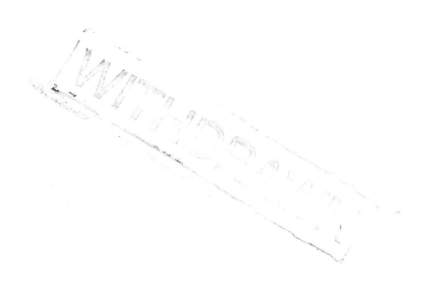

Advances in VIRUS RESEARCH

VOLUME 82

Bacteriophages, Part A

Advances in VIRUS RESEARCH

VOLUME **82**

Bacteriophages, Part A

Edited by

MAŁGORZATA ŁOBOCKA
Autonomous Department of Microbial Biology
Faculty of Agriculture and Biology
Warsaw University of Life Sciences
Nowoursynowska 159, Warsaw, Poland
Department of Microbial Biochemistry
Institute of Biochemistry and Biophysics
Polish Academy of Sciences
Pawińskiego 5A, Warsaw, Poland

WACŁAW T. SZYBALSKI
Professor Emeritus of Oncology
McArdle Laboratory for Cancer Research
University of Wisconsin Medical School
Madison, Wisconsin, USA

ELSEVIER

AMSTERDAM • BOSTON • HEIDELBERG • LONDON
NEW YORK • OXFORD • PARIS • SAN DIEGO
SAN FRANCISCO • SINGAPORE • SYDNEY • TOKYO
Academic Press is an imprint of Elsevier

Academic Press is an imprint of Elsevier

32 Jamestown Road, London, NW1 7BY, UK
Radarweg 29, PO Box 211, 1000 AE Amsterdam, The Netherlands
225 Wyman Street, Waltham, MA 02451, USA
525 B Street, Suite 1900, San Diego, CA 92101-4495, USA

First edition 2012

Library of Congress Cataloging-in-Publication Data

A catalog record for this book is available from the Library of Congress

British Library Cataloguing-in-Publication Data

A catalogue record for this book is available from the British Library

ISBN: 978-0-12-394621-8
ISSN: 0065-3527

For information on all Academic Press publications
visit our website at elsevierdirect.com

Printed and bound in USA
12 13 14 15 10 9 8 7 6 5 4 3 2 1

CONTENTS

CONTRIBUTORS

Hans-W. Ackermann
Department of Microbiology, Epidemiology and Infectiology, Faculty of Medicine, Laval University, Quebec, Canada
Email: ackermann4@gmail.com

Lindsay W. Black
Department of Biochemistry and Molecular Biology, University of Maryland Medical School, Baltimore, Maryland, USA
Email: LBlack@som.umaryland.edu

E. Fidelma Boyd
Department of Biological Sciences, University of Delaware, Newark, Delaware, USA
Email: fboyd@UDel.Edu

Mickaël Boyer
URMITE, Centre National de la Recherche Scientifique UMR IRD 6236, Faculté de Médecine, Aix-Marseille Université, Marseille Cedex 5, France
Email: mickboyer@free.fr

Christelle Desnues
URMITE, Centre National de la Recherche Scientifique UMR IRD 6236, Faculté de Médecine, Aix-Marseille Université, Marseille Cedex 5, France
Email: Christelle.DESNUES@univ-amu.fr

Patrick Forterre
Department of Microbiology, Institut Pasteur, Molecular Biology of the Gene in Extremophiles Unit, Paris, France
Email: forterre@pasteur.fr

Graham F. Hatfull
Department of Biological Sciences, Pittsburgh Bacteriophage Institute, University of Pittsburgh, Pittsburgh, Pennslyvania, USA
Email: gfh@pitt.edu

Mart Krupovic
Department of Microbiology, Institut Pasteur, Molecular Biology of the Gene in Extremophiles Unit, Paris, France
Email: krupovic@pasteur.fr

Katarzyna Licznerska
Department of Molecular Biology, University of Gdańsk, Gdańsk, Poland
Email: k.licznerska@biotech.ug.edu.pl

Marcin Łoś
Department of Molecular Biology, University of Gdańsk, Gdańsk, Poland; Institute of Physical Chemistry, Polish Academy of Sciences, Warsaw, Poland; Phage Consultants, Gdańsk, Poland
Email: mlos@biotech.ug.gda.pl

Daniel Muñoz-Espín
Instituto de Biología Molecular "Eladio Viñuela" (CSIC), Centro de Biología Molecular "Severo Ochoa" (CSIC-UAM), Universidad Autónoma, Madrid, Spain
Email: dmunoz@cnio.es

David Prangishvili
Department of Microbiology, Institut Pasteur, Molecular Biology of the Gene in Extremophiles Unit, Paris, France
Email: david.prangishvili@pasteur.fr

Venigalla B. Rao
Department of Biology, Catholic University of America, Washington DC, USA
Email: rao@cua.edu

Didier Raoult
URMITE, Centre National de la Recherche Scientifique UMR IRD 6236, Faculté de Médecine, Aix-Marseille Université, Marseille Cedex 5, France
Email: didier.raoult@gmail.com

Margarita Salas
Instituto de Biología Molecular "Eladio Viñuela" (CSIC), Centro de Biología Molecular "Severo Ochoa" (CSIC-UAM), Universidad Autónoma, Madrid, Spain
Email: msalas@cbm.uam.es

Gemma Serrano-Heras
Experimental Research Unit, General University Hospital of Albacete, Albacete, Spain
Email: gemmas@sescam.jccm.es

Agnieszka Szczepankowska
Department of Microbial Biochemistry, Institute of Biochemistry and Biophysics, Polish Academy of Sciences, Warsaw, Poland
Email: agaszczep@ibb.waw.pl

Alicja Węgrzyn
Laboratory of Molecular Biology (affiliated with the University of Gdańsk), Institute of Biochemistry and Biophysics, Polish Academy of Sciences, Gdańsk, Poland
Email: alawegrzyn@yahoo.com

Grzegorz Węgrzyn
Department of Molecular Biology, University of Gdańsk, Gdańsk, Poland
Email: wegrzyn@biotech.ug.gda.pl

Malcolm F. White
Biomedical Sciences Research Complex, University of St. Andrews, North Haugh, St. Andrews, Fife, United Kingdom
Email: mfw2@st-andrews.ac.uk

PREFACE

The history of studies on prokaryotic viruses that was initiated with the independent discoveries of bacteriophages by Hankin, Twort and d'Herelle at the junction between XIX and XX centuries has its ups and downs. Intuitively, since they were killing bacteria, phages were used initially as antibacterial agents. However, they were not enough standardized, and too variable, which did not permit to design highly reproducible experiments. Thus, clinical use of phages was more of an art than science. Therefore, in 1940's, bacteriophages were quickly supplanted by antibiotics and other antibacterial drugs, which acquired an aura of "miracle" medicines.

Although phages became "visualized" only after the discovery of electron microscope, their representatives soon acquired an important status in several research laboratories all over the world, as sophisticated genetic and molecular biology tools and as model "organisms" to study molecular mechanisms that underlie basic biological processes. They remained as such for decades contributing to numerous groundbreaking discoveries and to the development of genetic and molecular biology methods that enabled a rapid progress in several fields in biology and medical sciences. The discovery of restriction enzymes and the development of phage display-based drug discovery methods are just some examples.

When one of us (W.S.) was exposed for the first time to the Biology at the high school level in Lwów, Poland, in late 1930s, the smallest unit of life was the cell. Under the microscope, he had seen and admired the bacterial, yeast or larger eukaryotic cells. Moreover, he was told that there existed also viruses, that are even smaller than cells and that attack cells. He learned that viruses lysing bacterial cells were called bacteriophages. He saw the first electron micrographs of them, obtained from prof. Ruska in Berlin, when visiting the Institute of Prof. Rudolf Weigl in Lwów, at about 1940. As a student of chemical engineering, he immediately became mesmerized by these creatures, because of both scientific and practical reasons: (1) they seemed to look like hexagonal microcrystals, while appearing to be on the borderline between the "lifeless" chemical molecules and smallest units of life, (2) they seemed to be be a good candidate to unravel the chemical essence of life, (3) they might permit him to undertake the first time a chemical enzymatic/cathalytic synthesis of life (4) they were killing bacteria and thus they could cure the diseases. This is why he elected to become microbiologist in late 1940s and later joined the phage group of CSHL.

When the second one of us (M. L.) attended a high school in Warsaw, viruses, phages among them, were now included in the program of Biology classes. Descriptions of deliberate experiments on phages – at that time still considered to be physical structures, dominated the lectures on Genetics at the Warsaw University, and spurred the fascination of many students with this viruses. This fascination transformed into a laboratory practice, in 1970s and 1980s, at the time when phages were commonly used as simple models to elucidate mechanisms of basic biological processes, and as tools to study and engineer bacterial genomes.

The second half of XX century could be seen as a golden era of bacteriophage research. However, the glitter of this era was predicted by some to decline and be over at 1990's, when they claimed that ''all what was important was already discovered with prokaryotes" and that eukaryotic organisms would became the focus of attention. The development of cell culture-based techniques enabled a rapid progress in studies on eukaryotic cells and organisms. However, the discoveries of recent decades that pointed out the numerical dominance of viruses, especially phages, over all cellular organisms in most environments and enabled to get insight into the genomes of various phages and bacteria, revolutionized our understanding of the role of phages in the control of bacterial populations, in the adaptation of bacteria to new environmental niches, in the exchange of genetic information between them, and in the global circulation of matter. Thus, phages are now seen as key factors that shape our environment and again they attract the attention of more and more scientists. At the same time the emergence and spread of multiresistant bacterial pathogens revitalized the interest in the potential use of phages as antibacterial agents – an approach that was abandoned in Western countries with the introduction of antibiotics.

This book should reach the readers somewhat over a year after a crucial event – the first International Congress on Viruses of Microbes, at the Pasteur Institute in Paris. The congress attracted nearly a thousand participants, like no other phage-focused conference before, although the numerically smaller Phage Meetings initiated in CSHL have been held every year since 1950, as were other series of phage gathering including those in Olympia. WA and Salamanca, Spain.

No single book can reflect the full richness of phage world, with its estimated number of 10^{30} representatives. No book can cover in sufficient detail the topics of the First International Congress on Viruses of Microbes, which inspired us to initiate this undertaking. Thus, a reader will find here several topics that seem worth of more attention because of their novelty, importance or historical value. Most what we know about the prokaryotic viruses comes from bacteriophage studies. However, archeal viruses, whose only sparse representatives are known so far, may be even more diversified than bacteriophages, as is described in a chapter of this book by Mart Krupovic and coauthors. Clearly, there is still plenty of room for new discoveries. A few additional examples are the recently discovered new virus representatives, virophages, that propagate at the expense of other viruses in the "viral factories" of the latter. Although not sensu-stricto prokaryotic viruses, virophages resemble to some extent bacteriophages and thus, a chapter that concerns virophages is included in this book.

Małgorzata Łobocka
Warsaw, Poland
e-mail: malgorzata_lobocka@sggw.pl, lobocka@ibb.waw.pl
Wacław T. Szybalski
Wisconsin, USA
e-mail: szybalski@oncology.wisc.edu

Section 1
Historical, Ecological and Evolutionary Considerations

CHAPTER 1

Bacteriophage Electron Microscopy

Hans-W. Ackermann

Contents

Department of Microbiology, Epidemiology and Infectiology, Faculty of Medicine, Laval University, Quebec, Canada

Advances in Virus Research, Volume 82
ISSN 0065-3527, DOI: 10.1016/B978-0-12-394621-8.00017-0

Abstract

Since the advent of the electron microscope approximately 70 years ago, bacterial viruses and electron microscopy are inextricably linked. Electron microscopy proved that bacteriophages are particulate and viral in nature, are complex in size and shape, and have intracellular development cycles and assembly pathways. The principal contribution of electron microscopy to bacteriophage research is the technique of negative staining. Over 5500 bacterial viruses have so far been characterized by electron microscopy, making bacteriophages, at least on paper, the largest viral group in existence. Other notable contributions are cryoelectron microcopy and three-dimensional image reconstruction, particle counting, and immunoelectron microscopy. Scanning electron microscopy has had relatively little impact. Transmission electron microscopy has provided the basis for the recognition and establishment of bacteriophage families and is one of the essential criteria to classify novel viruses into families. It allows for instant diagnosis and is thus the fastest diagnostic technique in virology. The most recent major contribution of electron microscopy is the demonstration that the capsid of tailed phages is monophyletic in origin and that structural links exist between some bacteriophages and viruses of vertebrates and archaea. DNA sequencing cannot replace electron microscopy and vice versa.

I. INTRODUCTION

The discovery of bacterial viruses or bacteriophages, often called "phages," was one of the most momentous events in microbiology. Bacteriophages were discovered almost simultaneously by Frederick William Twort in England (1915) and Félix d'Herelle in France (1917). However, the first observation of their lytic activity was reported even earlier by British bacteriologist Ernst Hankin in 1896 (Hankin, 1896). The study of bacteriophages generated an enormous volume of scientific publications. Raettig's phage bibliography (1967) listed 11,405 articles, books, and book chapters from the years 1917–1965. It has been estimated from the author's personal bibliography that the number of phage publications is now near 50,000. This reflects the ever-increasing number of phage descriptions. In 2007, the astonishing number of more than 5500 prokaryote viruses, of which 99.6% were bacteriophages, had been examined in the electron microscope (Ackermann, 2007). Phages appear thus, at least theoretically, as the largest virus group in existence. There is no end in sight; on the contrary, it appears now that phages occur in astronomical numbers (10^{30} to 10^{32}) in the biosphere and are the most frequent biological entities on earth (Breitbart and Rohwer, 2005; Brüssow and Hendrix, 2002; Suttle, 2005).

D'Herelle coined the term "bacteriophage" or "bacteria eater" and postulated that his novel entities were viruses, analogous to the already known viruses of plants and vertebrates, for example, the tobacco mosaic and foot-and-mouth-disease viruses. He also postulated that they were particulate in nature and multiplied within bacterial cells (Herelle, 1921). Other scientists disputed this view and considered "phages" as enzymes, genes, or "transmissible autolysis." This controversy on the nature of phages was linked to a discussion on the priority of phage discovery (Summers, 1999).

The study of phages initiated the rise of molecular biology (Summers, 1999) and provided fundamental insights into virus replication and assembly. Phages made contributions to the epidemiology and understanding of infectious diseases, appeared responsible for faulty fermentations in the dairy industry, and were used for countless purposes ranging from bacterial diagnosis to the testing of air filters and condoms. At the present time, the most fertile fields of phage research are genomics (Brüssow and Kutter, 2005a,b; Hatfull, 2008), phage evolution (Brüssow, 2009; Hendrix, 2008), phage ecology (Abedon, 2008; Angly et *al.*, 2006), phage display (Hemminga *et al.*, 2010; Hertveldt *et al.*, 2009), and the discovery of new phages. Phage therapy, long neglected in Western countries after the advent of antibiotics, but practiced on a large scale in the former Soviet Union, Georgia, and Poland, is presently experiencing a comeback (Dublanchet, 2009; Kutter *et al.*, 2010; Sulakvelidze and Kutter, 2005).

This chapter focuses on the impact of electron microscopy on phage research and the role of phages in advancing electron microscopy. Indeed, bacterial viruses and electron microscopy have long been in a symbiotic relationship. The history of phage electron microscopy is one of cross-fertilization; that is, phages prompted the improvement of electron microscopes and the development of new techniques, and electron microscopy led to the discovery of new phages and a better understanding of phage biology. Electron microscopy is omnipresent in phage research. This chapter reviews the role of phages in electron microscopy and the importance of the latter in establishing the nature of phages, their classification, the description of novel phages, phage "life" and assembly within the infected cell, and phage ecology.

II. ELECTRON MICROSCOPY AND THE NATURE OF PHAGES

Electron microscopy was developed in the early 1930s (Haguenau *et al.*, 2003) and is much indebted to two brothers, Helmut and Ernst Ruska, both working in Berlin. The former focused on biological applications of electron microscopy and the second on electron optics. In 1938, two prototype electron microscopes located in a laboratory of the Siemens & Halske Company in Berlin were used for biological studies. Independently, electron microscopes were developed in Canada, Japan, and the United States (Haguenau *et al.*, 2003). The first bacteriophage micrographs, all of coliphages, were published in 1940 in Germany. In 1939, despite the worsening political situation in Europe, some of the micrographs had been sent to Professor Rudolf Weigl in Lwów, Poland and were justly perceived as sensational (W. Szybalski, personal communication). Phages appeared as round or elongated dark particles, which were or were not associated with bacteria (Pfankuch and Kausche, 1940; Ruska, 1940). These observations were noticed overseas despite World War II restrictions on scientific exchange and were soon followed by the first phage micrographs in the United States (Luria and Anderson, 1942; Luria *et al.*, 1943). Images showed phages of *Escherichia coli* and staphylococci. Two coliphages, apparently T4-like viruses, had long, thick tails and elongated heads. In Germany, Ruska (1942, 1943) observed cell destruction by phage-induced lysis of various enterobacteria and enterococci and, as early as 1943, was able to assemble a gallery of different phage morphotypes (Fig. 1). This corroborated the studies of Burnet (1933) who had already shown that enteric bacteriophages differed in size, antigenic properties, and inactivation by methylene blue, citrate, and urea. He showed that phages, contrary to an early postulate of d'Herelle (1921), were not a single entity with many races, but a diversified group of viruses. Most important, the observation of phage particles put the enzymatic theory of phage nature to a rest and established once and for all that

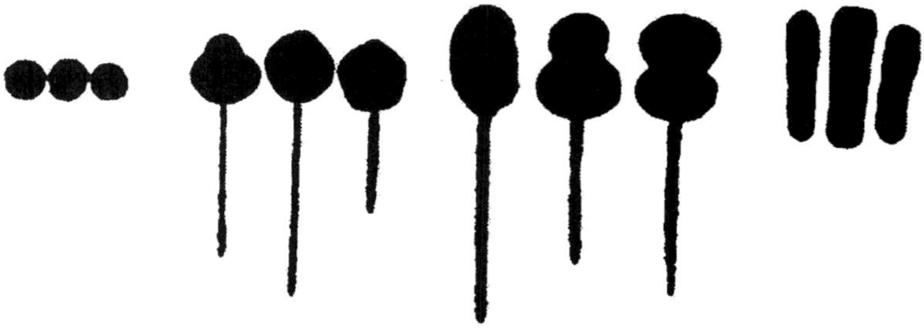

FIGURE 1 Types of tailed phages observed by H. Ruska; ink drawing (?) of unstained electron dense particles. The particles at left were probably T7-like phages. The visualization of short tails and very small phages, such as ϕX174 and MS2, was beyond the reach of early electron microscopes. The figure is the first graphical representation of any virus. Reproduced from Ruska, H. (1943) Ergebnisse der Bakteriophagenforschung und ihre Deutung nach morphologischen Befunden. *Ergeb Hyg Bakteriol Immunforsch Exp Ther* **25**:437–498. Copyright Springer 1943, with kind permission of Springer Science+Business Media.

bacteriophages are particulate and thus viruses. It is said that d'Herelle was on his death bed when the French scientist Hauduroy showed him an electron micrograph of a bacteriophage (Dubochet, 1988).

Biological electron microscopy progressed slowly by accretion of large and small improvements and knowledge (see the review of Holt and Beveridge, 1982). For example, the complex technique of sectioning depends on the design of ultramicrotomes and knifes, fixatives, buffers, embedding resins, and finally the mounting and staining of specimens. Present-day electron microscopes are the results of countless modifications. The latest major developments are the advent of digital electron microscopes and the increasing replacement of darkroom photography by charged-couple device (CCD) cameras and electronic image acquisition. Because most improvements developed over decades, it is difficult to define clear turning points. It was about 1990 that digital electron microscopes and CCD cameras slowly, but not completely, replaced conventional electron microscopes. Because bacteriophages, especially coliphages, are nonhazardous and usually easy to manipulate, they played a key role in the development of many electron microscopical techniques. Phages were among the first viruses to be examined in the electron microscope (Haguenau *et al.*, 2003).

Electron microscopy is now an increasingly large research field with countless applications in material research and biology. Its main branches are transmission (TEM), scanning (SEM), and high-voltage electron microscopy. The last two techniques have little or no impact on phage research, but TEM is of enormous importance in all of virology. It comprises such diverse techniques as staining of isolated viruses, thin sectioning, shadowing, autoradiography, immunoelectron microscopy (IEM), cryoelectron microscopy (cryoEM) with or without shadowing and three-dimensional image reconstruction, the replica technique, or enzymatic virus digestion on the grid. All these techniques have been applied to bacterial viruses, but most of them are rarely used and only in specialized laboratories. However, the negative staining of isolated particles is of universal importance in virology as an easy, fast, and inexpensive approach to the study and identification of viruses by means of a standard transmission electron microscope.

III. STUDYING THE VIRION

A. Shadowing and staining

The first phage images showed unstained particles. Phage particles were of variable size and sometimes provided with tails (Pfankuch and Kausche, 1940; Ruska, 1940, 1942). Filamentous and small isometric phages were not shown or not resolved. Phage particles were dark because, as we know today, heads of tailed phages contain a large amount of electron-dense, double-stranded DNA (dsDNA). Tails, which are proteinic in nature, were pale and much less visible. At the same time, it was observed that infected bacteria disappeared and were replaced by indistinct debris. Phages were much more contrasted than this debris. This confirmed d'Herelle's statement (1921) that phages were liberated by lysis of bacteria.

To improve contrast and resolve structural details, shadow casting was introduced by Williams and Wyckoff (1945). The specimen was coated with a chromium salt and evaporated in a vacuum and laterally from the specimen so that parts of the object were not completely covered by metal. This produced a few highly contrasted images with "shadows" of areas that were without a metal coat. Subsequently, chromium was replaced by other heavy or high-density metals such as gold or platinum. Coliphage T4 was shadowed in 1948 and was followed 5 years later by other coliphages of the T series (Williams and Fraser, 1953). Phages were air-dried or freeze-dried and then shadowed. Phages belonged to four morphotypes, represented by phages T1, T5, T3-T7, and T2-T4-T6. This study yielded a rich harvest of information on phage structure and dimensions and established that phages of one and the same strain *E. coli* could be of different morphology. Phage heads appeared as geometric bodies and tails were seen with unprecedented clarity. By 1952, the structure of such complicated viruses as coliphage T2 was essentially understood (Fig. 2) (Anderson *et al.*, 1953).

The staining of isolated, purified viruses was invented in 1955. The basic idea was to stain viruses with an electron-dense salt solution of high molecular weight and small molecular size. The technique was first applied to the tomato bushy stunt virus, an isometric plant RNA virus, and the rod-shaped tobacco mosaic virus (Hall, 1955). Viruses were stained with phosphotungstic and silicotungstic acid, osmium tetroxide, and various silver, platinum, thorium, and lanthanum salts. Particles that had taken up the stain were called "positively stained." In contrast, "negatively stained" viruses were simply surrounded by a stain and appeared white or gray on a dark background. Brenner and Horne (1959) developed the technique further using phage T2. They standardized experimental conditions and are generally credited with the introduction of negative staining. Phosphotungstic acid was now applied to other T-even phages (T4, T6) and produced images of great clarity (Brenner *et al.*, 1959), far better than many phage images published today. Uranyl acetate, already used to stain thin sections of tissue, was applied in 1960 to isolated viruses and was found to have a strong affinity for dsDNA. When applied to phages T2 and T7, it caused a strong positive staining of phage heads that appeared in deep black (Huxley and Zubay, 1961). Although many other chemicals have been tested on bacteriophages (Ackermann and DuBow, 1987; Hayat and Miller, 1990), phosphotungstates and uranyl acetate are still the most widely used stains. However, it is little known that uranyl acetate is a tricky substance that

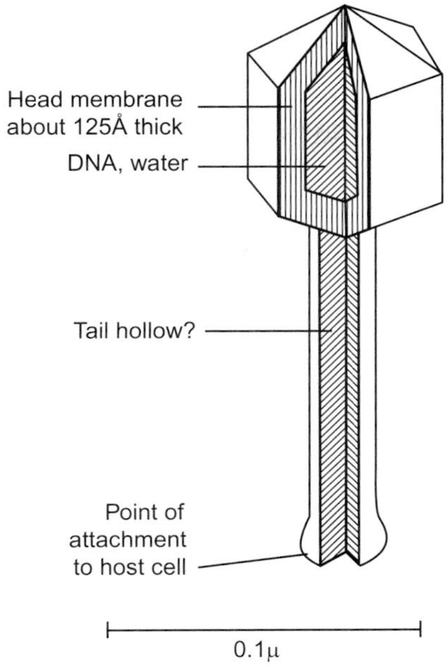

FIGURE 2 Schematical drawing of coliphage T2 by T.F. Anderson (1952). Reproduced from Anderson, T.F., Rappoport, C., Muscatine, C.A. (1953). On the structure and osmotic properties of phage particles. *Annales de l'Institut Pasteur* **84**:5–15. Copyright Elsevier–Masson 1953, with permission.

produces both negative and positive staining and numerous artifacts, notably shrinkage of positively stained capsids and swelling of proteinic structures (Ackermann *et al.*, 1974; Ackermann and DuBow, 1987). Negative staining, along with thin sectioning, is one of the two most important techniques in biological electron microscopy. Its usefulness for virus descriptions, structural studies, and classification cannot be overemphasized. Negative staining is irreplaceable for the investigation of phage gross morphology, dimensions, and fine structure. Moreover, it has revealed an astonishing array of structural details, especially in tailed phages such as T4 (capsomers, tail tubes and tail sheaths, base plates, collars and collar fibers, head and tail fibers) and morphological aberrations such as polyheads and polytails.

B. Scanning electron microscopy

Conceived and introduced as early as 1938 by Von Ardenne (Haguenau *et al.*, 2003) and first applied to various animal and plant viruses, SEM was applied in 1957 to bacteriophages. Coliphage P1 served as an experimental model because of its large size and distinctive morphology. This early work was mainly to define parameters such as fixation and drying techniques. After metal coating, phages could be seen adsorbed on the surface of bacterial cells, but no structural details were resolved (Wendelschafer-Crabb *et al.*, 1975). It is now possible to equip transmission electron microscopes with SEM detectors. A long-tailed *Staphylococcus* phage and coliphage T4 were studied, and tail striations could be resolved in phage T4 (Broers *et al.*, 1975). This work also illustrates the intrinsic and stringent limitations of SEM. In a general way, due to the necessity to dry specimens and to coat them with metal, the resolution of phage fine structure is poor and does not compare with that achieved by negative staining. However, technical improvements have shown that scanning transmission electron microscopy (STEM) has some limited potential. For example, STEM is suitable for mass measurements, while dark-field STEM produces high-contrast images of unstained, freeze-dried T4 phages with unfolded tail fibers. Negatively stained T4 tail fibers show globular and fibrous domains (Cerritelli *et al.*, 1996).

C. Cryoelectron microscopy and three-dimensional image reconstruction

The freeze-etching technique was developed to circumvent the chemical treatment of specimens and extensive loss of water. This technique consists of quick freezing of the specimen in nitrogen sludge or nitrogen cooled (Freon, propane, ethane) and cleaving followed by differential sublimation of the ice on the specimen surface ("etching"). The specimen is then contrasted with a metal (e.g., platinum) onto which carbon is evaporated. Originally developed for the study of plant viruses (Steere, 1957), the technique yielded excellent results in coliphage T2 (Bayer and Remsen, 1970). Phage heads, which so far had looked smooth and structureless, now appeared to be built of hollow capsomers with subunits, whereas extended and contracted tails showed a helical structure and tail fibers. When applied to phage λ, cryoEM showed skewed hollow capsomers corresponding to a triangulation number of T=21 (Bayer and Bocharov, 1973), which was later corrected to T=7 (Williams and Richards, 1974). Capsomers were also seen on abnormal giant heads of phage T2 (Bayer and Cummings, 1977). CryoEM without freeze fracture permitted the visualization of phage T4 replicative intermediates (Gogol *et al.*, 1992).

More recently, cryoEM became the basis for computer-based three-dimensional (3D) image reconstruction of virus structures. Technically, an electron microscopical grid with viruses is plunged into liquid ethane and frozen at around −160 °C. Vitrified T4 bacteriophages form arrays of highly contrasted and detailed virions (Dubochet *et al.*, 1985). The technique is suitable to show the arrangement of DNA within T4 and λ phage heads (Lepault *et al.*, 1987). Subsequently, vitrified viruses are photographed, the images are digitized, and 3D reconstructions are computed from thousands of photographs. A full description of the technique may be found in Dryden *et al.* (1993) and Morais *et al.* (2005). The inconvenience of cryoEM is that phage heads often appear rounded and never show transverse edges. Cryoelectron microscopic tomography has been used to elucidate the injection of coliphage T5 DNA into liposomes (Böhm *et al.*, 2001) and, very recently, the mode of DNA translocation in two T7-like podoviruses (Chang *et al.*, 2010; Liu *et al.*, 2010).

Three-dimensional image reconstruction has been used with spectacular results to study heads, head–tail connectors, tails, and base plates of tailed phages. In combination with X-ray crystallography, this technique is particularly valuable for the investigation of capsid symmetry (Steven *et al.*, 1997) and has indicated phylogenetic relationships between tailed phages that seem to have little in common, namely myoviruses (T4, ϕKZ, SPO1), siphoviruses (λ, HK97, T5), and podoviruses (ϕ29)(Duda *et al.*, 2006; Effantin *et al.*, 2006; Fokine *et al.*, 2004, 2005, 2007; Morais *et al.*, 2005; Tao *et al.*, 1998), thereby providing proofs for the basic unicity and monophyletic origin of tailed phage heads, thus the head–tail principle and the viral order Caudovirales. This unicity had long been postulated on the basis of phage assembly patterns and physiology, but without direct proof by electron microscopy. Even more far reaching is the discovery that the capsids of *E. coli* siphovirus HK97, *Bacillus* myovirus SPO1, and herpesviruses have structural relationships (Baker *et al.*, 2005; Duda *et al.*, 2006). A combination of transmission electron microscopy, genomics, sequence alignments, and nuclear magnetic resonance spectroscopy has shown that the tail tubes of myovirus and siphovirus bacteriophages and the bacterial type VI secretion system are related evolutionarily (Pell *et al.*, 2009).

The combination of cryoEM and 3D image reconstruction is suitable to determine triangulation numbers and to show hexagonal and pentagonal capsomers and decoration proteins. Together with X-ray crystallography, it has been applied with great success to tailless isometric prokaryote viruses, for example, novel *Sulfolobus* archaeal virus STIV (Khayat *et al.*, 2005), novel and unrelated archaeal virus SH1 (Jäälinoja *et al.*, 2008), corticovirus PM2 (Huiskonen *et al.*, 2004), microvirus SpV4 of *Spiroplasma* (Chipman *et al.*, 1998), and various tectiviruses (Rydman *et al.*, 1999). Investigation of STIV and tectiviruses indicates structural relationships among eukaryal, bacterial, and archaeal viruses, for example, between tectiviruses and adenoviruses (Huiskonen and Butcher, 2007).

D. Visualization of nucleic acids

Although now rarely used, this technique allows the observation of DNA or RNA filaments. It was introduced in 1962 with coliphage T2 as an experimental model (Kleinschmidt *et al.*, 1962). In its most common version, phage nucleic acid is extracted, spread on a cytochrome-formamide film on water, and shadowed or stained with uranium oxide (UO_2). Several variant techniques have been devised, for example, dark-field microscopy of unstained DNA or protein-free spreading (Dubochet *et al.*, 1971; Portmann *et al.*, 1974). Conformation of the nucleic acid, linear or circular, can be observed, and the length of molecules permits calculating their molecular weights. Other applications include the visualization of replicative forms, intracellular DNA, single-strand gaps, and DNA hybrids (heteroduplex analysis). For the latter, dsDNA of phage mutants or closely related phages is heated to separate DNA strands and cooled to associate (anneal) them again. Heteroduplex analysis is used for comparing mutants and closely related phages, for example, lambdoid phages (Fiandt *et al.*, 1971), T2, T4, and T6 (Kim and Davidson, 1974). The degree of homology provides a measure of phylogenetic and taxonomic relationships. Regions of nonhomology may form bubbles and loops, indicating sites of deletions, duplications, inversions, or insertions. In lambdoid phages, the method permits visualization of nonhomologous single-stranded DNA (ssDNA) regions within dsDNA molecules to "see genes" and to establish physical genomic maps (Hradecna and Szybalski, 1969; Westmoreland *et al.*, 1969). This was the most precise method of physical mapping until the advent of DNA sequencing. Electron microscopy is also able to physically map the binding of RNA polymerase to phage λ DNA (Vollenweider and Szybalski, 1978) and the localization, identification and comparison of IS insertion sequences (Fiandt *et al.*, 1972).

E. Virus counts

Electron microscopical particle counts were developed because the biological titration of animal and plant viruses was difficult at best. Even in bacteriophages, where titration of viable viruses is generally easy and accurate, titration does not account for defective viruses and viral debris. A breakthrough occurred with the use of latex spheres for reference. Briefly, known volumes of virus suspension and calibrated suspension of latex spheres are mixed, sprayed with a nebulizer onto a grid, and shadowed (Williams and Backus, 1949). The technique was refined with the help of coliphages T2 and λ by spraying viruses and latex spheres onto agar covered with a collodion film. The film was then cut out, mounted onto a grid, and shadowed (Kellenberger and Arber, 1957). Virus counts in aquatic environmental samples are based on the ability of uranyl acetate to induce positive staining of dsDNA-containing structures, namely phage heads and phycodnaviruses. This is explained in more detail in the section on phage ecology.

F. Immunoelectron microscopy

Already introduced in 1941 into virology and then applied to the tobacco mosaic virus (Anderson and Stanley), IEM is performed sporadically on bacterial viruses. In essence, this very simple technique consists of mixing native or diluted (1:40) antiserum with an antigen (virus) directly on a specimen grid or in a tube and applying it later to a grid. The allotted reaction time varies from 5 minutes to 6 hours. The grid is usually rinsed with a buffer and then stained. Excellent electron microscopical resolution is required. Phages or phage parts appear coated with a fur of antibodies. IEM may be used to investigate relationships between phages, such as T4-like viruses, by agglutination or to locate specific head or tail antigens (Yanagida and Ahmed-Zadeh, 1970). Antibodies or Fab antibody fragments may be conjugated with gold in order to visualize the location of specific antigens on phage heads or tails. Colloidal gold has been used to locate specific tail fiber sites of a relative of phage T4 and to visualize them by means of a field emission scanning electron microscope (Hermann *et al.*, 1991).

G. Electron holography

Off-axis electron holography of biological samples started in 1986 with the examination of ferritin and was extended to the tobaco mosic virus, a bacterial flagellum, bacterial cell wall components, the Semliki Forest arbovirus, and coliphage T5. Samples are examined unstained and carbon coated. A reference wave is directed through a hole in the substrate adjacent to the object. Reference and object exit waves are superimposed by a positively charged wire or "biprism" to produce a hologram at the detector plane. Phase images of phage T5 show edges on the T5 head and resemble those of a shadowed phage. The hologram is recorded by a CCD camera. The resolution is said to be 20 nm (Simon *et al.*, 2008). Although interesting, this technique has not provided useful insights into bacteriophage structure.

H. Atomic force microscopy

An atomic force microscope consists of a cantilever (silica or similar material) with a sharp tip to scan a specimen. The cantilever is deflected in proximity of the specimen surface. Cantilever motion is detected by a laser spot. Viruses are deposited on silica or mica wafers. Advantages of atomic force microscopy (AFM) are vaunted as three-dimensional image acquisition and the possibility of studying samples in different surroundings without pretreatment such as metal coating. AFM was applied to coliphage T4, a T4-like *Salmonella* phage, and lysed bacteria. Phages appeared as tailed blobs without details (Dubrovin *et al.*, 2008; Kolbe *et al.*, 1992). More recently, AFM was applied to a *Synechococcus* myovirus and, again, phage T4. Capsomers, tail striations, and extruded DNA were visualized (Kuznetzov *et al.*, 2010, 2011). Unfortunately, no images of negatively stained phages were presented for comparison. AFM is apparently unable to generate overviews of phage preparations. The most interesting application of AFM seems to be the visualization of capsomers.

IV. STUDYING PHAGE LIFE

A. Productive cycle

The life cycles of phages are called the productive or virulent cycle and the temperate or lysogenic cycle. The former generally ends in the production of novel virions and destruction of the host (lysis). Filamentous phages of the Inoviridae family constitute an exception as they are secreted continuously into the medium without lysis of the host. In the lysogenic state, the phage genome either is integrated into the host DNA or is free as a plasmid and becomes latent within the bacterium. When this equilibrium is broken, the phage genome initiates a phase of phage production and host lysis.

The life cycle of tailed virulent phages was pieced together from countless physiological experiments summarized in the classic book of Adams (1959). It appeared as a multistep process comprising an adsorption period, infection of the host cell by phage DNA, a mysterious period of intracellular multiplication, and final liberation of novel infectious phages. The role of electron microscopy in the investigation of phage reproductive cycles and intracellular multiplication was relatively minor and was used primarily to document and illustrate the steps of the phage life cycle. Phage T4 was almost always the workhorse of these studies.

A key observation already made by Ruska (1942, 1943) showed that phage infection led to abrupt burst and lysis of infected bacteria, leaving only virus particles behind. This confirmed d'Herelle's early contention that infected bacteria dissolved into a cloud of material (d'Herelle, 1921). Adsorption of phages to the cell wall of bacteria was documented as early as 1942. Masses of rod-shaped coliphages, later identified as short-tailed phages with long heads, were seen adsorbed to the outside of bacterial cells, forming a palisade around the bacterium (Kottmann, 1942). Various observations indicated that T-even-like phages also adsorbed to the bacterial cell wall and sometimes showed thickened tails and empty heads. These phages were called "ghosts." The meaning of these observations became clear when Hershey and Chase (1952) showed in a famous experiment that bacteria were infected by phage nucleic acid and not proteins. Tagged with the radioactive ^{32}P isotope, T2 phage nucleic acid was shown to enter the bacterium and initiate phage reproduction, while the phage protein coat, tagged with ^{35}S, remained outside. Infection was followed by the "eclipse" or latent period during which infectious phages disappeared from the medium until lysis of the infected bacterium. Simultaneously, new phages were liberated in one explosive event or "burst." In parallel, multiplication of temperate phages was shown to end in the production of full-fledged complete phage particles (Kellenberger and Kellenberger, cited by Adams, 1959). It was shown in the early 1960s that filamentous, ssDNA-containing phages (Inoviridae family) and male-specific, isometric, ssRNA-containing phages (Leviviridae family) adsorbed to bacterial pili (Hoffmann-Berling *et al.*, 1963). The extrusion of inoviruses was observed directly in the electron microscope (Hofschneider and Preuss, 1963). A crowning achievement of these studies on phage infection was the demonstration, presented in a series of stunning electron micrographs, that phages T2 and T4 adsorbed to bacterial cell walls by their tail fibers and injected a DNA filament into bacteria (Simon and Anderson, 1967). As shown by cryoEM tomography, phage T5 injects its DNA into liposomes by means of its central tail fiber (Böhm *et al.*, 2001). When applied to T7-like podoviruses, the same technique indicates that DNA release in *Prochlorococus* virus P-SSP7, a cyanophage, is triggered by the reorientation of tail fibers (Liu *et al.*, 2010), whereas DNA in *Salmonella* phage epsilon15 is ejected via a tunnel of core proteins or cellular components (Chang *et al.*, 2010).

B. Intracellular multiplication

Thin sectioning of cells developed after 1943 (Haguenau *et al.*, 2003) and was now applied to phage-infected bacteria. The Hershey–Chase experiment mentioned earlier had shown that the bacteria-infecting component of phage T2 was DNA. This was confirmed by the observation of phages fixed to the outside of bacteria and of "ghosts" or shadows of phages, which appeared flat and transparent and apparently had lost their inner content. Similar ghosts of T2 phage were produced by osmotic shock, leading to loss of nucleic acid (Levinthal and Fisher, 1953). The electron microscope now provided direct evidence for intracellular phage multiplication (Ackermann and DuBow, 1987; DeMars *et al.*, 1953; Kellenberger and Arber, 1957; Kellenberger and Wunderli-Allenspach, 1995) and showed that this takes place during the eclipse period. In the case of T-even phages, DNA-filled, dark phage heads are seen to appear at the periphery of the nucleoplasm at the end of the eclipse period; simultaneously, bacterial DNA is disrupted. Later in infection, tails become visible within the infected bacteria and appear fixed to the plasma membrane. In temperate phages, for example, coliphage λ, bacterial DNA is not disrupted. Intracellular viral crystals are sometimes seen in cells infected with short-tailed phages (Schito, 1974). They are the rule in leviviruses (Schwartz and Zinder, 1963), perhaps because the latter have no tails that could interfere with crystal formation mechanically.

C. Particle assembly

Intrigued by how bacteriophages came into being, early electron microscopists tried to find intermediate stages in phage morphogenesis. T2-infected bacteria were disrupted and shadowed before complete phage lysis. Incomplete phages, called "doughnuts," were indeed found (Levinthal and Fisher, 1953). Investigations could not be carried further with the electron microscopes of these times and without negative staining. In the sixties, the question was tackled again with conditionally lethal mutants of phage T4. Under "permissive" conditions, normal phages were produced. However, under "restrictive" conditions, infection was abortive and led to the accumulation of unassembled phage components. By mixing lysates of mutant-infected cells in various combinations and sequences, it was possible to reconstitute *in vitro* a fully infectious bacteriophage T4. It appeared that T4 had three separate assembly lines for the phage head, the tail, and tail fibers, respectively (Wood, 1992; Wood and Edgar, 1967). This approach was now extended to phages λ and HK97, T7, P22, and ϕ29 (Kellenberger and Wunderli-Allenspach, 1995). In this way, the morphogenetic pathway of representatives of all three tailed phage families (Myoviridae, Siphoviridae, Podoviridae) became known, and it was realized that tailed phages share features such as scaffolding proteins, proheads, and head–tail connectors.

Many phages, foremost the T-evens, produce abnormal particles of widely different size and shape. They are intermediate stages in phage synthesis or errors in phage assembly and include such structures as polyheads, polysheaths, phages with abnormally long tails, and multi-tailed phages, as well as dwarf, misshapen, or giant heads with or without DNA. Some phages, for example, coliphage P1, produce virions of different head size (Ackermann and DuBow, 1987; Kellenberger and Wunderli-Allenspach, 1995). The most frequent aberration, observed in siphoviruses and almost never in myoviruses, is the presence of abnormally long tails. These structures are the result of genetic defects or can be induced by growing phages on amino acid analogues (Cummings *et al.*, 1977). Although of little practical importance, these structures are the delight of the electron microscopist.

V. PHAGE CLASSIFICATION AND NOVEL VIRUSES

A. Classification into orders and families

Electron microscopy has provided a framework for high-level virus classification. This is certainly one of its main contributions to virology. Principal criteria are nature and conformation of nucleic acid and gross morphology, including the absence or presence of an envelope. Although nucleic acid is considered more important than morphology, the latter is generally easy to investigate so that morphological data abound and nucleic data are relatively scarce. This gives some microbiologists the false impression that virus classification is essentially morphological. Three orders and over 65 virus families have been established in this way. All bacterial and archaeal virus families have been individualized by electron microscopy. Bacteriophages are currently classified into one order with three families and seven additional families (Ackermann, 2005, 2006, 2007; Fauquet *et al.*, 2005). Although archaea possess a few myoviruses and siphoviruses, the divide between archaeal and bacterial viruses is generally sharp. The International Committee on Taxonomy of Viruses (ICTV) virus classification used by GenBank presents genome sequences and proteins in an orderly way.

Bacteriophage classification goes back to a seminal paper by Bradley (1967). He distinguished three groups of tailed phages and three types of isometric and filamentous phages, corresponding to present-day Myo-, Sipho-, Podo-, Micro-, Levi-, and Inoviridae families, respectively. Bacteriophages contain dsDNA, ssDNA, dsRNA, or ssRNA. Approximately 96% of phages are tailed. Polyhedral, filamentous, and pleomorphic phages are generally rare and have narrow host ranges (Ackermann, 2007). The general properties of basic phage groups are summarized later and in Table I. The morphology of important phage families is illustrated in Figure 3. The classification of bacteriophages is explained more fully elsewhere (Ackermann, 2005, 2006, 2007, Fauquet *et al.*, 1995). Phages may be roughly categorized by shape.

1. Tailed phages. They constitute the order Caudovirales, are ubiquitous, contain dsDNA, and comprise the families Myoviridae, Siphoviridae, and Podoviridae. All tailed phages have a head and a hollow, helical tail built of subunits. Its purpose is the transfer of DNA into a bacterium. The head or capsid is icosahedral or a more or less elongated derivative of this body. Most tails have fixation structures such as base plates, fibers, or spikes. The tail of myoviruses (24.5% of tailed phages) is contractile and consists of an axial needle surrounded by a contractile sheath, separated from the head by an empty space or "neck." Siphovirus (61%) tails are long and flexible or rigid tubes,

TABLE I Bacteriophage families[a]

Shape	Family	Example	Characteristics[b]	Number
Tailed	Myoviridae	T4	dsDNA, L, tail contractile	1320
	Siphoviridae	λ	dsDNA, L, tail long and noncontractile	3269
	Podoviridae	T7	dsDNA, L, tail short	771
Polyhedral	Microviridae	ϕX174	ssDNA, C, 12 capsomers, 30 nm	38
	Corticoviridae	PM2	dsDNA, C, complex capsid, lipids, 60 nm	3?
	Tectiviridae	PRD1	dsDNA, L, inner lipid vesicle, pseudo-tail, 60 nm	19
	Leviviridae	MS2	ssRNA, L, like poliovirus, 25 nm	38
	Cystoviridae	ϕ6	dsRNA, L, segmented, envelope, 70 nm	3
Filamentous	Inoviridae	fd	ssDNA, C, long filaments or short rods 90–1300 nm in length	66
Pleomorphic	Plasmaviridae	L2	dsDNA, C, envelope, no capsid, ~90 nm	5

[a] After Ackermann (2005,2006,2007).
[b] C, circular; L, linear.

whereas podovirus (14%) tails are short and generally 10–20 nm long. Most tails have fixation structures, such as base plates, fibers, or spikes. Phages may be virulent or temperate. All are liberated by burst of the infected bacterium.

2. Polyhedral bacterial viruses are icosahedra or quasi-icosahedral bodies. They are said to have "cubic symmetry" and comprise seven families of viruses, four of which contain lipids and two containing RNA. Microviridae (ssDNA) correspond to phage ϕX174 and its relatives. Interestingly, they are found not only in enterobacteria, but also in phylogenetically distant hosts such as *Bdellovibrio* and *Chlamydia*. Corticoviridae (dsDNA) have a multilayered capsid of alternating proteins and lipids. Tectiviridae (dsDNA) possess an icosahedral protein capsid that surrounds a lipid-containing vesicle. The latter has the unique property of transforming itself, for the purpose of infecting bacteria, into a tail-like tube of about 60 nm in length. Leviviridae (ssRNA) are small phages that resemble the poliovirus. Cystoviridae (dsRNA) have a flexible, lipid-containing

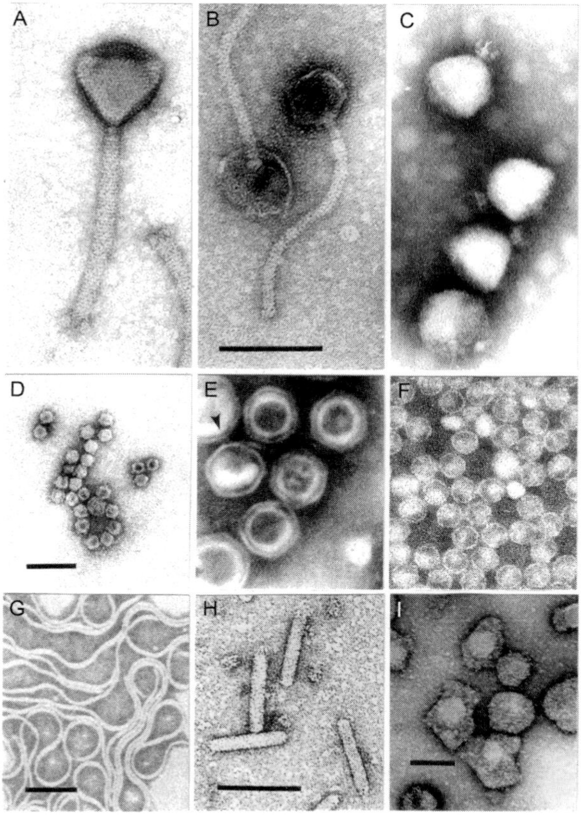

FIGURE 3 Bacteriophage morphology. A. Myovirus ϕBC6 of *B. cereus* with extended tail; UA. B. Siphovirus γ of *B. anthracis*, UA. The capsid of the right particle shows a pentagonal outline indicating an icosahedral shape. Phage is used for identification of *B. anthracis*. Podovirus P22 of *Salmonella typhimurium*, PT. D. Microvirus ΦX174 of *E. coli*, UA. E. Tectivirus 37–14 of *Thermus thermophilus* showing outer capsid and inner vesicle; PT. One particle at left (arrow) displays a full, deformed vesicle. F. Levivirus MS2 of *E. coli*, PT. G. Inovirus X of *E. coli*, showing unusual flexibility; PT. H. Plectrovirus MVL51 of *Acholeplasma laidlawii*, UA. I, Plasmavirus L2 of *A. laidlawii* after density gradient purification, UA. Corticoviruses and cystoviruses are not shown because of their superficial resemblance to tectiviruses. UA, uranyl acetate; PT, phosphotungtate. Bars indicate 100 nm. Final magnifications are × 297 000 (A–C, E, F), × 148 000 (D and E), × 183 000 (H), and × approximately 150 000 (I). Figs. H and I, respectively, are reproduced with kind permission of the ICTV Database (curator Dr. Cornelia Büchen–Osmond, Columbia University, New York, NY) and Dr. J. Maniloff, University of Rochester, Rochester, NY).

envelope surrounding an icosahedral capsid containing three pieces of dsRNA. All polyhedral phages are virulent and are liberated by burst.

3. Filamentous phages comprise Inoviridae (ssDNA), which include long filaments (the genus *Inovirus)* or short rods (genus *Plectrovirus)* and are probably heterogeneous. Plectroviruses are found in mycoplasmas only. Inoviruses are liberated by slow extrusion from the host bacterium.
4. Pleomorphic phages are represented by the Plasmaviridae family. They have lipoprotein envelopes and contain naked dsDNA without a capsid. Plasmaviridae appear as round particles that only infect mycoplasmas and are liberated by budding.

B. Temporal sequence of discoveries

The first tailed phages, and indeed the first of all viruses, described after the introduction of negative staining were coliphages T2, T4, and T6. Discoveries were sometimes simultaneous; for example, both Hall and colleagues and Sinsheimer described the morphology of coliphage ϕX174 independently in 1959 in the same periodical. Similarly, two tectiviruses, although representing a small and rare virus group, were described in 1974 (Table II). It appears that discoveries tumble over each other when conditions are ripe. However, the last phage family to be discovered, the Tectiviridae, was described in the early 1980s.

C. Classification into subfamilies, genera, and species

The role of electron microscopy is less obvious at lower taxonomical levels because classification depends here largely on sequencing and other molecular data. Electron microscopical and molecular data are complementary or corroborate each other and are of equal importance. However, electron microscopical data are mandatory for classification of a new taxon by the ICTV. Electron microscopy is particularly important for instant diagnosis and attribution of novel phages to morphospecies, which may or may not be subdivided by molecular criteria. For example, the morphospecies "T7" is now classified as the subfamily Autographivirinae and is subdivided by genomics into three genera and 15 possible "species" (Lavigne *et al.*, 2008). In tailed phages of enterobacteria alone, at least 35 morphospecies are recognizable by electron microscopy (Ackermann *et al.*, 1997). In phages of rare and characteristic morphology, electron microscopy may provide direct and immediate identification. However, no morphological diagnosis is absolute; for example, it will not detect genetic hybrids with individual genes from another taxon. This is to be expected in phylogenetically related viruses such as tailed phages. A recent example of this kind is that of a myovirus with podovirus tail spike genes (Walter *et al.*, 2008). The great value of diagnostic electron microscopy is evident in phage studies in the dairy industry. It is well known that phages may

TABLE II Electron microscopical discovery of bacterial virus families[a]

Year	Family or genus	Phage	Investigators	Host
1959	Myoviridae	T2, T4, T6	Brenner *et al.*	*Escherichia coli*
	Microviridae	ϕX174	Hall *et al.*; Sinsheimer	*E. coli*
1960	Siphoviridae	T1, T5	Bradley and Kay	*E. coli*
	Podoviridae	P22	Anderson	*Salmonella* typhimurium
		T7	Huxley and Zubay	*E. coli*
1961	Leviviridae	f2	Loeb and Zinder	*E. coli*
		MS2	Davis *et al.*	
1963	Inoviridae	fd	Marvin and Hoffmann-Berling	*E. coli*
		f1	Zinder *et al.*	
		M13	Hofschneider	
1968	Corticoviridae	PM2	Espejo and Canelo	*Pseudoalteromonas espejiana*
1971	Plasmaviridae	MVL1	Gourlay	*Acholeplasma laidlawii*
	Plectrovirus	MVL2	Gourlay *et al.*	*A. laidlawii*
1973	Cystoviridae	ϕ6	Vidaver *et al.*	*Pseudomonas phaseolicola*
1974	Tectiviridae	AP50	Nagy	*Bacillus anthracis*
		PRD1	Olsen *et al.*	Gram negatives

[a] Negatively stained viruses only.

interfere with cheese making, destroying starter cultures and causing faulty fermentations (Moineau and Lévesque, 2005). Electron microscopy was able to provide family and often morphospecies identification in 700 phages of *Lactococcus* (Ackermann, 2007; Jarvis *et al.*, 1991).

D. Novel phages

One of the greatest contributions of electron microscopy to bacteriophage science is the ongoing description of novel phages. This is a world-wide effort, carried out by hundreds of investigators in many countries. Novel phages are described at the rate of 100 per year (Ackermann, 2007). The large number of observations is largely due to the simplicity of negative staining and the fact that electron microscopy allows for instant diagnosis. Indeed, phage samples can be processed after 2–3 hours of purification by centrifugation and washing in buffer and even tap water. This can be done in a medium-sized centrifuge and a fixed-angle rotor at only 25,000 g. Staining is instantaneous, and an examination may take as little as 1–2 min in the hands of a skilled observer. This makes electron microscopy one of the fastest techniques in microbiology. Generally, even without a study of their nucleic acid, viruses can be recognized immediately as novel or at least ascribed to known families. This is paramount for the identification of industrial or commercial phages.

VI. PHAGE ECOLOGY

A. Cautionary remarks

Bacterial viruses are ubiquitous and occur in any place where their hosts are found. It is widely agreed that bacteriophages are the most abundant life forms on earth (see Section I). Aquatic viruses have been investigated to a considerable extent. Insofar as known, most are tailed phages of the cyanobacterial genera *Synechococcus* and *Prochlorococcus*. By lysing their hosts, they influence many geochemical and biological processes, including carbon cycling and bacterial proliferation (Fuhrman, 1999; Suttle, 2005). Metagenomics, the analysis of collective microbial genes contained in an environmental sample, indicates that there are "an estimated 5000 viral genotypes in 200 liters of seawater" (Breitbart and Rohwer, 2005), but three quarters to 90% of the sequences encountered are unrelated to any sequence in extant databases (Angly *et al.*, 2006; Breitbart *et al.*, 2003). However, metagenomics reveals that marine samples contain 2–3% of sequences that can be related to specific phages of Myoviridae, Siphoviridae, Podoviridae, or Microviridae families and even to "prophages" (Angly *et al.*, 2006; Breitbart *et al.*, 2003; Breitbart and Rohwer, 2005). These studies are not supported by phage isolation or electron microscopical evidence.

Most environmental phages have been isolated after enrichment and are therefore not representative of the environment. Electron microscopical studies of nonbiased, uncultured phage populations are much rarer and, generally and regrettably, unacceptably poor. All too often phages are shown at low magnification and positively stained, making comparison with known phages impossible. Their dimensions, if indicated at all, seem to have been obtained without magnification control or are, if measured on positively stained viruses, essentially worthless. Nevertheless, precious few publications exist with high-quality electron micrographs of phages from marine and freshwater environments (Demuth *et al.*, 1993; Suttle and Chan, 1993; Torrella and Morita, 1979). These publications show that most aquatic viruses are tailed phages of Myoviridae, Siphoviridae, and Podoviridae families, whereas icosahedral and filamentous phages seem to be rare in water.

The same considerations apply to investigations of phages in the gut, one of the most important habitats of bacteriophages. Metagenomics indicates that human feces contain an estimated 1200 viral genotypes (Breitbart *et al.*, 2003). A similar diversity is found in horse feces. Many of the phages present there are related to coliphages T4 and λ (Cann *et al.*, 2005). Electron microscopy confirms that equine feces contain indeed an enormous variety of phages, but micrographs are often poor and the investigators do not attempt to identify their phages. Most of them are tailed, but this may reflect an intrinsic difficulty of diagnosing small isometric viruses (Kulikov *et al.*, 2007).

B. Phage counts in water

The quantification of waterborne phages relies on the affinity of dsDNA for uranyl acetate. Viruses are centrifuged directly onto a grid and stained. The best and seemingly only practical technique requires an ultracentrifuge with a swinging-bucket rotor and tubes provided with a flat bottom of Epoxy resin onto which an electron microscopical grid is deposited. Samples are centrifuged at 80,000 g for 90 min. A water column of up to 60 mm height can be centrifuged this way (Borsheim *et al.*, 1990). Viruses are stained with uranyl acetate. Positively stained viruses, mostly tailed phage heads and occasional phycodnaviruses, appear deep black and are counted easily at relatively low magnification. Virus numbers are calculated in a final step (Bratbak and Heldal, 1993). This technique provided the first reliable data on total virus numbers in seawater (Bergh *et al.*, 1989) and showed that viruses, generally tailed phages, occur in enormous numbers in water and marine sediments. It is at the basis of our estimates of phage frequency in the biosphere (see Section I).

Viral (phage) abundance decreases with depth and distance from the shore. Direct phage counts have been used to estimate burst sizes and mortality rates of aquatic phages, providing unique and precious data on virus turnover in nature. However, the technique has four major shortcomings. (a) It depends on positive staining and is thus unsuitable for the detection of viruses that do not contain large compact masses of dsDNA, in particular filamentous and ssRNA viruses. (b) It does not lend itself to the identification of short-tailed phages because short tails are almost invisible in positively stained particles. (c) Viruses (phages) detected can rarely be identified. (d) Their dimensions, and thus estimates of DNA content, are unreliable.

C. New phages everywhere?

Metagenomic studies indicate that we know only a very small part of the bacteriophages present in the biosphere and may never know them all. For example, phage research is still centered on phages of γ-proteobacteria , and vast parts of the earth, for example, tropical Africa and Siberia, have never been investigated for their presence. The author's personal experience is that novel phages are easy to find by electron microscopy and that every water sample contains novel phages.

VII. CONCLUSIONS

A. Advantages of electron microscopy

The contributions of electron microscopy to phage research are many and are summarized in Table III. To state it briefly: (a) TEM is the fastest and most cost-effective virological technique in existence. (b) Images constitute a permanent record and are stored and exchanged easily. (c) TEM allows for instant diagnosis of virus families and morphospecies. In contrast, sequencing is still an expensive procedure that may take many months if unusual bases are present. (d) TEM provides precise information on dimensions and structure and 3D information on the assembly products of virus proteins. (e) TEM is a handy way and inescapable

TABLE III How bacteriophages benefited from electron microscopy

Domain	Insights or discoveries
General	Virus nature of phages
Virion structure	Complexity and variability, dimensions, presence of organelles, capsomers, and triangulation numbers in tailed phages
Phylogeny	Monophyletic nature of tailed phage capsids and possibly tails
	Links between (a) tailed phages and herpesviruses and (b) tectiviruses and adenoviruses
Replication and assembly	Visualization of phage adsorption, intracellular phages, replicative intermediates; reconstitution of assembly chains
Classification	Establishment of families and morphospecies
	Establishment of subfamilies, genera, and species (alone or jointly with sequence data)
	Description of over 5500 viruses
Ecology	Phages are the most frequent entities in the biosphere
	Diversity in nature
	Visualization of infected bacteria
Industrial microbiology	Visualization of harmful phages

requirement for preliminary characterization of phages used in therapy and biocontrol (Sulakvelidze and Kutter, 2005). (f) Electron microscopes have a long life span, provided that they are maintained properly. For example, the author is the happy and satisfied user of a 43-year-old Philips EM 300. (g) Electron microscopy often allows predicting properties by analogy with known viruses, for example, the nature, conformation, and molecular weight of nucleic acids, particle weight, and buoyant density. It gives us a view of a finished product, the complete virus, that is easy to classify. One may say in a succinct formula that "you see the particle and you know what it is."

B. Problems of electron microscopy

Electron microscopy always had problems of imaging and interpretation, but the rise of digital electron microscopy and CCD cameras in the 1990s created a novel situation. In a general way, it appears that the quality of phage electron microscopy has slipped and that many present-day phage electron micrographs are far inferior in quality to the first images of negatively stained phages taken in the late 1950s (Brenner *et al.*, 1959). A peak in phage electron microscopy was reached in the 1970s (see Dalton and Haguenau, 1973), but this seems to be forgotten. For example, in a personal survey of about 130 phage papers since 2006, which described novel phages by mostly digital TEM, 70 featured low-contrast, unsharp, astigmatic, poor to very poor pictures. Some "phage descriptions" reported neither phage dimensions nor stains and did not even specify the electron microscopes used. Only some 20 papers showed good-quality figures. The decline of phage electron microscopy may be linked to personal factors, namely the loss of great electron microscopists such as Eduard Kellenberger or Tom Anderson, their replacement by inexperienced investigators, and a perceived leniency even of reputed journals to accept substandard micrographs. Indeed, regardless of the electron microscope used, poor micrographs can be associated with an inadequate technique, whether in specimen processing or imaging.

Digital TEMs and CCD cameras are here to stay. CCD cameras have largely obviated darkroom photography and are wildly popular with inexperienced microscopists who fear work in the darkroom.

1. Compared to conventional TEMs, digital electron microscopes appear to be more expensive, cannot be maintained normally by users, and need expensive service contracts.
2. Their life span remains to be seen and they are more difficult to control than "manual" TEMs with respect to contrast and magnification. However, conventional photographic chemicals and papers may be difficult to find because the market has shrunk.

3. The relative quality of the various digital TEMs and CCD cameras is difficult to evaluate in the absence of comparative studies. It seems that present top-grade TEMs, whether produced by FEI, JEOL, or Hitachi, and concomitant CCD cameras are roughly equivalent with respect to resolution. The instruments are improved continuously. For example, TEMs manufactured by the FEI Company (Hillsboro, OR), which acquired the Philips Electron Optics Division, produce micrographs of striking quality.
4. With "manual" TEMs, contrast is controlled in the darkroom by means of graded filters and papers. In the case of digital TEMs, one can obtain high-resolution and high-contrast pictures by the adjustment of pixel intensities with CCD camera software (Tiekotter and Ackermann, 2009). It is unfortunate that the manufacturers of electron microscopes have seemingly neglected to issue guidelines for contrast enhancement, leaving users to fend for themselves.
5. With both "manual" and digital TEMs, magnification is controlled by means of test specimens, for example, catalase crystals (Luftig, 1967) or T4 phage tails. Latex spheres or diffraction grating replicas are suitable for low magnification only (10–30,000×). With "manual" microscopes, magnification can be corrected in the darkroom within minutes. The magnification of digital electron microscopes is normally set by the installer and cannot, or only with great difficulty, be adjusted by the user. To control magnification, the user must photograph test specimens and define correction factors by calculation.

Practically, it is recommended that

1. TEM manufacturers publish instructions for contrast enhancement.
2. All specimens be purified before examination. Crude lysates are to be banished. Purification is achieved most easily by differential centrifugation and washing in buffer.
3. Improve contrast of digital microscopes via Photoshop technology.
4. Control magnification regularly by means of test specimens.

C. Genomics vs electron microscopy

Can genomics replace electron microscopy? This might be suggested by the rise of rapid sequencing and the ensuing increased availability of completely sequenced virus genomes. It is indeed advocated in discussions by unconditional partisans of genomics. The answer is roundly "no." Genomics gives us the genome and genes, thus the elementary building blocks of a virus. It also gives gene order and direction of transcription, and it identifies genes coding for proteins with homology to known enzymes or virion components, restriction-modification enzymes, capsid protein size, or the length of tape measure proteins. Further, genomics indicates horizontal gene transfer or gene swapping, may indicate relationships between virus groups and individual viruses, and allows for quantification of relationships and the construction of phylogenetic trees. All this provides unprecedented insights into virus evolution and is a precious help in phage classification.

However, electron microscopy provides information on virion structure, while genomics does not show the whole virus, gives not a single dimension, provides no information on virus structure and physicochemical properties, does not identify unusual bases such as 5-hydroxymethylcytosine, and predicts only some biological properties, such as a lysogenic nature. No sequence can indicate simple things such as the size of phage capsids, their geometry, or the number of capsomers. If, as likely, the length of phage tails depends on the length of ruler protein genes (Katsura and Hendrix, 1984; Pedulla *et al.*, 2003), this must be ascertained by the measurement of many phage tails under strict magnification control. Unfortunately, this has not been the case. If, as pretended, a genome contains all information on a virus, we have not yet found the instruction manual to read it.

With respect to virus identification, genomics generally does not indicate to which virus family a tailed phage belongs; for example, there are no sequences specific to Myo-, Sipho-, or Podoviridae. Only in the case of small polyhedral or filamentous phages (Micro-, Levi, and Inoviridae) does genomics allow for an identification of virus families (Ackermann and Kropinski, 2007). Similarly, a *Bacillus* tectivirus from the earthworm gut was identified by genomics alone without the benefit of electron microscopy (Schuch *et al.*, 2010). However, in a general way, investigation of a complete virus sequence may take months and is infinitely slower and more labor-intensive than electron microscopy.

Can metagenomics replace electron microscopy? The answer is "no" again. For virus identification, metagenomics relies totally on known and identified genes and genomes, which, in turn, belong to viruses known and characterized by electron microscopy. In other terms, the vast majority of countless genes detected by metagenomics can be identified only to the extent as they belong to known sequences from known viruses. Further, metagenomics will not tell whether any detected sequences belong to complete, infectious virions or not. Can electron microscopy replace genomics? The answer is "yes," but only when it comes to the identification of high-level taxonomic categories. Clearly, electron microscopy and genomics (or metagenomics) are not alternatives, but complementary. Both of them answer different questions and appear as different fingers of the same hand.

REFERENCES

Abedon, S. T. (ed.) (2008). Bacteriophage Ecology. Population Growth, Evolution, and Impact of Bacterial Viruses. Cambridge University Press, Cambridge, UK.

Ackermann, H.-W. (2005). Bacteriophage classification. *In* ''Bacteriophages: Biology and Applications'' (E. Kutter and A. Sulakvelidze, eds.), pp. 67–89. CRC Press, Boca Raton, FL.

Ackermann, H.-W. (2006). Classification of bacteriophages. *In* ''The Bacteriophages'' (R. Calendar, ed.), 2nd edn. pp. 8–16. Oxford University Press, New York.

Ackermann, H.-W. (2007). 5500 Phages examined in the electron microscope. *Arch. Virol.* **152:**227–243.

Ackermann, H.-W., and DuBow, M. S. (1987). *Viruses of Prokaryotes* **Vol. I, pp. 67, 73, 116:**CRC Press, Boca Raton, FL.

Ackermann, H.-W., DuBow, M. S., Gershman, M., Karska-Wysocki, B., Kasatiya, S. S., Loessner, M. J., Mamet-Bratley, M. D., and Regué, M. (1997). Taxonomic changes in tailed phages of enterobacteria. *Arch. Virol.* **142:**1381–1390.

Ackermann, H.-W., Jolicoeur, P., and Berthiaume, L. (1974). Avantages et inconvénients de l'acétate d'uranyle en virologie comparée: Étude de quatre bactériophages caudés. *Can. J. Microbiol.* **20:**1093–1099.

Ackermann, H.-W., and Kropinski, A. M. (2007). Curated list of prokaryote viruses with fully sequenced genomes. *Res. Microbiol.* **158:**555–566.

Adams, M. H. (1959). Bacteriophages, pp. 38, 161–187. Interscience Publishers, New York.

Anderson, T. F. (1960). On the fine structure of the temperate bacteriophages P1, P2 and P22. *In* ''Proc. Eur. Reg. Conf. Electron Microscopy, Delft 1960'' (A. L. Houwink and B. Spit, eds.), Vol. 2, pp. 1008–1011. De Nederlandse Vereniging voor Elektronenmicroscopie, Delft.

Anderson, T. F., Rappaport, C., and Muscatine, N. A. (1953). On the structure and osmotic properties of phage particles. *In* ''Le Bactériophage, Premier Colloque International, Rouaumont, 1952'' (Institut Pasteur, ed.), *Ann. Inst Pasteur* **84:**5–14.

Anderson, T. F., and Stanley, W. M. (1941). A study by means of the electron microcope of the reaction between tobacco mosaic virus and its antiserum. *J. Biol. Chem.* **139:**339–344.

Angly, F. E., Felts, B., Breitbart, M., Salamon, P., Edwards, R. A., Carlson, C., Chan, A. M., Haynes, M., Kelley, S., Liu, H., Mahaffy, J. M., Mueller, J. E., *et al.* (2006). The marine viromes of four oceanic regions. *PloS Biol.* **4:**2121–2131 (e368).

Baker, M. L., Jiang, W., Rixon, F. J., and Chiu, W. (2005). Common ancestry of herpesviruses and tailed DNA bacteriophages. *J. Virol.* **79:**14967–14970.

Bayer, M. E., and Bocharov, A. F. (1973). The capsid structure of bacteriophage lambda. *Virology* **54:**465–475.

Bayer, M. E., and Cummings, D. J. (1977). Structural aberrations in T-even bacteriophage. VIII. Surface morphology of T4 lollipops. *Virology* **76:**767–780.

Bayer, M. E., and Remsen, C. C. (1970). Bacteriophage T2 as seen with the freeze-etching technique. *Virology* **40:**703–718.

Bergh, O., Borsheim, K. Y., Bratbak, G., and Heldal, M. (1989). High abundance of viruses found in aquatic environments. *Nature* **340:**467–468.

Böhm, J., Lambert, O., Frangakis, A. S., Letellier, L., Baumeister, W., and Rigaud, J. L. (2001). FhuA-mediated phage genome transfer into liposomes: A cryoelectron tomography study. *Curr. Biol.* **11:**1168–1175.

Borsheim, K. Y., Bratbak, G., and Heldal, M. (1990). Enumeration and biomass estimation of planktonic bacteria and viruses by transmission electron microscopy. *Appl. Environ. Microbiol.* **56:**352–356.

Bradley, D. E. (1967). Ultrastructure of bacteriophages and bacteriocins. *J. Bacteriol.* **31:**230–314.

Bradley, D. E., and Kay, D. (1960). The fine structure of bacteriophages. *J. Gen. Microbiol.* **23:**553–563.

Bratbak, G., and Heldal, M. (1993). Total count of viruses in aquatic environments. *In* ''Handbook of Methods in Aquatic Microbial Ecology'' (P. F. Kemp, B. F. Sherr, and J. J. Cole, eds.), pp. 135–138. CRC Press, Boca Raton, FL.

Breitbart, M., Hewson, I., Felts, B., Mahaffy, J. M., Nulton, J., Salamon, P., and Rohwer, F. (2003). Metagenomic analysis of an uncultured viral community from human feces. *J. Bacteriol.* **185:**6220–6223.

Breitbart, M., and Rohwer, F. (2005). Here a virus, there a virus, everywhere the same virus? *Trends Microbiol.* **13:**278–284.

Brenner, S., and Horne, R. W. (1959). A negative staining method for high resolution electron microscopy of viruses. *Biochim. Biophys. Acta* **34:**103–110.

Brenner, S., Streisinger, G., Horne, R. W., Champe, S. P., Barnett, L., Benzer, S., and Rees, M. W. (1959). Structural components of bacteriophage. *J. Mol. Biol.* **1:**281–292.

Broers, A. N., Panessa, B. J., and Gennaro, J. F. (1975). High-resolution scanning electron microscopy of bacteriophages 3C and T4. *Science* **189:**637–639.

Brüssow, H. (2009). The not so universal tree of life *or* the place of viruses in the living world. *Philos. Transact. Royal Soc. B* **364:**2263–2274.

Brüssow, H., and Hendrix, R. W. (2002). Phage genomics: Small is beautiful. *Cell* **108:**13–16.

Brüssow, H., and Kutter, E. (2005a). Genomics and evolution of tailed phages. *In* ''Bacteriophages: Biology and Applications'' (E. Kutter and A. Sulakvelidze, eds.), pp. 91–128. CRC Press, Boca Raton, FL.

Brüssow, H., and Kutter, E. (2005b). Phage ecology. *In* ''Bacteriophages: Biology and Applications'' (E. Kutter and A. Sulakvelidze, eds.), pp. 129–163. CRC Press, Boca Raton, FL.

Burnet, F. M. (1933). The classification of dysentery-coli bacteriophages. *III. A correlation of the serological classification with certain biochemical tests. J. Pathol. Bacteriol.* **37:**179–184.

Cann, A. J., Fandrich, S. E., and Heaphy, S. (2005). Analysis of the virus population present in equine faeces indicates the presence of hundreds of uncharacterized virus genomes. *Virus Genes* **30:**151–156.

Cerritelli, M. E., Wall, J. S., Simon, M. N., Conway, J. F., and Steven, A. C. (1996). Stoichiometry and domainal organization of the long tail-fiber of bacteriophage T4: A hinged viral adhesin. *J. Mol. Biol.* **260:**767–780.

Chang, J. T., Schmid, M. F., Haase-Pettingell, C., Weigele, P. R., King, J. A., and Chiu, W. (2010). Visualizing the structural changes of bacteriophage epsilon15 and its *Salmonella* host during infection. *J. Mol. Biol.* **402:**731–740.

Chipman, P. R., Agbandje-McKenna, M., Renaudin, J., Baker, T., and McKenna, R. (1998). Structural analysis of the spiroplasma virus, SpV4: Implications for evolutionary variation to obtain host diversity among the *Microviridae. Structure* **6:**135–145.

Cummings, D. J., Chapman, V. A., and DeLong, S. S. (1977). Structural aberrations in T-even bacteriophage. *IX. Effect of mixed infection on the production of giant bacteriophage. J. Virol.* **22:**489–499.

Davis, J. E., Strauss, J. H., and Sinsheimer, R. L. (1961). Bacteriophage MS2: Another RNA phage. *Science* **134:**1427.

DeMars, R. I., Luria, S. E., Fisher, H., and Levinthal, C. (1953). The production of incomplete phage particles by the action of proflavine and the properties of incomplete particles. *Ann. Inst. Pasteur* **84:**113–128.

Demuth, J., Neve, H., and Witzel, K.-P. (1993). Direct electron microscopy study on the morphological diversity of bacteriophage populations in Lake Plussee. *Appl. Environ. Microbiol.* **59:**3378–3384.

D'Herelle, F. (1917). Sur un microbe invisible antagoniste des bacilles dysentériques. *C.R. Hebd. Seances Acad. Sci. D* **165:**373–375.

D'Herelle, F. (1921). Le bactériophage: Son comportement dans l'immunité. Masson, Paris 112 and 374.

Dryden, K. A., Wang, G., Yeager, M., Nibert, M. L., Coombs, K. M., Furlong, D. B., Fields, B. N., and Baker, T. S. (1993). Early steps in reovirus infection are associated with dramatic changes in supramolecular structure and protein conformation: Analysis of virions and subviral particles by cryoelectron microscopy and image reconstruction. *J. Cell Biol.* **122:**1023–1041.

Dublanchet, A. (2009). ''Des virus pour combattre les infections. Renouveau, La phagothérapie d'un traitement au secours des antibiotiques''. Editions Favre, Lausanne.

Dubochet, J. (1988). The contribution to society from electron microscopy in the life sciences. *In* ''The Contribution of Electron Microscopy to Society''. Philips Electron Optics Bull. Special Issue 128, pp. 17–20. Philips Analytical, Eindhoven, The Netherlands.

Dubochet, J., Adrian, M., Lepault, J., and McDowall, A. W. (1985). Cryo-electron microscopy of vitrified biological specimens. *TIBS* **10:**143–146.

Dubochet, J., Ducommun, M., Zollinger, M., and Kellenberger, E. (1971). A new preparation method for dark-field electron microscopy of biomacromolecules. *J. Ultrastruct. Res.* **35:**147–167.

Dubrovin, E. V., Voloshin, A. G., Kraevsky, S. V., Ignatyuk, T. E., Abramchuk, S. S., Yaminsky, I. V., and Ignatov, S. G. (2008). Atomic force microscopy investigation of phage infection of bacteria. *Langmuir* **24:**13068–13074.

Duda, R. L., Hendrix, R. W., Huang, W. M., and Conway, J. F. (2006). Shared architecture of bacteriophage SPO1 and herpesvirus capsids. *Curr. Biol.* **16:**R11–R13 16, 440 (Addendum).

Effantin, G., Boulanger, P., Neumann, E., Letellier, L., and Conway, J. F. (2006). Bacteriophage T5 structure reveals similarities with HK97 and T4 suggesting evolutionary relationships. *J. Mol. Biol.* **361:**993–1002.

Espejo, R. T., and Canelo, E. S. (1968). Properties of bacteriophage PM2: A lipid-containing bacterial virus. *Virology* **34:**738–747.

Fauquet, C. M., Mayo, M. A., Maniloff, J., Desselberger, U., and Ball, L. A. (2005). Virus Taxonomy. VIIIth Report of the International Committee on Taxonomy of Viruses. Elsevier Academic Press.

Fiandt, M., Hradecna, Z., Lozeron, H. A., and Szybalski, W. (1971). Electron micrographic mapping of deletions, insertions, inversions, and homologies in the DNAs of coliphages lambda and ϕ80. *In* ''The Bacteriophage Lambda'' (A. D. Hershey, ed.), pp. 329–354. Cold Spring Harbor Laboratory Press, Cold Spring Harbor, NY.

Fiandt, M., Szybalski, W., and Malamy, M. H. (1972). Polar mutations in *lac*, *gal* and phage λ consist of a few IS-DNA sequences inserted with either orientation. *Mol. Gen. Genet.* **119:**223–231.

Fokine, A., Battisti, A. J., Bowman, V. D., Efimov, A. V., Kurochkina, L. P., Chipman, P. R., Mesyanzhinov, V. V., and Rossmann, M. G. (2007). Cryo-EM study of the *Pseudomonas* bacteriophage ϕKZ. *Structure* **15:**1099–1104.

Fokine, A., Chipman, P. R., Leiman, P. G., Mesyanzhinov, V. V., Rao, V. B., and Rossmann, M. G. (2004). Molecular architecture of the prolate head of bacteriophage T4. *Proc. Natl. Acad. Sci. USA* **101:**6003–6008.

Fokine, A., Leiman, P. G., Shneider, M. M., Ahvazi, B., Boeshans, K. M., Steven, A. C., Black, L. W., Mesyanzhinov, V. V., and Rossmann, M. G. (2005). Structural and functional similarities between the capsid proteins of bacteriophages T4 and HK97 point to a common ancestry. *Proc. Natl. Acad. Sci. USA* **102:**7128–7168.

Fuhrman, J. A. (1999). Marine viruses and their biogeochemical and ecological effects. *Nature* **399:**541–548.

Gogol, E. P., Young, M. C., Kubasek, W. L., Jarvis, T. C., and Von Hippel, P. H. (1992). Cryoelectron microscopic visualization of functional subassemblies of the bacteriophage T4 DNA replication complex. *J. Mol. Biol.* **224:**395–412.

Gourlay, R. N. (1971). Mycoplasmatales virus-laidlawii 2, a new virus isolated from *Acholeplasma laidlawii*. *J. Gen. Virol.* **12:**65–67.

Gourlay, R. N., Bruce, J., and Garwes, D. J. (1971). Characterization of Mycoplasmatales virus laidlawii 1. *Nature New Biol.* **229:**118–119.

Haguenau, F., Hawkes, P. W., Hutchison, J. L., Satiat-Jeunemaître, B., Simon, G. T., and Williams, D. B. (2003). Key events in the history of electron microscopy. *Microsc. Microanal.* **9:**96–138.

Hall, C. E. (1955). Electron densitometry of stained virus particles. *J. Biophys. Biochem. Cytol.* **1:**1–12.

Hall, C. E., Maclean, E. C., and Tessman, I. (1959). Structure and dimensions of bacteriophage ϕX174 from electron microscopy. *J. Mol. Biol.* **1:**192–194.

Hankin, E. H. (1896). L'action bactericide des eaux de la Jumna et du Gange sur le vibrion du cholera. *Ann. Inst. Pasteur* **10:**511.

Hatfull, G. F. (2008). Bacteriophage genomics. *Curr. Opin. Microbiol.* **11:**447–453.

Hayat, M. A., and Miller, S. E. (1990). Negative Staining. McGraw-Hill, New York.

Hemminga, M. A., Vos, W. L., Nazarov, P. V., Koehorst, R. B. M., Wolfs, C. J. A. M., Spruijt, R. B., and Stopar, D. (2010). Viruses: Incredible nanomachines. *New advances with filamentous phages. Eur. Biophys. J.* **39:**541–550.

Hendrix, R. W. (2008). Phage evolution. *In* "Bacteriophage Ecology. Population Growth, Evolution, and Impact of Bacterial Viruses" (S. T. Abedon, ed.), pp. 177–194. Cambridge University Press, Cambridge, UK.

Hermann, R., Schwarz, H., and Müller, M. (1991). High precision immunoscanning electron microscopy using Fab fragments coupled to ultra-small colloidal gold. *J. Struct. Biol.* **107:**38–47.

Hertveldt, K., Beliën, T., and Volckaert, G. (2009). General M13 phage display: M13 phage display in identification and characterization of protein-protein interactions. *In* "Bacteriophages, Methods and Protocols" (M. R. J. Clokie and A. M. Kropinski, eds.), Vol. 2, pp. 321–339. Methods in Molecular Biology, 502. Humana Press, Clifton, NJ.

Hershey, A. D., and Chase, M. (1952). Independent functions of viral protein and nucleic acid in growth of bacteriophages. *J. Gen. Physiol.* **36:**39–56.

Hoffmann-Berling, H., Dürwald, H., and Beulke, I. (1963). Ein fädiger DNS-Phage (fd) und ein sphärischer RNS-Phage (fr) wirtsspezifisch für männliche Stämme von *E. coli*. III. Biologisches Verhalten von fd und fr. *Zeitschr. Naturforsch.* **18B:**893–898.

Hofschneider, P. H. (1963). Untersuchungen über 'kleine' E. coli K-12 Bakteriophagen. 1. Die Isolierung und einige Eigenschaften der 'kleinen' Bacteriophagen M12, M13 und M20. *Zeitschr. Naturforsch.* **18B:**203–205.

Hofschneider, P. H., and Preuss, A. (1963). M13 bacteriophage liberation from intact bacteria as revealed by the electron microscope. *J. Mol. Biol.* **7:**450–451.

Holt, S. C., and Beveridge, T. J. (1982). Electron microscopy: Its development and application to microbiology. *Can. J. Microbiol.* **28:**1–53.

Hradecna, Z., and Szybalski, W. (1969). Electron micrographic maps of deletions and substitutions in the genomes of transducing coliphages *λdg* and *λbio*. *Virology* **38:**473–477.

Huiskonen, J. T., and Butcher, S. J. (2007). Membrane-containing viruses with icosahedrally symmetric capsids. *Curr. Opin. Struct. Biol.* **17:**229–236.

Huiskonen, J. T., Kivelä, H. M., Bamford, D. H., and Butcher, S. J. (2004). The PM2 virion has a novel organization with an internal membrane and pentameric receptor binding. *Nat. Struct. Mol. Biol.* **11:**850–856.

Huxley, H. E., and Zubay, G. (1960). Fixation and staining of nucleic acids for electron microscopy. *In* "Proc Eur. Reg. Conf. Electron Microscopy, Delft 1960" (A. L. Houwink and B. J. Spit, eds.), Vol. II, pp. 699–702. Nederlandse Vereniging voor Electronenmicroscopie, Delft, The Netherlands.

Huxley, H. E., and Zubay, G. (1961). Preferential staining of nucleic acid-constaining structures for electron microscopy. *J. Biophys. Biochem. Cytol.* **11:**273–296.

Jäälinoja, H. T., Roine, E., Laurinmäki, P., Kivelä, H. M., Bamford, D. H., and Butcher, S. J. (2008). Structure and host-cell interaction of SH1, a membrane-containing, halophilic euryarchaeal virus. *Proc. Natl. Acad. Sci. USA* **105:**8008–8013.

Jarvis, A. W., Fitzgerald, G. F., Mata, M., Mercenier, A., Neve, H., Powell, I. B., Ronda, C., Saxelin, M., and Teuber, M. (1991). Species and type phages of lactococcal bacteriophages. *Intervirology* **32:**2–9.

Katsura, I., and Hendrix, R. W. (1984). Length determination in bacteriophage lambda tails. *Cell* **39:**691–698.

Kellenberger, E., and Arber, W. (1957). Electron microscopical studies of phage multiplication. *I. A method for quantitative analysis of particle suspensions. Virology* **3:**245–255.

Kellenberger, E., and Wunderli-Allenspach, H. (1995). Electron microscopic studies on intracellular phage development: History and perspectives. *Micron* **26:**213–245.

Khayat, R., Tang, L., Larson, E. T., Lawrence, M. C., Young, M., and Johnson, J. E. (2005). Structure of an archaeal virus capsid protein reveals common ancestry to eukaryotic and bacterial viruses. *Proc. Natl. Acad. Sci. USA* **102:**18944–18949.

Kim, J.-S., and Davidson, N. (1974). Electron microscope heteroduplex study of sequence relations of T2, T4, and T6 bacteriophage DNAs. *Virology* **57:**93–111.

Kleinschmidt, A. K., Lang, D., Jacherts, D., and Zahn, R. K. (1962). Darstellungen und Längenmessungen des gesamten Desoxyribonucleinsäure-Inhaltes von T2-Bakteriophagen. *Biochim. Biophys. Acta* **61:**857–864.

Kolbe, W. F., Ogletree, D. F., and Salmeron, M. B. (1992). Atomic force microscopy imaging of T4 bacteriophages on silicon substrates. *Ultramicroscopy* **42–44:**1113–1117.

Kottmann, U. (1942). Morphologische Befunde aus taches vierges von Coliculturen. *Arch. Ges. Virusforsch.* **2:**388–396.

Kulikov, E. E., Isaeva, A. S., Rotkina, A. S., Manykin, A. A., and Letarov, A. V. (2007). Diversity and dynamics of bacteriophages in equine feces (Russian). *Mikrobiologiya* **76:**271–278.

Kutter, E. M., DeVos, D., Gvasalia, G., Alavidze, Z., Gogokhia, L., Kuhl, S., and Abedon, S. T. (2010). Phage therapy in clinical practice: Treatment of human infection. *Curr. Pharmaceut. Biotechnol.* **11:**58–86.

Kuznetzov, Yu.G., Chang, S.-C., and McPherson, A. (2011). Investigation of bacteriophage T4 by atomic force microscopy. *Bacteriophage* **1:**165–173.

Kuznetzov, Yu.G., Martiny, J. B. H., and McPherson, A. (2010). Structural analysis of a Syechococcus myovirus S-CAM4 and infected cells by atomic force microscopy. *J. Gen. Virol.* **91:**3095–3104.

Lavigne, R., Seto, D., Mahadeva, P., Ackermann, H.-W., and Kropinski, A. M. (2008). Unifying classical and molecular taxonomic classification: Analysis of the *Podoviridae* using BLASTP-based tools. *Res. Microbiol.* **159:**406–414.

Lepault, J., Dubochet, J., Baschong, W., and Kellenberger, E. (1987). Organization of double-stranded DNA in bacteriophages: A study by cryo-electron microscopy of vitrified samples. *EMBO J.* **6:**1507–1512.

Levinthal, C., and Fisher, H. W. (1953). Maturation of phages: The evidence of phage precursors. *Cold Spring Harbor Symp. Quant. Biol.* **18:**29–33.

Liu, X., Zhang, Q., Murata, K., Baker, M. L., Sullivan, M. B., Fu, C., Dougherty, M. T., Schmid, M. F., Osburne, M. S., Chisholm, S. W., and Chiu, W. (2010). Structural changes in a marine podovirus associated with release of its genome into *Prochlorococcus. Nat. Struct. Mol. Biol.* **17:**830–836.

Loeb, T., and Zinder, N. D. (1961). A bacteriophage containing RNA. *Proc. Natl. Acad. Sci. USA* **47:**282–289.

Luftig, R. B. (1967). An accurate measurement of the catalase crystal period and its use as an internal marker for electron microscopy. *J. Ultrastruct. Res.* **20:**91–102.

Luria, S. E., and Anderson, T. F. (1942). The identification and characterization of bacteriophages with the electron microscope. *Proc. Natl. Acad. Sci. USA* **27:**127–130.

Luria, S. E., Delbrück, M., and Anderson, T. F. (1943). Electron microscope studies of bacterial viruses. *J. Bacteriol.* **46:**57–58.

Marvin, D. A., and Hoffmann-Berling, H. (1963). Physical and chemical properties of two new small bacteriophages. *Nature* **197:**517–518.

Moineau, S., and Lévesque, C. (2005). Control of bacteriophages in industrial fermentations. *In* "Bacteriophages: Biology and Applications" (E. Kutter and A. Sulakvelidze, eds.), pp. 285–295. CRC Press, Boca Raton, FL.

Morais, M. C., Choi, K. H., Koti, J. S., Chipman, P. R., Anderson, D. L., and Rossmann, M. G. (2005). Conservation of the capsid structure in tailed dsDNA bacteriophages: The pseudoatomic structure of ϕ29. *Mol. Cell* **18:**149–159.

Nagy, E. (1974). A highly specific phage attacking *Bacillus anthracis* strain Sterne. *Acta Microbiol. Acad. Sci. Hung.* **21:**257–263.

Olsen, R. H., Siak, J.-S., and Gray, R. H. (1974). Characteristics of PRD1, a plasmid-dependent broad host-range DNA bacteriophage. *J. Virol.* **14:**689–699.

Pedulla, M. L., Ford, M. E., Houtz, J. M., Karthikeyan, T., Wadsworth, C., Lewis, J. A., Jacobs-Sera, D., Falbo, J., Gross, J., Pannunzio, N. R., Brucker, W., Kumar, V., *et al.* (2003). Origin of highly mosaic mycobacteriophage genomes. *Cell* **113:**171–182.

Pell, L. G., Kanelis, V., Donaldson, L. W., Howell, P. L., and Davidson, A. R. (2009). The phage λ major tail protein structure reveals a common evolution for long-tailed phages and the type VI bacterial secretion system. *Proc. Natl. Acad. Sci. USA* **106:**4160–4165.

Pfankuch, E., and Kausche, G. A. (1940). Isolierung und übermikroskopische Abbildung eines Bakteriophagen. *Naturwissenschaften* **28:**46.

Portmann, R., Sogo, J. M., Koller, T., and Zillig, W. (1974). Binding sites of *E. coli* RNA polymerase on T7-DNA as determined by electron microscopy. *FEBS Lett* **45:**64–67.

Raettig, H. (1967). *Bakteriophagie 1957–1965.* **Vol. II:** Gustav Fischer, Stuttgart, Germany.

Ruska, H. (1940). Über die Sichtbarmachung der bakteriophagen Lyse im Übermikroskop. *Naturwissenschaften* **28:**45–46.

Ruska, H. (1942). Morphologische Befunde bei der bakteriophagen Lyse. *Arch. Ges. Virusforsch.* **2:**345–387.

Ruska, H. (1943). Ergebnisse der Bakteriophagenforschung und ihre Deutung nach morphologischen Befunden. *Ergeb. Hyg. Bakteriol. Immunforsch. Exp. Ther.* **25:**437–498.

Rydman, P. S., Caldentey, J., Butcher, S. J., Fuller, S. D., Rutten, T., and Bamford, D. H. (1999). Bacteriophage PRD1 contains a labile receptor-binding structure at each vertex. *J. Mol. Biol.* **291:**575–587.

Schito, G. C. (1974). Development of coliphage N4: Ultrastructural studies. *J. Virol.* **13:**186–196.

Schuch, R., Pelzek, A. J., Kan, S., and Fischetti, V. A. (2010). Prevalence of *Bacillus anthracis*-like organisms and bacteriophages in the intestinal tract of the earthworm *Eisenia fetida*. *Appl. Environ. Microbiol.* **76:**2286–2294.

Schwartz, F. M., and Zinder, N. D. (1963). Crystalline aggregates in bacterial cells infected with the RNA phage f2. *Virology* **21:**276–278.

Simon, L. D., and Anderson, T. F. (1967). The infection of *Escherichia coli* by T2 and T4 bacteriophages as seen in the electron microscope. *I. Attachment and penetration. Virology* **32:**279–297.

Simon, P., Lichte, H., Formanek, P., Lehmann, M., Huhle, R., Carillo-Cabrera, W., Harscher, A., and Ehrlich, H. (2008). Electron holography of biological samples. *Micron* **39:**229–256.

Sinsheimer, R. L. (1959). A single-stranded deoxyribonucleic acid from bacteriophage ϕX174. *J. Mol. Biol.* **1:**43–53.

Steere, R. L. (1957). Electron microscopy of structural details in frozen biological specimens. *J. Biophys. Biochem. Cytol.* **3:**45–59.

Steven, A. C., Trus, B. L., Booy, F. P., Cheng, N., Zlotnick, A., Caston, J. R., and Conway, J. F. (1997). The making and breaking of symmetry in virus capsid assembly: Glimpses of capsid biology from cryoelectron microscopy. *FASEB J.* **11:**733–742.

Sulakvelidze, A., and Kutter, E. (2005). Bacteriophage therapy in humans. *In* ''Bacteriophages: Biology and Applications'' (E. Kutter and A. Sulakvelidze, eds.), pp. 381–436. CRC Press, Boca Raton, FL.

Summers, W. C. (1999). Felix d'Herelle and the Origins of Molecular Biology. pp. 60–81. Yale University Press, New Haven, CT.

Suttle, C. A. (2005). Viruses in the sea. *Nature* **437:**356–361.

Suttle, C. A., and Chan, A. M. (1993). Marine cyanophages infecting oceanic and coastal strains of *Synechococcus*: Abundance, morphology, cross-infectivity, and growth characteristics. *Mar. Ecol. Prog. Ser.* **92:**99–109.

Tao, Y., Olson, N. H., Xu, W., Anderson, D. L., Rossmann, M. G., and Baker, T. S. (1998). Assembly of a tailed bacterial virus and its genome release studied in three dimensions. *Cell* **95:**431–437.

Tiekotter, K. L., and Ackermann, H.-W. (2009). High-quality virus images obtained by TEM and CCD technology. *J. Virol. Meth.* **159:**87–92.

Torrella, F., and Morita, R. Y. (1979). Evidence by electron micrographs for a high incidence of bacteriophage particles in the waters of Yaquina bay, Oregon: Ecological and taxonomical implications. *Appl. Environ. Microbiol.* **37:**774–778.

Twort, F. W. (1915). An investigation on the nature of ultra-microscopic viruses. *Lancet ii*1241–1243.

Vidaver, A. K., Koski, R. K., and Van Etten, J. L. (1973). Bacteriophage ϕ6: A lipid-containing virus of *Pseudomonas phaseolicola*. *J. Virol.* **11:**799–805.

Vollenweider, H. J., and Szybalski, W. (1978). Electron microscopic mapping of RNA polymerase binding to coliphage lambda DNA. *J. Mol. Biol.* **123:**485–498.

Walter, M., Fiedler, C., Grassl, R., Biebl, M., Rachel, R., Hermo-Parrado, X. L., Llamas-Saiz, A. L., Seckler, R., Miller, S., and Van Raaij, M. J. (2008). Structure of the receptor-binding protein of bacteriophage Det7: A podoviral tail spike in a myovirus. *J. Virol.* **82:**2265–2273.

Wendelschafer-Crabb, G., Erlandsen, S. L., and Walker, D. H. (1975). Conditions critical for optimal vizualization of bacteriophage adsorbed to bacterial surfaces by scanning electron microscopy. *J. Virol.* **15:**1498–1503.

Westmoreland, B. C., Szybalski, W., and Ris, H. (1969). Mapping of deletions and substitutions in heteroduplex DNA molecules of bacteriophage lambda by electron microscopy. *Science* **163:**1343–1348.

Williams, R. C., and Backus, R. C. (1949). Macromolecular weights determined by direct particle counting. *I. The weight of bushy stunt virus particles. J. Am. Chem. Soc.* **71:**40–52.

Williams, R. C., and Fraser, D. (1953). Morphology of the seven T-bacteriophages. *J. Bacteriol.* **66:**458–464.

Williams, R. C., and Richards, K. E. (1974). Capsid structure of bacteriophage lambda. *J. Mol. Biol.* **88:**547–550.

Williams, R. C., and Wyckoff, R. W. G. (1945). Electron shadow-micrography of virus particles. *Proc. Soc. Exp. Biol. Med.* **58:**265–270.

Wood, W. B. (1992). Assembly of a complex bacteriophage *in vitro*. *BioEssays* **14:**635–640.

Wood, W. B., and Edgar, R. S. (1967). Building a bacterial virus. *Sci. Am.* **217:**60–74.

Yanagida, M., and Ahmad-Zadeh, C. (1970). Determination of gene product positions in bacteriophage T4 by specific antibody association. *J. Mol. Biol.* **51:**411–421.

Zinder, N. D., Valentine, R. C., Roger, M., and Stoeckenius, W. (1963). f1, a rod-shaped male-specific bacteriophage that contains DNA. *Virology* **20:**638–640.

CHAPTER 2

Postcards from the Edge: Structural Genomics of Archaeal Viruses

Mart Krupovic,* Malcolm F. White,† Patrick Forterre,* and **David Prangishvili***

Contents

Abstract

Ever since their discovery, archaeal viruses have fascinated biologists with their unusual virion morphotypes and their ability to thrive in extreme environments. Attempts to understand the biology of these viruses through genome sequence analysis were not efficient.

* Department of Microbiology, Institut Pasteur, Molecular Biology of the Gene in Extremophiles Unit, Paris, France
† Biomedical Sciences Research Complex, University of St. Andrews, North Haugh, St. Andrews, Fife, United Kingdom

Advances in Virus Research, Volume 82
ISSN 0065-3527, DOI: 10.1016/B978-0-12-394621-8.00012-1

Genomes of archaeoviruses proved to be *terra incognita* with only a few genes with predictable functions but uncertain provenance. In order to facilitate functional characterization of archaeal virus proteins, several research groups undertook a structural genomics approach. This chapter summarizes the outcome of these efforts. High-resolution structures of 30 proteins encoded by archaeal viruses have been solved so far. Some of these proteins possess new structural folds, whereas others display previously known topologies, albeit without detectable sequence similarity to their structural homologues. Structures of the major capsid proteins have illuminated intriguing evolutionary connections between viruses infecting hosts from different domains of life and also revealed new structural folds not yet observed in currently known bacterial and eukaryotic viruses. Structural studies, discussed here, have advanced our understanding of the archaeal virosphere and provided precious information on different aspects of biology of archaeal viruses and evolution of viruses in general.

LIST OF ABBREVIATIONS

AAA^+	ATPases associated with diverse cellular activities
AFV1	Acidianus filamentous virus 1
ATV	*Acidianus* two-tailed virus
ds	double-stranded
HHPV-1	*Haloarcula hispanica* pleomorphic virus 1
HRPV-1	*Halorubrum* pleomorphic virus 1
(w)HTH	(winged) helix-turn-helix
ITR	inverted terminal repeats
MCM	minichromosome maintenance helicase
MCP	major capsid protein
PBCV-1	*Paramecium bursaria Chlorella* virus type 1
PSV	*Pyrobaculum* spherical virus
RCR	rolling-circle replication
REP	Replication protein
RHH	ribbon-helix-helix
ROP	repressor of primer
SIFV	*Sulfolobus islandicus* filamentous virus
SIRV	*Sulfolobus islandicus* rod-shaped virus
ss	single-stranded
SSV1	*Sulfolobus* spindle-shaped virus 1
SSV-RH	*Sulfolobus* spindle-shaped virus Ragged Hills
STIV	*Sulfolobus* turreted icosahedral virus
STSV1	*Sulfolobus tengchongensis* spindle-shaped virus 1
TYLCV	Tomato yellow leaf curl virus

I. INTRODUCTION

The discovery of archaea as one of the three domains of life in addition to the domains bacteria and eukarya has stimulated strong interest in revealing special features of members of this domain, including the nature of the associated virosphere. Considerable progress has been made in the last two decades in isolating viruses that infect hyperthermophilic and extremely halophilic archaea from extreme geothermal and hypersaline environments. At present (May 2011), 45 such viruses have been characterized. Despite their modest number, the morphological diversity of these viruses is extraordinary and comprises morphotypes that have not been previously observed in nature. The double-stranded (ds) DNA viruses of hyperthermophilic hosts from the phylum Crenarchaota are exceptionally diverse both morphologically and genomically. Due to their unique properties, eight novel families have been established by the International Committee on Taxonomy of Viruses for their classification, including rod-shaped Rudiviridae, filamentous Lipothrixviridae, spindle-shaped Fuselloviridae, bottle-shaped Amplullaviridae, two-tailed Bicaudaviridae, spherical Globuloviridae, droplet-shaped Guttaviridae, and bacilliform Clavaviridae (Table I). The virions of members of these families are shown in Figure 1.

Viruses of another phylum of the archaeal domain, the Euryarchaeota, are less diverse morphologically. A few of them have been assigned to the families Myoviridae and Siphoviridae (which also include bacterial head–tail viruses) and two comprise the genus *Salterprovirus*, whereas four are still awaiting taxonomical assessments (Table I). All these viruses except one have double-stranded DNA genomes. *Halorubrum* pleomorphic virus 1 (HRPV-1) is the only known archaeal virus with a single-stranded DNA genome (Pietilä et al., 2009). No RNA viruses from the archaeal domain have been described.

II. GENOMICS OF ARCHAEAL VIRUSES

Along with an exceptional diversity of morphotypes, another remarkable property of archaeal viruses is a very low proportion of genes with recognizable functions and homologues. Comparative genomic analysis, even pushed to the limits of significance, revealed only a few proteins with detectable homologues in public sequence databases (Prangishvili *et al.*, 2006b). Consequently, the wealth of genomic and functional information available for bacterial and eukaryotic viruses is of little assistance when trying to understand the biology of archaeal viruses. Perhaps the only exception to this general trend is the relationship between bacterial

TABLE I Representatives of families of archaeal viruses and unclassified species

Family, genus, species	Host	dsDNA size, bp	Genome sequence accession #	Reference
Viruses of Crenarchaeota				
Family Rudiviridae				
Sulfolobus islandicus rod-shaped virus 1, SIRV1	*Sulfolobus*	Linear, 32,308	AJ414696	Prangishvili *et al.*, 1999
S. *islandicus* rod-shaped virus 2, SIRV2	*Sulfolobus*	Linear, 35,450	AJ344259	Peng *et al.*, 2001
Family Lipothrixviridae				
Genus *Alphalipothrixvirus*				
Thermoproteus tenax virus 1, TTV1	*Thermoproteus*	Linear, 15,900	X14855	Janekovic *et al.*, 1983
Genus *Betalipothrixvirus*				
Sulfolobus islandicus filementous virus, SIFV	*Sulfolobus*	Linear, 40,852	AF440571	Arnold *et al.*, 2000b
Genus *Gammalipothrixvirus*				
Acidianus filamentous virus 1, AFV1	*Acidianus*	Linear, 21,080	AJ567472	Bettstetter *et al.*, 2003
Genus *Deltalipotrixvirus*				
Acidianus filamentous virus 2, AFV2	*Acidianus*	Linear, 31,1787	AJ854042	Häring *et al.*, 2005
Family Globuloviridae				
Pyrobaculum spherical virus, PSV	*Pyrobaculum* *Thermoproteus*	Linear, 28,337	AJ635162	Häring *et al.*, 2004
Family Guttaviridae				
Sulfolobus newzealandicus droplet-shaped virus, SNDV	*Sulfolobus*	Circular, 20,000	nd[a]	Arnold *et al.*, 2000a
Family Fuselloviridae				
Sulfolobus spindle-shaped virus 1, SSV1	*Sulfolobus*	Circular, 15,465	XO7234	Schleper *et al.*, 1992

Sulfolobus spindle-shapeed virus 6, SSV6	*Sulfolobus*	Circular, 15,684	FJ870915	Redder *et al.*, 2009
Family Bicaudaviridae				
Acidianus two-tailed virus, ATV	*Acidianus*	Circular, 62,730	AJ888457	Prangishvili *et al.*, 2006c
Family Ampullaviridae				
Acidianus bottle-shaped virus, ABV	*Acidianus*	Linear, 23,814	EF432053	Peng *et al.*, 2007
Family Clavaviridae				
Aeropyrum pernix bacilliform virus 1, APBV1	*Aeropyrum*	Circular, 5278	AB537968	Mochizuki *et al.*, 2010
Unclassified, virion morphology: isometric				
Sulfolobus turreted icosahedral virus, STIV	*Sulfolobus*	Circular, 17,663	AY569307	Rice *et al.*, 2004
Sulfolobus turreted icosahedral virus 2, STIV2	*Sulfolobus*	Circular, 16,622	GU0803365	Happonen *et al.*, 2010
Unclassified, virion morphology: spindle shaped				
Sulfolbus technogensis spindle-shaped virus 1, STSV1	*Sulfolobus*	Circular, 75,294	AJ783769	Xiang *et al.*, 2005
Viruses of Euryarchaeota				
Family Myoviridae				
Genus ΦH-like viruses				
Natrialba phage ΦCh1	*Natrialba*	Linear, 58,498	AF440695	Klein *et al.*, 2002
Unassigned				
Halorubrum phage HF2	*Halorubrum*	Linear, 77,670	AF222060	Tang *et al.*, 2004
Family Siphoviridae				
Genus ψM1-like viruses				
Methanobacterium phage ψM2	*Methanothermobacter*	Linear, 30,400	AF065412	Pfister *et al.*, 1998
Genus *Salterprovirus*				
His1 virus	*Haloarcula*	Linear, 14,464	AF191796	Bath and Dyall-Smith, 1998

(continued)

TABLE I (*continued*)

Family, genus, species	Host	dsDNA size, bp	Genome sequence accession #	Reference
His2 virus		Linear, 16,067	AF191797	Bath *et al.*, 2006
Unclassified, virion morphology: spindle shaped				
Pyrococcus abyssi virus 1, PAV1	*Pyrococcus*	Circular, 18,098	EF071488	Geslin *et al.*, 2007
Unclassified, virion morphology: isometric,				
Halovirus SH1	*Haloarcula*	Linear, 30,898	AY950802	Bamford *et al.*, 2005
Unclassified, virion morphology: pleomorphic				
Haloarcula hispanica pleomorphic virus 1, HHPV-1	*Haloarcula*	Circular, 8082	GU321093	Roine *et al.*, 2010
Unclassified, virion morphology: pleomorphic				
Halorubrum pleomorphic virus 1, HRPV1	*Halorubrum*	Circular, 7048[b]	FJ685651	Pietilä *et al.*, 2009

[a] Not determined.
[b] The virus DNA is single stranded.

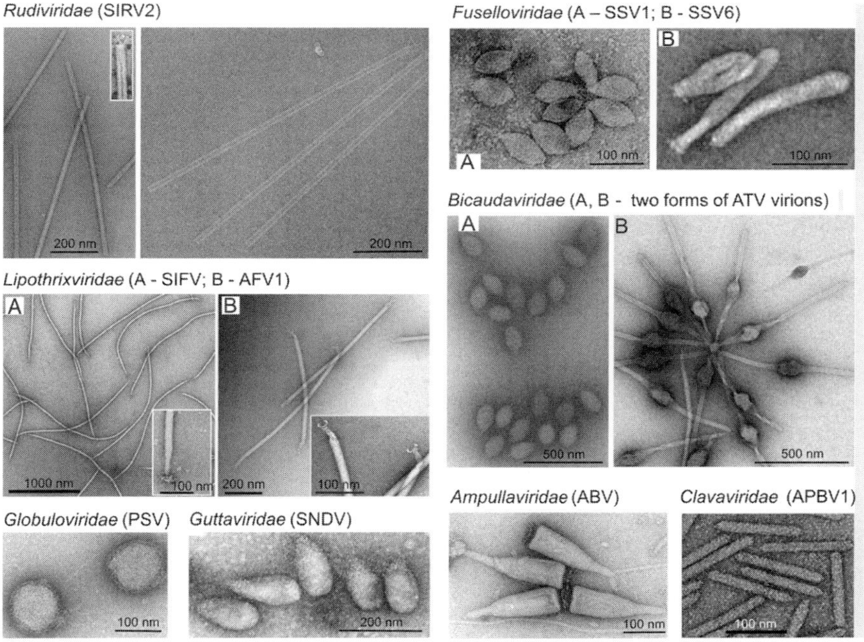

FIGURE 1 Transmission electron micrographs of representative members of eight families of viruses of the Crenarchaeota. Genome accession numbers for viruses depicted in this figure can be found in Table I.

and archaeal viruses of the order Caudovirales, belonging to families Myoviridae (contractile tails) and Siphoviridae (long, noncontractile tails). Viruses from the two domains are not only remarkably similar in their overall morphology (Prangishvili *et al.*, 2006a), but also the genomic relationship is readily recognizable (Krupovic *et al.*, 2010a, 2011b). Comparative genomic analysis of these two viral groups revealed that principles of virion assembly and maturation, as well as the genome packaging mechanism, which have been studied extensively in bacterial tailed viruses (Rao and Feiss, 2008; Steven *et al.*, 2005), are also common to tailed viruses of archaea (Krupovic *et al.*, 2010a).

Comprehensive genomic analysis revealed that only a small pool of genes is shared among distinct groups of archaeal viruses, as well as between these viruses and their hosts (Prangishvili *et al.*, 2006b). This common pool includes genes for predicted transcriptional regulators, P-loop ATPases implicated in viral DNA replication and packaging, enzymes for nucleic acid metabolism and modification, and glycosylases. However, for the majority of archaeal virus genes, no functional annotation could be offered. An example of a unique gene ensemble in an archaeal virus genome has been reported in *Aeropyrum pernix* bacilliform virus 1 (family Clavaviridae), where none of the 14 putative genes displayed significant similarity to sequences in the public databases (Mochizuki *et al.*, 2010).

The three-dimensional protein structure is generally conserved over longer periods of time when compared to the primary one. Therefore, to guide functional characterization of archaeal virus proteins, with the goal to gain insights into the biology of these viruses, several research groups undertook a structural genomics approach. As a result, a number of X-ray structures have been solved during the past few years for different proteins of archaeal viruses. The next section summarizes progress in this line of research and highlights (sadly, the few) cases where structural information was sufficient to guide the functional characterization of these mysterious viral proteins.

III. STRUCTURAL GENOMICS AND ARCHAEAL VIRUSES

High-resolution structures are currently available for archaeal viruses belonging to the families Rudiviridae, Lipothrixviridae, Globuloviridae, Fuselloviridae, Bicaudaviridae, and the unclassified *Sulfolobus* turreted icosahedral virus (STIV; Table II). As of today, not a single high-resolution structure is available for proteins encoded by euryarchaeal viruses. For some families, proteins from a single representative member have been characterized structurally, while in the case of other families, structures have been solved for proteins encoded by several members of the same family (Table II).

Structural studies on fuselloviruses and STIV have been summarized previously (Lawrence *et al.*, 2009). Seven X-ray structures have been determined for proteins encoded by fuselloviruses (five from *Sulfolobus* spindle-shaped virus 1, SSV1, and two from *Sulfolobus* spindle-shaped virus Ragged Hills, SSV-RH) (Fig. 2), and four structures are available for STIV proteins (Table II). X-ray structures for five proteins encoded by *Pyrobaculum* spherical virus (PSV; Globuloviridae) have been solved in the framework of the structural genomics initiative carried out by The Scottish Structural Proteomics Facility (Oke *et al.*, 2010; Fig. 3). The largest number of high-resolution structures is currently available for filamentous and rod-shaped crenarchaeal viruses (families Lipothrixviridae and Rudiviridae, respectively). As a result of the collective effort of several laboratories, seven structures are available for lipothrixviruses and six for rudiviruses (Fig. 4). In addition, an X-ray structure has been determined for a protein encoded by the *Acidianus* two-tailed virus (ATV; Goulet *et al.*, 2010b). These viral proteins are discussed according to their functional category, which in some cases became apparent only through structural analysis.

TABLE II Proteins of archaeal viruses with available high-resolution structures

Family/virus	Protein name	Function /feature	Accession number	PDB ID	Disulfide bonds	Homologues in other viruses	Reference
Rudiviridae							
SIRV1	ORF56a	DNA binding (?)[a]; HTH	NP_666589	2X48		—	Oke *et al.*, 2010
	ORF56b (SvtR)	DNA binding; RHH	NP_666596	2KEL		Lipothrix-	Guillière *et al.*, 2009
	ORF114	DNA binding (?)	NP_666617	2X4I	Cys33-Cys58	Lipothrix-, Bicauda-, STIV	Oke *et al.*, 2010
	ORF119	Replication initiation protein	NP_666597	2X3G		Gemini-, Circo-, Micro-, etc.[b]	Oke *et al.*, 2010
	ORF131		NP_666598	2X5T		—	Oke *et al.*, 2010
Yellowstone SIRV	ORF134	Major capsid protein	NP_666607 (SIRV1)	3F2E		Lipothrix-	Szymczyna *et al.*, 2009
Lipothrixviridae							
AFV1	ORF102		YP_003753	2WB6	Cys56-Cys62	—	Keller *et al.*, 2009a
	ORF99		YP_003728	3DJW		Rudi-	Goulet *et al.*, 2009c
	ORF132	Major capsid protein 1	YP_003749	3FBL		Rudi-	Goulet *et al.*, 2009a
	ORF140	Major capsid protein 2	YP_003750	3FBZ		—	Goulet *et al.*, 2009a
	ORF157	Nuclease	YP_003730	3II3		Fusello- (SSV-RH)	Goulet *et al.*, 2010a
AFV3	ORF109	DNA-binding	YP_001604358	2J6B		Rudi-, Bicauda-, STIV	Keller *et al.*, 2007

(*continued*)

TABLE II *(continued)*

Family/virus	Protein name	Function /feature	Accession number	PDB ID	Disulfide bonds	Homologues in other viruses	Reference
SIFV	ORF14		NP_445679	2H36		Rudi-, Fusello- (SSV6)	Goulet *et al.*, 2009b
Globuloviridae							
PSV	ORF126	C2C2 Zn-finger	YP_015568	2X5R		—	Oke *et al.*, 2010
	ORF131		YP_015553	2X5C		—	Oke *et al.*, 2010
	ORF137	SM-like RNA binding motif	YP_015530	2X4J	Cys55-Cys133	—	Oke *et al.*, 2010
	ORF165a	DNA binding (?); wHTH	YP_015525	2VXZ	Cys24-Cys66	—	Oke *et al.*, 2010
	ORF239		YP_015532	2X3M		—	Oke *et al.*, 2010
Fuselloviridae							
SSV1	B129	DNA binding (?); C2H2	NP_039795	2WBT	Cys121-Cys127	—	Lawrence *et al.*, 2009
	D63	ROP-like adaptor protein (?)	NP_039786	1SKV		—	Kraft *et al.*, 2004a
	F93	DNA binding; wHTH	NP_039783	1TBX		STIV[b]	Kraft *et al.*, 2004b
	F112	DNA binding; wHTH	NP_039787	2VQC	Cys51-Cys58	Bicauda-	Menon *et al.*, 2008
	E96		NP_039785	—		—	Lawrence *et al.*, 2009
SSV-RH	D212	Nuclease fold; PD-(D/E)XK	NP_963934	2W8M		—	Menon *et al.*, 2010
	E73	DNA binding (?); RHH	NP_963940	4A1Q		—	Schlenker *et al.*, 2009

Bicaudaviridae							
ATV	P131	Major structural protein	YP_319893	3FAJ		STSV1	Goulet *et al.*, 2010b
Unassigned							
STIV	A197	Glycosyltransferase (?); GT-A fold	YP_024997	2C0N	Cys86-Cys116	—	Larson *et al.*, 2006
	B116	DNA binding	YP_025003	2J85	Cys33-Cys62	Rudi-, Bicauda-, Lipothrix-,	Larson *et al.*, 2007b
	B345	Major capsid protein	YP_025022	2BBD		Tecti-, Cortico-, Phycodna-, etc.[b]	Khayat *et al.*, 2005
	F93	DNA binding; wHTH	YP_025013	2CO5	Cys93-Cys93′	Fusello-[b]	Larson *et al.*, 2007a

[a] Question mark indicates that activity has not been demonstrated experimentally.
[b] Homology is evident only through structural comparison.

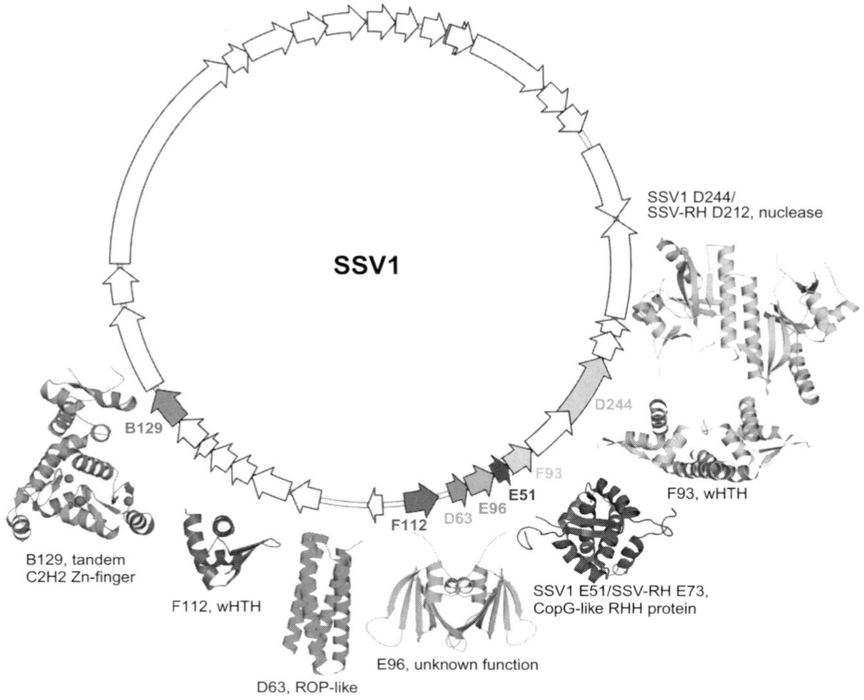

FIGURE 2 X-ray and NMR structures of fusellovirus proteins. Available SSV1 and SSV-RH protein X-ray and NMR structures are shown beneath the SSV1 genome map and are shaded with distinct colors that follow the shading of their corresponding genes. PDB accession numbers and references for the shown structures can be found in Table II. The figure is modified from Lawrence *et al.* (2009).

The presence of intracellular disulfide bonds is a recognized feature of proteins from hyperthermophilic crenarchaea and is thought to contribute toward protein thermostability in these organisms (Beeby *et al.*, 2005). Nine of the 30 virus protein crystal structures summarized in Table II have disulfide bonds. This represents 30% of the total and, if representative, suggests that up to one-third of all proteins from crenarchaeal viruses may be stabilized in this manner. Analysis of the positions of the disulfides in these nine proteins shows that three pairs of cysteines are found within eight amino acids of one another and link local regions of structure (SSV1 B129, F112 and AFV1 ORF102), four are at least 25 residues apart and thus link between subdomains of proteins (PSV ORF137, ORF165a; SIRV1 ORF114; STIV A197 and B116), and one links the two subunits of a dimeric protein (STIV F93). It has also been suggested that stabilizing disulfide bonds are more prevalent in intracellular viral

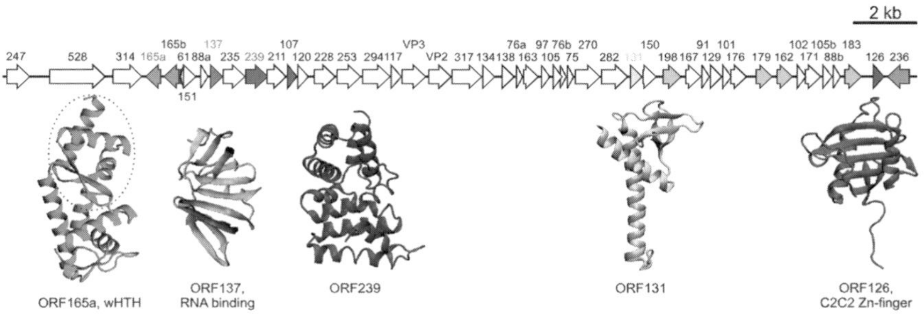

FIGURE 3 Structural genomics of globulovirus PSV. Available PSV protein X-ray structures are shown beneath the genome map and are shaded with distinct colors that follow the shading of their corresponding genes. PDB accession numbers and references for the shown X-ray structures can be found in Table II. Paralogous genes are colored identically. The wHTH domain of ORF165a is circled.

proteins (e.g., transcriptional factors) than in structural proteins involved in virion formation (Larson *et al.*, 2007a; Menon *et al.*, 2008).

A. Transcription regulators and other DNA-binding proteins

In silico analysis has illuminated the diversity of transcriptional regulators encoded by different archaeal viruses (Prangishvili *et al.*, 2006b). Among the most prevalent structural motifs found in these proteins are the helix–turn–helix (HTH) and the ribbon–helix–helix (RHH). Proteins with the latter domain are encoded by nearly all crenarchaeal viruses (Prangishvili *et al.*, 2006b) and some euryarchaeal viruses (Geslin *et al.*, 2007). In addition, some archaeal viruses encode proteins with looped–hinged helix and Zn-finger domains (Prangishvili *et al.*, 2006b).

Despite their abundance in archaeal virus genomes and their crucial role in the regulation of viral genome expression, only one archaeoviral transcription factor, protein SvtR (also known as ORF56b) encoded by SIRV1, has been studied in appreciable detail (Guillière *et al.*, 2009). The nuclear magnetic resonance structure of the protein revealed a typical RHH fold (Fig. 4). The protein was found to form a dimer and bind DNA with its β-sheet face. Two regions within the SIRV1 genome were pinpointed as the SvtR-binding sites; the protein was found to act as a repressor of its own gene as well as the gene for the viral structural protein gp30 (ORF1070; Guillière *et al.*, 2009). An NMR structure of the homodimeric protein E73 from the fusellovirus SSV-RH also revealed an RHH fold (Schlenker *et al.*, 2009). However, the role of this protein in transcription regulation remains to be verified.

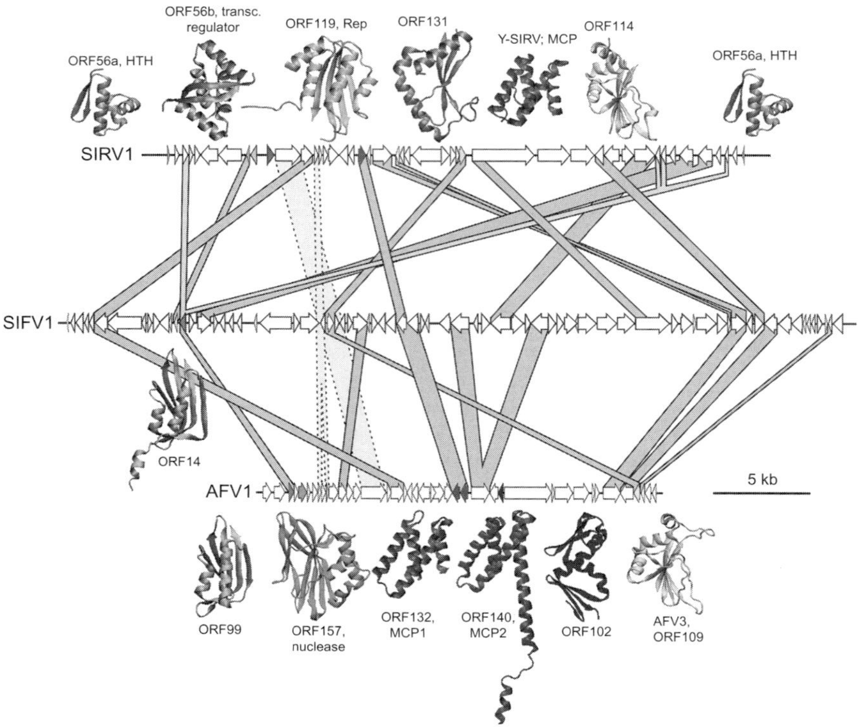

FIGURE 4 Structural genomics of linear dsDNA viruses of Rudiviridae and Lipothrixviridae families. The available protein structures are shown next to the genome maps and are shaded with distinct colors that follow the shading of their corresponding genes. PDB accession numbers and references for the shown X-ray and NMR structures can be found in Table II. Genes shared by *Sulfolobus islandicus* rod-shaped virus 1 (SIRV1; Rudiviridae) and lipothrixviruses *S. islandicus* filamentous virus (SIFV) and *Acidianus* filamentous virus 1 (AFV1) are connected via gray shading. (See Page 1 in Color Section at the back of the book.)

High-resolution structural information is available for two additional SIRV1 proteins that may be involved in transcription regulation or DNA binding. The first protein is ORF56a, which is encoded within the inverted terminal repeats, with the consequence of two identical gene copies being present in the SIRV1 genome (Fig. 4). The X-ray structure of ORF56a revealed a HTH fold (Oke *et al.*, 2010), which is most similar to the DNA-binding domain of the transposase encoded by the Tc3 transposon (Tc1/*mariner* family) of *Caenorhabditis elegans* (van Pouderoyen *et al.*, 1997). However, the ability of ORF56a to interact with DNA has yet to be demonstrated. The second protein, ORF114 (Fig. 4), is a member of a protein family common to crenarchaeal viruses of the families Rudiviridae, Lipothrixviridae, Bicaudaviridae, and the unclassified virus STIV, as well as to several bacterial proviruses (Keller *et al.*, 2007; Prangishvili *et al.*, 2006b). Currently, three members of this protein family have been characterized structurally (Table II). In addition to SIRV1 ORF114, X-ray structures are available for proteins ORF109 of AFV3 (Lipothrixviridae; Keller *et al.*, 2007) and B116 of STIV (Larson *et al.*, 2007b). Proteins in this family possess a unique fold consisting of a five-stranded β sheet flanked on one side by three α helices. The protein forms a dimer with two conserved loops, containing positively charged residues, pointing away from the protein core. The distance between the two loops was found to be equivalent to the spacing of the major grooves in B-form DNA, suggesting that the protein might be involved in DNA binding. *In vitro* assays performed with STIV B116 and AFV3 ORF114 supported such a hypothesis (Keller *et al.*, 2007; Larson *et al.*, 2007b). However, the exact role of these proteins in the viral infection cycle remains to be determined.

In addition, X-ray structures of four winged HTH (wHTH; ORF165a from PSV, F93 from STIV, F93 and F112 from SSV1) and two Zn-finger (B129 from SSV1 and ORF126 from PSV) domain-containing proteins from crenarchaeal viruses have been determined (Figs. 2 and 3, Table II). Interestingly, whereas SSV1 encodes a C2H2 Zn-finger protein (Fig. 1; Lawrence *et al.*, 2009), PSV encodes two paralogous C2C2 Zn-binding proteins (Fig. 3). Nonspecific binding of SSV1 F112 and STIV F93 to dsDNA has been demonstrated using an electrophoretic mobility shift assay (Larson *et al.*, 2007a; Menon *et al.*, 2008). The proteins were reasoned to be likely involved in transcription regulation, but the identification of their target sites has not been attempted.

B. RNA-binding proteins

Structural analysis has revealed two putative RNA-binding proteins encoded by crenarchaeal viruses (Table II). SSV1 protein D63 displays structural similarity to bacterial ROP-like adaptor proteins (Kraft *et al.*, 2004a). The ROP (repressor of primer) protein of bacterial plasmid ColE1 controls the plasmid copy number by increasing the affinity between two complementary RNAs (RNA I and RNA II), thereby preventing primer formation and initiation of DNA replication (Tomizawa and Som, 1984). The genome copy number control of SSV1 is not well understood, but the structural similarity between D63 and ROP suggests that the mechanism may be similar to that of ColE1 (Menon *et al.*, 2008).

Another potential RNA-interacting protein encoded by archaeal viruses has been illuminated though structural analysis of the PSV protein ORF137 (Oke *et al.*, 2010). The protein displays unexpected structural similarity to SM-like RNA-binding proteins such as bacterial Hfq.

SM proteins play a crucial role during ribonucleoprotein assembly and are required for cellular RNA processing, including tRNA and rRNA processing, mRNA decapping and decay, and intron splicing in pre-mRNA (Wilusz and Wilusz, 2005). The SM fold consists of an N-terminal α-helix followed by a highly twisted five-strand β sheet. Bacterial Hfq proteins form a doughnut-shaped homohexamer, whereas eukaryotic counterparts form heteroheptamers (Wilusz and Wilusz, 2005). Unexpectedly, the structure of PSV ORF137 revealed a twisted 10-stranded β sheet, which is equivalent to a dimer of canonical SM proteins. It is tempting to speculate that ORF137 arose as a result of duplication of the ancestral SM-like protein-coding gene. This is in line with the observation that paralogous copies for several genes other than *orf137* are present in the PSV genome, suggesting that gene duplication did indeed shape the PSV genome (Fig. 3). The precise role of ORF137 during the PSV infection cycle is still not known; it is likely to be involved in the metabolism of viral or host RNA molecules, as is the case for cellular SM-like proteins.

C. Viral nucleases

Structures for two distinct nuclease-like proteins encoded by crenarchaeal viruses have been reported (Table II). The X-ray structure of SSV-RH D212, a protein conserved in fuselloviruses, has unexpectedly revealed a typical nuclease fold of the PD-(D/E)XK superfamily (Fig. 2), despite the lack of any detectable sequence similarity to other nucleases (Menon *et al.*, 2010). Even though all known active site residues characteristic to PD-(D/E)XK nucleases were found to be conserved in D212, its biochemical activity could not be confirmed experimentally.

The X-ray structure of ORF157 encoded by AFV1 displayed a novel fold (Fig. 4), remotely related to nucleotidyltransferases (Goulet *et al.*, 2010a). Subsequent structural information-guided *in vitro* assays not only revealed that the protein displays nuclease activity toward linear dsDNA, but also implicated Glu86 as a residue essential for catalytic activity. Interestingly, the only homologue of AFV1 ORF157 is encoded by fusellovirus SSV-RH; no other lipothrixviruses or fuselloviruses encode ORF157-like proteins. The two proteins share ~50 % identity, suggesting that AFV1 and SSV-RH were relatively recently engaged in horizontal gene exchange.

D. Replication proteins

Genome replication of archaeal viruses has not been extensively studied experimentally, and knowledge of this fundamental process is mainly based on sequence analyses. Only three taxonomic groups of archaeal viruses encode their own DNA polymerases. Protein-primed type B DNA polymerases are encoded by spindle-shaped viruses His1 and His2 (genus *Salterprovirus*) infecting halophilic euryarchaea (Bath *et al.*, 2006) and also by the crenarchaeal *Acidianus* bottle-shaped virus (Ampullaviridae; Peng *et al.*, 2007). Haloarchaeal- tailed viruses HF1 and HF2 (Myoviridae), however, encode typical RNA-primed type B DNA polymerases (Tang *et al.*, 2004). It should be noted, however, that the majority of tailed euryarchaeal viruses appear to rely on the minichromosome maintenance helicases (MCM) for genome replication initiation (Krupovic *et al.*, 2010b). Interestingly, phylogenetic analysis revealed that *mcm* genes were recruited by these viruses from their respective hosts on multiple independent occasions (Krupovic *et al.*, 2010b).

Identifiable genome replication proteins are also encoded by unclassified pleomorphic haloarchaeal viruses HRPV-1 and HHPV-1 (Pietilä *et al.*, 2009; Roine *et al.*, 2010). Despite the fact that the two viruses possess genomes of different nucleic acid types (ssDNA and dsDNA, respectively), they are clearly evolutionary related based on the synteny of their genomes and amino acid sequence similarity between their protein products (Roine *et al.*, 2010). On their circular genomes, HRPV-1 and HHPV-1 both encode proteins that are recognizably similar to rolling-circle replication (RCR) initiation proteins and are therefore expected to replicate via RCR mechanism.

In addition to the DNA polymerase of ABV, genome replication proteins from crenarchaeal viruses of two other families were predicted. Fuselloviruses and the *Acidianus* two-tailed virus (Bicaudaviridae) encode AAA^{+} ATPases, which were suggested to be involved in the initiation of DNA replication (Koonin, 1992; Prangishvili *et al.*, 2006c). ATPases encoded by fuselloviruses and ATV are related to DnaA-like and Cdc48-like proteins, respectively (Prangishvili *et al.*, 2006b). However, the role of these proteins in viral genome replication has not been investigated.

Structural analysis of the ORF119 protein from SIRV1 (Oke *et al.*, 2010, 2011), which is conserved in all rudiviruses, unexpectedly revealed that its topology is remarkably similar to that of RCR-initiating REP proteins encoded by diverse viruses and plasmids. The catalytic domain of the plant-infecting tomato yellow leaf curl virus (Geminiviridae; Campos-Olivas *et al.*, 2002) was identified as the closest structural homologue of ORF119 (Fig. 5A). Structure-based alignment of the two proteins facilitated identification of the three conserved motifs characteristic to RCR proteins (Oke *et al.*, 2011). The protein was found to function as a dimer (Fig. 5B), and its nicking and joining activities, as well as the nicking target site within the SIRV1 genome, have been subsequently demonstrated experimentally (Oke *et al.*, 2011). Whereas nicking and joining REP proteins are typically responsible for replication initiation of circular DNA molecules, the genome of SIRV1 (and other rudiviruses) is a linear dsDNA molecule with covalently closed ends and inverted terminal repeats (ITR; Blum *et al.*, 2001).

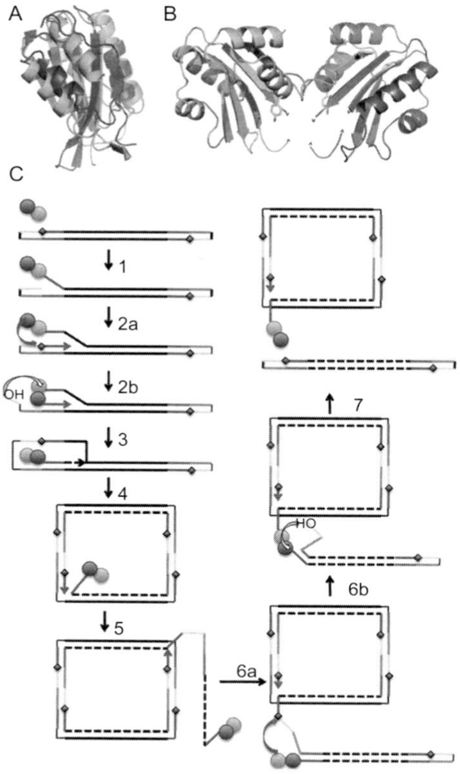

FIGURE 5 Replication protein (REP) of *Sulfolobus islandicus* rod-shaped virus 1 (SIRV1). (A) Superposition of the SIRV1 replication protein monomer (cyan) with the rolling-circle replication initiation protein from geminivirus TLYCV (purple). (B) The dimer of SIRV1 REP protein. Subunit A is colored in cyan, and subunit B is in slate. The Tyr residue, which forms the covalent link to the DNA, is shown in sticks for each subunit. Carets mark disordered loops. Each subunit exchanges a β strand (marked with an asterisk) with the other. (C) Proposed model for DNA replication in rudiviruses. See main text for details on each of the depicted steps. The figure is adapted from Oke *et al*. (2011). (See Page 2 in Color Section at the back of the book.)

The structural and biochemical characterization of SIRV1 REP allowed a model for rudivirus genome replication to be proposed (Fig. 5C). According to this model, SIRV1 genome replication is initiated by the REP protein, which nicks one strand of the viral genome at the *ori* site within the ITRs (Fig. 5C, step 1), releasing a 3′ DNA end, which can be used as a primer to initiate DNA replication. As a result, a covalent adduct between the dimeric replication protein and the newly created 5′ end is formed. Replication quickly regenerates the *ori* site, which is attacked by the second subunit of REP to form a dual-adducted REP dimer (Fig. 5C, step 2a). This species is likely to be transient, as the new 3′ DNA end generated in subunit 2 can flip over to attack the tyrosyl-phosphoester in subunit 1, forming a new contiguous DNA strand (Fig. 5C, step 2b). Displacement replication is then used to replicate the rest of the genome, which generates a dsDNA circle (Fig. 5C, step 4). As replication continues around the circle, the previously copied viral DNA is displaced and can fold up into the linear DNA structure found in the virion (Fig. 5C, step 5). This folding would ensure that REP was suitably positioned to attack the newly displaced *ori* site shown in step 6a, generating a transient double-adducted REP dimer and a new 3′ end that can immediately attack the other REP subunit (Fig. 5C, step 6b), releasing a covalently closed linear viral genome and leaving REP attached to the emerging DNA strand ready for another round of replication (Fig. 5C, step 7; Oke *et al.*, 2011).

E. Structural proteins of archaeal viruses

The ability to form a virion is the main feature that distinguishes viruses from other types of mobile genetic elements, such as plasmids or transposons (Krupovic and Bamford, 2010). In other words, a virion is a hallmark of a virus (Raoult and Forterre, 2008). Furthermore, whereas genome replication proteins are often being exchanged horizontally between unrelated mobile genetic elements (e.g., viruses and plasmids or unrelated groups of viruses), the virion assembly principles and the structure of the major components of the virion tend to remain conserved within the group of viruses sharing a common ancestor (Krupovic and Bamford, 2007, 2009). Consequently, structural studies directed at virion proteins are often very informative in that they not only provide information on the assembly and organization of viral particles, but may also reveal deep evolutionary connections between distantly related viruses (Bamford *et al.*, 2002; Krupovic and Bamford, 2008b). This point is well illustrated by the structural characterization of virion proteins from several archaeal viruses. Structures of five such proteins have been determined (Table II).

1. Rudiviridae and Lipothrixviridae—The Ligamenvirales

Linear viruses of archaea belong to two distinct families: Rudiviridae and Lipothrixviridae (Prangishvili *et al.*, 2006a). At least two proteins were found to constitute the rigid rod-shaped virions of rudiviruses. The highly basic major capsid protein (MCP) associates with the genomic DNA to form a helical body of the virion, whereas the minor capsid protein is implicated in the formation of terminal structures present at both ends of the linear virions (Steinmetz *et al.*, 2008). The structure of the

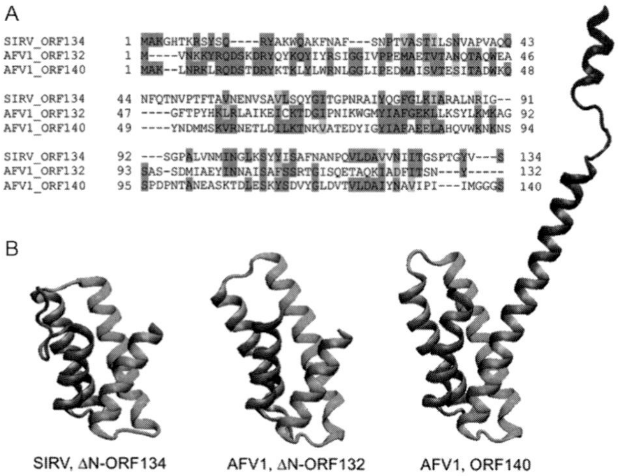

FIGURE 6 Major capsid proteins (MCP) of linear dsDNA viruses infecting archaea. (A) Sequence alignment of the MCP of *Sulfolobus islandicus* rod-shaped virus (SIRV; ORF134) with the two MCPs (ORF132 and ORF140) of *Acidianus* filamentous virus 1 (AFV1). The alignment is colored according to sequence conservation (BLOSUM62 matrix). (B) Comparison of the N-terminally truncated MCP of SIRV with the two MCPs of AFV1. Structures are colored using a rainbow color gradient from the N terminus (blue) to the C terminus (red). PDB accession numbers and references for the shown X-ray structures can be found in Table II. Figure is adapted from Prangishvili and Krupovic, 2012. (See Page 2 in Color Section at the back of the book.)

C-terminal domain of the MCP from *Sulfolobus islandicus* rod-shaped virus (SIRV) revealed a novel four-helix bundle topology (Fig. 6), not observed previously in capsid proteins of other viruses (Szymczyna *et al.*, 2009). Notably, arrangement of the four helices is different from that observed in the capsid protein of the ssRNA genome-containing tobacco mosaic virus (Fig. 7; Goulet *et al.*, 2009a), suggesting an independent origin for these archaeal and plant virus capsid proteins.

Unlike rudiviruses, filamentous virions of lipothrixviruses are flexible and possess a lipid envelope (Fig. 1; Prangishvili *et al.*, 2006a). Furthermore, lipothrixvirus virions are composed of two major capsid proteins (MCP1 and MCP2), not one as in the case of rudiviruses (Goulet *et al.*, 2009a). Both MCPs were shown to interact with dsDNA and form virion-like filaments *in vitro*. The structures of the two MCPs from lipothrixvirus AFV1 have been determined by X-ray crystallography, revealing that they are structurally related to each other (Fig. 6; Goulet *et al.*, 2009a). However, the two proteins display a distinct hydrophobicity profile, which allowed the topological model of the two proteins in the AFV1 virion to be proposed. According to this model, the basic MCP1 protein forms a core around which the genomic dsDNA is wrapped, whereas MCP2 interacts

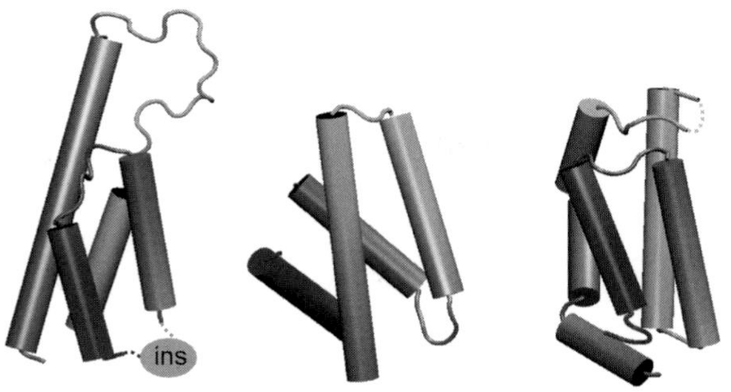

FIGURE 7 Arrangement of α-helixes in viral four-helix bundle proteins. Major capsid proteins of tobacco mosaic virus (TMV; PDB ID:1EI7), *Acidianus* filamentous virus 1 (AFV1; PDB ID:3FBL), and *Acidianus* two-tailed virus (ATV; PBD ID:3FAJ) are colored using a rainbow color gradient from the N terminus (blue) to the C terminus (red). Insertion (ins) between the second and the third α-helixes in the TMV protein was omitted for easier comparison. The figure is modified from Krupovic and Bamford (2011). (See Page 3 in Color Section at the back of the book.)

with the genome with its basic N-terminal region and the hydrophilic C-terminal domain is embedded into the lipid envelope (Goulet *et al.*, 2009a). Strikingly, MCP1 (and MCP2) of AFV1 is structurally remarkably similar to the MCP of rudiviruses (Fig. 6), despite very low pair-wise sequence similarity (17 % identity between the MCP1 of AFV1 and MCP of SIRV).

Comparative genomics of rudiviruses and lipothrixviruses has suggested previously that the two groups of viruses are related evolutionarily (Krupovic *et al.*, 2011a; Prangishvili *et al.*, 2006b). Indeed, on the genome level, some lipothrixviruses are no more similar to other members of the Lipothrixviridae than they are to rudiviruses (Fig. 4). For example, lipothrixviruses AFV1 and *S. islandicus* filamentous virus (SIFV; Arnold *et al.*, 2000b) share 10 homologous genes. The same number of genes is also common to SIFV and rudivirus SIRV1 (Krupovic *et al.*, 2011a). The two sets of genes (AFV1–SIFV and SIFV–SIRV) do overlap but are not identical (Fig. 4). Structural similarity between MCPs of rudiviruses and lipothrixviruses further reinforces the hypothesis that rigid rod-shaped and flexible filamentous viruses have arisen from a common ancestor. Based on structural relatedness of lipothrixviral and rudiviral MCPs, it has been envisioned that lipothrixviruses may have evolved from a "simpler" nonenveloped rudivirus-like ancestor (Goulet *et al.*, 2009a). Accordingly, in order to reflect the evolutionary relationship between linear viruses of the two families, a new taxonomic order, the "Ligamenvirales" (from the Latin *ligamen*, for string, thread) has been proposed (Prangishvili and Krupovic, 2012).

2. Major capsid protein of the *Acidianus* two-tailed virus

The major capsid protein structure has also been reported for the ATV (Goulet *et al.*, 2010b; Prangishvili *et al.*, 2006c). The X-ray structure revealed a unique fold (Goulet *et al.*, 2010b), not clearly related to other major capsid proteins for which high-resolution structures are available. The structure of ATV MCP consists of six α-helixes and two short β strands (Fig. 8A). Notably, packing of the six α-helixes resembles the four-helix bundle fold in which the fourth α-helix is kinked at the termini. Indeed, the DALI server (Holm and Rosenstrom, 2010) identifies four-helix bundle proteins, such as the helical histidine phosphotransferase domain of CheA (PDB ID: 1TQG; DALI Z score = 8.5), as closest structural relatives of the ATV MCP (Krupovic and Bamford, 2011). It should be mentioned that the topology of the pseudo-four-helix bundle fold of the ATV MCP is radically different from that of the lipothrixviral and rudiviral MCPs (Fig. 7). The only identifiable homologue of ATV MCP is encoded by the unclassified *Sulfolobus tengchongensis* spindle-shaped virus 1 (STSV1), where the homologue of the ATV MCP (37 % identity; Fig. 8B) was also found to be the major structural protein of the virion (Xiang *et al.*, 2005). STSV1 virion morphology resembles that of ATV in that both virions have a spindle-shaped body (Fig. 8C). However, unlike in ATV, tails are present only at one end of the spindle in STSV1 (Xiang *et al.*, 2005). It should be noted, however, that tails in STSV1 were found to be of variable size; it therefore cannot be ruled out that STSV1 virions develop two tails under certain condition, as is the case for

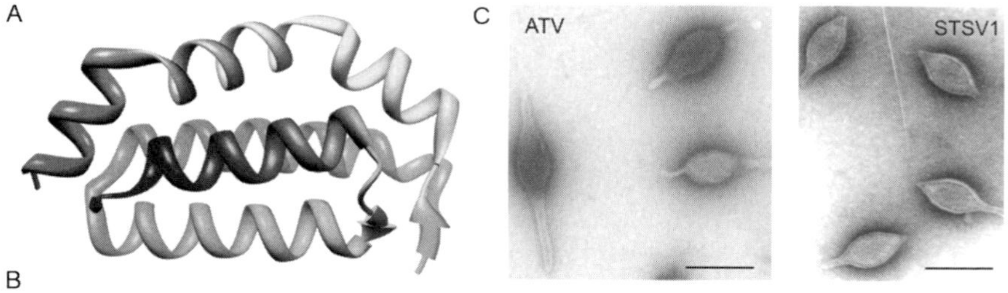

B

```
ATV_ORF131    1 MAKYEPKKGDYAGGAVKILDMFENGQLGYPEVTLKLAGEEANARRAGDERTKEAIHAIVKMISDAMKPYRNKGS 74
STSV1_ORF40   1 MAREEPYKGDYVGGVAKILQGYFANYYGFPNVSLRLAGEEANLSKTGHANAKAIVHEMIKVIKEASKPLR-RGK 73

ATV_ORF131   75 GFQS-----QPIPGEVIAQVTSNPE--YQQAKAFLASPATQVRNIEREEVLSKGAKKLAQAMAS---------  131
STSV1_ORF40  74 GFKEAWVYFSQVPENAPPNSTALPPNIAQQVKAKLEALYQQL--LATRAAASEAAKAPAYTNPTPGLTMQASM 144
```

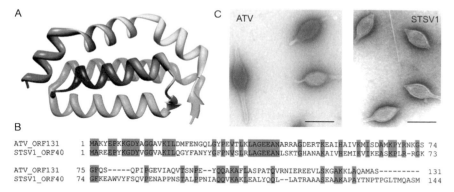

FIGURE 8 (A) X-ray structure of the major capsid protein of *Acidianus* two-tailed virus (ATV). The structure is colored using a rainbow color gradient from the N terminus (blue) to the C terminus (red). (B) Sequence alignment of the ATV MCP to that of the *Sulfolobus tengchongensis* spindle-shaped virus (STSV1). (C) Transmission electron micrographs showing morphological similarity between the two-tailed virions of ATV and the one-tailed virions of STSV1. Bars: 200nm. ATV micrograph is a courtesy of Dr. Soizick Lucas-Staat, Institut Pasteur. The STSV1 micrograph is reproduced from Xiang *et al.* (2005). (See Page 3 in Color Section at the back of the book.)

ATV (Prangishvili *et al.*, 2006c). Furthermore, both viruses have circular dsDNA genomes of comparable size (62.7 and 75.3kb for ATV and STSV1, respectively) and share a set of nine genes. These observations, along with the fact that MCPs of ATV and STSV1 are homologous, suggest that the two viruses share a common ancestor and should be probably classified together into the family Bicaudaviridae, where ATV is currently the sole member.

3. Double jelly-roll viruses

The high-resolution X-ray structure of the MCP of STIV was another milestone toward our understanding of viral origin and evolution. The STIV virion consists of an icosahedral protein capsid surrounding a lipid membrane vesicle, which encloses a circular dsDNA genome (Rice *et al.*, 2004). The MCP of STIV was found to display a double jelly-roll topology (Khayat *et al.*, 2005), a structural fold consisting of two eight-stranded antiparallel β barrels joined by a linker region (Krupovic and Bamford, 2008b). The same MCP topology was observed previously in bacterial tectivirus PRD1 (Benson *et al.*, 1999), algae-infecting *Paramecium bursaria Chlorella* virus type 1 (Nandhagopal *et al.*, 2002), and human adenovirus (Roberts *et al.*, 1986). Figure 9 shows examples of double jelly-roll capsid proteins from representative viruses infecting hosts within all three

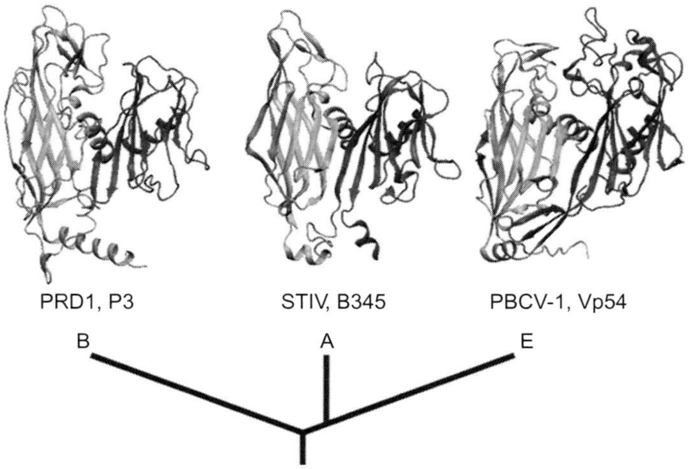

FIGURE 9 Double jelly-roll major capsid proteins from viruses infecting hosts in the three domains of life. Comparison of the double jelly-roll MCPs from bacterial (B) tectivirus PRD1 (PDB ID:1HX6), archaeal (A) virus STIV (PDB ID:2BBD), and eukaryotic phycodnavirus *Paramecium bursaria Chlorella* virus type 1 (PBCV-1; PDB ID:1J5Q). The two eight-stranded β barrels constituting the double jelly-roll fold are shown in green and red, respectively.

domains of life. Importantly, similarity between double jelly-roll capsid viruses extends beyond the structural similarity of their MCPs and includes common virion assembly and genome packaging principles (discussed in Krupovic and Bamford, 2008b). Identification of a double jelly-roll MCP-containing virus-infecting archaea provided strong support for the viral lineage hypothesis (Bamford *et al.*, 2002), which predicts a common origin for viruses that, despite infecting hosts from different domain of life, share the same capsid architecture. In other words, common principles for virion assembly and architecture in such viruses are inherited from a common viral ancestor that existed prior to diversification of the last universal cellular ancestor.

Electron cryomicroscopy and bioinformatic studies further expanded the double jelly-roll viral lineage to include nine officially recognized virus families and three additional viruses that have not yet been assigned to a family (Krupovic and Bamford, 2010). In addition to STIV and its close relative STIV2 (Happonen *et al.*, 2010), both infecting a hyperthermophilic acidophilic crenarchaeon *Sulfolobus solfataricus*, no other double jelly-roll archaeal viruses have been isolated. However, two putative proviruses encoding homologues of the STIV MCP and putative genome packaging ATPase have been identified in the genomes of two euryarchaeal species, *Thermococcus kodakarensis* KOD1 and *Methanococcus voltae* A3 (Krupovic and Bamford, 2008a). This suggests that double jelly-roll viruses might not be restricted to crenarchaeal hosts, but are (or were) also infecting organisms in the other major archaeal phylum, the Euryarchaeota.

F. Viral glycosyltransferases

The majority of characterized crenarchaeal viruses encode at least one glycosyltransferase (SIFV encodes five!) per genome (Prangishvili *et al.*, 2006b). Glycosyltransferases may be involved in modification of a variety of cellular or viral targets, such as virion structural proteins, viral DNA, cellular proteins, or host cell envelope (Markine-Goriaynoff *et al.*, 2004). Indeed, virion proteins of rudiviruses (MCP; Vestergaard *et al.*, 2005), STIV (MCP; Maaty *et al.*, 2006), haloarchaeal pleomorphic virus HRPV-1 (putative receptor-binding protein; Pietilä *et al.*, 2010), and *Aeropyrum pernix* bacilliform virus 1 (MCP; Mochizuki *et al.*, 2010) were found to be glycosylated.

A high-resolution structure is available only for one archaeal virus glycosyltransferase. X-ray crystallographic analysis of the STIV protein A197 revealed a GT-A fold that is common to many members of the glycosyltransferase superfamily (Larson *et al.*, 2006). Notably, the similarity between A197 and other described glycosyltransferases could not be detected through sequence-based approaches due to very low similarity between these proteins (15% identity in a structure-based alignment)

(Larson *et al.*, 2006; Prangishvili *et al.*, 2006b). However, structure-based alignment of A197 and other glycosyltransferases helped locating the DXD motif as well as the catalytic Asp residue conserved in GT-A glucosyltransferases. Based on structural information, it was suggested that A197 is responsible for glycosylation of the STIV MCP (Larson *et al.*, 2006). However, the identity of the substrate and the glycosylation activity itself are yet to be demonstrated for this protein.

G. Proteins without structural homologues and predictable functions

It has been noted that viruses and plasmids often encode proteins that have no homologues in cellular genomes; in many cases, these proteins display novel structural folds (Keller *et al.*, 2009b). The same is true for archaeal viruses. X-ray structures of seven proteins from different archaeal viruses show neither sequence nor structural similarity to previously described proteins (Table II). These proteins could be involved in unknown steps of the viral cycle or provide new solutions to such processes as genome replication, modulation of host defence systems, virion entry, assembly and egress, etc. Functional characterization of proteins without currently known structural homologues is therefore a research priority.

IV. CONCLUDING REMARKS

As pointed out correctly by Koonin (2010): "the structures do no magic." The structures of novel, uncharacterized proteins rarely provide direct insights into their functions. However, if distantly related homologues from other organisms have been characterized structurally and functionally, structural analysis may be very informative in terms of revealing deep evolutionary connections that often hint about the function of the protein under study. This point is well illustrated by functional insights obtained from the structural characterization of the replication initiation protein from SIRV1, nucleases from SIRV1 and SSV-RH, glycosyltransferase from STIV, and SM-like RNA-binding protein from PSV (Table II). Although experimental confirmation of the predicted functions for some of these proteins is yet to be obtained, the structures provide a clear guidance for functional analysis. It is also true that once the function of a protein is uncovered, complete characterization is impossible without high-resolution structural information.

The available results of structural genomics projects on archaeal viruses, discussed in this chapter, definitely provide insights into the biology of this unique group of viruses and warrant further structural characterization of their proteins.

ACKNOWLEDGMENTS

This work was supported by the European Molecular Biology Organization (Long-Term Fellowship ALTF 347–2010 to MK) and Agence Nationale de la Recherche (program Blanc to DP). Work in MFW's laboratory is supported by a grant from the Biotechnology and Biological Sciences Research Council (reference BB/G011400).

REFERENCES

Arnold, H. P., Ziese, U., and Zillig, W. (2000a). SNDV, a novel virus of the extremely thermophilic and acidophilic archaeon Sulfolobus. *Virology* **272**:409–416.

Arnold, H. P., Zillig, W., Ziese, U., Holz, I., Crosby, M., Utterback, T., Weidmann, J. F., Kristjanson, J. K., Klenk, H. P., Nelson, K. E., and Fraser, C. M. (2000b). A novel lipothrixvirus, SIFV, of the extremely thermophilic crenarchaeon *Sulfolobus*. *Virology* **267**:252–266.

Bamford, D. H., Burnett, R. M., and Stuart, D. I. (2002). Evolution of viral structure. *Theor. Popul. Biol.* **61**:461–470.

Bamford, D. H., Ravantti, J. J., Ronnholm, G., Laurinavicius, S., Kukkaro, P., Dyall-Smith, M., Somerharju, P., Kalkkinen, N., and Bamford, J. K. (2005). Constituents of SH1, a novel lipid-containing virus infecting the halophilic euryarchaeon Haloarcula hispanica. *J Virol* **79**:9097–9107.

Bath, C., Cukalac, T., Porter, K., and Dyall-Smith, M. L. (2006). His1 and His2 are distantly related, spindle-shaped haloviruses belonging to the novel virus group, *Salterprovirus*. *Virology* **350**:228–239.

Bath, C., and Dyall-Smith, M. L. (1998). His1, an archaeal virus of the *Fuselloviridae* family that infects *Haloarcula hispanica*. *J. Virol.* **72**:9392–9395.

Beeby, M., O'Connor, B. D., Ryttersgaard, C., Boutz, D. R., Perry, L. J., and Yeates, T. O. (2005). The genomics of disulfide bonding and protein stabilization in thermophiles. *PLoS Biol.* **3**:e309.

Benson, S. D., Bamford, J. K., Bamford, D. H., and Burnett, R. M. (1999). Viral evolution revealed by bacteriophage PRD1 and human adenovirus coat protein structures. *Cell* **98**:825–833.

Bettstetter, M., Peng, X., Garrett, R. A., and Prangishvili, D. (2003). AFV1, a novel virus infecting hyperthermophilic archaea of the genus acidianus. *Virology* **315**:68–79.

Blum, H., Zillig, W., Mallok, S., Domdey, H., and Prangishvili, D. (2001). The genome of the archaeal virus SIRV1 has features in common with genomes of eukaryal viruses. *Virology* **281**:6–9.

Campos-Olivas, R., Louis, J. M., Clerot, D., Gronenborn, B., and Gronenborn, A. M. (2002). The structure of a replication initiator unites diverse aspects of nucleic acid metabolism. *Proc. Natl. Acad. Sci. USA* **99**:10310–10315.

Geslin, C., Gaillard, M., Flament, D., Rouault, K., Le Romancer, M., Prieur, D., and Erauso, G. (2007). Analysis of the first genome of a hyperthermophilic marine virus-like particle, PAV1, isolated from *Pyrococcus abyssi*. *J. Bacteriol.* **189**:4510–4519.

Goulet, A., Blangy, S., Redder, P., Prangishvili, D., Felisberto-Rodrigues, C., Forterre, P., Campanacci, V., and Cambillau, C. (2009a). *Acidianus* filamentous virus 1 coat proteins display a helical fold spanning the filamentous archaeal viruses lineage. *Proc. Natl. Acad. Sci. USA* **106**:21155–21160.

Goulet, A., Pina, M., Redder, P., Prangishvili, D., Vera, L., Lichiere, J., Leulliot, N., van Tilbeurgh, H., Ortiz-Lombardia, M., Campanacci, V., and Cambillau, C. (2010a). ORF157 from the archaeal virus *Acidianus* filamentous virus 1 defines a new class of nuclease. *J. Virol.* **84**:5025–5031.

Goulet, A., Spinelli, S., Blangy, S., van Tilbeurgh, H., Leulliot, N., Basta, T., Prangishvili, D., Cambillau, C., and Campanacci, V. (2009b). The crystal structure of ORF14 from *Sulfolobus islandicus* filamentous virus. *Proteins* **76:**1020–1022.

Goulet, A., Spinelli, S., Blangy, S., van Tilbeurgh, H., Leulliot, N., Basta, T., Prangishvili, D., Cambillau, C., and Campanacci, V. (2009c). The thermo- and acido-stable ORF-99 from the archaeal virus AFV1. *Protein Sci.* **18:**1316–1320.

Goulet, A., Vestergaard, G., Felisberto-Rodrigues, C., Campanacci, V., Garrett, R. A., Cambillau, C., and Ortiz-Lombardia, M. (2010b). Getting the best out of long-wavelength X-rays: De novo chlorine/sulfur SAD phasing of a structural protein from ATV. *Acta Crystallogr. D Biol. Crystallogr.* **66:**304–308.

Guillière, F., Peixeiro, N., Kessler, A., Raynal, B., Desnoues, N., Keller, J., Delepierre, M., Prangishvili, D., Sezonov, G., and Guijarro, J. I. (2009). Structure, function, and targets of the transcriptional regulator SvtR from the hyperthermophilic archaeal virus SIRV1. *J. Biol. Chem.* **284:**22222–22237.

Happonen, L. J., Redder, P., Peng, X., Reigstad, L. J., Prangishvili, D., and Butcher, S. J. (2010). Familial relationships in hyperthermo- and acidophilic archaeal viruses. *J. Virol.* **84:**4747–4754.

Häring, M., Peng, X., Brugger, K., Rachel, R., Stetter, K. O., Garrett, R. A., and Prangishvili, D. (2004). Morphology and genome organization of the virus PSV of the hyperthermophilic archaeal genera *Pyrobaculum* and *Thermoproteus*: A novel virus family, the *Globuloviridae. Virology* **323:**233–242.

Häring, M., Vestergaard, G., Brugger, K., Rachel, R., Garrett, R. A., and Prangishvili, D. (2005). Structure and genome organization of AFV2, a novel archaeal lipothrixvirus with unusual terminal and core structures. *J. Bacteriol.* **187:**3855–3858.

Holm, L., and Rosenstrom, P. (2010). Dali server: Conservation mapping in 3D. *Nucleic Acids Res.* **38:**W545–W549.

Janekovic, D., Wunderl, S., Holz, I., Zillig, W., Gierl, A., H. N., and Neumann, H. (1983). TTV1, TTV2 and TTV3, a family of viruses of extremely thermophilic, anaerobic sulfur-reducing archaebacterium *Thermoproteus tenax. Mol. Gen. Genet.* **192:**39–45.

Keller, J., Leulliot, N., Cambillau, C., Campanacci, V., Porciero, S., Prangishvilli, D., Forterre, P., Cortez, D., Quevillon-Cheruel, S., and van Tilbeurgh, H. (2007). Crystal structure of AFV3-109, a highly conserved protein from crenarchaeal viruses. *Virol. J.* **4:**12.

Keller, J., Leulliot, N., Collinet, B., Campanacci, V., Cambillau, C., Pranghisvilli, D., and van Tilbeurgh, H. (2009a). Crystal structure of AFV1-102, a protein from the acidianus filamentous virus 1. *Protein Sci.* **18:**845–849.

Keller, J., Leulliot, N., Soler, N., Collinet, B., Vincentelli, R., Forterre, P., and van Tilbeurgh, H. (2009b). A protein encoded by a new family of mobile elements from Euryarchaea exhibits three domains with novel folds. *Protein Sci.* **18:**850–855.

Khayat, R., Tang, L., Larson, E. T., Lawrence, C. M., Young, M., and Johnson, J. E. (2005). Structure of an archaeal virus capsid protein reveals a common ancestry to eukaryotic and bacterial viruses. *Proc. Natl. Acad. Sci. USA* **102:**18944–18949.

Klein, R., Baranyi, U., Rossler, N., Greineder, B., Scholz, H., and Witte, A. (2002). *Natrialba magadii* virus phiCh1: First complete nucleotide sequence and functional organization of a virus infecting a haloalkaliphilic archaeon. *Mol. Microbiol.* **45:**851–863.

Koonin, E. V. (1992). Archaebacterial virus SSV1 encodes a putative DnaA-like protein. *Nucleic Acids Res.* **20:**1143.

Koonin, E. V. (2010). New variants of known folds: Do they bring new biology? *Acta Crystallogr. Sect. F Struct. Biol. Cryst. Commun.* **66:**1226–1229.

Kraft, P., Kummel, D., Oeckinghaus, A., Gauss, G. H., Wiedenheft, B., Young, M., and Lawrence, C. M. (2004a). Structure of D-63 from *Sulfolobus* spindle-shaped virus 1: Surface properties of the dimeric four-helix bundle suggest an adaptor protein function. *J. Virol.* **78:**7438–7442.

Kraft, P., Oeckinghaus, A., Kummel, D., Gauss, G. H., Gilmore, J., Wiedenheft, B., Young, M., and Lawrence, C. M. (2004b). Crystal structure of F-93 from Sulfolobus spindle-shaped virus 1, a winged-helix DNA binding protein. *J. Virol.* **78:**11544–11550.

Krupovic, M., and Bamford, D. H. (2011). Double-stranded DNA viruses: 20 families and only five different architectural principles for virion assembly. *Curr. Opin. Virol.* **1:**118–124.

Krupovic, M., and Bamford, D. H. (2007). Putative prophages related to lytic tailless marine dsDNA phage PM2 are widespread in the genomes of aquatic bacteria. *BMC Genomics* **8:**236.

Krupovic, M., and Bamford, D. H. (2008a). Archaeal proviruses TKV4 and MVV extend the PRD1-adenovirus lineage to the phylum *Euryarchaeota*. *Virology* **375:**292–300.

Krupovic, M., and Bamford, D. H. (2008b). Virus evolution: How far does the double beta-barrel viral lineage extend? *Nat. Rev. Microbiol.* **6:**941–948.

Krupovic, M., and Bamford, D. H. (2009). Does the evolution of viral polymerases reflect the origin and evolution of viruses? *Nat. Rev. Microbiol.* **7:**250.

Krupovic, M., and Bamford, D. H. (2010). Order to the viral universe. *J. Virol.* **84:**12476–12479.

Krupovic, M., Forterre, P., and Bamford, D. H. (2010a). Comparative analysis of the mosaic genomes of tailed archaeal viruses and proviruses suggests common themes for virion architecture and assembly with tailed viruses of bacteria. *J. Mol. Biol.* **397:**144–160.

Krupovic, M., Gribaldo, S., Bamford, D. H., and Forterre, P. (2010b). The evolutionary history of archaeal MCM helicases: A case study of vertical evolution combined with hitchhiking of mobile genetic elements. *Mol. Biol. Evol.* **27:**2716–2732.

Krupovic, M., Prangishvili, D., Hendrix, R. W., and Bamford, D. H. (2011a). Genomics of bacterial and archaeal viruses: dynamics within the prokaryotic virosphere. *Microbiol. Mol. Biol. Rev.* **75:**610–635.

Krupovic, M., Spang, A., Gribaldo, S., Forterre, P., and Schleper, C. (2011b). A thaumarchaeal provirus testifies for an ancient association of tailed viruses with archaea. *Biochem. Soc. Trans.* **39:**82–88.

Larson, E. T., Eilers, B., Menon, S., Reiter, D., Ortmann, A., Young, M. J., and Lawrence, C. M. (2007a). A winged-helix protein from *Sulfolobus* turreted icosahedral virus points toward stabilizing disulfide bonds in the intracellular proteins of a hyperthermophilic virus. *Virology* **368:**249–261.

Larson, E. T., Eilers, B. J., Reiter, D., Ortmann, A. C., Young, M. J., and Lawrence, C. M. (2007b). A new DNA binding protein highly conserved in diverse crenarchaeal viruses. *Virology* **363:**387–396.

Larson, E. T., Reiter, D., Young, M., and Lawrence, C. M. (2006). Structure of A197 from *Sulfolobus* turreted icosahedral virus: A crenarchaeal viral glycosyltransferase exhibiting the GT-A fold. *J. Virol.* **80:**7636–7644.

Lawrence, C. M., Menon, S., Eilers, B. J., Bothner, B., Khayat, R., Douglas, T., and Young, M. J. (2009). Structural and functional studies of archaeal viruses. *J. Biol. Chem.* **284:**12599–12603.

Maaty, W. S., Ortmann, A. C., Dlakic, M., Schulstad, K., Hilmer, J. K., Liepold, L., Weidenheft, B., Khayat, R., Douglas, T., Young, M. J., and Bothner, B. (2006). Characterization of the archaeal thermophile Sulfolobus turreted icosahedral virus validates an evolutionary link among double-stranded DNA viruses from all domains of life. *J. Virol.* **80:**7625–7635.

Markine-Goriaynoff, N., Gillet, L., Van Etten, J. L., Korres, H., Verma, N., and Vanderplasschen, A. (2004). Glycosyltransferases encoded by viruses. *J. Gen. Virol.* **85:**2741–2754.

Menon, S. K., Eilers, B. J., Young, M. J., and Lawrence, C. M. (2010). The crystal structure of D212 from *Sulfolobus* spindle-shaped virus ragged hills reveals a new member of the PD-(D/E)XK nuclease superfamily. *J. Virol.* **84:**5890–5897.

Menon, S. K., Maaty, W. S., Corn, G. J., Kwok, S. C., Eilers, B. J., Kraft, P., Gillitzer, E., Young, M. J., Bothner, B., and Lawrence, C. M. (2008). Cysteine usage in *Sulfolobus* spindle-shaped virus 1 and extension to hyperthermophilic viruses in general. *Virology* **376:**270–278.

Mochizuki, T., Yoshida, T., Tanaka, R., Forterre, P., Sako, Y., and Prangishvili, D. (2010). Diversity of viruses of the hyperthermophilic archaeal genus *Aeropyrum*, and isolation of the *Aeropyrum pernix* bacilliform virus 1, APBV1, the first representative of the family *Clavaviridae*. *Virology* **402:**347–354.

Nandhagopal, N., Simpson, A. A., Gurnon, J. R., Yan, X., Baker, T. S., Graves, M. V., Van Etten, J. L., and Rossmann, M. G. (2002). The structure and evolution of the major capsid protein of a large, lipid-containing DNA virus. *Proc. Natl. Acad. Sci. USA* **99:**14758–14763.

Oke, M., Carter, L. G., Johnson, K. A., Liu, H., McMahon, S. A., Yan, X., Kerou, M., Weikart, N. D., Kadi, N., Sheikh, M. A., Schmelz, S., Dorward,, S., *et al.* (2010). The Scottish Structural Proteomics Facility: Targets, methods and outputs. *J. Struct. Funct. Genomics* **11:**167–180.

Oke, M., Kerou, M., Liu, H., Peng, X., Garrett, R. A., Prangishvili, D., Naismith, J. H., and White, M. F. (2011). A dimeric Rep protein initiates replication of a linear archaeal virus genome: implications for the Rep mechanism and viral replication. *J. Virol.* **85:**925–931.

Peng, X., Basta, T., Haring, M., Garrett, R. A., and Prangishvili, D. (2007). Genome of the *Acidianus* bottle-shaped virus and insights into the replication and packaging mechanisms. *Virology* **364:**237–243.

Peng, X., Blum, H., She, Q., Mallok, S., Brugger, K., Garrett, R. A., Zillig, W., and Prangishvili, D. (2001). Sequences and replication of genomes of the archaeal rudiviruses SIRV1 and SIRV2: Relationships to the archaeal lipothrixvirus SIFV and some eukaryal viruses. *Virology* **291:**226–234.

Pfister, P., Wasserfallen, A., Stettler, R., and Leisinger, T. (1998). Molecular analysis of *Methanobacterium* phage psiM2. *Mol. Microbiol.* **30:**233–244.

Pietilä, M. K., Laurinavicius, S., Sund, J., Roine, E., and Bamford, D. H. (2010). The single-stranded DNA genome of novel archaeal virus *Halorubrum* pleomorphic virus 1 is enclosed in the envelope decorated with glycoprotein spikes. *J. Virol.* **84:**788–798.

Pietilä, M. K., Roine, E., Paulin, L., Kalkkinen, N., and Bamford, D. H. (2009). An ssDNA virus infecting archaea: A new lineage of viruses with a membrane envelope. *Mol. Microbiol.* **72:**307–319.

Prangishvili, D., Arnold, H. P., Gotz, D., Ziese, U., Holz, I., Kristjansson, J. K., and Zillig, W. (1999). A novel virus family, the *Rudiviridae*: Structure, virus-host interactions and genome variability of the *Sulfolobus* viruses SIRV1 and SIRV2. *Genetics* **152:**1387–1396.

Prangishvili, D., Forterre, P., and Garrett, R. A. (2006a). Viruses of the Archaea: A unifying view. *Nat. Rev. Microbiol.* **4:**837–848.

Prangishvili, D., Garrett, R. A., and Koonin, E. V. (2006b). Evolutionary genomics of archaeal viruses: Unique viral genomes in the third domain of life. *Virus Res.* **117:**52–67.

Prangishvili, D., and Krupovic, M. (2012). A new proposed taxon for double-stranded DNA viruses, the order "Ligamenvirales". *Arch. Virol.* doi: 10.1007/s00705-012-1229-7.

Prangishvili, D., Vestergaard, G., Häring, M., Aramayo, R., Basta, T., Rachel, R., and Garrett, R. A. (2006c). Structural and genomic properties of the hyperthermophilic archaeal virus ATV with an extracellular stage of the reproductive cycle. *J. Mol. Biol.* **359:**1203–1216.

Rao, V. B., and Feiss, M. (2008). The bacteriophage DNA packaging motor. *Annu. Rev. Genet.* **42:**647–681.

Raoult, D., and Forterre, P. (2008). Redefining viruses: Lessons from Mimivirus. *Nat. Rev. Microbiol.* **6:**315–319.

Redder, P., Peng, X., Brugger, K., Shah, S. A., Roesch, F., Greve, B., She, Q., Schleper, C., Forterre, P., Garrett, R. A., and Prangishvili, D. (2009). Four newly isolated fuselloviruses from extreme geothermal environments reveal unusual morphologies and a possible interviral recombination mechanism. *Environ. Microbiol.* **11:**2849–2862.

Rice, G., Tang, L., Stedman, K., Roberto, F., Spuhler, J., Gillitzer, E., Johnson, J. E., Douglas, T., and Young, M. (2004). The structure of a thermophilic archaeal virus shows a double-stranded DNA viral capsid type that spans all domains of life. *Proc. Natl. Acad. Sci. USA* **101:**7716–7720.

Roberts, M. M., White, J. L., Grutter, M. G., and Burnett, R. M. (1986). Three-dimensional structure of the adenovirus major coat protein hexon. *Science* **232:**1148–1151.

Roine, E., Kukkaro, P., Paulin, L., Laurinavičius, S., Domanska, A., Somerharju, P., and Bamford, D. H. (2010). New, closely related haloarchaeal viral elements with different nucleic acid types. *J. Virol.* **84:**3682–3689.

Schlenker, C., Menon, S., Lawrence, C. M., and Copié, V. (2009). (1)H, (13)C, (15)N backbone and side chain NMR resonance assignments for E73 from Sulfolobus spindle-shaped virus ragged hills, a hyperthermophilic crenarchaeal virus from Yellowstone National Park. *Biomol. NMR Assign.* **3:**219–222.

Schleper, C., Kubo, K., and Zillig, W. (1992). The particle SSV1 from the extremely thermophilic archaeon *Sulfolobus* is a virus: Demonstration of infectivity and of transfection with viral DNA. *Proc. Natl. Acad. Sci. USA* **89:**7645–7649.

Steinmetz, N. F., Bize, A., Findlay, R. C., Lomonossoff, G. P., Manchester, M., Evans, D. J., and Prangishvili, D. (2008). Site-specific and spatially controlled addressability of a new viral nanobuilding block: *Sulfolobus islandicus* rod-shaped virus 2. *Adv. Funct. Mater.* **18:**3478–3486.

Steven, A. C., Heymann, J. B., Cheng, N., Trus, B. L., and Conway, J. F. (2005). Virus maturation: Dynamics and mechanism of a stabilizing structural transition that leads to infectivity. *Curr. Opin. Struct. Biol.* **15:**227–236.

Szymczyna, B. R., Taurog, R. E., Young, M. J., Snyder, J. C., Johnson, J. E., and Williamson, J. R. (2009). Synergy of NMR, computation, and X-ray crystallography for structural biology. *Structure* **17:**499–507.

Tang, S. L., Nuttall, S., and Dyall-Smith, M. (2004). Haloviruses HF1 and HF2: Evidence for a recent and large recombination event. *J. Bacteriol.* **186:**2810–2817.

Tomizawa, J., and Som, T. (1984). Control of ColE1 plasmid replication: Enhancement of binding of RNA I to the primer transcript by the Rom protein. *Cell* **38:**871–878.

van Pouderoyen, G., Ketting, R. F., Perrakis, A., Plasterk, R. H., and Sixma, T. K. (1997). Crystal structure of the specific DNA-binding domain of Tc3 transposase of *C. elegans* in complex with transposon DNA. *EMBO J* **16:**6044–6054.

Vestergaard, G., Häring, M., Peng, X., Rachel, R., Garrett, R. A., and Prangishvili, D. (2005). A novel rudivirus, ARV1, of the hyperthermophilic archaeal genus *Acidianus*. *Virology* **336:**83–92.

Wilusz, C. J., and Wilusz, J. (2005). Eukaryotic Lsm proteins: Lessons from bacteria. *Nat. Struct. Mol. Biol.* **12:**1031–1036.

Xiang, X., Chen, L., Huang, X., Luo, Y., She, Q., and Huang, L. (2005). *Sulfolobus tengchongensis* spindle-shaped virus STSV1: Virus-host interactions and genomic features. *J. Virol.* **79:**8677–8686.

CHAPTER 3

Sputnik, a Virophage Infecting the Viral Domain of Life

Christelle Desnues,* Mickaël Boyer,* and **Didier Raoult**

Contents

* These authors contributed equally.

The authors declare no conflict of interest.

URMITE, Centre National de la Recherche Scientifique UMR IRD 6236, Faculté de Médecine, Aix-Marseille Université, Marseille Cedex 5, France

Advances in Virus Research, Volume 82
ISSN 0065-3527, DOI: 10.1016/B978-0-12-394621-8.00013-3

Abstract This chapter discusses the astonishing discovery of the Sputnik virophage, a new virus infecting giant viruses of the genera Mimivirus and Mamavirus. While other virophages have also since been described, this chapter focuses mainly on Sputnik, which is the best described. We detail the general properties of the virophage life cycle, as well as its hosts, genomic characteristics, ecology, and origin. In addition to genetic, phylogenetic, and structural evidence, the existence of virophages has deeply altered our view of the tripartite division of life to include the addition of a fourth domain constituted of the nucleocytoplasmic large DNA viruses, an important point that is discussed.

LIST OF ABBREVIATIONS

2D gel electrophoresis	two dimensional gel electrophoresis
AAV	adeno-associated virus
APMV	*Acanthamoeba polyphaga* Mimivirus
BBH	best BLAST hit
BLAST	basic local aligment search tool
CBPSV	chronic bee paralysis satellite virus
CEO	capsid encoding organisms
CIV	Chilo iridescent virus
COG	cluster of orthologous group
CroV	Cafeteria roenbergensis virus
Cryo-EM	cryo-electron microscopy
DAPI	4′,6-diamidino-2-phenylindole
dsDNA	double-stranded DNA
Env_nr	Environmental non redundant database
HGT	horizontal gene transfer
IFF	immunofluorescence
MALDI-MS	matrix-assisted laser desorption/ionization mass spectrometry
MCP	major capsid protein
MGE	mobile genetic element
NCLDV	nucleocytoplasmic large DNA virus
ORF	open reading frame
PBCV-1	*Paramecium bursaria* Chlorella virus 1

PFGE	pulse field gel electrophoresis
p.i.	post-infection
PMSV	panicum mosaic satellite virus
PolB	DNA polymerase B
REO	ribosome-encoding organism
SF	superfamily
ssDNA	single-stranded DNA
ssRNA	single-stranded RNA
TEM	transmission electron microscopy
TNSV	tobacco necrosis satellite virus
TNV	tobacco necrosis virus

I. THE MIMIVIRIDAE FAMILY AND THE HISTORY OF SPUTNIK

A. Mimivirus, Mamavirus, and Sputnik

The first member of the Mimiviridae family, *Acanthamoeba Polyphaga Mimivirus* (APMV) or Mimivirus, was described in 2003 (La Scola *et al.*, 2003). Mimivirus was initially isolated in 1992 from the water of a cooling tower during a pneumonia outbreak in Bradford, England, by T. J. Rowbotham. At that time, it was mistakenly thought to be a type of small Gram-positive cocci and was accordingly named "*Bradford* coccus." Eleven years later, electron microscopy performed on a *Bradford* coccus suspension demonstrated the presence of nonenveloped icosahedral particles (approximately 500 nm in diameter) covered by fibrils instead of bacteria (La Scola *et al.*, 2003). The presence of a typical viral eclipse phase and further genome sequencing confirmed that Mimivirus was indeed a virus (La Scola *et al.*, 2003; Raoult *et al.*, 2004). The infection cycle of Mimivirus occurs entirely within the cytoplasm of *Acanthamoeba polyphaga*, a free-living amoeba that is ubiquitous in the environment (Mutsafi *et al.*, 2010). Once viral DNA is delivered into the cytoplasm of the amoeba, replication and transcription take place within the viral core, and early mRNAs accumulate rapidly at localized sites adjacent to the replication site. This process is similar to replication and transcription observed in poxviruses (Mutsafi *et al.*, 2010; Schramm and Locker, 2005).

The Mimivirus genome was published in 2004. While it was predicted to be ~0.8Mb from preliminary pulse field gel electrophoresis experiments, the Mimivirus genome was ultimately shown to be 1.2Mb after sequencing. It was the largest viral genome ever published, more than twice the genome size of *Bacillus megaterium* phage G, which is ~0.670Mb (complete chromosome sequence). The Mimivirus genome was initially predicted to contain 911 open reading frames (ORFs), and 75 new ORFs have recently been added (Legendre *et al.*, 2010; Raoult *et al.*, 2004). While most of these ORFs have not been found in any other viruses, Mimivirus can still be confidently related to other nucleocytoplasmic large DNA viruses (NCLDVs). NCLDVs are a group of large DNA viruses infecting various eukaryotic hosts, including iridoviruses, ascoviruses, poxviruses, the asfarvirus, phycodnaviruses, and the marseillevirus (Boyer *et al.*, 2009). Indeed, a concatenated phylogenetic tree constructed using the nine ORFs belonging to class I core genes conserved among all NCLDVs places Mimivirus as one of the six families of this viral group (Koonin and Yutin, 2010; Raoult *et al.*, 2004; Yutin *et al.*, 2009).

Of the 986 ORFs detected in the Mimivirus genome, almost half are considered to be ORFans (i.e., ORFs lacking homologues in the current databases) (Boyer *et al.*, 2010a), and only 24% have an inferred function (Colson and Raoult, 2010). Many of these genes are involved in cellular processes and are not expected to be found in parasitic organisms, which are thought to depend entirely on host machinery. For instance, several components of the translational apparatus are detected and expressed during the Mimivirus replication cycle (Legendre *et al.*, 2010). It has been argued that the presence of ORFs related to cellular function resulted from extensive horizontal gene transfer (HGT), largely from the amoebal host, but also from bacteria and archaea to the virus (Moreira and Brochier-Armanet, 2008; Moreira and Lopez-Garcia, 2009). From this point of view, Mimivirus has been defined as a mere "bag of genes" stolen from different sources (Moreira and Brochier-Armanet, 2008; Moreira and Lopez-Garcia, 2009). However, reanalysis of these studies has demonstrated that the number of genes acquired from the host appears to be overestimated and may account for <5% of the genome (Forterre, 2010), confirming that unlike other NCLDVs, the number of host-related genes in the Mimivirus genome is not correlated with its size (Filee and Chandler, 2010; Filee *et al.*, 2008). In fact, most of the genes of foreign origin in the Mimivirus genome are related to bacteria (Filee and Chandler, 2010; Filee *et al.*, 2007). Because amoebas feed on bacteria, they may constitute an ecological niche favorable to HGT between bacteria and Mimivirus (Filee and Chandler, 2010; Raoult and Boyer, 2010; Thomas and Greub, 2010). Several mobile genetic elements (MGEs), such as transposases and endonucleases, are found among the transferred genes, and these genetic elements have been hypothesized to promote horizontal gene transfer (HGT) and recombination (Filee and Chandler, 2010).

Five years after the discovery of Mimivirus, a new giant virus called Mamavirus was discovered in the waters of a cooling tower in Les Halles (Paris, France)(La Scola *et al.*, 2008). It was found that Mamavirus resembles Mimivirus in many aspects, including size and genome content. However, early experiments demonstrated reduced infectivity of Mamavirus toward the amoeba *Acanthamoeba castellanii*. When observed by transmission electron microscopy (TEM), Mamavirus viral factories appeared to be colonized at one pole by a small virus (~50 nm in diameter) that we named Sputnik (Fig. 1). By analogy to bacteriophages, we

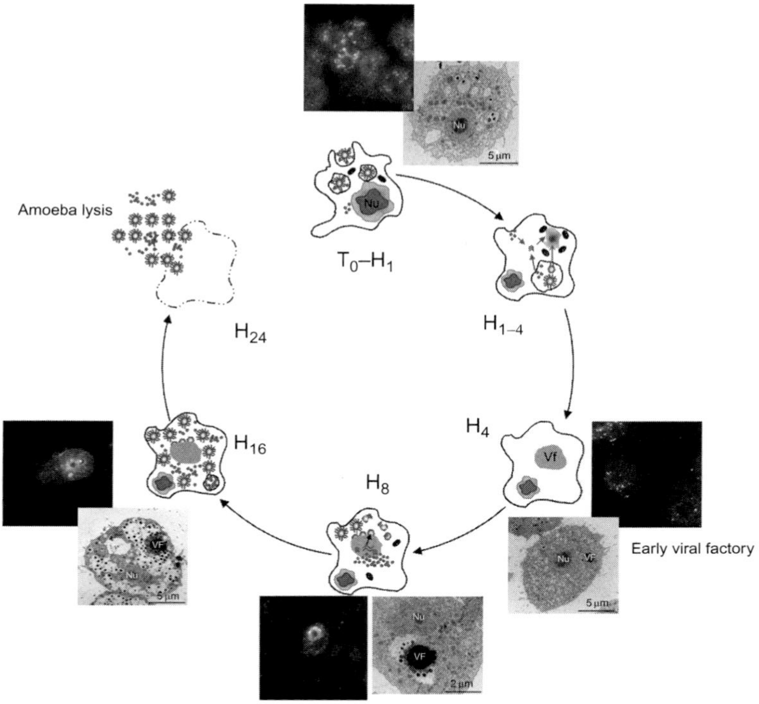

FIGURE 1 Life cycle of Mamavirus and Sputnik at T_0-H_1, H_4, H_8, H_{16}, and H_{24}. Schematic representation (center), immunofluorescence (IFF), and transmission electron microscopy (TEM) images are shown. IFF was performed with rabbit anti-Mimivirus (red) and mouse anti-Sputnik (green) antibodies. Nucleic acids were stained with 4′,6-diamidino-2-phenylindole. Viral factory (Vf) and nucleus (Nu) are indicated. Modified with permission from Desnues and Raoult (2010). (See Page 4 in Color Section at the back of the book.)

proposed Sputnik to be the first virophage (i.e., the first virus infecting a virus) described (La Scola *et al.*, 2008).

B. Other Mimi-like viruses associated with amoebas and a second virophage

Using a mixture of selected antibiotics to prevent bacterial growth, La Scola *et al.* (2010) were able to isolate new giant viruses from various sources (La Scola *et al.*, 2010). The authors tested 105 samples of environmental origin (e.g., freshwater from rivers and lakes, seawater, and soil), human construction-associated freshwater (e.g., decorative fountains, cooling towers, tap water, and hospital water), and contact lens fluid from a patient with keratitis. Nineteen new giant viruses were isolated, and 14 of them have a capsid size $\geq$400 nm. The genomes of four of these viruses have been sequenced (Courdo11, Moumou, Terra 1, and Terra 2), and preliminary phylogenetic analysis of their complete DNA polymerase B (PolB) sequences indicates that they are new Mimivirus-like viruses. These results are also consistent with results observed using matrix-assisted laser desorption/ionization mass spectrometry (MALDI-MS) (La Scola *et al.*, 2010). Further, a second virophage has also been isolated. This virophage was found in association with a giant virus (called CL) recovered from the liquid used to rinse contact lenses from a patient with keratitis. The genome of this second strain of virophage is almost identical to the Sputnik genome (data not shown).

C. The marine *Cafeteria roenbergensis* virus and its virophage Mavirus

Cafeteria roenbergensis virus (CroV) is a giant virus (~300 nm in diameter) that infects a marine heterotrophic flagellate (*C. roenbergensis* strain VENT1). CroV was isolated in the early 1990s from Texas marine waters (Garza and Suttle, 1995), and its genome has been published (Fischer *et al.*, 2010a). Its double-stranded DNA (dsDNA) genome is approximately 730 kb. The central coding region contains 544 ORFs, and highly repetitive sequences flank both strands of the genome. Phylogenetic analysis of conserved regions of PolB indicates that CroV is a bona fide NCLDV in the Mimiviridae group, although based on its capsid and genome size, CroV has largely diverged from Mimivirus. However, >30% of the CroV genome displays significant similarity to the Mimivirus genome (Fischer *et al.*, 2010a). CroV protein-encoding genes are involved in DNA replication, transcription, translation, and DNA repair, and each of these genes is under the control of early or late promoters. A 38-kb region of the CroV genome has not been associated with any promoter and may have been acquired through horizontal transfer from bacteria. This region contains genes involved in a pathway for the biosynthesis of 3-deoxy-D-manno-octulosonate, an essential component of the lipolysaccharide membrane of Gram-negative bacteria, along with other protein-encoding genes involved in carbohydrate metabolism.

A phylogenetic tree based on the concatenated alignment of four universal clusters of orthologous groups of proteins from NCLDVs (i.e., primase-helicase, DNA polymerase, packaging ATPase, and the A2L-like transcription factor) places CroV as a new subfamily within the Mimiviridae (Colson *et al.*, 2011). As with other members of the Mimiviridae, such as Mamavirus and CL, this new giant virus was also associated with a virophage (Fischer *et al.*, 2010b). This virophage, called Mavirus, has similarities to the Maverick family of DNA transposons found in many eukaryotes (Fischer *et al.*, 2010b).

II. SPUTNIK STRUCTURE: MORPHOLOGY, CHEMICAL COMPOSITION, AND PROTEIN COMPONENTS

Structural studies of Sputnik using cryoelectron microscopy (cryo-EM) have been performed, allowing the first three-dimensional reconstruction of the Sputnik virion to be proposed (Sun *et al.*, 2009). Sputnik particles contain an icosahedral capsid of about 740 Å in diameter. The protein encoded by ORF 20, the most abundant protein in Sputnik particles (La Scola *et al.*, 2008), is therefore likely used as the major capsid protein (MCP) of the hexagonal surface lattice of the particle, characterized by a $T = 27$ triangulation number ($h = 3$; $k = 3$). Capsomers are formed by trimeric molecules of MCP assembling into pseudohexameric and pentameric structural units that compose the external capsid shell of the virion. It appears that the particle surface is covered with 55-Å protrusions containing a triangular head protruding from the center of each pseudohexameric unit (Fig. 2). Other viruses, notably NCLDVs, such as PBCV-1 (Cherrier *et al.*, 2009), CIV (Yan *et al.*, 2009), and Mimivirus (Xiao *et al.*, 2009), as well as bacteriophages such as T4 (Fokine *et al.*, 2004) and phi29

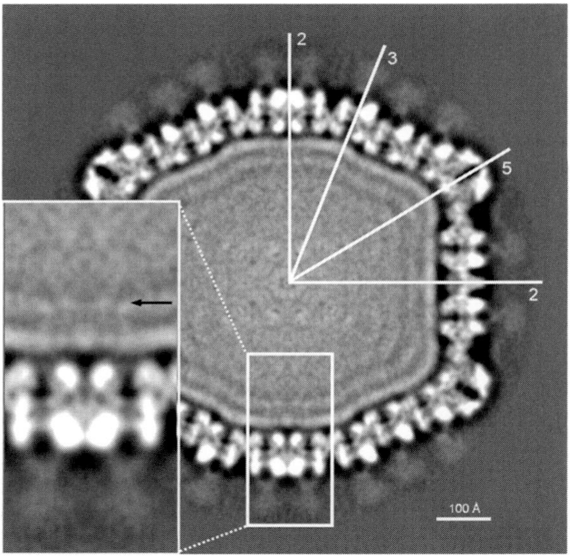

FIGURE 2 Cross section through a Sputnik cryo-EM image. Orientations of icosahedral (2-, 3-, and 5-fold) axes are shown with white lines. Note the putative lipid bilayer under the protein capsid and protusions on the trimeric capsomers. In the magnified view, a black arrow points toward possible transmembrane protein densities. Copyright © American Society for Microbiology, *Journal of Virology*, Vol. 84, 2010, pp. 894–897, doi:10.1128/JVI.01957-09.

(Tao *et al.*, 1998), have protusions on their surface, but their function remains uncertain. Likewise, adenovirus particles include fibers that emanate from pentameric vertices and are involved in host recognition (Balakireva *et al.*, 2003). Although the function of Sputnik protusions remains unknown, we speculate that they play a role in the recognition and adhesion of Sputnik to the giant virus particles, allowing the virophage to enter host cells.

The pentameric units located at the vertices of the capsid shell do not contain protusions but instead exhibit a type of cavity at the center of the pentamer. As described previously for other viruses, particularly bacteriophages (Cherrier *et al.*, 2009; Leiman *et al.*, 2004), these cavities may serve as a portal for DNA entry or exit (Xiao *et al.*, 2009; Zauberman *et al.*, 2008).

Inside the capsid shell, a cryo-EM cross section of the virion suggests the presence of a lipid bilayer beneath the protein capsid. Further analysis of the lipid fraction of Sputnik using mass spectrometry indicated that Sputnik samples contain between 12 and 24% lipid by weight and revealed that phosphatidylserine is the major lipid component of the virion. These observations are consistent with the presence of a lipid membrane within Sputnik. This membrane encloses the Sputnik DNA genome (18,343bp) in a volume of 3.6×10^7 Å^3. Thus, the density of the packaged DNA is approximately 1.996 Å^3/bp, which is comparable to other DNA viruses, such as T4 (168,903 bp), phiKZ (280,334bp), PRD1 (14,927bp), and adenovirus, which have densities of 2907, 2100, 2148, and 2057 Å^3/bp, respectively.

The thickness of the Sputnik capsomer (75 Å) and the distance between adjacent capsomers (75 Å) suggest that the Sputnik MCP exhibits comparable dimensions to viruses of the PRD1–adenovirus lineage. This lineage is composed of icosahedral dsDNA viruses, including adenovirus, bacteriophage PRD1, Sulfolobus turreted icosahedral virus, the marine bacteriophage PM2, and NCLDVs [including Mimivirus and *Paramecium bursaria* Chlorella virus 1 (PBCV-1)] (Krupovic and Bamford, 2008). All of these viruses have MCPs whose polypeptides, in some cases, such as the NCLDVs, have significant sequence similarity. MCP structures of the aforementioned viruses that have been determined to atomic resolution contain a similar fold that includes two consecutive "jelly roll" domains (double jelly roll fold). A jelly roll domain is an antiparallel β barrel consisting of eight β strands. Typically, MCPs are organized into capsomers with a thickness of ~75 Å, a diameter varying between 74 and 85 Å, and contain three monomers with a double jelly roll fold. It has been shown that the crystal structure of PBCV-1 MCP fits the cryo-EM map of Sputnik well (Sun *et al.*, 2009). Thus, the MCP of Sputnik likely also contains a double jelly roll fold as seen in viruses belonging to the PRD1–adenovirus lineage. However, the absence of significant sequence similarity between the MCP of Sputnik and other members of the PRD1–adenovirus lineage suggests that Sputnik is a new branch in this lineage.

III. LIFE CYCLE: HOST CELLS, ENTRY, UNCOATING, DNA REPLICATION, TRANSCRIPTION, TRANSLATION, ASSEMBLY, MATURATION, AND RELEASE

The predominant Sputnik host virus is Mamavirus. However, Sputnik can successfully coinfect *A. castellanii* along with Mimivirus. While the development cycles of Sputnik within Mimi and Mamaviruses have the same kinetics, Sputnik has a lower affinity for Mimivirus. For convenience, we use either Mimivirus or Mamavirus to designate the Mimi-like host virus associated with Sputnik.

A. Entry in the amoeba

Infections of amoebas by Sputnik particles have been performed either with or without Mamavirus (La Scola *et al.*, 2008). In the absence of Mamavirus, no amoebal cell lysis was observed, even at 7days postinfection (p.i.)(La Scola *et al.*, 2008). Additionally, no Sputnik particles were found inside the amoeba by TEM or immunofluorescence. The pathway by which Mimi-like viruses enter amoebas is unknown. It has been shown that Mimivirus invades human and mouse macrophages by phagocytosis, a mechanism usually used to engulf bacteria and parasites larger than 0.5 μm (Ghigo *et al.*, 2008). The surface fibers of Mimivirus are glycosylated proteins, ~140 nm in length, with a diameter of 1.4 nm (Kuznetsov *et al.*, 2010), contributing to its exceptional size. These fibers are capped with an approximately 25-kDa protein head and are anchored to the capsid by a protein layer displaying multiple fiber anchor sites (Kuznetsov *et al.*, 2010). The dense covering of fibers surrounding the capsid forms a protective layer that is resistant to proteases unless first treated with lysozyme (Xiao *et al.*, 2009). This covering is reminiscent of the peptidoglycan found in bacteria and may contribute to the bacterial-like entry mechanism of Mimivirus into amoebas (Claverie *et al.*, 2006; Desnues and Raoult, 2010; Kuznetsov *et al.*, 2010; Raoult *et al.*, 2004; Xiao *et al.*, 2009).

Sputnik has been observed frequently in association with the surface fibers of Mamavirus, and it has been hypothesized that Sputnik may be internalized together with Mamavirus in the same endocytic vacuole by the amoeba (Desnues and Raoult, 2010). Three proteins (R135, L725, and L829) were identified from purified Mimivirus fibers by two-dimensional (2D) gel electrophoresis coupled with MALDI-MS (Boyer *et al.*, 2011). The R135 protein is a putative GMC-type oxydoreductase and has been recovered from the protein pattern of purified Sputnik particles (La Scola *et al.*, 2008). This protein is likely involved in the adhesion between Sputnik and Mamavirus because Sputnik replication was not detected during coinfection with a bald form of Mimivirus (Boyer *et al.*, 2011).

B. Virophage hijacking of the viral factory

Immunofluorescence analysis using mouse anti-Sputnik antibodies allowed the detection of Sputnik particles inside amoebas at T0, corresponding to 30min p.i. The colocalization of Sputnik and Mimivirus signals further confirms the hypothesis that the two viruses share the same endocytic vacuoles. The mechanism by which Sputnik releases its genome into the cytoplasm of the amoeba is currently unknown but likely thrives on Mimivirus genome delivery. Once in the endosome, Mimivirus genome delivery occurs via a tunnel formed by fusion of the membrane of the endocytic vacuole and the internal viral membrane (Zauberman *et al.*, 2008). Delivery is preceded by a large-scale rearrangement of the Mimivirus capsid, leading to the massive opening of a "stargate" located at a unique vertex of the portal system (Zauberman *et al.*, 2008). As shown by electron microscopy, the Mimivirus genome is enclosed within a vesicle (the Mimivirus core), likely derived from the inner membrane of the virus when it is released (Mutsafi *et al.*, 2010; Zauberman *et al.*, 2008). Transcription is initiated inside the viral core, and newly synthesized mRNAs accumulate in regions localized near the core. A genome replication factory is observed by electron microscopy as a dense region at the periphery of the viral core (Mutsafi *et al.*, 2010). Multiple cores can be observed in close vicinity and may fuse to form a multilobular early viral factory. At 4h p.i., the early viral factory can be observed with epifluorescence microscopy as a strongly 4′,6-diamidino-2-phenylindole-stained patch. Sputnik and Mimivirus particles are not observed at this time, reflecting the eclipse phase of the viruses.

C. Production and release of progeny virions

Between 6 and 8h p.i., the viral factory expands (Fig. 1), and Sputnik progeny virions begin to be produced at one pole of the Mimivirus viral factory before the production of newly synthesized Mimivirus. The unipolarity of Sputnik production may result from the presence of a distinct packaging zone. Mimivirus factories displaying Sputnik production only are observed frequently at that time (La Scola *et al.*, 2008). Occasionally, the viral factory appears to be split into two: one filled with Sputnik progeny and the other producing Mimivirus (Desnues and Raoult, 2010). At 16h p.i., the cell is completely filled with newly formed Sputnik and Mimivirus virions. Sputnik particles may be detected either free in the cytoplasm of the amoeba or accumulated within vacuoles (Fig. 1).

One effect of Sputnik on the viability of Mamavirus-infected amoebas has been noted (La Scola *et al.*, 2008). Approximately 92% of amoeba cells are lysed after 24h of culture with Mamavirus alone, while only 79% are lysed following Mamavirus and Sputnik coinfection. Thus, the presence of Sputnik results in a 13% reduction of amoeba cell lysis and also results in about a one-half log reduction of the Mamavirus infectious titer (La Scola *et al.*, 2008).

D. Sputnik coinfection with other viruses

The effect of Sputnik on the growth of other amoeba-associated viruses has been tested previously. It appears that Sputnik is host-genus specific. For instance, Sputnik does not replicate along with Marseillevirus (Desnues and Raoult, 2010), another NCLDV found in amoeba. Other Marseillevirus-like viruses were tested, and similar results were obtained (unpublished data). In contrast, Sputnik can grow in association with all of the Mimivirus-like viruses that were tested, although it has higher infectious titers with Mamavirus (unpublished data).

IV. GENOMICS: GENE CONTENT, SPECIFIC GENES, LATERALLY TRANSFERRED GENES, ORFANS, GENE EXPRESSION, AND METAGENOMICS

A. Genome organization

The Sputnik genome is a 18,343-bp circular dsDNA. Twenty-one protein-encoding genes have been predicted, with a size ranging from 88 to 779 amino acids (La Scola *et al.*, 2008) (Table I). The Sputnik start and stop codons are predominantly AUG and UAA. Further, organization of the Sputnik genome is similar to the organization of other viral genomes, including a tightly spaced arrangement with little overlap of the ORFs (except for adjacent ORFs 18/19, which present an overlap of 22 nucleotide residues). The average distance between two predicted ORFs is ~190 nucleotide residues. The distribution of Sputnik protein-coding genes exhibits a strand bias toward the positive strand (17 ORFs). The high A+T content (73%) of the Sputnik genome is similar to that of Mamavirus and Mimivirus. Sputnik also shares the codon bias toward AT-rich codons.

B. Gene content and sources of Sputnik genes

Seven of the predicted Sputnik proteins share similarity with protein sequences in GenBank (five with E $<10^{-4}$, Table I). The others are considered to be ORFans. Best BLAST hit analysis demonstrated that Sputnik homologues are derived from Mimi/Mamavirus (ORFs 6, 12, and 13), bacteriophages (ORF 13), and archaea viruses (ORF 10)(Table I); these

TABLE I Best BLAST hit analysis of Sputnik proteins searched against GenBank nr (BLASTP, E $<10^{-4}$) database and Environmental nr database (Env_nr, TBLASTN, E $<10^{-4}$)[a]

Gene (size, aa)	Best BLAST Hit in GenBank nr (accession #, % identity/alignment length/E value)	Best BLAST Hit in env_nr (% identity/alignment length/E value)	Domain architecture/ protein family/predicted activity	Predicted function in virophage replication
ORF 1 (144)	—	—	Unknown	Unknown
ORF 2 (114)	—	—	Unknown	Unknown
ORF 3 (245)	—	Microbial mat metagenome (56%/187/3e-54)	FtsK-HerA superfamily ATPase	DNA packaging
ORF 4 (139)	—	—	Zn-ribbon-containing protein	Transcription regulation
ORF 5 (119)	—	—	Unknown	Unknown
ORF 6 (310)	collagen triple helix repeat-containing protein [*Acanthamoeba castellanii* Mamavirus] (EU827539.1, 75%/251/2e-90) *MIMIR196*	Marine metagenome 1091140559032 (58%/73/ 1e-12)	Collagen triple helix repeat-containing protein	Protein–protein interactions in factories
ORF 7 (236)	Collagen-like protein V6 [Sputnik virophage] (YP_002122367 , 65%, 106, 4e-24)	—	Collagen triple helix repeat-containing protein	Protein–protein interactions in factories

ORF 8 (184)	—	—	Unknown	Minor virion protein
ORF 9 (175)	—	Marine metagenome ctg_1101668235028 (67%, 100, 3e-42)	Unknown	Unknown
ORF 10 (226)	Phage integrase family protein [*Methanococcus aeolicus* Nankai-3] (YP_001324883, 32%/166/6e-13)	Hot springs metagenome ctg_1106445187876 (29%/164/1e-10)	Tyr recombinase family integrase	Integration of virophage into Mimivirus genome
ORF 11 (162)	—	—	Unknown	Unknown
ORF 12 (152)	*MAMA_R546 [Acanthamoeba castellanii* Mamavirus (Q5UR26, 62%/119/2e-35)	—	Unknown	Unknown
ORF 13 (779)	Putative DNA-polymerase or DNA-primase [*Lactobacillus* phage phiadh] (NP_050131.1, 29%/171/4e-12) *MIMI L207/206*	Marine metagenome 1096626676013 (32%/409/2e-46)	Primase–helicase	DNA replication
	—	Marine metagenome ctg_1101668410675 (32%/298/3e-24)		
ORF 14 (114)	—	—	Zn ribbon-containing protein	Transcription regulation
ORF 15 (109)	—	—	Membrane protein	

(continued)

TABLE I *(continued)*

Gene (size, aa)	Best BLAST Hit in GenBank nr (accession #, % identity/alignment length/E value)	Best BLAST Hit in env_nr (% identity/alignment length/E value)	Domain architecture/protein family/predicted activity	Predicted function in virophage replication
				Modification of APM membrane?
ORF 16 (130)	—	—	Unknown	Unknown
ORF 17 (88)	—	—	IS3 family transposase A protein	DNA-binding protein
ORF 18 (167)	—	Marine metagenome ctg_1101668235028 (43%/174/3e-29)	Unknown	Unknown
ORF 19 (218)	—	Marine metagenome ctg_1101668242663 (31%/230/2e-13)	Unknown	Minor virion protein
ORF20 (595)	—	Marine metagenome ctg_1101668040111 (30%/266/1e-27)	Unknown	Major capsid protein
ORF21 (438)	—	Marine metagenome ctg_1101668119947 (30%/330/6e-28)	Unknown	Unknown

[a] Modified from La Scola *et al.* (2008).

results were further confirmed by constructing phylogenetic trees (La Scola *et al.*, 2008).

Open reading frames 6 and 7 contain a highly conserved collagen triple helix motif (Rasmussen *et al.*, 2003). This motif is characteristic of extracellular structural proteins involved in matrix formation and/or adhesion processes (Rasmussen et al., 2003). The ORF 12 protein, which shares similarities with ORF R546 from Mamavirus and Mimivirus, has no assigned function. The Sputnik ORF 12 protein is truncated at the C terminus and lacks end repeats compared to the Mamavirus homologue. The predicted ORF 13 coding sequence can be divided in two domains involved in DNA replication. The first domain, located in the C-terminal portion of the protein, is a superfamily 3 helicase, and this SF3 domain is highly conserved in bacteriophage DNA primase/helicase proteins and in NCLDV D5 proteins. The N-terminal portion of the protein shows no significant similarities to sequences in the GenBank nr database. In contrast, two domains of ORF 13 have highly significant matches ($E = 2e^{-46}$ and $E = 3e^{-24}$ for the C-terminal and the N-terminal domains, respectively) with sequences in the environmental nr database (Table I). Based on sequence signatures, it has been hypothesized that the N-terminal domain of this ORF may represent a highly divergent version of the archaeoeukaryotic primase (La Scola *et al.*, 2008).

Open reading frame 10 displays significant similarity with integrases belonging to the tyrosine recombinase family. The Sputnik integrase, like all 130 members of the tyrosine recombinase family, contains the invariant RHRY amino acid tetrad in the C-terminal domain of the protein (La Scola *et al.*, 2008). The closest homologue of the Sputnik integrase is found in archaea viruses and proviruses (Table I). Because archaea are found frequently in extreme environments, sequences related to Sputnik ORF 10 recovered from the hot spring metagenome (Table I) may correspond to either archaea viruses or Sputnik relatives able to thrive under such adverse conditions.

In the initial search for Sputnik homologues in the environmental nr database, few sequences were recovered using BLASTP (La Scola *et al.*, 2008). However, six additional Meta-ORFans (i.e., ORFans with homologues found in the environmental nr database) (Boyer *et al.*, 2010a) were found during reanalysis using TBLASTN (ORFs 9, 10, 18, 19, 20, and 21, Table I).

C. Gene expression

Because the Sputnik genome does not encode its own DNA-dependent RNA polymerase, it may use that of Mimivirus for transcription.This possibility is supported by both the detection of the unique Mimivirus hairpin signal for polyadenylation (Byrne *et al.*, 2009; Claverie and

Abergel, 2009) and the presence of a Mimivirus late promoter in the Sputnik genome (Legendre *et al.*, 2010).

As with other eukaryotic viruses, Mimivirus protein-encoding genes are expressed as polyadenylated transcripts. More than 80% of the analyzed mature mRNA 3′ ends contain palindromic sequences, allowing for perfect pairing of at least 13 successive nucleotides into hairpin-like structures (Byrne *et al.*, 2009). This signal, which is absent in the amoeba genome, governs the polyadenylation of Mimivirus transcripts. The Sputnik genome displays 16 Mimivirus-like putative hairpin structures (Claverie and Abergel, 2009). Further, the distribution of these signals is not random because 14 of them are found in intergenic regions, whereas only 2 are located within a gene.

Additionally, a study of mRNA transcripts in Mimivirus revealed that gene transcription is regulated through early, intermediate, and late promoters (Legendre *et al.*, 2010). A search for these promoters in the Sputnik genome demonstrated the presence of late promoter motifs upstream of 12 different genes. The distance between the motif and the beginning of the ORF is $\sim 29 \pm 2.5$ nucleotide residues and is similar to that observed in Mimivirus ($D = 21 \pm 5$).

D. Proteomics

Proteins from purified Sputnik particles were analyzed using 2D gel electrophoresis (La Scola *et al.*, 2008). The most abundant protein, encoded by ORF 20, was identified in 55 out of the 60 detected spots. This protein corresponds to the major virion coat protein and has a mass of approximately 65kDa. Two other proteins, encoded by ORF 8 and ORF 19, were also identified and may correspond to minor virion proteins. These two proteins exhibited *N*-acetylation of their N-terminal amino acids, a modification that is common in eukaryotes. Western blotting performed with anti-Sputnik mouse serum confirmed that spots detected in the 2D gels correspond to Sputnik proteins.

It has been shown previously that Sputnik viral particles contain all viral RNAs except ORF 17, which encodes for a transposase (Desnues and Raoult, 2010). However, ORF 17 mRNA was detected at 4h p.i., showing that this gene is functional (Desnues and Raoult, 2010).

E. Ecology of Sputnik and Sputnik ORFs in metagenomic data sets

In addition to homologues found in the environmental nr database (Table I), Sputnik-like sequences were further analyzed by BLASTP searching the environmental database CAMERA (Seshadri *et al.*, 2007) (Fig. 3). As of December 2010, CAMERA contains 951 metagenomes (454- and shotgun-sequenced metagenomes) from 75 projects. BLASTP was

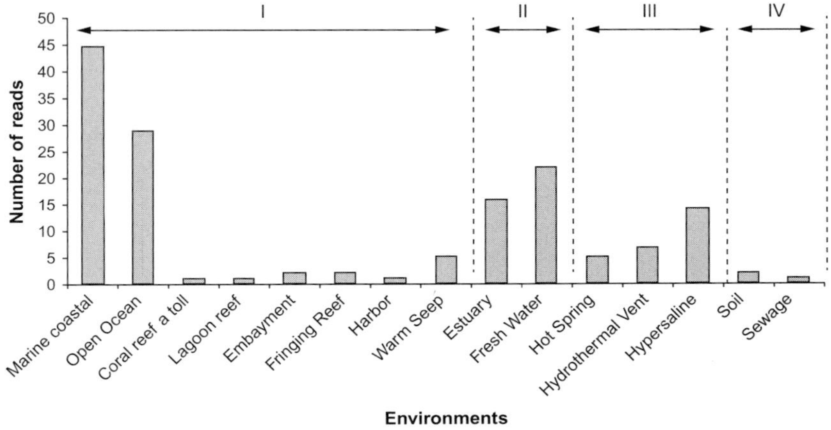

FIGURE 3 Repartitioning of Sputnik reads among environmental metagenomes. Environments are grouped into marine (I), freshwater (II), extreme (III), and terrestrial (IV) categories.

TABLE II BLASTP results (best BLAST hit analysis) of Sputnik protein sequences against the Lake Gatun metagenome

ORF number and annotation	E value (for the best Blast hit)
ORF3_FtsK-HerA-ATPase	7.64E-35
ORF6_collagen_protein	6.77E-09
ORF9_unknown	1.02E-41
ORF18_unknown	5.73E-29
ORF_19_minor_virion_protein	5.71E-13
ORF20_major_capsid_protein	5.76E-03
ORF21_unknown	9.55E-05

performed against all metagenomic ORF peptides, which excluded 454 sequencing data. Using no minimal e-value and an alignment of 25 reads per query, 153 hits were obtained from 43 different metagenomes further grouped into marine, freshwater, extreme, and terrestrial (Fig. 3). Indeed, 15 of the 21 Sputnik ORFs were recovered from the metagenomes (ORFs 3, 4, 6, 8, 9, 10, 11, 13, 14, 16, 17, 18, 19, 20, and 21).

The Lake Gatun data set is the metagenome displaying the largest number of hits (22 reads). Seven ORFs were recovered with very high confidence (Table II). Lake Gatun is a large artificial freshwater lake created in the Republic of Panama at the beginning of the 20th century. This metagenome data set was generated from a 0.1- to 0.8-μm fraction collected 2 m below the water level, and the water temperature was 28.6°C at the time of sampling. Numerous Mimivirus-related sequences are also found in the Lake Gatun metagenome (561 reads with an e-value $<10^{-4}$ among 228 different ORFs, data not shown), suggesting that both the virophage and its host are common in this environment.

Sputnik-related sequences have been found in environmental metagenomes derived from very diverse habitats, such as marine waters, fresh water, extreme environments, sewage, and soil (Tables I and III, Fig. 3). Thus, Sputnik and its relatives may represent an unknown family of viruses able to survive in various environments.

Numerous Sputnik-like sequences were also recovered from marine environments but must be normalized according to the number of metagenomes. The presence of a high number of very similar Sputnik-like and Mimivirus-like sequences in the metagenome of Lake Gatun (Table II) strongly suggests that close relatives of Sputnik and Mimivirus are common in this freshwater environment.

V. VIROPHAGE VS SATELLITE VIRUS

Satellites are subviral agents that depend on another virus (called the helper virus) for their multiplication. There are two classes of satellite agents: satellite viruses whose genome encodes their own capsid proteins and satellite nucleic acids whose genome does not encode any structural proteins (Fauquet *et al.*, 2005). Because Sputnik requires the presence of Mimivirus to accomplish its replication cycle and possesses its own capsid-encoding gene, it may be considered a satellite virus. The term satellite was first used in 1962 by Kassanis to describe the relationship between the tobacco necrosis virus (TNV) and its small, accompanying virus (the tobacco necrosis satellite virus, TNSV)(Kassanis, 1962). TNSV is a viral particle (17 nm in diameter) found in association with some isolates of the 26-nm-diameter TNV. The smaller particle is dependent on TNV for its multiplication but is serologically unrelated to TNV. Thus far, few other satellite viruses have been described, which are reported in Table III.

Based on their nucleic acids (ssDNA or ssRNA), two classes of satellite viruses can be differentiated (Table III). The first is represented by the human adeno-associated virus (AAV), which was discovered in 1965 in adenovirus preparations (Atchison *et al.*, 1965). AAV is a small, nonenveloped virus with an icosahedral capsid ranging from 20 to 24 nm in diameter. AAV have been classified as a *Dependovirus* in the Parvoviridae family and are not included in satellite virus group of the International Committee on Taxonomy of Viruses (ICTV) (Fauquet *et al.*, 2005). AAV is highly seroprevalent in the human population, but it has not been associated with any human pathology. AAV has either a positive or a negative linear ssDNA genome of approximately 4.7 kb in size. The genome contains the *rep* (replication) and *cap*

TABLE III Satellite viruses, modified from Palukaitis *et al.* (2008)

Helper/ satellite virus	Nucleic acid	Particle (nm)	Genome size (nt)	MCP[a] (mass in kDa)	Host	References
Adenoviruses or herpesviruses/ adeno-associated virus (Dependovirus, Parvoviridae)	ssDNA	20–24	4700	87, 73, and 62	Vertebrate	Berns and Giraud (1996)
Subgroup 1						
Chronic bee-paralysis virus (CBPV)/ CBPV-associated satellite virus	ssRNA	17	~1100	15	Animal	Bailey *et al.* (1980)
Subgroup 2						
Necrovirus, tobacco necrosis virus (TNV)/satellite TNV	ssRNA	17	1239	21.6	Plant	Reichmann (1964)
Sobemovirus, Panicum mosaic virus (PMV)/satellite PMV	ssRNA	16	824–826	~17.5	Plant	Masuta *et al.* (1987)
Tobamovirus, tobacco mosaic virus (TMV)/satellite TMV	ssRNA	17	1059	~17.5	Plant	Valverde and Dodds (1986)
Nodavirus, *Macrobranchium rosenbergii* nodavirus (MrNV)/ extra small virus	ssRNA	15	796	~17	Plant	Qian *et al.* (2003)
Unassigned, Maize white line mosaic virus (MWLMV)/ satellite MWLMV	ssRNA	17	1168	24	Plant	Zhang *et al.* (1991)

[a] MCP, major capsid protein.

(capsid) genes encoding nonstructural and structural proteins, respectively. Both ends of the genome consist of a 145-nucleotide-long palindromic terminal repeat that can form a hairpin structure and is required for integration into the host chromosome (Wang *et al.*, 1995). In the absence of a helper virus (adenovirus or herpes simplex virus), AAV can integrate into the human host chromosome (Kotin *et al.*, 1990). Excision of the provirus is observed in cases of super-infection with the helper virus or, with lower efficiency, when the host cell is subjected to cellular genotoxic stress (Yalkinoglu *et al.*, 1988). In some cases, AAV is also able to replicate autonomously in the absence of helper viruses or genotoxic agents (Meyers *et al.*, 2000).

The second class of satellite viruses (which includes all the satellite viruses classified as such by the ICTV) is composed of ssRNA viruses infecting insects (subgroup 1) and plants (subgroup 2). The chronic bee paralysis satellite virus (CBPSV), first described in 1980, is the only member of the chronic bee paralysis virus-associated satellite subgroup (Fauquet *et al.*, 2005). This particle is ~17 nm in diameter and has a single capsid protein of ~15kDa (Bailey *et al.*, 1980). The CBPSV genome consists of three ssRNA fragments of equal size (~1100 nucleotide residues each) but different secondary structures. However, it is unclear whether CBPSV is a true satellite virus (dependent on CBPV for replication) or an abortive particle (Ribière *et al.*, 2010). Currently, the most abundant isolated satellite viruses are satellites that resemble TNSV (subgroup 2), all of which depend on plant viruses for their replication (Table III). Satellite viruses of this subgroup contain an ssRNA of approximately 800–1200 nucleotide residues. Particles have a capsid size ranging from 15 to 17 nm in diameter and are composed of 60 protein subunits. Interference with accumulation of the helper virus has been described for TNSV. The panicum mosaic satellite virus (PMSV) can also enhance the mild symptoms of PMV to cause a severe plant disease.

Because Sputnik requires the presence of Mamavirus to accomplish its life cycle, it can be considered a satellite virus. Thus, Sputnik would be the first described dsDNA satellite virus, with a genome (18.3kb), a capsid size (50 nm), and a major capsid protein molecular mass (65kDa) much larger than that of currently described satellite viruses (Table III). In contrast to AAV, Sputnik cannot infect the amoeba host in the absence of Mamavirus. Indeed, it has been hypothesized that Sputnik multiplication relies on Mamavirus machinery. In addition to the results showing that Sputnik probably uses the transcription machinery of its viral host (Legendre et al., 2010), preliminary results of DNA–DNA *in situ* hybridization using a Sputnik-specific probe demonstrate that Sputnik genome replication occurs within the Mamavirus viral factory (unpublished data).

Attenuation of virulence, mostly associated with reduced replication, is frequently observed in association with satellites viruses, while deleterious morphological effects are not observed. Sputnik differs from satellite viruses because it leads to abnormal Mamavirus particle production, including an excess of membrane layers and incomplete assemblage of fibrils at the periphery of the particles. Some Sputnik particles are even observed within wide Mamavirus particles.

VI. GIANT VIRUSES, VIROPHAGES, AND THE FOURTH DOMAIN OF LIFE

Classification of living organisms based on ribosomal genes necessarily excludes viruses from the universal tree of life. Viruses are able to infect organisms of the three canonical domains of life (eukaryotes, bacteria, and archaea), suggesting that viral lineages have diversified over a long period of time. However, recent developments in evolutionary genomics using sequence-based approaches support a monophyletic origin of several eukaryotic viral lineages. Indeed, it has been demonstrated that diversification of a picornavirus-like superfamily of RNA viruses occurred at an early stage of eukaryogenesis, before the divergence of supergroups of eukaryotes (Koonin *et al.*, 2008). Likewise, another study demonstrated that NCLDVs had conserved a repertoire of genes also present in the genomes of organisms belonging to the three domains of life, which are involved in DNA biosynthesis and processing (nucleotide biosynthesis, DNA replication, repair, recombination, and transcription) (Boyer *et al.*, 2010b). In this study, phyletic and phylogenetic approaches based on proteins encoded by this set of genes demonstrated that NCLDVs form a clade that likely represents a fourth domain of life, supporting a common origin of the NCLDVs (Fig. 4). Moreover, these data suggest that NCLDV diversification likely occurred concomitantly with that of eukaryotes. Overall, these evolutionary genomic studies support the hypothesis that DNA and RNA viruses have an ancestral origin, as do cellular organisms.

Furthermore, structural studies have been informative with regard to virus evolution; it has been argued that structural information is more conserved than genetic information during virus evolution (Krupovic and Bamford, 2008). The double jelly roll motif consisting of the β barrel MCP found in the capsids of viruses infecting archaea, eukarya, and bacteria is proposed to be evidence for the ancestry of some viruses. Indeed it has been suggested that the extent of similarity between the virion organization of viruses with the β barrel structural fold enables the establishment of a higher level taxonomy for nine officially recognized viruses families (including NCLDVs) and three additional viruses (including the virophage Sputnik) that have not yet been assigned to a family (Krupovic and Bamford, 2010). Thus, it was proposed that the structural information shared between different viral families allows the grouping of

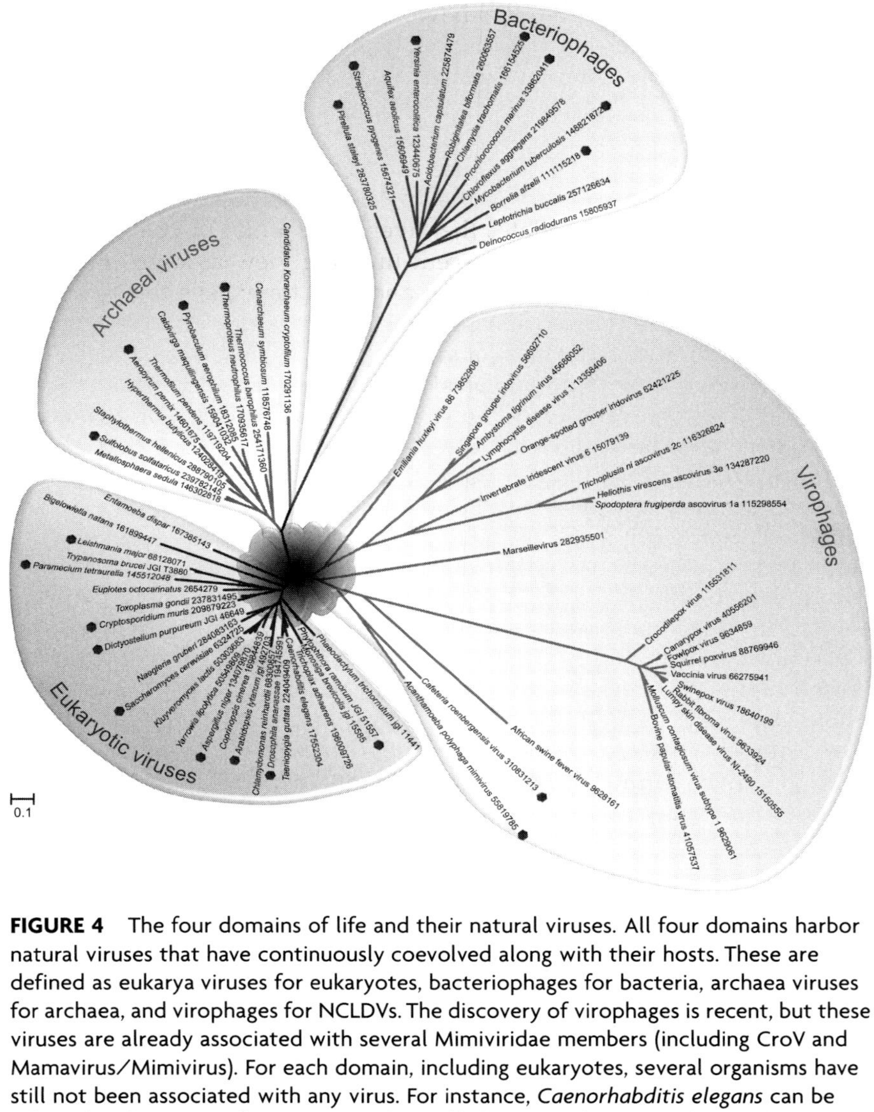

FIGURE 4 The four domains of life and their natural viruses. All four domains harbor natural viruses that have continuously coevolved along with their hosts. These are defined as eukarya viruses for eukaryotes, bacteriophages for bacteria, archaea viruses for archaea, and virophages for NCLDVs. The discovery of virophages is recent, but these viruses are already associated with several Mimiviridae members (including CroV and Mamavirus/Mimivirus). For each domain, including eukaryotes, several organisms have still not been associated with any virus. For instance, *Caenorhabditis elegans* can be infected with viruses infecting mammalian cells, but natural viruses infecting this species remain to be discovered (Shaham, 2006). Similarly, several bacterial or archaeal species have never been associated with virus-like particles, even though some harbor complete or defective prophages in their genome (Ackermann, 2007). Maximum-likelihood phylogenetic tree of RNA polymerase II generated from the curated alignment of 73 protein sequences from Eukarya (blue), bacteria (purple), archaea (green), and NCLDVs (red). Genera infected by identified natural viruses are indicated with a small viral icon. The scale bar represents the number of estimated changes per position for a unit of branch length. More details can be found in Boyer *et al.* (2010b). (See Page 5 in Color Section at the back of the book.)

Villarreal, L. P. (2009). Origin of Group Identity: Viruses, Addiction and Cooperation. Springer, New York.
Wang, X.-S., Ponnazhagan, S., and Srivastava, A. (1995). Rescue and replication signals of the adeno-associated virus 2 genome. *J. Mol. Biol.* **250:**573–580.
Xiao, C., Kuznetsov, Y. G., Sun, S., Hafenstein, S. L., Kostyuchenko, V. A., Chipman, P. R., Suzan-Monti, M., Raoult, D., McPherson, A., and Rossmann, M. G. (2009). Structural studies of the giant mimivirus. *PLoS Biol.* **7:**e92.
Yalkinoglu, A. O., Heilbronn, R., Bürkle, A., Schlehofer, J. R., and zur Hausen, H. (1988). DNA amplification of adeno-associated virus as a response to cellular genotoxic stress. *Cancer Res.* **48:**3123–3129.
Yan, X., Yu, Z., Zhang, P., Battisti, A. J., Holdaway, H. A., Chipman, P. R., Bajaj, C., Bergoin, M., Rossmann, M. G., and Baker, T. S. (2009). The capsid proteins of a large, icosahedral dsDNA virus. *J. Mol. Biol.* **385:**1287–1299.
Yutin, N., Wolf, Y. I., Raoult, D., and Koonin, E. V. (2009). Eukaryotic large nucleo-cytoplasmic DNA viruses: Clusters of orthologous genes and reconstruction of viral genome evolution. *Virol. J.* **6:**223.
Zauberman, N., Mutsafi, Y., Halevy, D. B., Shimoni, E., Klein, E., Xiao, C., Sun, S., and Minsky, A. (2008). Distinct DNA exit and packaging portals in the virus Acanthamoeba polyphaga mimivirus. *PLoS Biol.* **6:**e114.
Zhang, L., Zitter, T. A., and Palukaitis, P. (1991). Helper virus-dependent replication, nucleotide sequence and genome organization of the satellite virus of maize white line mosaic virus. *Virology* **180:**467–473.

Section 2
Genomics and Molecular Biology

CHAPTER 4

Bacteriophage-Encoded Bacterial Virulence Factors and Phage–Pathogenicity Island Interactions

E. Fidelma Boyd

Contents

Abstract

The role of bacteriophages as natural vectors for some of the most potent bacterial toxins is well recognized and includes classical type I membrane-acting superantigens, type II pore-forming lysins, and type III exotoxins, such as diphtheria and botulinum toxins. Among Gram-negative pathogens, a novel class of bacterial virulence factors called effector proteins (EPs) are phage encoded

Department of Biological Sciences, University of Delaware, Newark, Delaware, USA

Advances in Virus Research, Volume 82
ISSN 0065-3527, DOI: 10.1016/B978-0-12-394621-8.00014-5

among pathovars of *Escherichia coli*, *Shigella* spp., and *Salmonella enterica*. This chapter gives an overview of the different types of virulence factors encoded within phage genomes based on their role in bacterial pathogenesis. It also discusses phage–pathogenicity island interactions uncovered from studies of phage-encoded EPs. A detailed examination of the filamentous phage CTXϕ that encodes cholera toxin is given as the sole example to date of a single-stranded DNA phage that encodes a bacterial toxin.

LIST OF ABBREVIATIONS

CT	cholera toxin
CTXϕ	cholera toxin phage
Ds	double-stranded
EHEC	enterohaemorrhagic *E. coli*
EPs	effector proteins
EPEC	enteropathogenic *E. coli*
ETA	epidermolytic toxin exfoliative toxin A
ICEs	integrative and conjugative element
LPS	lipopolysaccharide
LEE	the locus of enterocyte effacement
MIGE	mobile and integrative genetic elements
PAIs	pathogenicity islands
SPaI	*S. aureus* pathogenicity island
STEC	shiga toxin *E. coli*
SPI-1	*Salmonella* pathogenicity island-1
SCV	*Salmonella* containing vacuole
Ss	single-stranded
T3SSs	type III secretion systems
TSST-1	toxic shock syndrome toxin 1
SaPI-1	*S. aureus* pathogenicity island-1
Stx	shiga-toxins
SIF	*Salmonella* inducing filament
ROS	reactive oxygen species

I. GENERAL BACKGROUND

Virulence factors for a range of human pathogens are encoded on mobile and integrative genetic elements (MIGEs), such as plasmids, bacteriophages, conjugative transposons, integrative and conjugative elements (ICEs), and pathogenicity islands (PAIs). Virulence factors encoded on MIGEs are diverse and are involved in all the different stages of bacterial pathogenesis. Bacterial pathogens must first enter their host using a range of entry points. For example, in the case of gastrointestinal pathogens, entry occurs by ingestion of contaminated food or water; in the case of respiratory tract pathogens, entry usually occurs by inhalation of airborne infectious droplets; among urinary and reproduction tract pathogens, uptake usually occurs by sexual transmission; and blood borne infections generally occur via cuts or abrasions or other assaults on the natural physical barrier of the skin. Once inside the host, the pathogen must then negotiate the internal environment to reach its site of attachment and proliferation, toxin elaboration, or its specific intracellular host target (Fig. 1A). For many of these steps (Figs. 1A1 to 1A4), the bacterial functions required and the toxins produced are encoded by phage DNA that reside in the chromosome of the bacterium as temperate, helper, or satellite (defective) prophages. Bacteria have adapted successfully to colonize all available niches on earth, and the human body is just another eukaryotic host that requires factors to protect and shield as well as aid in survival and proliferation of the bacterium (Brussow *et al.*, 2004). Over short and large-scale evolutionary time, bacteria have adapted to new niches, such as the human host, by either adapting mechanisms required for survival in their original niche and/or acquiring new traits by horizontal transfer of genes encoded on MIGEs such as phages.

Historically, phages are the most studied and well-recognized MIGEs known to play a key role in the emergence of pathogenic bacteria. In classical terms, phage conversion is a process whereby phage-encoded virulence factors convert their bacterial host from a nonvirulent strain to a virulent strain. Many phages form stable relationships (lysogens) with their bacterial host, averting bacterial lysis to permit both vertical and horizontal transmission within a bacterial population. Several different functional categories of bacterial virulence factors (Fig. 1B) are encoded on a diversity of lysogenic phages, mostly double-stranded (dsDNA) phages; the next sections discuss traits encoded by phages that enable bacterial colonization and survival in the human host (Table I).

II. PHAGE-ENCODED EFFECTOR PROTEINS (EPS)

A. Phage-encoded EPs and pathogenicity island (PAI)-encoded type three secretion systems

Intracellular pathogens have to gain entry into their eukaryotic host cell. To do this, the pathogen must first attach to the host cell surface and then manipulate the host cell to achieve internalization (Fig. 2). Type III secretion systems (T3SSs) are bacterial cell wall-spanning structures found in a range of pathogens that inject proteins called effectors from the bacterial

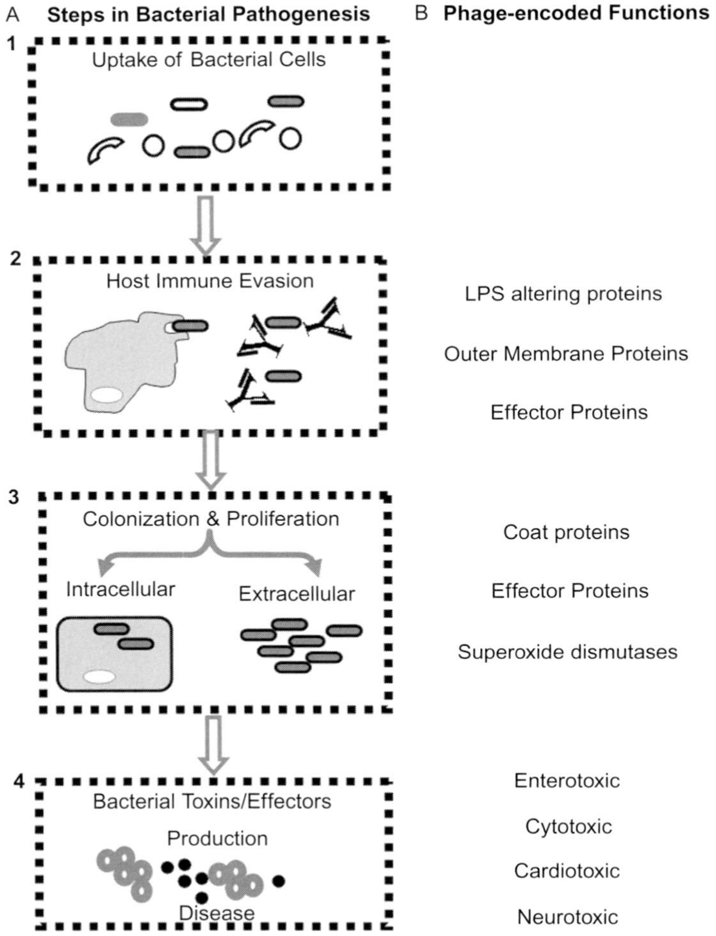

FIGURE 1 Stages in bacterial pathogenesis and virulence factors required. (A) Outline of major steps (rectangular boxes) in bacterial pathogenesis and (B) categories of phage-encoded virulence factors involved in each step. (A1) Uptake of different pathogens (circles, rods, commas) can occur via oral, nasal, or urogenital tracts. (A2) A number of pathogens avoid the host immune system cells (phagocytosis, antibodies) utilizing phage-encoded modification proteins such as those listed in the right column. (A3) Once the pathogen is at its site of activity (tissue/cell), it can either remain extracellular and elaborate toxins that are taken up by the host target cell or, in the case of intracellular pathogens, are internalized within the target eukaryotic cell. A range of virulence factors that are phage encoded are required for uptake and survival within the host cell and are listed in the right-hand column. (A4) Phage-encoded bacterial toxins and effector proteins (open and closed circles) can then cause a range of effects as listed in the right-hand column.

TABLE I Phage-encoded bacterial virulence factors

Bacterial species	Phage	Gene/function	Reference
Escherichia coli	lambdoid	*lom*/cell attachment	Barondess and Beckwith, 1990
E. coli	lambdoid	*bor*/cellular survival	Barondess and Beckwith, 1990
E. coli EHEC	H-19B	*stx1, stx2*/Shiga toxins	O'Brien *et al.*, 1984
E. coli EHEC	lambdoid	*tccP*/effector protein (EP)	Perna *et al.*, 2001
E. coli EHEC	lambdoid	*nleA-F*/EPs	Creuzburg *et al.*, 2005
E. coli EHEC	SpLE3-like	*espL2*/EP	Miyahara *et al.*, 2009
E. coli EPEC	lambdoid	*cdt1*/toxin	Asakura *et al.*, 2007
E. coli EPEC	lambdoid	*cif*/EP	Loukiadis *et al.*, 2008
E. coli EPEC	lambdoid	*nleH*/EP	Hemrajani *et al.*, 2008
E. coli	lambdoid	*eib*/cellular survival	Sandt *et al.*, 2002
E. coli EHEC	Sp4, 10	*sodC*/cellular survival	Ohnishi *et al.*, 2001
E. coli	phiFC3208	*hly2*/enterohaemolysin	Beutin *et al.*, 1993
Salmonella enterica	SopEphi	*sopE*/EP	Mirold *et al.*, 1999
S. enterica	ST64B	*sseK3*/EP	Brown *et al.*, 2011
S. enterica	*Gifsy*-1	*gogB*/EP	Figueroa-Bossi *et al.*, 2001
S. enterica	*Gifsy*-1	*gipA*/invasion	Figueroa-Bossi *et al.*, 2001
S. enterica	*Gifsy*-2	*sseI(gtgB)*/EP	Figueroa-Bossi *et al.*, 2001
S. enterica	*Gifsy*-3	*sspH1*/EP	Figueroa-Bossi *et al.*, 2001
S. enterica	*Gifsy*-2	*sodCI*/cellular survival	Figueroa -Bossi *et al.*, 2001
S. enterica	*Fels-1*	*sodCIII*/cellular survival	Figueroa-Bossi *et al.*, 2001
S. enterica	*Gifsy*-1,2	*ailT, ailF*/serum resistance	McClelland, 2001
S. enterica	eplison34	*rfb*/glucosylation antigenicity	Wright, 1971
S. enterica	P22	*gtr*/glucosylation antigenicity	Allison and Verma, 2000
S. enterica	*Fels-1*	*nanH*/neuraminidase	Figueroa-Bossi *et al.*, 2001
S. enterica	*Gifsy*	*grvA*/antivirulence gene	Ho and Slauch, 2001

(*continued*)

TABLE I *(continued)*

Bacterial species	Phage	Gene/function	Reference
Shigella flexneri	Sf6	*oac*/O-antigen acetylase	Clark *et al.*, 1991; Verma *et al.*, 1991
S. flexneri	SfII,V,X	*gtrII*/antigenicity	Allison and Verma, 2000
Shigella dysenteriae	lambdoid	*stx1,2*/Stx	Strockbine *et al.*, 1988
Neisseria meningitidis	Mu-like OMPs	Antigenicity	Masignani *et al.*, 2001
Pseudomonas aeruginosa ϕCTX	*ctx*/cytotoxin		Nakayama *et al.*, 1999
Vibrio cholerae	CTXϕ	*ctxAB*/CT	Waldor and Mekalanos, 1996
V. cholerae	CTXϕ	*ace* /enterotoxin	Trucksis *et al.*, 1993,
V. cholerae	CTXϕ	*zot*/enterotoxin	Fasano *et al.*, 1991
V. cholerae	K139	*glo*/G-protein-like	Reidl and Mekalanos, 1995
Clostridium botulinum	phage C1	C1/botulium toxin	Barksdale and Arden, 1974
Corynebacterium diphtheriae	β-phage	*tox*/diphtheria toxin	Freeman, 1951
Staphylococcus aureus	NA	*see*, *sel*/enterotoxin	Betley and Mekalanos, 1985
S. aureus	ϕN315	*sep*/enterotoxin P	Kuroda *et al.*, 2001
S. aureus	ϕ13	*entA*/enterotoxin A	Coleman *et al.*, 1989
S. aureus	ϕMu50A	*sea*/enterotoxin A	Kuroda *et al.*, 2001
S. aureus	ϕ13	*sak*/staphylokinase	Coleman *et al.*, 1989
S. aureus	ϕETA	*eta*/exfoliative toxin A	Yamaguchi *et al.*, 2000
S. aureus	Various	*tst*/TSST-1	Lindsay *et al.*, 1998
Streptococcus pyogenes	T12	*speA*/toxin type A	Weeks and Ferretti, 1984
S. pyogenes	CS112	*speC*/toxin type C	Goshorn and Schlievert, 1989
S. pyogenes	H4489A	*hylP*/hyaluronidase	Hynes and Ferretti, 1989
S. mitis	SM1	*pblA*,*pblB*/cell attachment	Bensing *et al.*, 2001b
Mycoplasma arthritidis	MAV1	*vir*/cell attachment	Voelker and Dybvig, 1999

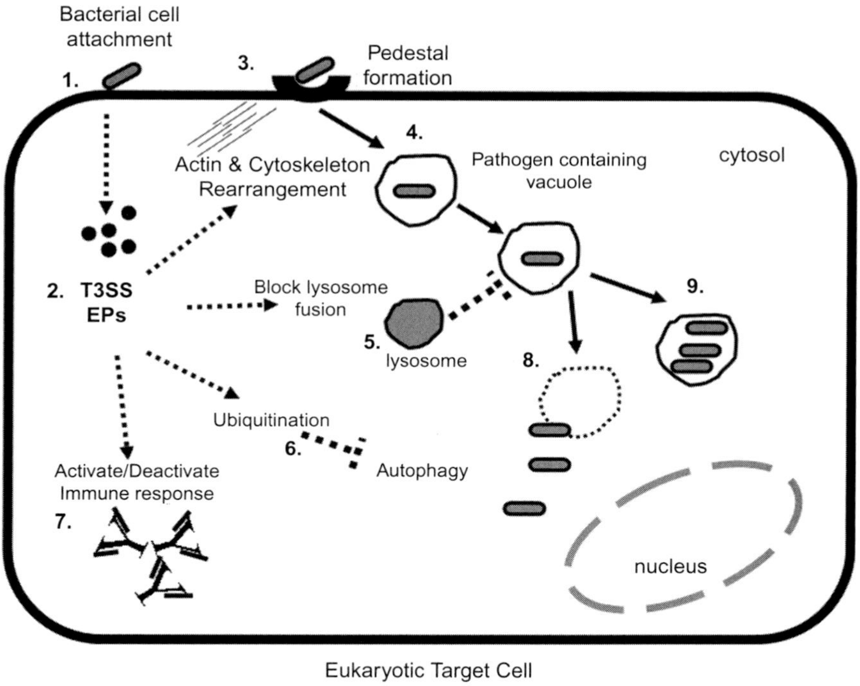

FIGURE 2 A simplified schematic of some of the roles of T3SS effector proteins (EPs) in pathogen uptake and survival within its target eukaryotic cell (dotted arrows). (1) Initial bacterial cell attachment to target cell (2) with subsequent secretion of EPs into target cell by T3SS, (3) which in a number of pathogens results in actin microfilament rearrangement with pseudopod or pedestal formation (4) followed by internalization within a vacuole or phagosome. (5) Some EPs are believed to prevent lysosome fusion with the pathogen containing vacuole. (6) Other EPs are known to act as either ubiquitin ligases or deubiquitinases to disrupt ubiquitin tagging to prevent autophagy. (7) A number of EPs are known to activate and deactivate the host cell immune response. Intracellular pathogens either (8) escape the vacuole and multiple with the cell cytosol or (9) remain within the vacuole and replicate there; EPs have also been identified that aid in each of these processes.

cell directly into the cytosol of their target eukaryotic cell (Hansen-Wester and Hensel, 2001; Hansen-Wester *et al.*, 2002a,b). EPs are multifunctional proteins that once inside the eukaryotic cell aid in bacterial invasion and survival, usually by redirecting eukaryotic cell signaling pathways, reorganizing the host cytoskeleton and disrupting the host immune responses (Fig. 2)(Dean, 2011; Miyahara *et al.*, 2009; Yen *et al.*, 2010). EPs are multidomain proteins that, depending on the effector, have diverse subversive functions. The N-terminal region of the EP contains the sequence signal for the secretion sequence signal, and the C-terminal region contains the biological activity function of the EP [for a comprehensive review, see Dean (2011)]. Enterohemorrhagic *Escherichia coli* (EHEC), a zoonosis pathogen, encodes a T3SS that injects an array of EPs into epithelial cells of their human host, primarily to promote cell attachment and colonization (Miyahara *et al.*, 2009; Ogura *et al.*, 2009; Tobe *et al.*, 2006). At least 39 different EPs are transported by the EHEC T3SS; 7 EPs are encoded on the locus of the enterocyte effacement (LEE) pathogenicity island and the remainder are encoded throughout the genome, many within prophage genomes (Ogura *et al.*, 2009; Tobe *et al.*, 2006). In EHEC O157:H7 strains, at least 9 different lambdoid prophages are present and proposed to encode between two to eight different effector proteins (Tobe *et al.*, 2006). Furthermore, depending on the EHEC strain examined and the prophage content, the repertoire of EPs can be very different. The presence of different EPs among strains may explain differences in pathogenesis and cell tropism (Ogura *et al.*, 2009). In enteropathogenic *E. coli* (EPEC) strains, the lysogenic lambdoid phage CDT-1Φ encodes a homologue of Cif, a T3SS EPs, which is transmitted into the host cell via a PAI-encoded T3SS (Asakura *et al.*, 2007). It has been shown that EHEC, as well as EPEC, strains contain a LEE island-encoded Cif protein that is widespread among isolates (Loukiadis *et al.*, 2008). Shiga toxin *E. coli* (STEC) strains also contain a T3SS encoded within the LEE island whose effector proteins cause the characteristic attaching and effacing lesions. In STEC strains, the NleA (non-LEE-encoded effector A) protein is present on a prophage, and the expression of this protein is regulated by LEE-encoded regulators (Creuzburg *et al.*, 2005; Gruenheid *et al.*, 2004; Schwidder *et al.*, 2011).

Salmonella enterica subspecies I isolates can cause infections ranging from mild and self-limiting enterocolitis to the systemic disease typhoid. Similar to what is known in *E. coli*, strain differences and diversity in *S. enterica* can be based in large part on the presence and absence of prophage genomes (Fig. 3). *S. enterica* is a prototypical intracellular pathogen and carries two specialized T3SSs encoded on *Salmonella* pathogenicity island-1 (SPI-1) and SPI-2 that function at distinct steps in pathogenesis (Hansen-Wester and Hensel, 2001, 2002a,b). As with *E. coli* EPEC, EHEC, and STEC strains, in *S. enterica*, specific EPs are translocated by specific T3SSs (either T3SS-1 encoded on SPI-1 or T3SS-2 encoded on SPI-2), and many of these effector proteins are encoded on unlinked chromosomal sites that were acquired separately (Fig. 3). At least 30 different EPs have been identified in *S. enterica*; approximately 20 are secreted by T3SS-1 and 15 by T3SS-2. Two of these EPs, named SopE2 and SspH2, are encoded adjacent to phage-like sequences, and five EPs, SopE, GogB, SseI, SspH1, and SseK3, are found on prophages SopEphi, Gifsy-1, Gifsy-3, and SM64B (Table II, Fig. 3)(Brown *et al.*, 2011; Ehrbar *et al.*, 2002, 2003; Ehrbar and Hardt, 2005; Figueroa-Bossi *et al.*, 1997, 2001;

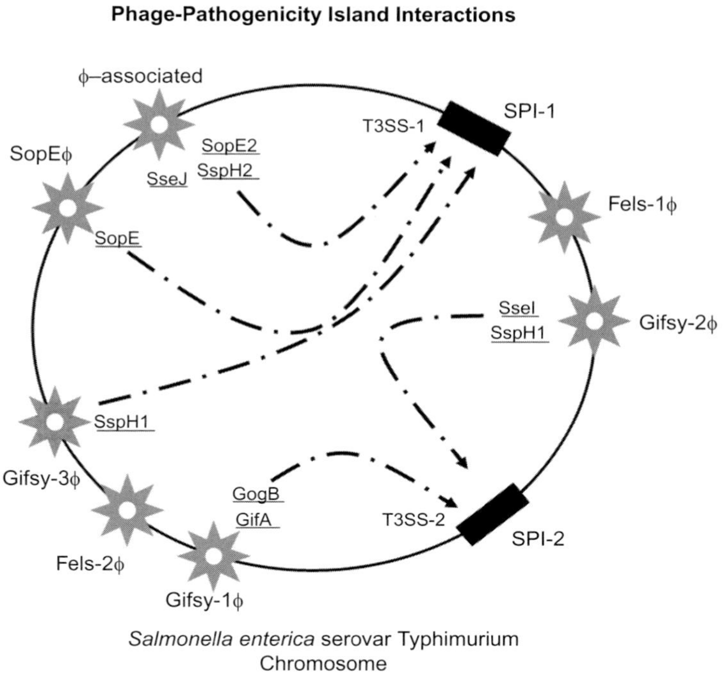

FIGURE 3 Schematic of *S. enterica* genome with key phages (designated as stars) and PAIs (rectangular boxes) marked. Depiction of EPs, their corresponding phages, and dotted arrows show the T3SS that translocates them into the eukaryotic target cell. Effector proteins are underlined.

Figueroa-Bossi and Bossi, 1999; Friebel *et al.*, 2001; Mirold *et al.*, 2001). Similar to what is known in *E. coli*, distribution of these effectors varies among strains (Brown *et al.*, 2011).

B. Phages and PAIs are distinct unrelated mobile and integrative genetic elements (MIGEs)

It should be noted at this point that phages and PAIs are not closely related to one another as some have suggested (Dobrindt *et al.*, 2004; Oelschlaeger *et al.*, 2002). Pathogenicity islands were hypothesized to be degenerate forms of phages or other MIGEs due mainly to the presence of phage-like integrases and because some regions had features associated with other MIGEs (Dobrindt *et al.*, 2004). However, the similarity of PAI integrases to phage integrases is a historical artifact. Phage integrases belonging to either tyrosine or serine recombinase families were the first integrases to be identified and characterized; thus many different MIGEs

TABLE II Phage-encoded bacterial toxins and their eukaryotic cell effect

Category		
Toxin type	Activity and effect where known	
Type I toxins (cell membrane acting toxin)		
Superantigens	Binding to MHC class II and Vβ/Vγ T-cell receptors	T-cell activation Cytokine secretion
SpeB	Cysteine protease	Alteration Ig-binding properties
Type II toxins (membrane pore-forming toxin)		
Streptolysin O	Cell membrane permeabilization	Cell death
Perfringolysin O		
α toxin		
Leukotoxin		
Aerolysin		
Type III toxins (extracellular toxin internalized by target cells)		
Diphtheria toxin	ADP-ribosylation EF2	Cell death
Shiga toxin	*N*-Glycosidase activity 28S rRNA	Cell death
Cholera toxin	ADP-ribosylation of Gs	cAMP increase, water loss
Botulinum toxin	Cleavage VAMP/synaptobrevin	Neurotoxin flaccid paralysis
Cytolethal distending toxin	DNase I activity	Cyclomodulins
Type IV toxins [effector proteins (EP) injected directly by T3SS into target cells]		
E. coli EPs		
TccP	Actin remodeling	
EspL2	Actin remodeling	Pseudopod formation
EspJ	Transmission	

NleG	Ubiquitin ligases	Proinflammatory inactivation
Cif	Ubiquitin inhibitor	Cyclomodulin/cell cycle arrest
NleH	Effector	
S. enterica EPs		
SopE/SpoE2 (T3SS-1)	Guanine exchange factor mimic	Membrane ruffling activation
	Cytoskeletal reorganization	Proinflammatory activation
SseK3 (T3SS-2)	Invasion	
GogB (T3SS-1)	E3 ubiquitin ligase	Proinflammatory inactivation
GipA/	Invasion	
SseI/gtgB (T3SS-2)	Inhibits actin polymerization	Remodels SCV
SspH1/SspH2 (T3SS-1)	E3 ubiquitin ligase	Proinflammatory inactivation
SspH1/SspH2 (T3SS-2)	Inhibits actin polymerization	Remodels SCV

that integrate into the host chromosome via site-specific recombination encode a "phage-like integrase" by BLAST analysis.

Pathogenicity islands as originally defined in *E. coli* are large chromosomal regions that encode multiple virulence factors, are present in pathogenic isolates and absent from nonpathogenic isolates, encode an integrase, and insert at a tRNA locus. Most PAIs have aberrant base composition, such as percent guanine-cytosine and dinucleotide frequency compared to that of their host genome (Hacker and Kaper, 2000; Hacker *et al.*, 1997). Additional features of PAIs include nonself-mobilization, ability to integrate and excise from host chromosome using an integrase, an attachment site, and a recombination directionality factor/excisionase (Boyd *et al.*, 2009; Napolitano *et al.*, 2011). The recombination module is characteristic of many MIGEs such as phages and ICEs that insert into the host genome via site-specific recombination of a circular intermediate. Evolutionary relationships have been determined among PAIs and prophages from *E. coli* strains based on a phylogenetic analysis of their encoded integrase (Napolitano *et al.*, 2011). The phylogenetic trees constructed showed that all integrases encoded within PAIs clustered together, separate and distinct from the phage-encoded integrases. Our conclusion is that the PAIs analyzed were evolutionarily distinct from prophages (Boyd *et al.*, 2009; Napolitano *et al.*, 2011). Thus, findings that specific T3SSs encoded within PAIs are required for the transport of specific effectors encoded within diverse phages are quite unexpected.

III. SURVIVAL IN EUKARYOTIC HOST CELLS

Intracellular pathogens are internalized via a membrane-bound vacuole triggered, in the case of *Salmonella* spp., *Shigella* spp., and some pathogenic *E. coli*, by EPs translocated into the cell by T3SSs discussed earlier (Fig. 2). Some bacteria survive within the vacuole without proliferation, whereas others, such as *Shigella* spp., *Burkholderia pseudomallei*, and *Listeria monocytogenes*, among others, escape the vacuole to gain access to the cytoplasm and replicate there. Once internalized within a host cell, *Salmonella enterica* survives with a modified phagosome known as the *Salmonella*-containing vacuole (SCV)[for a short review on this topic, see Steele-Mortimer (2008)]. In either case, the bacterium has to survive the internal environment of the host cell, such as damage from oxidative stress, innate immune responses, lysosome fusion, and autophagy. Two *S. enterica* phages, Gifsy-2 and Fels-1, encode a periplasmic copper- and zinc-cofactor superoxide dismutase (Cu,Zn-SodC) that catalyzes the conversion of superoxide to hydrogen peroxide and molecular oxygen protecting the bacterium from reactive oxygen species (ROS)(Figueroa-Bossi *et al.*, 2001). In *S. enterica*, three Cu, Zn-SodC-encoding genes, *sodCI*, *sodCII*, and *sodCIII*, have been identified (Figueroa-

Bossi and Bossi, 1999; Figueroa-Bossi *et al.*, 1997, 2001). The *sodCI* gene is carried on phage Gifsy-2 and is confined to subspecies I isolates, and the *sodCIII* gene is within the genome of phage Fels-1, in a location similar to that of *sodCI* in Gifsy-2 (Figueroa-Bossi *et al.*, 2001). The *sodCII* gene is chromosomally encoded (Figueroa-Bossi *et al.*, 2001). A *sodC* gene has been identified in two lambdoid phages in *E. coli* O157 strain Sakai; however, the functional significance has not been elucidated to date. In *S. enterica*, establishment and survival within modified phagosome SCV are dependent on effector proteins that are translocated across the SCV membrane specifically by T3SS-2 encoded on SPI-2 (McGhie *et al.*, 2009). Although the precise role of each individual EP involved in this process is not known, given the redundancy of EPs, many of these EPs are phage encoded. One possible role of these EPs may be the ability to block SCV-lysosome fusion, which would otherwise be detrimental to the bacterium, in addition to their known role in actin and microtubule rearrangements and *Salmonella*-inducing filament (Sif) formation. *S. enterica* can also protect itself from the toxic effects of ROS by preventing recruitment of oxidase to the SCV, which is thought to be mediated by the secretion of effector proteins by T3SS-2 also [for a comprehensive review of *S. enterica* effector function, see McGhie *et al.* (2009)].

IV. ATTACHMENT TO HOST EUKARYOTIC CELLS

Bacteria attachment to eukaryotic cells can require a number of different bacterial cell surface structures, such as pili, fimbriae, and adhesins, as well as the capsule and lipopolysaccharide (LPS). None of these bacterial cell surface structures have been shown to be encoded within phages. Examples of phage-encoded proteins involved in bacterial attachment are limited, and thus the proposal that phage-encoded genes in *Streptococcus mitis* are central to the attachment to platelets, an early step in infectious endocarditis, is a novel and unique finding (Bensing *et al.*, 2001a,b; Siboo *et al.*, 2003). It has been proposed that two bacterial surface proteins, PblA and PblB, encoded within the lysogenic phage SM1 from *S. mitis* promote the binding to human platelets through their interaction with the membrane ganglioside GD3 (Bensing *et al.*, 2001a,b; Siboo *et al.*, 2003). The Sullam group has shown that phage lysin also mediates the binding of *S. mitis* to human platelets via an interaction with fibrinogen (Seo *et al.*, 2010).

V. EVASION OF HOST IMMUNE CELLS

The host immune system has evolved to resist infection by pathogens using both nonspecific innate responses and specific adaptive immune responses. In turn, bacteria have evolved mechanisms to circumvent and sometimes co-opt these immune responses, such as autophagy and the inflammasome for their own benefit. The effector protein SspH1 in *S. enterica*, which is encoded by the phage Gifsy-3 and is translocated by both T3SS-1 and T3SS-2, has been shown to inhibit both NF-κB activity and proinflammatory cytokine secretion. An EP IcsB secreted by a T3SS in *Shigella* spp. hides the bacterium from the autophagy pathway. The inflammasome can be activated by the presence of a number of different molecules, which include the bacterial cell wall LPS. The O-antigen of bacterial cell wall LPS is a major antigenic determinant. The O-antigen modification genes encoding bacterial LPS O-antigen glucosylating, acetylating, and transferase proteins are important in many bacterial pathogens in allowing the bacterium to redecorate its outer surface to disguise it for the immune system, preventing, for example, phagocytosis by macrophage and/or inflammasome activation. O-antigen switching has been demonstrated experimentally in *S. enterica*, *E. coli*, *V. cholerae*, *S. pneumonia*, and *Neisseria meningitides* to name but a few. Genes involved can be located in the genomes of several morphologically diverse phages and, in many cases, genes are located immediately downstream of the phage chromosomal attachment site *attP*, the DNA sequence required for site-specific recombination between the phage and bacterium (Boyd and Brussow, 2002; Brussow *et al.*, 2004). Interestingly, it is proposed that expression of the glycosyltransferase operon encoded by phage P22 in *S. enterica* is under the control of phase variation enabled by a novel epigenetic mechanism requiring OxyR in conjunction with the DNA methyltransferase Dam (Broadbent *et al.*, 2010). This is an example of how phages have integrated into the regulatory circuits of their host. The ability to switch O-antigen, although important directly in pathogenesis, also has a serious indirect effect on bacterial pathogenesis in the area of bacterial pathogen vaccine development and efficacy. In 1992, a novel *V. cholerae* serogroup emerged in southeast Asia via a horizontal transfer of O-antigen genes that resulted in replacement of the dominant O1 serogroup by a novel O139 serogroup (Bik *et al.*, 1995; Waldor *et al.*, 1994). Initially, *V. cholerae* serogroup O139 strains infected many more adults than O1 serogroup strains, which was due to a lack of immunity in the adult population (Faruque *et al.*, 2003). Research demonstrated that the pathogenic O139 serogroup strain was derived from a *V. cholerae* O1 serogroup strain by the acquisition of O139 antigen genes (Bik *et al.*, 1995; Waldor *et al.*, 1994). Indeed, evidence shows that several non-O1/non-O139 *V. cholerae* strains with pathogenic potential have arisen by exchange of O-antigen biosynthesis regions in O1 serogroup isolates (Li *et al.*, 2002; O'Shea *et al.*, 2004). O-antigen switching in *V. cholerae* has been proposed to be mediated by a bacteriophage, as has been found with *E. coli* and *S. enterica* (Mooi and Bik, 1997; Stroeher *et al.*, 1997). It has been demonstrated that O-antigen switching in *V. cholerae* can also result from chitin-induced transformation (Blokesch and Schoolnik, 2008; Meibom *et al.*, 2005; Miller *et al.*, 2007).

VI. EXTRACELLULAR TOXINS

A large variety of bacterial toxins are phage encoded in both Gram-negative and Gram-positive bacteria (Table I). Among Gram-positive bacteria, these include some of the most potent bacterial toxins ever described, such as the neurotoxic botulinum toxin and cardio and neuro-toxic diphtheria toxin encoded within *Clostridium botulinum* and *Corynebacterium diphtheriae,* respectively, to cause diseases ranging from flaccid paralysis to mild gastrointestinal disease to life-threatening toxemia and sepsis. Streptococcal and staphylococcal isolates cause a wide range of infections, including fasciitis, rheumatic fever, pharyngitis, and pyoderma. These infectious diseases are due in part to superantigen toxins such as enterotoxin B,C,K,L,Q, and toxic shock syndrome toxin 1 (TSST-1), epidermolytic toxin exfoliative toxin A (ETA), enterotoxin P, hyaluronidase, and staphylokinase, virulence factors all encoded by phages.

A. Toxins encoded by phage-related regions

Staphylococcus aureus strains can elicit a range of illnesses, many with severe clinical outcomes that result from phage-encoded virulence factors; TSST-1 toxin that causes toxic shock syndrome, enterotoxins that cause acute gastroenteritis, leucocidin that elicits leukocytolysis, SPEA toxin that causes scarlet fever and tissue necrosis, and ETA causing scalded-skin syndrome. It was proposed that some of these toxins are not phage encoded but are present on *S. aureus* pathogenicity islands named SaPIs, similar to the PAIs first described in *E. coli* pathogenic strains (Lindsay *et al.*, 1998; Ruzin *et al.*, 2001). Originally, it was proposed that in *S. aureus* strain RN4282, the *tst* gene, which encodes the TSST-1 toxin, was present within a 15-kb genetic element named SaPI-1 (Lindsay *et al.*, 1998; Ruzin *et al.*, 2001). However, SaPI-1 does not conform to the definition of a PAI and has more similarities to phages. SaPI-1 has a G+C content of 31%, which is similar to the rest of the *S. aureus* chromosome, and SaPI-1 encodes an integrase homologous to the staphylococcal bacteriophage ΦPVL integrase and integrates at a 17-bp site-specific *att* site in the *tyrB* gene. It was shown that *S. aureus* phages 13 and 80α can encapsulate and transduce SaPI-1 to recipient strains, and in strain RN4282, SaPI-1 is induced to excise and replicate specifically by 80α (Lindsay *et al.*, 1998; Ruzin *et al.*, 2001). We suggested that these features and interactions of SaPI-1 are more akin to P2 and P4 interactions in *E. coli.* The P2 and P4 interaction is an example of mobilization of a defective/satellite phage P4 by a helper phage P2 (Boyd *et al.*, 2001). More recently, Novick and co-workers (2010) have described these regions as phage-related chromosomal islands. Evolutionary and functional analyses of SaPIs suggest that these regions encode indeed satellite phages that co-opt the morphogenesis genes of other indigenous phages for particle formation (Novick *et al.*, 2010).

B. Toxins encoded by lambdoid phages

Escherichia coli can cause enteric diseases, such as severe watery diarrhea, dysentery, and hemorrhagic colitis, as well as extraintestinal infections such as cystitis, septicemia, and meningitis. It has been shown that the cytolethal distending toxin Cdt-1, encoded by *cdtA*, *cdtB*, and *cdtC* produced by EPEC strains, is encoded on a lysogenic lambdoid phage named CDT-1Φ (Asakura *et al.*, 2007). CDTs are cyclomodulins that interfere with the eukaryotic cell cycle; the CdtB subunit enters into the cell and is trafficked to the nucleus to cause DNA damage through its DNase I activity.

Enterohemorrhagic *E. coli* strains, STEC strains, and *Shigella dysenteriae* type1 strains cause severe hemorrhagic colitis and hemolytic uremic syndrome due to the production of Shiga toxins (Stx), Stx1 and Stx2, and enterohemolysins (Hly), all of which are encoded by lambdoid phages (Herold *et al.*, 2004; Schmidt, 2001). Stx is an AB_5 toxin with one active A subunit and five identical B subunits. B subunits bind to glycolipids on the host cells, and the A subunit is taken up by the cell and retrograde trafficked to its active site. Depending on the strain, *stx* genes can be encoded on unrelated phages and produce very different amounts of toxin (Herold *et al.*, 2004; Schmidt, 2001). Genomic sequences of *E. coli* O157 strains Sakai and EDL933 demonstrated the presence of 18 and 12 prophages, respectively, many of which are lambdoid that encode genes related to virulence (Ohnishi *et al.*, 2002). Other phage-encoded virulence genes present in *E. coli* strains include the Bor and Lom proteins, which confer serum resistance and cell adhesion (Ohnishi *et al.*, 2002).

C. Toxins encoded by filamentous phages

Cholera toxin (CT), an AB_5 enterotoxin, is the cause of explosive watery diarrhea characteristic of cholera and is the only known toxin encoded by a filamentous phage (Waldor and Mekalanos, 1996). The CT-encoding phage named CTXϕ is present in *V. cholerae* serogroup O1 and O139 strains predominantly (Waldor and Mekalanos, 1996). The *ctxAB* genes that encode the AB subunits share extensive sequence homology with heat-labile enterotoxins produced by *E. coli* EPEC strains. The *ctxAB* genes reside at the 3′ end of the integrated CTXϕ and are not required for phage

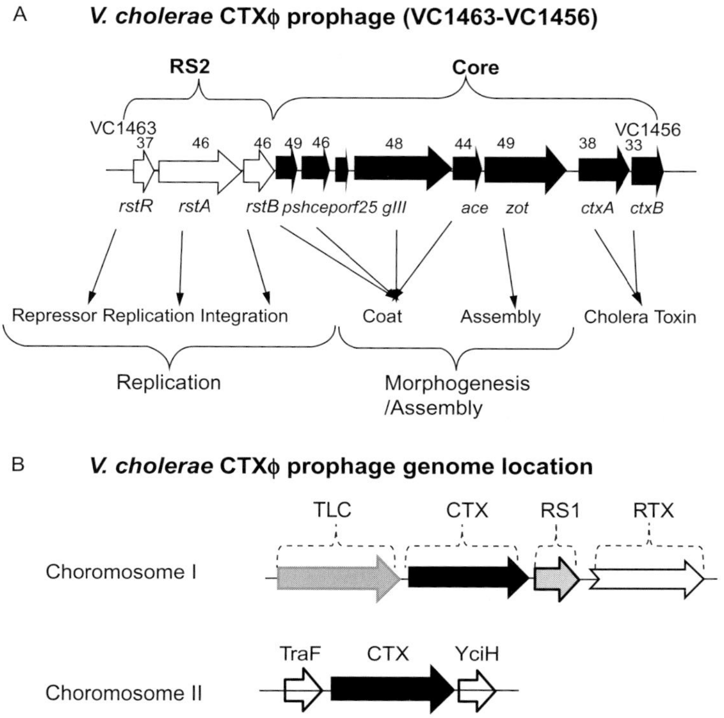

FIGURE 4 CTXϕ prophage structure and genome arrangement. (A) Schematic representation of the CTXϕ prophage. Arrows indicate open reading frames and the direction of transciption. Two functional modules make up the CTXϕ genome, RS2 region (white arrows), and Core region (black arrows). VC1463–VC1456 inclusive represent open reading frame (ORF) numbers from the *V. cholerae* N16961 genome sequence. Numbers above ORFs indicate percent GC of each gene within the CTXϕ genome. Four letter designations below the ORFs represent gene designations. (B) *V. cholerae* isolates may contain a copy of CTXϕ in each, both, or none of the two chromosomes. Among El Tor isolates, CTXϕ is usually only present in chromosome I. In chromosome I, CTXϕ, when present, is always located between the TLC cluster, which encodes a satellite phage, and the RTX cluster, which encodes the Repeat in toxin proteins. In specific strains, an additional satellite phage is present that may flank an end or both ends of the CTXϕ prophage. RS1ϕ prophages are identical to the RS2 region of CTXϕ but contain an additional gene *rstC*. Among classical strains, CTXϕ is always present on both chromosomes. In chromosome II, CTXϕ is always located between ORF TraF and YciH.

replication or production (Fig. 4) (Waldor and Mekalanos, 1996). CTXϕ is a lysogenic filamentous phage similar to *E. coli* filamentous phages M13, Ff, and f1 in size, structure, and gene order (Fig. 4A). Apart from *V. cholerae* O1 and related isolates, the distribution of CTXϕ is sporadic among non-O1/non-O139 serogroup isolates, and the CTXϕ integrates into either or both of the two *V. cholerae* chromosomes (Fig. 4B)(Faruque *et al.*, 1998a,b, 2003; O'Shea *et al.*, 2004). The limited distribution of CTXϕ is likely due to restricted distribution of the *V. cholerae* cell surface receptor, the type IV pilus TCP, which is encoded on the *Vibrio* pathogenicity island (VPI) or TCP island (Karaolis *et al.*, 1998; Taylor *et al.*, 1987; Waldor and Mekalanos, 1996). The VPI region is also present predominately in O1 and O139 isolates (Faruque *et al.*, 1998a,b, 2003; O'Shea *et al.*, 2004). Thus, only strains that have first acquired the VPI region can subsequently acquire CTXϕ, an example of phage–PAI interactions.

1. CTXϕ genome

The 7-kb CTXϕ genome has two functionally distinct modules: (a) the DNA replication module named repeat sequence 2 (RS2) that contains *rstR*, *rstA*, and *rstB* genes and (b) the core region containing *psh*, *cep*, *orfU* (*gIII*), *ace*, *zot*, and *ctxAB* genes (Fig. 4A)(Waldor and Mekalanos, 1996). The *rstA* gene is required for phage replication, whereas the *rstR* gene encodes a repressor protein of *rstA* and *rstB*. The protein product of the latter is required for phage integration into the host chromosomes (Waldor *et al.*, 1997). It should be noted here that the *rstR* gene is the most divergent sequence between CTXϕ genomes and is used to differentiate phages among strains (Davis *et al.*, 1999). For example, whereas all other genes on the CTXϕ genome between biotypes classical and El Tor share nearly 95% nucleotide identity, *rstR* genes share only 44% nucleotide identity (Davis *et al.*, 1999). Thus, RstR is a biotype-specific repressor of its cognate *rstA* gene. The core genes *cep*, *orfU*, *ace*, and *zot* correspond to *gVIII*, *gIII*, *gVI*, and *gI* of the *E. coli* Ff phage (Waldor and Mekalanos, 1996). The Psh, Cep, gpOrfU, and Ace proteins are phage structural proteins, and Zot is an assembly protein (Waldor and Mekalanos, 1996). The Ace and Zot proteins were demonstrated to have an enterotoxic effect on cells (Baudry *et al.*, 1992; Fasano *et al.*, 1991; Trucksis *et al.*, 1993). Unlike other prophages, CTXϕ particles are generated from a chromosomally integrated phage by a process analogous to rolling circle replication (Davis and Waldor, 2000). The CTXϕ genome lacks a homologue of *gIV* from the Ff phage, which encodes a protein required for phage secretion. Instead in *V. cholerae*, convergence of the secretory pathways for CT and CTXϕ has occurred (Davis *et al.*, 2000a). The outer membrane component EspD of the type-II general secretion system is required for both CT and CTXϕ release from *V. cholerae* cells (Davis *et al.*, 2000a; Waldor and Mekalanos, 1996).

2. CTXϕ copy number and chromosomal location

In *V. cholerae*, classical strains in chromosome I, which are believed to be responsible for the first six pandemics of cholera, CTXϕ is present in the intergenic region between the Toxin Linked Cryptic (TLC) gene cluster (ORF VC1465) and Repeat in the (RTX)(ORF VC1451) gene cluster, designated as the El Tor site, and in chromosome II in the intergenic region between VCA0569 and VCA0570, designated as the classical site (Fig. 4B)(Davis and Waldor, 2000). The CTXϕ in *V. cholerae* O1 serogroup El Tor biotype and O139 serogroup isolates also contain an adjacent region named RS1 and is a defective phage. It encodes RstR, RstA, and RstB, which are identical to the three RS2 proteins, as well as an RstC, an antirepressor that allows CTXϕ gene expression (Davis *et al.*, 2002; Waldor *et al.*, 1997). Different CTXϕ and RS1ϕ arrangements have been observed among different biotypes and strains of *V. cholerae* (Fig. 5). The

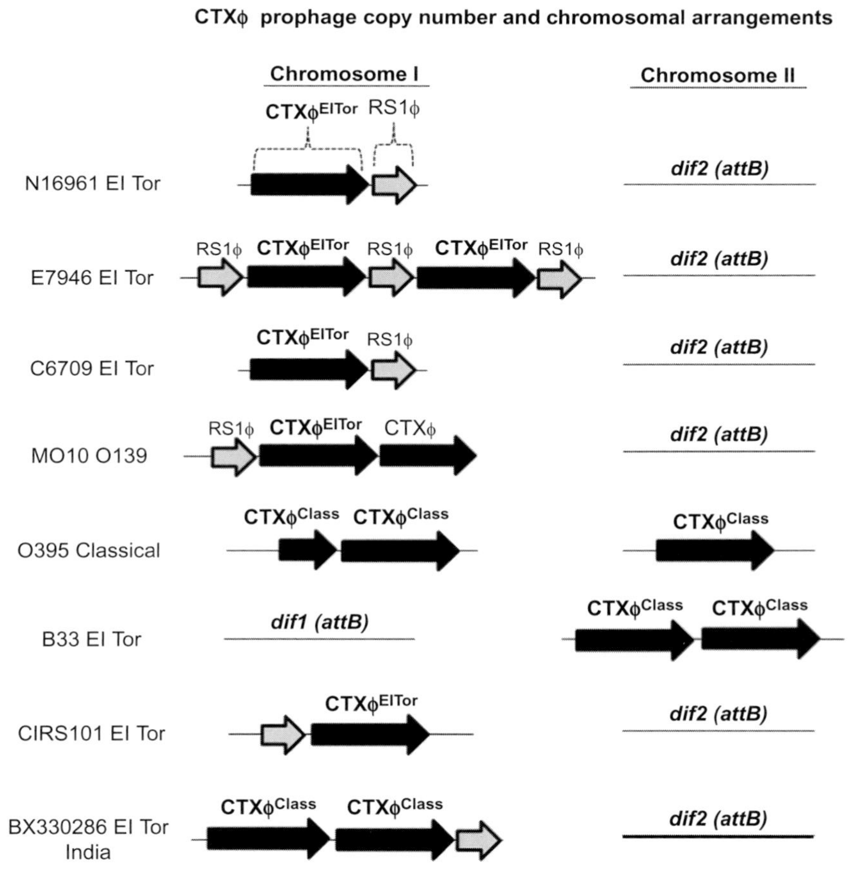

FIGURE 5 Schematic representation of some known CTXϕ arrangements among *V. cholerae*-sequenced isolates. Black arrows represent CTXϕ prophage and light greay arrows represent RS1ϕ satellite prophage. *dif1* and *dif2* represent the *attB* site for CTXϕ integration on each of the two *V. cholerae* chromosomes. Strain N16961 was the first *V. cholerae* strain to be fully sequenced and contains a single copy of CTXϕ at the *dif1* attachment site and no copy at the *dif2* attachment site in chromosome II.

arrangement of CTXϕ in the host genome varies depending on whether it is integrated in chromosome I, chromosome II, or both (Fig. 5). Among *V. cholerae* El Tor isolates, the cause of the ongoing seventh pandemic of cholera, CTXϕ^{ET}, is in chromosome I either as a CTXϕ^{ET}-CTXϕ^{ET} array or as a CTXϕ^{ET}-RS1ϕ array (Davis and Waldor, 2000; Mekalanos, 1983; Waldor and Mekalanos, 1994). In *V. cholerae* El Tor strain E7946, a RS1-CTXϕ^{ET}-RS1ϕ-CTXϕ^{ET}-RS1ϕ arrangement is present in chromosome I, and in strain C6709, isolated in Peru in 1991, a CTXϕ^{ET}-RS1ϕ arrangement is present in chromosome I (Fig. 5)(Mekalanos *et al.*, 1983; Waldor and Mekalanos, 1994). *V. cholerae* classical biotype isolates, the cause of earlier pandemics, CTXϕ is integrated in both chromosomes and never contains an RS1ϕ element (Fig. 5)(Davis *et al.*, 2000b). Within classical strains O395, GP12, and C33, in chromosome I, CTXϕ^{class} is present as an array of truncated, fused prophages, and in chromosome II, a solitary CTXϕ^{class} prophage is present (Davis *et al.*, 2000b). In strains CA401, C1, C14, C21, and C34, a single CTXϕ^{class} prophage is present in chromosome I and a truncated CTXϕ^{class} array is present in chromosome II (Davis *et al.*, 2000b). It has been proposed that the production of pCTX, the extrachromosomal copy of CTXϕ, requires the presence of either two CTXϕ-CTXϕ or a CTXϕ-RS1ϕ array (Davis and Waldor, 2000). Classical isolates are unable to produce pCTXϕ^{class} and CTXϕ^{class} (Davis *et al.*, 2000b).

Among *V. cholerae* O139 serogroup strains, which are derived from El Tor biotype strains, some unusual CTXϕ rearrangements are found (Fig. 5)(Bhadra *et al.*, 1995; Davis and Waldor, 2000; Davis *et al.*, 1999, 2000b). *V. cholerae* O139 strain SG24 contains a CTXϕ^{ET}-RS1ϕ-CTXϕ^{ET} array in chromosome I, and strains AS197 and AS209 contain a RS1ϕ-CTXϕ^{ET}-CTXϕ^{Calc}-CTXϕ^{Calc} array in chromosome I (Fig. 5)(Bhadra *et al.*, 1995; Davis *et al.*, 1999, 2000b; Davis and Waldor, 2000). More recently, a number of novel CTXϕ arrangements and types have been identified in *V. cholerae* El Tor strains. A CTXϕ-CTXϕ array in chromosome II, which includes a CTXϕ similar to El Tor CTXϕ^{ET} and a novel CTXϕ similar to CTXϕ^{Calc} isolated from O139 isolates, was identified. A 2004 Mozambique El Tor *V. cholerae* strain B-33 contained a CTXϕ^{class} with a CTXϕ^{class}-CTXϕ^{class} array in chromosome II (Faruque *et al.*, 2007; Grim *et al.*, 2010). The basis of specificity and efficiency of integration of CTXϕ and the mechanism of emergence of strains with novel CTXϕ chromosomal arrangements until recently was much debated and is discussed later.

3. Specificity of CTXϕ chromosomal integration

CTXϕ has a positive single-stranded DNA [(+) ssDNA)] genome that integrates site specifically in *V. cholerae* chromosomes to form stable lysogens, ensuring stable vertical transmission from parents to offspring. CTXϕ integration requires phage-encoded and host-encoded factors (Huber and Waldor, 2002; McLeod and Waldor, 2004). An integrase that catalyzed the recombination between host and donor DNA is absent from the CTXϕ genome, and the phage co-ops the host tyrosine recombinases, XerC and XerD, which ordinarily serve to resolve chromosome dimers during cell division (Barre *et al.*, 2001; Sherratt *et al.*, 2004). In *E. coli*, XerCD bind and catalyze recombination at homologous 28-bp *dif* sites, which are made up of two 12-bp binding sites for XerCD separated by a 6-bp spacer or overlap region, which allows for XerC–XerD interactions that ensure stable synapsis (Barre *et al.*, 2001; Sherratt *et al.*, 2004). The *V. cholerae* genome consists of two nonhomologous circular chromosomes I and II (Trucksis *et al.*, 1998), and therefore two *dif* sites are present named *dif1* and *dif2* in chromosome I and II, respectively, that coordinate dimer resolution on each chromosome with cell division (Val *et al.*, 2005). The two *V. cholerae* 28-bp *dif* sites differ from one another at four polymorphic sites, one of which is located in the XerC-binding site and the other three sites are located in the 6-bp spacer region (Huber and Waldor, 2002; Val *et al.*, 2005). CTXϕ integrates at one or two of the chromosome dimer resolution sites, *dif1* and *dif2*.

The substrate for recombination and integration between the dsDNA *V. cholerae* chromosome and CTXϕ could be potentially a ssDNA or double-stranded DNA (dsDNA) form of CTXϕ (McLeod and Waldor, 2004; Val *et al.*, 2005). Originally it was proposed that CTXϕ integration in an El Tor isolate resulted from recombination between a 200-bp intergenic region (*attP*) on the replicative dsDNA form of CTXϕ and *dif1* (*attB*) in chromosome I of *V. cholerae* (Huber and Waldor, 2002). It was demonstrated that the recombination reaction was catalyzed by XerC and XerD, did not require the cell division protein FtsK, and, unlike other phage site-specific integration mechanisms, integration was irreversible (Huber and Waldor, 2002; McLeod and Waldor, 2004). An alternative model for CTXϕ integration was proposed where the ssDNA genome was the substrate for integration at *dif1* by XerCD (Val *et al.*, 2005). Val and colleagues (2005) demonstrated the presence of a double-forked hairpin structure within the *attP* sequence in the ssDNA CTXϕ that creates an additional *attP+* site in the stem of the secondary structure. They showed that XerC catalyzes a single pair of strand exchanges between this target, *attP+*, and *dif1* in the presence of XerD (Val *et al.*, 2005). CTXϕ integration resulted upon conversion of the ensuing Holliday junction by repair and/or replication similar to earlier findings by McLeod and Waldor (2004) and Val *et al.* (2005).

Das and colleagues proposed a mechanism of CTXϕ integration that accounted for the ability of variant CTXϕ genomes to integrate at chromosomal site-specific *dif* sites in chromosomes I and II (Das *et al.*, 2010; Huber and Waldor, 2002; Waldor and Mekalanos, 1996). The Barre group demonstrated how and why there is specific integration between $CTX\phi^{ET}$ at *dif1* and none at *dif2*, in contrast to the integration of $CTX\phi^{class}$ at both *dif1* and *dif2* (Das *et al.*, 2010). They showed that integration of CTXϕ^{class} is due to two base changes in the overlap region of *attP2* in this phage, allowing XerCD recombination between CTXϕ^{class} *attP+* with *dif1* and *dif2*. Their data determined that the specificity of integration of the different CTXϕ variants is governed by the potential of the ssDNA CTXϕ to form base pair interactions that stabilize strand exchanges (Das *et al.*, 2010).

Interestingly, other filamentous phages, such as VGJϕ, use a dsDNA genome as a template for integration. It was propose that the *dif1* site could be the *attB* site for a larger family of filamentous vibriophages, such as the previously described VGJϕ, VEJϕ, VSK, VSKK, fs2, f237, and Vf33 phages (Das *et al.*, 2011). These phages differed from CTXϕ and related phages in that they contain a single *attP* Xer-binding site and are found as a single copy on the host chromosome (Das *et al.*, 2011). Barre and colleagues proposed that VGJϕ and related phages hijack the host XerCD recombinase system for integration at *dif1* but use the dsDNA form of the phage as the template for integration (Das *et al.*, 2011). How CTXϕ^{class} from biotype classical strains, which cannot produce CTXϕ infectious particles, are transferred between natural isolates and how recently emerged *V. cholerae* El Tor strains containing a CTXϕ^{class} prophage are formed is unresolved (Davis *et al.*, 2000b; Faruque *et al.*, 2007; Udden *et al.*, 2008). It is possible that these novel strains acquired CTXϕ^{class} via an alternative mechanism, such as by generalized transducing phage and/or transformation mechanisms, although the *in vitro* efficiencies of these events were shown to be quite low (Boyd and Waldor, 1999; Campos *et al.*, 2010; Udden *et al.*, 2008).

ACKNOWLEDGMENTS

I thank members of my group for their enthusiasm and hard work: Megan R. Carpenter, Nityananda Chowdhury, Brandy Haines Menges, Sai Kalburge, Joseph J. Kingston, Jean Bernard Lubin, Serge Ongagna-Yhombi, and W. Brian Whitaker. Work in the Boyd group is supported by National Science Foundation CAREER Grant DEB-0844409, NSF Grant IOS-0918429, and funding from the U.S. Department of Agriculture, NRI CSREES Grant 2008-01198. Literature citations have been limited due to space limitations, and I apologize to those authors whose research is not included or cited.

REFERENCES

Allison, G. E., and Verma, N. K. (2000). Serotype-converting bacteriophages and O-antigen modification in *Shigella flexneri*. *Trends Microbiol.* **8**:17–23.

Asakura, M., Hinenoya, A., Alam, M. S., Shima, K., Zahid, S. H., Shi, L., Sugimoto, N., Ghosh, A. N., Ramamurthy, T., Faruque, S. M., Nair, G. B., and Yamasaki, S. (2007). An inducible lambdoid prophage encoding cytolethal distending toxin (Cdt-I) and a type III effector protein in enteropathogenic *Escherichia coli*. *Proc. Natl. Acad. Sci. USA* **104**:14483–14488.

Barondess, J. J., and Beckwith, J. (1990). A bacterial virulence determinant encoded by lysogenic coliphage lambda. *Nature* **346:**871–874.

Barre, F. X., Soballe, B., Michel, B., Aroyo, M., Robertson, M., and Sherratt, D. (2001). Circles: The replication-recombination-chromosome segregation connection *Proc. Natl. Acad. Sci. USA* **98:**8189–8195.

Baudry, B., Fasano, A., Ketley, J., and Kaper, J. B. (1992). Cloning of a gene (*zot*) encoding a new toxin produced by *Vibrio cholerae. Infect. Immun.* **60:**428–434.

Bensing, B. A., Rubens, C. E., and Sullam, P. M. (2001a). Genetic loci of *Streptococcus mitis* that mediate binding to human platelets. *Infect. Immun.* **69:**1373–1380.

Bensing, B. A., Siboo, I. R., and Sullam, P. M. (2001b). Proteins PblA and PblB of *Streptococcus mitis,* which promote binding to human platelets, are encoded within a lysogenic bacteriophage. *Infect. Immun.* **69:**6186–6192.

Beutin, L., Stroeher, U. H., and Manning, P. A. (1993). Isolation of enterohemolysin (Ehly2)-associated sequences encoded on temperate phages of *Escherichia coli. Gene* **132:**95–99.

Bhadra, R. K., Roychoudhury, S., Banerjee, R. K., Kar, S., Majumdar, R., Sengupta, S., Chatterjee, S., Khetawat, G., and Das, J. (1995). Cholera toxin (CTX) genetic element in *Vibrio cholerae* O139. *Microbiology* **141:**1977–1983.

Bik, E. M., Bunschoten, A. E., Gouw, R. D., and Mooi, F. R. (1995). Genesis of the novel epidemic *Vibrio cholerae* O139 strain: Evidence for horizontal transfer of genes involved in polysaccharide synthesis. *EMBO J.* **14:**209–216.

Blokesch, M., and Schoolnik, G. K. (2008). The extracellular nuclease Dns and its role in natural transformation of *Vibrio cholerae. J. Bacteriol.* **190:**7232–7240.

Boyd, E. F., Almagro-Moreno, S., and Parent, M. A. (2009). Genomic islands are dynamic, ancient integrative elements in bacterial evolution. *Trends Microbiol.* **17:**47–53.

Boyd, E. F., and Brussow, H. (2002). Common themes among bacteriophage-encoded virulence factors and diversity among the bacteriophages involved. *Trends Microbiol.* **10:**521–529.

Boyd, E. F., Davis, B. M., and Hochhut, B. (2001). Bacteriophage-bacteriophage interactions in the evolution of pathogenic bacteria. *Trends Microbiol.* **9:**137–144.

Boyd, E. F., and Waldor, M. K. (1999). Alternative mechanism of cholera toxin acquisition by *Vibrio cholerae*: Generalized transduction of CTXϕ by bacteriophage CP-T1. *Infect. Immun.* **67:**5898–5905.

Broadbent, S. E., Davies, M. R., and van der Woude, M. W. (2010). Phase variation controls expression of *Salmonella* lipopolysaccharide modification genes by a DNA methylation-dependent mechanism. *Mol. Microbiol.* **77:**337–353.

Brown, N., Coombes, B. K., Bishop, J. L., Wickham, M. E., Lowden, M. J., Gal-Mor, O., Goode, D. L., Boyle, E. C., Sanderson, K. L., and Finlay, B. B. (2011). *Salmonella* phage ST64B encodes a member of the SseK/NleB effector family. *PLoS One* **6:**e17824.

Brussow, H., Canchaya, C., and Hardt, W. D. (2004). Phages and the evolution of bacterial pathogens: From genomic rearrangements to lysogenic conversion. *Microbiol. Mol. Biol. Rev.* **68:**560–602.

Campos, J., Martínez, E., Izquierdo, Y., and Fando, R. (2010). VEJϕ, a novel filamentous phage of *Vibrio cholerae* able to transduce the cholera toxin genes. *Microbiology* **156:**108–115.

Clark, C. A., Beltrame, J., and Manning, P. A. (1991). The *oac* gene encoding a lipopolysaccharide O-antigen acetylase maps adjacent to the integrase-encoding gene on the genome of *Shigella flexneri* bacteriophage Sf6. *Gene* **107:**43–52.

Coleman, D. C., Sullivan, D. J., Russell, R. J., Arbuthnott, J. P., Carey, B. F., and Pomeroy, H. M. (1989). *Staphylococcus aureus* bacteriophages mediating the simultaneous lysogenic conversion of beta-lysin, staphylokinase and enterotoxin A: Molecular mechanism of triple conversion. *J. Gen. Microbiol.* **135:**1679–1697.

Creuzburg, K., Recktenwald, J., Kuhle, V., Herold, S., Hensel, M., and Schmidt, H. (2005). The Shiga toxin 1-converting bacteriophage BP-4795 encodes an NleA-like type III effector protein. *J. Bacteriol.* **187:**8494–8498.

Das, B., Bischerour, J., and Barre, F. (2011). VGJphi integration and excision mechanisms contribute to the genetic diversity of *Vibrio cholerae* epidemic strains. *Proc. Natl. Acad. Sci. USA* **108:**2516–2521.

Das, B., Bischerour, J., Val, M., and Barre, F.-X. (2010). Molecular keys of the tropism of integration of the cholera toxin phage. *Proc. Natl. Acad. Sci. USA* **107**(9)**:**4377–4382.

Davis, B. M., Kimsey, H. H., Chang, W., and Waldor, M. K. (1999). The *Vibrio cholerae* O139 calcutta CTXF is infectious and encodes a novel repressor. *J. Bacteriol* **181:**6779–6787.

Davis, B. M., Kimsey, H. H., Kane, A. V., and Waldor, M. K. (2002). A satellite phage-encoded antirepressor induces repressor aggregation and cholera toxin gene transfer. *EMBO J.* **21:**4240–4249.

Davis, B. M., Lawson, E. H., Sandkvist, M., Ali, A., Sozhamannan, S., and Waldor, M. K. (2000a). Convergence of the secretory pathways for cholera toxin and the filamentous phage, CTXphi. *Science* **288:**333–335.

Davis, B. M., Moyer, K. E., Boyd, E. F., and Waldor, M. K. (2000b). CTX prophages in classical biotype *Vibrio cholerae*: Functional phage genes but dysfunctional phage genomes. *J. Bacteriol.* **182:**6992–6998.

Davis, B. M., and Waldor, M. K. (2000). CTXf contains a hybrid genome derived from tandemly integrated elements. *Proc. Natl. Acad. Sci. USA* **97:**8572–8577.

Dean, P. (2011). Functional domains and motifs of bacterial type III effector proteins and their roles in infection. *FEMS Microbiol. Rev.* **35:**1100–1125.

Dobrindt, U., Hochhut, B., Hentschel, U., and Hacker, J. (2004). Genomic islands in pathogenic and environmental microorganisms. *Nat. Rev. Microbiol.* **2:**414–424.

Ehrbar, K., Friebel, A., Miller, S. I., and Hardt, W. D. (2003). Role of the *Salmonella* pathogenicity island 1 (SPI-1) protein InvB in type III secretion of SopE and SopE2, two *Salmonella* effector proteins encoded outside of SPI-1. *J. Bacteriol.* **185:**6950–6967.

Ehrbar, K., and Hardt, W. D. (2005). Bacteriophage-encoded type III effectors in *Salmonella enterica* subspecies 1 serovar Typhimurium. *Infect. Genet. Evol.* **5:**1–9.

Ehrbar, K., Mirold, S., Friebel, A., Stender, S., and Hardt, W. D. (2002). Characterization of effector proteins translocated via the SPI1 type III secretion system of *Salmonella typhimurium*. *Int. J. Med. Microbiol.* **291:**479–485.

Faruque, S. M., Albert, M. J., and Mekalanos, J. J. (1998a). Epidemiology, genetics, and ecology of toxigenic *Vibrio cholerae*. *Microbiol. Mol. Biol. Rev.* **62:**1301–1314.

Faruque, S. M., Asadulghani, Saha, M. N., Alim, A. R., Albert, M. J., Islam, K. M., and Mekalanos, J. J. (1998b). Analysis of clinical and environmental strains of nontoxigenic *Vibrio cholerae* for susceptibility to CTXΦ: Molecular basis for origination of new strains with epidemic potential. *Infect. Immun* **66:**5819–5825.

Faruque, S. M., Sack, D. A., Sack, R. B., Colwell, R. R., Takeda, Y., and Nair, G. B. (2003). Emergence and evolution of *Vibrio cholerae* O139. *Proc. Natl. Acad. Sci. USA* **100:**1304–1309.

Faruque, S. M., Tam, V. C., Chowdhury, N., Diraphat, P., Dziejman, M., Heidelberg, J. F., Clemens, J. D., Mekalanos, J. J., and Nair, G. B. (2007). Genomic analysis of the Mozambique strain of *Vibrio cholerae* O1 reveals the origin of El Tor strains carrying classical CTX prophage. *Proc. Natl. Acad. Sci. USA* **104:**5151–5156.

Fasano, A., Baudry, B., Pumplin, D. W., Wasserman, S. S., Tall, B. D., Ketley, J., and Kaper, J. B. (1991). *Vibrio cholerae* produces a second enterotoxin, which affects intestinal tight junctions. *Proc. Natl. Acad. Sci. USA* **88:**5242–5246.

Figueroa-Bossi, N., and Bossi, L. (1999). Inducible prophages contribute to *Salmonella* virulence in mice. *Mol. Microbiol.* **33:**167–176.

Figueroa-Bossi, N., Coissac, E., Netter, P., and Bossi, L. (1997). Unsuspected prophage-like elements in *Salmonella typhimurium*. *Mol. Microbiol.* **25:**161–173.

Figueroa-Bossi, N., Uzzau, S., Maloriol, D., and Bossi, L. (2001). Variable assortment of prophages provides a transferable repertoire of pathogenic determinants in *Salmonella. Mol. Microbiol.* **39:**260–271.

Freeman, V. J. (1951). Studies on the virulence of bacteriophage-infected strains of *Corynebacterium diptheriae. J. Bacteriol.* **61:**675–688.

Friebel, A., Ilchmann, H., Aepfelbacher, M., Ehrbar, K., Machleidt, W., and Hardt, W. D. (2001). SopE and SopE2 from *Salmonella typhimurium* activate different sets of RhoGTPases of the host cell. *J. Biol. Chem.* **276:**34035–34040.

Goshorn, S. C., and Schlievert, P. M. (1989). Bacteriophage association of streptococcal pyrogenic exotoxin type C. *J. Bacteriol.* **171:**3068–3073.

Grim, C., Hasan, N. A., Taviani, E., Haley, B., Chun, J., Brettin, T. S., Bruce, D. C., Detter, J. C., Han, C. S., Chertkov, O., Challacombe, J., Huq, A., *et al.* (2010). Genome sequence of hybrid *Vibrio cholerae* O1 MJ-1236, B-33, and CIRS101 and comparative genomics with *V. cholerae. J. Bacteriol.* **192:**3524–3533.

Gruenheid, S., Sekirov, I., Thomas, N. A., Deng, W., O'Donnell, P., Goode, D., Li, Y., Frey, E. A., Brown, N. F., Metalnikov, P., Pawson, T., Ashman, K., *et al.* (2004). Identification and characterization of NleA, a non-LEE-encoded type III translocated virulence factor of enterohaemorrhagic *Escherichia coli* O157:H7. *Mol. Microbiol.* **51:**1233–1249.

Hacker, J., Blum-Oehler, G., Muhldorfer, I., and Tschape, H. (1997). Pathogenicity islands of virulent bacteria: structure, function and impact on microbial evolution. *Mol. Microbiol.* **23:**1089–1097.

Hacker, J., and Kaper, J. B. (2000). Pathogenicity islands and the evolution of microbes. *Annu. Rev. Microbiol.* **54:**641–679.

Hansen-Wester, I., and Hensel, M. (2001). *Salmonella* pathogenicity islands encoding type III secretion systems. *Microbes Infect.* **3:**549–559.

Hansen-Wester, I., Stecher, B., and Hensel, M. (2002a). Type III secretion of *Salmonella enterica* serovar Typhimurium translocated effectors and SseFG. *Infect. Immun.* **70:**1403–1409.

Hansen-Wester, I., Stecher, B., and Hensel, M. (2002b). Analyses of the evolutionary distribution of *Salmonella* translocated effectors. *Infect. Immun.* **70:**1619–1622.

Hemrajani, C., Marches, O., Wiles, S., Girard, F., Dennis, A., Dziva, F., Best, A., Phillips, A. D., Berger, C. N., Mousnier, A., Crepin, V. F., Kruidenier, L., *et al.* (2008). Role of NleH, a type III secreted effector from attaching and effacing pathogens, in colonization of the bovine, ovine, and murine gut. *Infect. Immun.* **276**(11)**:**4804–4813.

Herold, S., Karch, H., and Schmidt, H. (2004). Shiga toxin-encoding bacteriophages: Genomes in motion. *Int. J. Med. Microbiol.* **294:**115–121.

Ho, T. D., and Slauch, J. M. (2001). Characterization of *grvA*, an antivirulence gene on the Gifsy-2 phage in *Salmonella enterica* serovar Typhimurium. *J. Bacteriol.* **183:**611–620.

Huber, K. E., and Waldor, M. K. (2002). Filamentous phage integration requires the host recombinases XerC and XerD. *Nature* **417:**656–659.

Hynes, W. L., and Ferretti, J. J. (1989). Sequence analysis and expression in *Escherichia coli* of the hyaluronidase gene of *Streptococcus pyogenes* bacteriophage H4489A. *Infect. Immun.* **57:**533–539.

Karaolis, D. K., Johnson, J. A., Bailey, C. C., Boedeker, E. C., Kaper, J. B., and Reeves, P. R. (1998). A *Vibrio cholerae* pathogenicity island associated with epidemic and pandemic strains. *Proc. Natl. Acad. Sci. USA* **95:**3134–3139.

Kuroda, M., Ohta, T., Uchiyama, I., Baba, T., Yuzawa, H., Kobayashi, I., Cui, L., Oguchi, A., Aoki, K., Nagai, Y., Lian, J., Ito, T., *et al.* (2001). Whole genome sequencing of methicillin-resistant *Staphylococcus aureus. Lancet* **357:**1225–1240.

Li, M., Shimada, T., Morris, J. G., Jr., Sulakvelidze, A., and Sozhamannan, S. (2002). Evidence for the emergence of non-O1 and non-O139 Vibrio cholerae strains with pathogenic potential by exchange of O-antigen biosynthesis regions. *Infect. Immun.* **70:**2441–2453.

Lindsay, J. A., Ruzin, A., Ross, H. F., Kurepina, N., and Novick, R. P. (1998). The gene for toxic shock toxin is carried by a family of mobile pathogenicity islands in *Staphylococcus aureus*. *Mol. Microbiol.* **29:**527–543.

Loukiadis, E., Nobe, R., Herold, S., Tramuta, C., Ogura, Y., Ooka, T., Morabito, S., Kérourédan, M., Brugère, H., Schmidt, H., Hayashi, T., and Oswald, E. (2008). Distribution, functional expression, and genetic organization of Cif, a phage-encoded type III-secreted effector from enteropathogenic and enterohemorrhagic *Escherichia coli*. *J. Bacteriol.* **190:**275–285.

Masignani, V., Giuliani, M. M., Tettelin, H., Comanducci, M., Rappuoli, R., and Scarlato, V. (2001). Mu-like prophage in serogroup B *Neisseria meningitidis* coding for surface-exposed antigens. *Infect. Immun.* **69:**2580–2588.

McClelland, M., Sanderson, K. E., Spieth, J., Clifton, S. W., Latreille, P., Courtney, L., Porwollik, S., Ali, J., Dante, M., Du, F., Hou, S., Layman, D., *et al.* (2001). Complete genome sequence of *Salmonella enterica* serovar Typhimurium LT2. *Nature* **413:**852–856.

McGhie, E., Brawn, L. C., Hume, P. J., Humphreys, D., and Koronakis, V. (2009). *Salmonella* takes control: Effector-driven manipulation of the host. *Curr. Opin. Microbiol.* **12:**117–124.

McLeod, S. M., and Waldor, M. K. (2004). Characterization of XerC- and XerD-dependent CTX phage integration in *Vibrio cholerae*. *Mol. Microbiol.* **54:**935–947.

Meibom, K. L., Blokesch, M., Dolganov, N. A., Wu, C. Y., and Schoolnik, G. K. (2005). Chitin induces natural competence in *Vibrio cholerae*. *Science* **310:**1824–1827.

Mekalanos, J. J. (1983). Duplication and amplification of toxin genes in *Vibrio cholerae*. *Cell* **35:**253–263.

Mekalanos, J. J., Swartz, D. J., Pearson, G. D., Harford, N., Groyne, F., and de Wilde, M. (1983). Cholera toxin genes: Nucleotide sequence, deletion analysis and vaccine development. *Nature* **306:**551–557.

Miller, M. C., Keymer, D. P., Avelar, A., Boehm, A. B., and Schoolnik, G. K. (2007). Detection and transformation of genome segments that differ within a coastal population of *Vibrio cholerae* strains. *Appl. Environ. Microbiol.* **73:**3695–3704.

Mirold, S., Ehrbar, K., Weissmuller, A., Prager, R., Tschape, H., Russmann, H., and Hardt, W. D. (2001). *Salmonella* host cell invasion emerged by acquisition of a mosaic of separate genetic elements, including *Salmonella* pathogenicity island 1 (SPI1), SPI5, and sopE2. *J. Bacteriol.* **183:**2348–2358.

Mirold, S., Rabsch, W., Rohde, M., Stender, S., Tschape, H., Russmann, H., Igwe, E., and Hardt, W. D. (1999). Isolation of a temperate bacteriophage encoding the type III effector protein SopE from an epidemic *Salmonella typhimurium* strain. *Proc. Natl. Acad. Sci. USA* **96:**9845–9850.

Miyahara, A., Nakanishi, N., Ooka, T., Hayashi, T., Sugimoto, N., and Tobe, T. (2009). Enterohemorrhagic *Escherichia coli* effector EspL2 induces actin microfilament aggregation through annexin 2 activation. *Cell Microbiol.* **11:**337–350.

Mooi, F. R., and Bik, E. M. (1997). The evolution of epidemic *Vibrio cholerae* strains. *Trends Microbiol.* **5:**161–165.

Napolitano, M. G., Almagro-Moreno, S., and Boyd, E. F. (2011). Dichotomy in the evolution of pathogenicity island and bacteriophage encoded integrases from pathogenic *Escherichia coli* strains. *Infect. Genet. Evol.* **11:**423–436.

Novick, R., Christie, G. E., and Penades, J. R. (2010). The phage-related chromosomal islands of Gram-positive bacteria. *Nat. Rev. Microbiol.* **8:**541–551.

O'Brien, A. D., Newland, J. W., Miller, S. F., Holmes, R. K., Smith, H. W., and Formal, S. B. (1984). Shiga-like toxin-converting phages from *Escherichia coli* strains that cause hemorrhagic colitis or infantile diarrhea. *Science* **226:**694–696.

Oelschlaeger, T., Dobrindt, U., and Hacker, J. (2002). Pathogenicity islands of uropathogenic *E. coli* and the evolution of virulence. *Int. J. Antimicrob. Agents* **19:**517–521.

Ogura, Y., Ooka, T., Iguchi, A., Toh, H., Asadulghani, M., Oshima, K., Kodama, T., Abe, H., Nakayama, K., Kurokawa, K., Tobe, T., Hattori, M., *et al.* (2009). Comparative genomics reveal the mechanism of the parallel evolution of O157 and non-O157 enterohemorrhagic *Escherichia coli*. *Proc. Natl. Acad. Sci. USA* **106:**17939–17944.

Ohnishi, M., Terajima, J., Kurokawa, K., Nakayama, K., Murata, T., Tamura, K., Ogura, Y., Watanabe, H., and Hayashi, T. (2002). Genomic diversity of enterohemorrhagic *Escherichia coli* O157 revealed by whole genome PCR scanning. *Proc. Natl. Acad. Sci. USA* **99:**17043–17048.

O'Shea, Y. A., Reen, F. J., Quirke, A. M., and Boyd, E. F. (2004). Evolutionary genetic analysis of the emergence of epidemic *Vibrio cholerae* isolates based on comparative nucleotide sequence analysis and multilocus virulence gene profiles. *J. Clin. Microbiol.* **42:**4657–4671.

Perna, N. T., Plunkett, G., 3rd, Burland, V., Mau, B., Glasner, J. D., Rose, D. J., Mayhew, G. F., Evans, P. S., Gregor, J., Kirkpatrick, H. A., Posfai, G., Hackett, J., *et al.* (2001). Genome sequence of enterohaemorrhagic *Escherichia coli* O157:H7. *Nature* **409:**529–533.

Ruzin, A., Lindsay, J., and Novick, R. P. (2001). Molecular genetics of SaPI1: A mobile pathogenicity island in *Staphylococcus aureus*. *Mol. Microbiol.* **41:**365–377.

Sandt, C. H., Hopper, J. E., and Hill, C. W. (2002). Activation of prophage *eib* genes for immunoglobin-binding proteins from the IbrAB genetic island of *Escherichia coli* ECOR-9. *J. Bacteriol.* **184:**3640–3648.

Schmidt, H. (2001). Shiga-toxin-converting bacteriophages. *Res. Microbiol.* **152:**687–695.

Schwidder, M., Hensel, M., and Schmidt, H. (2011). Regulation of *nleA* in Shiga toxin-producing *Escherichia coli* O84:H4 strain 4795/97. *J. Bacteriol.* **193:**832–841.

Seo, H., Xiong, Y. Q., Mitchell, J., Seepersaud, R., Bayer, A. S., and Sullam, P. M. (2010). Bacteriophage lysin mediates the binding of *Streptococcus mitis* to human platelets through interaction with fibrinogen. *PLoS Pathog.* **6:**e1001047.

Sherratt, D. J., Soballe, B., Barre, F. X., Filipe, S., Lau, I., Massey, T., and Yates, J. (2004). Recombination and chromosome segregation. *Philos. Trans. R. Soc. Lond. B Biol. Sci.* **359:**61–69.

Siboo, I. R., Bensing, B. A., and Sullam, P. M. (2003). Genomic organization and molecular characterization of SM1, a temperate bacteriophage of *Streptococcus mitis*. *J. Bacteriol.* **185:**6968–6975.

Strockbine, N. A., Jackson, M. P., Sung, L. M., Holmes, R. K., and O'Brien, A. D. (1988). Cloning and sequencing of the genes for Shiga toxin from *Shigella dysenteriae* type 1. *J. Bacteriol.* **170**(3)**:**1116–1122.

Stroeher, U. H., Parasivam, G., Dredge, B. K., and Manning, P. A. (1997). Novel *Vibrio cholerae* O139 genes involved in lipopolysaccharide biosynthesis. *J. Bacteriol.* **179:**2740–2747.

Taylor, R. K., Miller, V. L., Furlong, D. B., and Mekalanos, J. J. (1987). Use of *phoA* gene fusions to identify a pilus colonization factor coordinately regulated with cholera toxin. *Proc. Natl. Acad. Sci. USA* **84:**2833–2837.

Tobe, T., Beatson, S. A., Taniguchi, H., Abe, H., Bailey, C. M., Fivian, A., Younis, R., Matthews, S., Marches, O., Frankel, G., Hayashi, T., and Pallen, M. J. (2006). An extensive repertoire of type III secretion effectors in *Escherichia coli* O157 and the role of lambdoid phages in their dissemination. *Proc. Natl. Acad. Sci. USA* **103:**14941–14946.

Trucksis, M., Galen, J. E., Michalski, J., Fasano, A., and Kaper, J. B. (1993). Accessory cholera enterotoxin (Ace), the third toxin of a *Vibrio cholerae* virulence cassette. *Proc. Natl. Acad. Sci. USA* **90:**5267–5271.

Trucksis, M., Michalski, J., Deng, Y. K., and Kaper, J. B. (1998). The *Vibrio cholerae* genome contains two unique circular chromosomes. *Proc. Natl. Acad. Sci. USA* **95:**14464–14469.

Udden, S. M., Zahid, M. S., Biswas, K., Ahmad, Q. S., Cravioto, A., Nair, G. B., Mekalanos, J. J., and Faruque, S. M. (2008). Acquisition of classical CTX prophage from *Vibrio cholerae* O141 by El Tor strains aided by lytic phages and chitin-induced competence. *Proc. Natl. Acad. Sci. USA* **105:**11951–11956.

Val, M. E., Bouvier, M., Campos, J., Sherratt, D., Cornet, F., Mazel, D., and Barre, F. X. (2005). The single-stranded genome of phage CTX is the form used for integration into the genome of *Vibrio cholerae*. *Mol. Cell.* **19:**559–566.

Verma, N. K., Brandt, J. M., Verma, D. J., and Lindberg, A. A. (1991). Molecular characterization of the O-acetyl transferase gene of converting bacteriophage SF6 that adds group antigen 6 to *Shigella flexneri*. *Mol. Microbiol.* **5:**71–75.

Voelker, L. L., and Dybvig, K. (1999). Sequence analysis of the *Mycoplasma arthritidis* bacteriophage MAV1 genome identifies the putative virulence factor. *Gene* **11:**101–107.

Waldor, M. K., Colwell, R., and Mekalanos, J. J. (1994). The *Vibrio cholerae* O139 serogroup antigen includes an O-antigen capsule and lipopolysaccharide virulence determinants. *Proc. Natl. Acad. Sci. USA* **91:**11388–11392.

Waldor, M. K., and Mekalanos, J. J. (1994). Emergence of a new cholera pandemic: Molecular analysis of virulence determinants in *Vibrio cholerae* O139 and development of a live vaccine prototype. *J. Infect. Dis.* **170:**278–283.

Waldor, M. K., and Mekalanos, J. J. (1996). Lysogenic conversion by a filamentous phage encoding cholera toxin. *Science* **272:**1910–1914.

Waldor, M. K., Rubin, E. J., Pearson, G. D., Kimsey, H., and Mekalanos, J. J. (1997). Regulation, replication, and integration functions of the *Vibrio cholerae* CTXΦ are encoded by region RS2. *Mol. Microbiol.* **24:**917–926.

Weeks, C. R., and Ferretti, J. J. (1984). The gene for type A streptococcal exotoxin (erythrogenic toxin) is located in bacteriophage T12. *Infect. Immun.* **46:**531–536.

Wright, A. (1971). Mechanism of conversion of the *Salmonella* O antigen by bacteriophage epsilon 34. *J. Bacteriol.* **105:**927–936.

Yamaguchi, T., Hayashi, T., Takami, H., Nakasone, K., Ohnishi, M., Nakayama, K., Yamada, S., Komatsuzawa, H., and Sugai, M. (2000). Phage conversion of exfoliative toxin A production in *Staphylococcus aureus*. *Mol. Microbiol.* **38:**694–705.

Yen, H., Ooka, T., Iguchi, A., Hayashi, T., Sugimoto, N., and Tobe, T. (2010). NleC, a type III secretion protease, compromises NF-kappaB activation by targeting p65/RelA. *PLoS Pathog.* **6:**e1001231.

CHAPTER 5

Structure, Assembly, and DNA Packaging of the Bacteriophage T4 Head

Lindsay W. Black* and **Venigalla B. Rao**[†]

Contents

Abstract

The bacteriophage T4 head is an elongated icosahedron packed with 172 kb of linear double-stranded DNA and numerous proteins. The capsid is built from three essential proteins: gp23*, which forms the hexagonal capsid lattice; gp24*, which forms pentamers at 11 of the 12 vertices; and gp20, which forms the unique dodecameric

* Department of Biochemistry and Molecular Biology, University of Maryland Medical School, Baltimore, Maryland, USA
† Department of Biology, Catholic University of America, Washington DC, USA

Advances in Virus Research, Volume 82
ISSN 0065-3527, DOI: 10.1016/B978-0-12-394621-8.00018-2

portal vertex through which DNA enters during packaging and exits during infection.

Intensive work over more than half a century has led to a deep understanding of the phage T4 head. The atomic structure of gp24 has been determined. A structural model built for gp23 using its similarity to gp24 showed that the phage T4 major capsid protein has the same fold as numerous other icosahedral bacteriophages. However, phage T4 displays an unusual membrane and portal initiated assembly of a shape determining self-sufficient scaffolding core. Folding of gp23 requires the assistance of two chaperones, the *Escherichia coli* chaperone GroEL acting with the phage-coded gp23-specific cochaperone, gp31. The capsid also contains two nonessential outer capsid proteins, Hoc and Soc, which decorate the capsid surface. Through binding to adjacent gp23 subunits, Soc reinforces the capsid structure. Hoc and Soc have been used extensively in bipartite peptide display libraries and to display pathogen antigens, including those from human immunodeficiency virus (HIV), *Neisseria meningitides*, *Bacillus anthracis*, and foot and mouth disease virus. The structure of Ipl*, one of a number of multiple (>100) copy proteins packed and injected with DNA from the full head, shows it to be an inhibitor of one specific restriction endonuclease specifically targeting glycosylated hydroxymethyl cytosine DNA. Extensive mutagenesis, combined with atomic structures of the DNA packaging/terminase proteins gp16 and gp17, elucidated the ATPase and nuclease functional motifs involved in DNA translocation and headful DNA cutting. The **cryoelectron microscopy** structure of the T4 packaging machine showed a pentameric motor assembled with gp17 subunits on the portal vertex. Single molecule optical tweezers and fluorescence studies showed that the T4 motor packages DNA at the highest rate known and can package multiple segments. Förster resonance energy transfer–fluorescence correlation spectroscopy studies indicate that DNA gets compressed in the stalled motor and that the terminase-to-portal distance changes during translocation. Current evidence suggests a linear two-component (large terminase plus portal) translocation motor in which electrostatic forces generated by ATP hydrolysis drive DNA translocation by alternating the motor between tensed and relaxed states.

LIST OF ABBREVIATIONS

aa	amino acids
gp23*	mature gp21 protease processed form of gp23 etc.
EF	edema factor
EM	electron microscopy
FCS	fluorescence correlation spectroscopy

FMDV	foot and mouth disease virus
FRET	Förster resonance energy transfer
gp	gene product
HIV	human immunodeficiency virus
Hoc	highly antigenic outer capsid protein
IP	internal protein
LF	lethal factor
PA	protective antigen
Soc	small outer capsid protein
kb	kilobase
bp	base pair

I. INTRODUCTION

The T4-type bacteriophages are ubiquitously distributed in nature and occupy environmental niches ranging from mammalian gut to soil, sewage, and oceans. More than 130 such viruses showing morphological features similar to phage T4 have been described; from the T4 superfamily, ~1400 major capsid protein sequences have been correlated to its three-dimensional structure (Desplats and Krisch, 2003; Krisch and Comeau, 2008; Tetart *et al.*, 1998). Features include a large elongated (prolate) head, contractile tail, and complex base plate with six long, kinked tail fibers emanating from it radially. Phage T4 continues to serve as an excellent model to elucidate the mechanisms of head structure, assembly, and DNA packaging not only of myoviridae but of other large icosahedral viruses such as herpes viruses, as well as providing a valuable biotechnology platform with diverse current applications. This chapter focuses on advances on the basic understanding of phage T4 head structure, assembly, and mechanism of DNA packaging. Application of some of this knowledge to develop phage T4 as a surface display and vaccine platform is also discussed. The reader is referred to the comprehensive review by Black *et al.* (1994) for earlier work (references through 1993) on T4 head assembly.

II. STRUCTURE AND ASSEMBLY OF PHAGE T4 CAPSID

The overall architecture and dimensions of the phage T4 head (Fokine *et al.*, 2004) are displayed in Fig. 1. The width and length of the elongated prolate icosahedron are $T_{end} = 13$ laevo and $T_{mid} = 20$ (86 nm wide and 120 nm long), and the copy numbers of gp23 (major capsid protein), gp24 (vertex protein), gp20 (portal protein), and nonessential display proteins Hoc and Soc are 930, 55, 12, 155, and 870, respectively (Fig. 1). Also, over

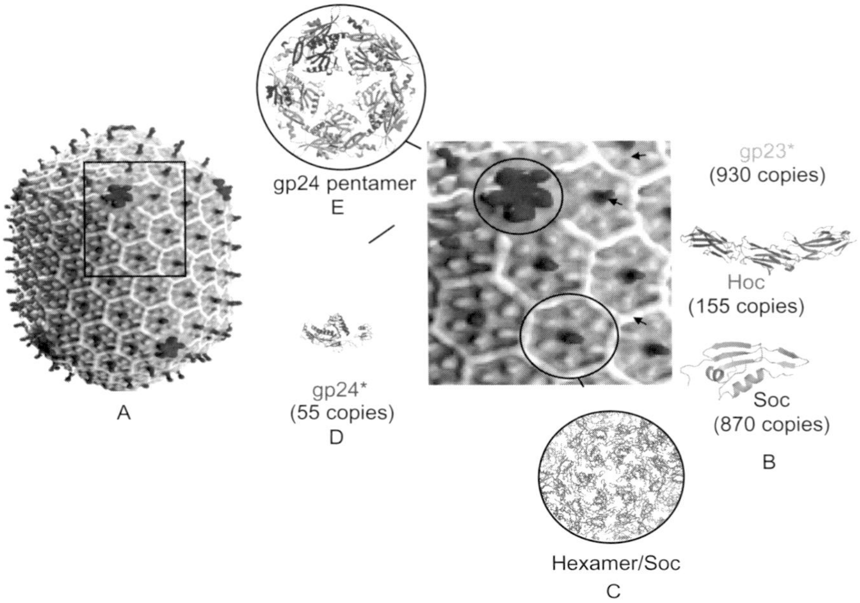

FIGURE 1 Structure of the bacteriophage T4 head. (A) CryoEM reconstruction of phage T4 capsid (Fokine *et al.*, 2004); the square block shows enlarged view showing gp23 (yellow subunits), gp24 (purple subunits), Hoc (red subunits), and Soc (white subunits). (B) Structures of RB49 Hoc (red) and RB69Soc (gray). (C) Structural model showing one gp23 hexamer (blue) surrounded by six Soc trimers (red). Neighboring gp23 hexamers are shown in green, black, and magenta (Qin *et al.*, 2009). (D) Structure of gp24 (Fokine *et al.*, 2005). (E) Structural model of gp24 pentameric vertex. (See Page 6 in Color Section at the back of the book.)

1000 internal proteins are packed together with the ~170-kb linear double-stranded DNA (dsDNA) within the capsid.

Assembly of the phage T4 prohead (Fig. 2A) is remarkable in (i) being wholly dependent on a membrane-bound portal initiator; (ii) ability to form a correct capsid size scaffolding core in the absence of the major capsid protein; (iii) extensive processing of nearly all the capsid proteins (only the portal protein is *not* processed) at consensus IP_2E- or LP_2E-processing sites by the gp21 morphogenetic protease; cleavage distal to >5000 E- peptide bonds is followed by (iv) detachment of the fully processed prohead from the host membrane for DNA packaging in the cytoplasm. According to bioinformatics classification, the gp21 morphogenetic protease is thought to occupy an isolated niche among phage and viral prohead protease genes classified as serine proteases (Cheng *et al.*, 2004; Liu and Mushegian, 2004). It should be noted that DNA packaging is initiated into the released processed procapsids already containing the

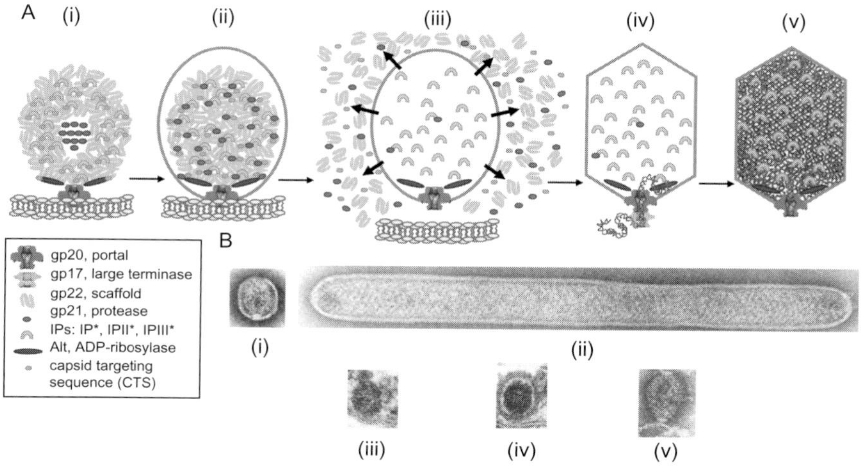

FIGURE 2 (A) Assembly of T4 prohead and head. (i) Membrane-bound portal initiates scaffolding core assembly; (ii) full prohead assembled on membrane; (iii) gp21 processing of core and procapsid components followed by detachment from membrane; (iv) DNA packaging into processed prohead; and (v) expanded fully packaged head. (B) Assembly intermediates from Black *et al.* (1994): (i) portal bound scaffolding core with procapsid; (ii) giant procapsid and core; (iii) *in vivo*-assembled naked core on membrane is precursor to (iv) full prohead; and (v) portal initiator assembles size-determining scaffolding core *in vitro*. (See Page 6 in Color Section at the back of the book.)

internal proteins to be ejected (IPI, IPII, IPIII, and Alt) along with DNA upon infection. Following packaging, the head is "plugged" by attachment of the head completion ("neck") proteins (gp13, gp14, and gp15) to the portal followed by assembly of the tail machine to form an infectious virus particle.

A gallery of unprocessed early prohead intermediates emphasizes a number of fundamental properties of T4 prohead assembly (Fig. 2B): (i) prominent and distinct internal prohead scaffolding core attached to the portal, (ii) aberrant giant proheads contain a giant core, (iii) "naked" cores without a procapsid shell are attached to the membrane-bound portal, (iv) can serve as prohead intermediates, and (v) core architecture is mainly imposed by assembly of the major scaffolding protein gp22, broken down to small peptides by the gp21 protease maturation. Mutations may also allow accumulation of other active intermediates such as vertex and distal cap-lacking proheads, showing that prohead assembly can diverge from a strictly linear pathway (Black *et al.*, 1994).

Assembly of the membrane-bound portal initiator is dependent on a viral chaperone (gp40), and gp20 is found at the membrane and can be cross-linked to Tig, DnaK, and YidC proteins. It is likely that DnaK transports the protein to the membrane, while YidC may function as a membrane-associated chaperone binding gp20 to the surface of the lipid bilayer until prehead assembly is completed (Quinten and Kuhn, personal communication). Following processing, the "empty" prohead maintains its early small procapsid precursor structure (unexpanded prohead, or empty small particle or esp) until irreversible expansion (~50% volume increase) leads to the stable elp (expanded prohead or empty large particle) conformation; as discussed later, packaging can be initiated into either type particle *in vitro*.

A significant advance was the crystal structure of the vertex protein gp24 and, by inference, the structure of its close relative, the major capsid protein gp23 (Fokine *et al.*, 2005). The gp24 and inferred gp23 structures are closely related to the structure of the major capsid protein of bacteriophage HK97, most probably also the same protein fold as the majority of tailed dsDNA bacteriophage major capsid proteins (Wikoff *et al.*, 2000). This ~0.3-nm resolution structure permits rationalization of head length mutations in the major capsid protein, as well as of mutations allowing bypass of the vertex protein. The former map to the periphery of the capsomer and the latter within the capsomer. It is likely that the special gp24 vertex protein of phage T4 is a relatively recent evolutionary addition as judged by the ease with which it can be bypassed. Cryoelectron (cryoEM) microscopy showed that in bypass mutants that substitute pentamers of the major capsid protein at the vertex, additional Soc decoration protein subunits surround these gp23* molecules, which does not occur in the gp23*–gp24* interfaces of the wild-type capsid (Fokine *et al.*, 2006). Nevertheless, despite the rationalization of the major capsid protein affecting head size mutations, it should be noted that these divert only a relatively small fraction of the capsids to altered and variable sizes. The primary determinant of the normally invariant prohead shape is thought to be the internal gp22-based major scaffolding core, which normally grows concurrently with the shell (Black *et al.*, 1994), although other core proteins (IPIII and Alt) also make a contribution to shape determination (Fig. 2). However, little progress has been made in establishing the exact mechanism of correct capsid size determination. Thus numerous "recent" T-even isolates are found to display increased and apparently invariant capsid prolate icosahedra; for example, KVP40, 254 kb, apparently has a single T_{mid} greater than the 170-kb T4 $T_{mid} = 20$)(Miller *et al.*, 2003). However, despite providing interesting material bearing on the T-even head size determination mechanism, few if any in-depth studies have been carried out on these phages to determine whether the morphogenetic core, the major capsid protein, or other factors are responsible for the different and precisely determined volumes of their capsids.

A. Folding of the major capsid protein gp23

Folding and assembly of the phage T4 major capsid protein gp23 into the prohead require special utilization of the GroEL chaperonin system and an essential phage cochaperonin, gp31. gp31 replaces the GroES cochaperonin utilized for folding the 10–15% of *E. coli* proteins that require folding by the GroEL folding chamber. Although T4 gp31 and the closely related RB49 cochaperonin CocO have been demonstrated to replace the GroES function for all essential *E. coli* protein folding, the GroES-gp31 relationship is not reciprocal; that is, GroES cannot replace gp31 to fold gp23 because of special folding requirements of the latter protein (Andreadis and Black, 1998; Keppel *et al.*, 2002). The N terminus of gp23 that is removed by protease processing appears to strongly target associated fusion proteins to the GroEL chaperonin (Bakkes *et al.*, 2005; Clare *et al.*, 2006; Snyder and Tarkowski, 2005). Especially perplexing, given this dedicated gp23 folding system, is the ability of mutant gp23 to fold without the gp31–GroEL system; four GroEL bypass gp23 amino acid missense mutations additively confer almost 50% folding efficiency and phage production without the GroEL chaperone. Binding of gp23 to the GroEL folding cage shows features distinct from those of most bound *E. coli* proteins. However, it has been shown that gp23 is able to interact with the GroEL–GroES complex, although not productively (Calmat *et al.*, 2009). Unlike substrates such as RUBISCO, gp23 occupies both chambers of the GroEL folding cage, and only gp31 is able to promote efficient capped single "*cis*" chamber folding, apparently by creating a larger folding chamber (Clare *et al.*, 2009). On the basis of the gp24 inferred structure of gp23 and structures of the GroES and gp31 complexed GroEL folding chambers, support for a critical increased chamber size to accommodate gp23 has been advanced as the explanation for the gp31 specificity (Clare *et al.*, 2006). However, because comparable size T-even phage gp31 homologues display preference for folding their own gp23s, more subtle features of the various T-even phage structured folding cages may also determine specificity. Fluorescence experiments support earlier evidence that gp23 adds to the prohead as a hexamer rather than as a monomer; however, the structure of the early hexamer or its precursors is unknown but likely differs substantially from the mature capsid gp24-derived gp23 structure, as gp23 undergoes complex structural transformations from early to mature capsid (Black *et al.*, 1994; Stortelder *et al.*, 2006).

The novel additional phage T4 GroEL-related gene 39.2 has been discovered and shown to restore GroEL function in certain hosts and to certain GroEL mutations. This small protein (58 residues) is speculated to function to strengthen the GroEL interaction with various host cochaperones by favoring the open GroEL state over the closed state (D. Ang and C. Georgopoulos, personal communication).

III. STRUCTURE OF THE PHAGE T4 HEAD

A. Packaged DNA

Packaged phage T4 DNA shares a number of general features with other tailed dsDNA phages: 2.5-nm side-to-side packing of predominantly B-form duplex DNA condensed to ~500 mg/ml. However, other features differ among phages; for example, T4 DNA is packed in an orientation parallel to the prolate and giant head tail axis together with ~1000 molecules of imbedded and mobile internal proteins, unlike the DNA arrangement that traverses the head–tail axis and is arranged around an internal protein core as seen in phage T7 (Cerritelli *et al.*, 1997). Use of the capsid-targeting sequence of internal proteins allows encapsidation of foreign proteins such as **green fluorescent protein** (GFP) and staphylococcal nuclease within the DNA of active virus (Mullaney and Black, 1998; Mullaney *et al.*, 2000). Digestion by the latter nuclease upon addition of calcium yields a pattern of short DNA fragments, predominantly a 160-bp repeat (Mullaney and Black, 1998). This pattern supports a discontinuous pattern of DNA packing such as in the icosahedral-bend, liquid-crystal, or spiral-fold models among a number of proposed models for phage T4 (Fig. 3). Experimental evidence bearing on these models was summarized in Mullaney *et al.* (2000) and in a review of a large-tailed, dsDNA-containing bacteriophage-condensed genome structure that supports markedly different inner capsid structures among such phages, with the condensed genome structure of a single phage type unlikely to be precisely determined or to fit a single general structure (Black and Thomas, 2012).

B. Packaged proteins

In addition to the uncertain arrangement at the nucleotide level of packaged phage DNA, the structure of other internal components is poorly understood in comparison to surface capsid proteins. The internal protein I* (IPI*) of phage T4 is injected to protect the DNA from a two subunit GmrS+GmrD glucose modified restriction endonuclease of a pathogenic *E. coli* that digests glucosylated hydroxymethylcytosine DNA of T-even phages (Bair and Black, 2007; Bair *et al.*, 2007). The 76 residue proteolyzed mature form of the protein has a novel compact protein fold consisting of two β sheets flanked with N- and C-terminal α helices, a structure that is required for its inhibitor activity that is apparently due to binding the GmrS/GmrD proteins (Fig. 4) (Rifat *et al.*, 2008). A single chain GmrS/GmrD homologue enzyme with 90% identity in its sequence to the two subunit enzyme has evolved IPI* inhibitor immunity. It thus appears that the phage T-evens have coevolved with their hosts, a diverse and highly specific set of internal proteins to counter the hmC

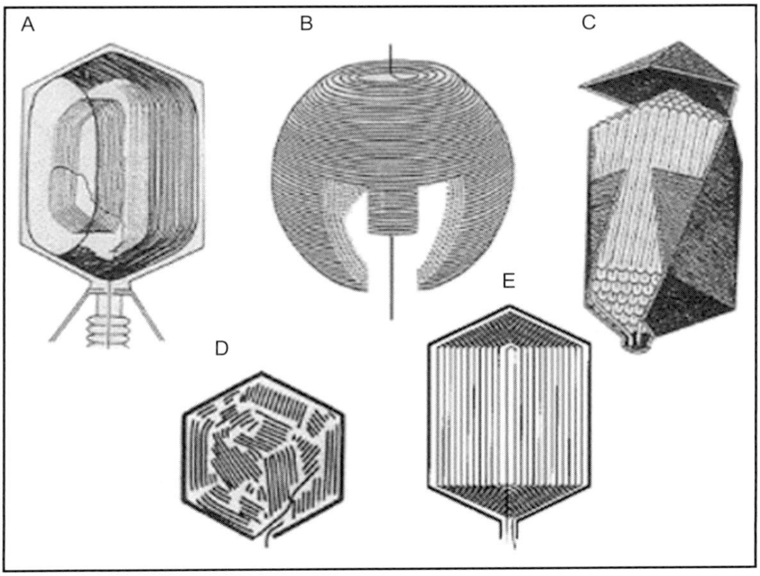

FIGURE 3 Models of packaged T4 DNA structure. (a) T4 DNA is packed longitudinally to the head–tail axis (Earnshaw *et al.*, 1978), unlike the transverse packaging in T7 capsids (Cerritelli *et al.*, 1997) (b). Other models shown include spiral fold (c), liquid crystal (d), and icosahedral bend (e). Both packaged T4 DNA ends are located in the portal (Ray *et al.*, 2010a). For references and evidence bearing on packaged models see (Mullaney and Black, 1998).

modification-dependent restriction endonucleases. Consequently, the internal protein components of the T-even phages are a highly diverse set of defense proteins against diverse attack enzymes with only a conserved capsid targeting sequence to encapsidate the proteins into the precursor scaffolding core (Repoila *et al.*, 1994).

Genes 2 and 4 of phage T4 likely are associated in function, and gp2 was shown previously by Goldberg and co-workers to be able to protect the ends of mature T4 DNA from the *recBCD* exonuclease V, likely by binding to DNA termini (Lipińska *et al.*, 1989). The gp2 protein has not been identified within the phage head because of its low abundance, but evidence for its presence in the head comes from the fact that gp2 can be added to gp2-deficient full heads to confer exonuclease V protection. Thus gp2 affects head–tail joining as well as protecting the DNA ends likely with as few as two copies per particle binding the two DNA ends (Wang *et al.*, 2000). It is unknown whether following packaging a DNA end descends part way into the tail poised for ejection as in some other phages. Neck proteins complete the head and are part of an extended family of phage components (Cardarelli *et al.*, 2010).

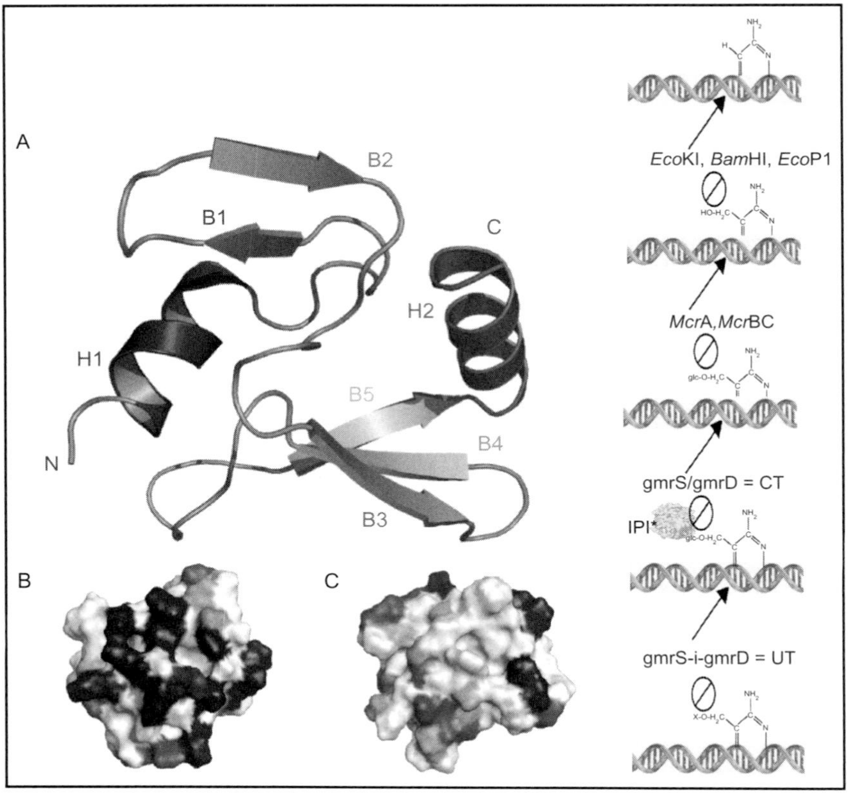

FIGURE 4 Structure and function of T4 internal protein I*. **Nuclear magnetic resonance** structure of IPI*, a highly specific inhibitor of the two-subunit CT (GmrS/GmrD) glucosyl-hmC DNA directed restriction endonuclease (right); DNA modifications blocking such enzymes are shown. The IPI* structure is compact with an asymmetric charge distribution on the faces (blue are basic residues) that may allow rapid DNA-bound ejection through the portal and tail without unfolding–refolding. (See Page 7 in Color Section at the back of the book.)

Solid-state **nuclear magnetic resonance** analysis of phage T4 particle shows that DNA is largely B form and allows its electrostatic interactions to be tabulated (Yu and Schaefer, 2008). This study reveals high-resolution interactions bearing on the internal structure of the phage T4 head. The DNA phosphate negative charge is balanced among lysyl amines, polyamines, and mono and divalent cations. Interestingly, among positively charged amino acids, only lysine residues of the internal proteins were seen to be in close contact with the DNA phosphates, arguing for specific internal protein DNA structures. Electrostatic contributions from internal proteins and interactions of polyamines with DNA entering the prohead to the packaging motor were proposed to account for the higher packaging rates achieved by the phage T4 packaging machine when compared to that of *Bacillus subtiis* phi29 and *E. coli* λ phages.

IV. DISPLAY ON CAPSID USING HOC AND SOC PROTEINS

In addition to the essential capsid proteins, gp23, gp24, and gp20, the T4 capsid is decorated with two nonessential outer capsid proteins: Hoc (highly antigenic outer capsid protein), a dumbbell-shaped monomer at the center of each gp23 hexon, up to 155 copies per capsid (39 kDa; red subunits); and Soc (small outer capsid protein), a rod-shaped molecule that binds between gp23 hexons, up to 870 copies per capsid (10 kDa; white subunits)(Fig. 1). Both Hoc and Soc are dispensable and bind to the capsid after the completion of capsid assembly (Ishii and Yanagida, 1977; Ishii *et al.*, 1978). Null (amber or deletion) mutations in either or both the genes do not affect phage production, viability, or infectivity.

The structure of Soc has been determined (Qin *et al.*, 2009). The 77 amino acid RB69 Soc is a tadpole-shaped molecule with two binding sites for gp23*. Interaction of Soc to the two gp23 molecules glues adjacent hexons. Trimerization of bound Soc molecules results in clamping of three hexons, and 270 such clamps form a cage reinforcing the capsid structure (Fig. 1). Soc assembly thus provides great stability to phage T4 to survive under hostile environments such as extreme pH (pH 11), high temperature (60°C), and a host of denaturing agents. Soc-minus phage lose viability at pH 10.6, and the addition of Soc enhances its survival by $\sim 10^4$-fold.

Hoc does not provide significant additional stability. A monomer of Hoc is present at the center of each hexameric gp23 capsomer. Hoc consists of a string of four domains, three immunoglobulin (Ig)-like domains and one non-Ig domain at the C terminus. The capsid-binding site is localized to the C-terminal 25 amino acids, which are well conserved in T4-related bacteriophages (Sathaliyawala *et al.*, 2010). The crystal structure of the first three Ig domains of RB49 Hoc shows a linear arrangement of the Ig domains (Fokine *et al.*, 2011). The Ig-like fold of each domain resembles ones found frequently in cell attachment molecules of higher organisms. Decorating the virus with long fibers of Hoc molecules may provide survival advantages to the virus. Hoc might be able to attach the phage capsids loosely to bacterial surface molecules, allowing the virus to stay attached to the cell while its tail fibers find their receptors.

The aforementioned properties of Hoc and Soc are uniquely suited to engineer the T4 capsid surface by arraying pathogen antigens. Recombinant vectors have been developed that allow fusion of pathogen antigens to the N or C termini of Hoc and Soc (Jiang *et al.*, 1997; Ren and Black, 1998; Ren *et al.*, 1996, 1997). The fusion proteins were expressed in *E. coli* and, upon infection with *hoc⁻soc⁻* phage, the fusion proteins assembled on the capsid. Phages purified from the infected extracts are decorated with the pathogen antigens. Alternatively, the fused gene can be transferred into the T4 genome by recombinational marker rescue, and infection with the recombinant phage expresses and assembles the fusion protein on the capsid as part of the infection process. Short peptides or protein domains from a variety of pathogens—*Neisseria meningitides* (Jiang *et al.*, 1997), polio virus (Ren *et al.*, 1996), human immunodeficiency virus (HIV) (Sathaliyawala *et al.*, 2006), swine fever virus (Wu *et al.*, 2007), and foot and mouth disease virus (Ren *et al.*, 2008)—have been displayed on the T4 capsid using this approach. The T4 system can be adapted to prepare bipartite libraries of randomized short peptides displayed on T4 capsid Hoc and Soc and use these libraries to "fish out" peptides that interact with the protein of interest (Malys *et al.*, 2002). Biopanning of libraries by the T4 large packaging protein gp17 selected peptides that match with the sequences of proteins thought to interact with gp17. Of particular interest was the selection of a peptide that matched with the T4 late sigma factor, gp55. The gp55-deficient extracts packaged concatemeric DNA about 100-fold less efficiently, suggesting that the gp17 interaction with gp55 helps in loading the packaging terminase onto the viral genome (Black and Peng, 2006; Malys *et al.*, 2002).

An *in vitro* display system has been developed that takes advantage of the high-affinity interactions between Hoc or Soc and the capsid (Fig. 5) (Li *et al.*, 2007; Shivachandra *et al.*, 2006). In this system, the pathogen antigen fused to Hoc or Soc with a hexa-histidine tag was overexpressed in *E. coli* and purified. The purified protein was assembled on the *hoc⁻soc⁻* phage by simply mixing the purified components. This system has certain advantages over the *in vivo* display: (i) a functionally well-characterized and conformationally homogeneous antigen is displayed on the capsid; (ii) the copy number of the displayed antigen can be controlled by altering the ratio of antigen to capsid binding sites; and (iii) multiple antigens can be displayed on the same capsid. This system was used to display full-length antigens from HIV (Sathaliyawala *et al.*, 2006) and anthrax (Li *et al.*, 2007; Shivachandra *et al.*, 2006) that are as large as 90 kDa. All 155 Hoc-binding sites can be filled with anthrax toxin antigens, protective antigen (PA, 83 kDa), lethal factor (LF, 89 kDa), or edema factor (EF, 90 kDa; Shivachandra *et al.*, 2007). Fusion to the N terminus of Hoc did not affect the apparent binding constant (K_d) or the copy number per capsid (B_{max}), but fusion to the C terminus reduced the K_d by 500-fold (Jiang *et al.*, 1997; Shivachandra *et al.*, 2007). All 870 copies of Soc-binding sites can be filled with Soc-fused antigens but the size of the fused antigen must be ~30 kDa or less; otherwise, the copy number is reduced significantly (Li *et al.*, 2007). For example, the 20-kDa PA domain-4 and the 30-kDa LFn domain fused to Soc can be displayed to full capacity. An insoluble Soc-HIV gp120

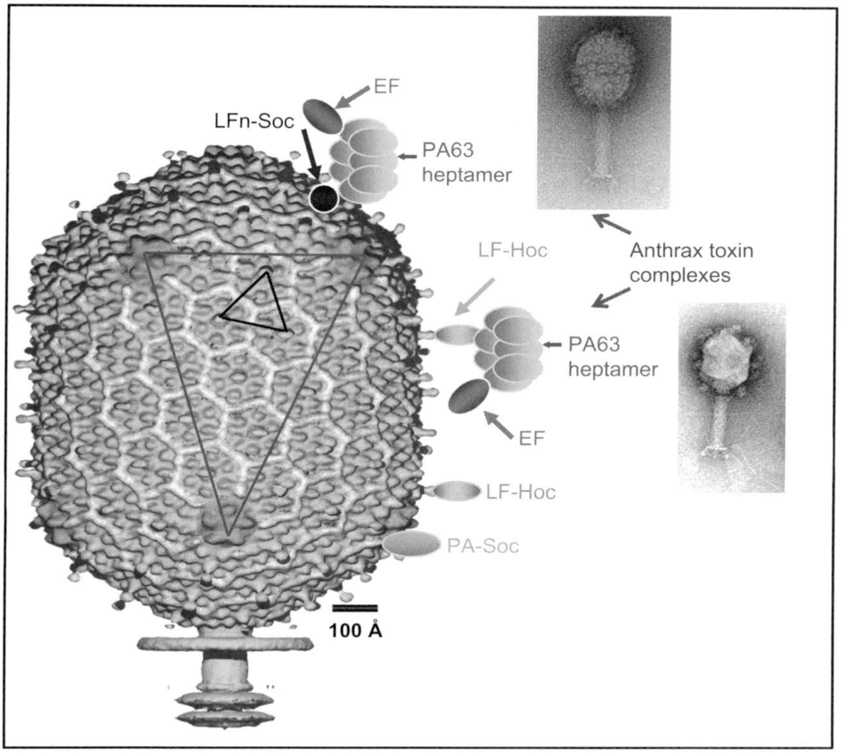

FIGURE 5 *In vitro* display of antigens on bacteriophage T4 capsid. Schematic representation of the T4 capsid decorated with large antigens, PA (83 kDa) and LF (89 kDa), or hetero-oligomeric anthrax toxin complexes through either Hoc or Soc binding (Li *et al.*, 2006, 2007). See text for details. (Insets) Electron micrographs of T4 phage with anthrax toxin complexes displayed through Soc (top) or Hoc (bottom). Note that the copy number of the complexes is lower with the Hoc display than with the Soc display. (See Page 8 in Color Section at the back of the book.)

V3 loop domain fusion protein with a 43 amino acid C-terminal addition could be refolded and bound with ~100% occupancy to mature phage head type polyheads (Ren *et al.*, 1996). Large 90-kDa anthrax toxins can also be displayed but the B_{max} is reduced to about 300 presumably due to steric constraints. Antigens can be fused to the N terminus, the C terminus, or both termini of Soc simultaneously, without significantly affecting the K_d or B_{max}. Thus, as many as 1895 antigen molecules or domains can be attached to each capsid using both Hoc and Soc (Li *et al.*, 2007).

The *in vitro* system offers novel avenues to display macromolecular complexes through specific interactions with the already attached antigens (Li *et al.*, 2006). Sequential assembly was performed by first attaching LF-Hoc and/or LFn-Soc to hoc–soc–phage and exposing the N-domain of LF on the surface. Heptamers of PA were then assembled through interactions between the LFn domain and the N-domain of cleaved PA (domain 1′ of PA63). EF was then attached to the PA63 heptamers, completing assembly of the ~700-kDa anthrax toxin complex on phage T4 capsid (Fig. 5). CryoEM reconstruction shows that native PA63(7)–LFn(3) complexes are assembled in which three adjacent capsid-bound LFn "legs" support the PA63 heptamers (Fokine *et al.*, 2007). Additional layers of proteins can be built on the capsid through interactions with the respective partners.

One of the main applications of T4 antigen particles is their potential use in vaccine delivery. A number of independent studies showed that the T4-displayed particulate antigens without any added adjuvant elicit strong antibody responses and, to a lesser extent, cellular responses. The 43 amino acid residue V3 loop of HIV gp120 fused to Soc displayed on the T4 phage was highly immunogenic in mice and induced anti-gp120 antibodies (Ren *et al.*,1996); Soc-displayed IgG anti-EWL was active (Ren and Black, 1998). The Hoc fused 183 amino acid N-terminal portion of the HIV CD4 receptor protein is displayed in active form. Strong anthrax lethal-toxin neutralization titers were elicited upon immunization of mice and rabbits with phage T4-displayed PA either through Hoc or through Soc (Rao, unpublished data; Shivachandra *et al.*, 2006, 2007). When multiple anthrax antigens were displayed, immune responses against all the displayed antigens were elicited (Shivachandra *et al.*, 2007). T4 particles displaying PA and LF or those displaying the major antigenic determinant cluster mE2 (123 amino acids) and the primary antigen E2 (371 amino acids) of the classical swine fever virus elicited strong antibody titers (Wu *et al.*, 2007). Furthermore, rhesus macaques immunized with T4-displayed PA were completely protected against aerosol challenge with lethal anthrax Ames spores (Rao *et al.*, 2011). Mice immunized with the Soc displayed foot and mouth disease virus (FMDV) capsid precursor polyprotein (P1, 755 amino acids) and proteinase 3C (213 amino acids) were completely protected upon challenge with a lethal dose of FMDV (Ren *et al.*, 2008; Wu *et al.*, 2007). Pigs immunized with a mixture of T4-P1 and T4-3C particles were also protected when these animals were cohoused with FMDV-infected pigs. In another type of application, a T4-displayed mouse Flt4 tumor antigen elicited anti-Flt4 antibodies and broke immune tolerance to self-antigens. These antibodies provided antitumor and antimetastasis immunity in mice (Ren *et al.*, 2009).

Other applications have shown that heavy single chain shark antibodies fused to Hoc and Soc bind tightly to phage T4 heads. These tightly arrayed antibodies display high antigen affinity and are attractive as biosensors and as antitoxins to ricin and other lethal agents (Archer and Liu, 2009; Robertson *et al.*, 2011). Moreover, dye labeled DNA containing heads are taken up into cells efficiently through clathrin-mediated endocytosis and show little cytotoxicity (J. L. Liu, personal communication).

The aforementioned studies provide abundant evidence that the phage T4 nanoparticle platform has the potential to engineer human as well as veterinary vaccines, as well as to act in transfer and as a biosensor platform.

V. PACKAGING PROTEINS

Two nonstructural terminase proteins, gp16 (18 kDa) and gp17 (70 kDa), link head assembly and genome processing (Black, 1989; Rao and Black, 2005; Rao and Feiss, 2008). These proteins are thought to form a hetero-oligomeric complex, which recognizes the concatemeric DNA and makes an endonucleolytic cut (hence the name "terminase"). The terminase–DNA complex docks on the prohead through gp17 interactions with the special portal vertex formed by the dodecameric gp20, thus assembling a DNA-packaging machine. The ATP-fueled machine translocates DNA into the capsid until the head is full, equivalent to about 1.02 times the genome length (171 kb). The terminase dissociates from the packaged head, makes a second cut to terminate DNA packaging, and attaches the concatemeric DNA to another empty head to continue translocation in a processive fashion. Structural and functional analyses of key parts of the machine—gp16, gp17, and gp20—as described here, led to models for the packaging mechanism.

A. Small terminase gp16

gp16, the 18-kDa small terminase subunit, is dispensable for packaging linear DNA *in vitro* but is essential *in vivo*; amber mutations in gene 16 accumulate mostly empty proheads and some partially filled heads (Black *et al.*, 1994; Kondabagil and Rao, 2006).

Mutational and biochemical analyses suggest that gp16 is involved in the recognition of viral DNA (Lin *et al.*, 1997; Lin and Black, 1998) and regulation of gp17 functions (Al-Zahrani *et al.*, 2009). gp16 is predicted to contain three domains: a central domain important for oligomerization and N- and C-terminal domains important for DNA binding, ATP binding, and/or gp17-ATPase stimulation (Al-Zahrani *et al.*, 2009; Mitchell *et al.*, 2002). At least one of the major specificity determinants for gp17 interaction resides in the C-terminal domain. Swapping of this region between T4 and RB49 small terminases leads to switching the ATPase

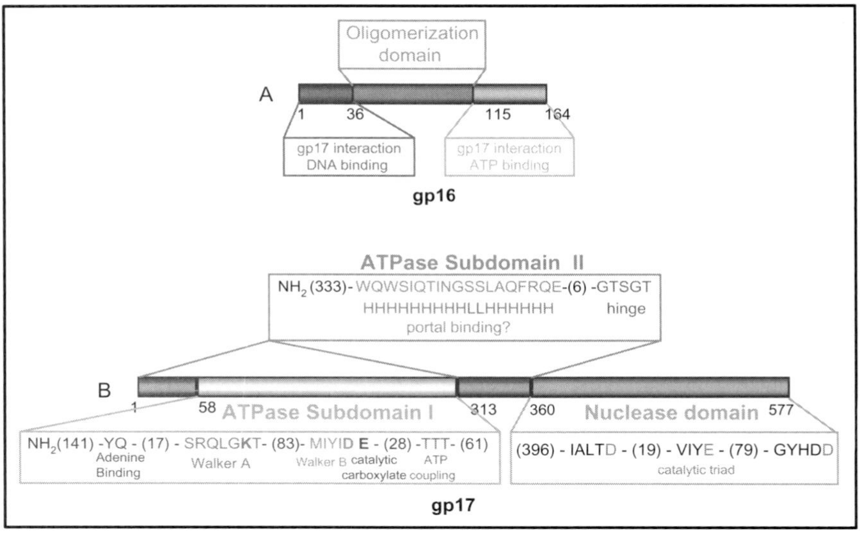

FIGURE 6 Domains and motifs in phage T4 terminase proteins. Schematic representation of domains and motifs in small terminase protein gp16 (A) and large terminase protein gp17 (B). Functionally critical amino acids are shown in bold. Numbers represent the number of amino acids in the respective coding sequence. For further detailed explanations of the functional motifs, refer to Rao and Feiss (2008) and Al-Zahrani *et al.* (2009).

stimulation specificity from T4 gp17 to RB49 gp17 and vice versa (Gao and Rao, 2011; Fig. 6).

Gp16 forms oligomeric single and side-by-side double rings, with each ring having a diameter of ~8 nm with a ~2-nm central channel (Lin *et al.*, 1997). Mass spectrometry determination shows that the single and double rings are 11- and 22-mers, respectively (Duijn, 2010). A number of *pac* site phages produce comparable small terminase subunit multimeric ring structures. Sequence analyses predict two to three coiled coil motifs in gp16 (Kondabagil and Rao, 2006). All T4 family gp16s, as well as other phage small terminases, consist of one or more coiled-coil motifs, consistent with their propensity to form stable oligomers. Oligomerization presumably occurs through parallel coiled-coil interactions between neighboring subunits. Mutations in the long central α-helix of T4 gp16 that perturb coiled-coil interactions lose the ability to oligomerize (Kondabagil and Rao, 2006). Indeed, the X-ray structure shows 11- and 12-mers formed through extensive hydrophobic and electrostatic interactions between the two predicted long helices of the central domain (Sun *et al.*, 2012).

Gp16 oligomerizes, forming a platform for the assembly of the large terminase gp17. This could presumably occur following interaction with the viral DNA concatemer. A predicted helix-turn-helix in the N-terminal domain is thought to be involved in DNA binding (Lin *et al.*, 1997; Mitchell *et al.*, 2002). The corresponding motif in the phage λ small terminase protein, gpNu1, has been well characterized and demonstrated to bind DNA. *In vivo* genetic studies and *in vitro* DNA-binding studies show that a 200-bp 3′ end sequence of gene 16 is a preferred "*pac*" site for gp16 interaction (Lin *et al.*, 1997; Lin and Black, 1998). About 200 bp pieces of DNA are tightly associated with T4 gp16 purified from overexpressed *E. coli* culture (Kondabagil and Rao, unpublished observations). It was proposed that the stable gp16 double rings were two-turn lock washers that constituted the structural basis for synapsis of two *pac* site DNAs. This could promote the gp16-dependent gene amplifications observed around the *pac* site that can be selected in *alt*- mutants that package more DNA; such synapsis could function as a gauge of DNA concatemer maturation (Black, 1995; Wu and Black, 1995; Wu *et al.*, 1995).

gp16 stimulates the gp17-ATPase activity by >50-fold (Leffers and Rao, 2000), but a high ATPase gp17 multimer does not require continued gp16 association (Baumann and Black, 2003). gp16 also stimulates *in vitro* DNA packaging activity in the crude system where phage-infected extracts containing all DNA replication/transcription/recombination proteins are present (Leffers and Rao, 2000; Rao and Black, 1988), but inhibits the packaging of linear DNA in the defined system where only two purified components, proheads and gp17, are present (Black and Peng, 2006; Kondabagil *et al.*, 2006). It stimulates *in vivo* gp17-nuclease activity when T4 transcription factors are also present. It also stimulates gp17-nuclease *in vitro* in the presence of ATP but inhibits the nuclease in the absence of ATP (Al-Zahrani *et al.*, 2009). gp16 also inhibits the binding of gp17 to DNA (Alam and Rao, 2008). Both N- and C-domains are required for ATPase stimulation or nuclease inhibition (Al-Zahrani *et al.*, 2009). Maximum effects were observed at a ratio of approximately one gp16 oligomer to one gp17 (Kanamaru *et al.*, 2004).

gp16 contains an ATP-binding site with broad nucleotide specificity (Lin *et al.*, 1997; Al-Zahrani *et al.*, 2009); however, it lacks the canonical nucleotide-binding signatures such as Walker A and Walker B (Mitchell *et al.*, 2002). No correlation was evident between nucleotide binding and gp17-ATPase stimulation or gp17-nuclease inhibition. Thus it is unclear what role ATP binding plays in gp16 function. Evidence thus far suggests that gp16 is a regulator of the DNA packaging machine, modulating the nuclease, ATPase, and translocase activities of gp17 for efficient packaging initiation. In one model, a packaging initiation complex consisting of gp16, gp17, and DNA forms at a putative *pac* site, makes a cut, and delivers the end to the portal. gp16 then stimulates gp17 ATPase and DNA translocation activities to initiate DNA packaging.

B. Large terminase gp17

gp17 is the 70-kDa large subunit of the terminase holoenzyme and the motor protein of the DNA packaging machine. gp17 consists of two functional domains (Fig. 6): an N-terminal ATPase domain having classic ATPase signatures, such as Walker A, Walker B, and catalytic carboxylate, and a C-terminal nuclease domain having a catalytic metal cluster with conserved aspartic and glutamic acid residues coordinating with Mg^{2+} (Kanamaru *et al.*, 2004).

1. ATPase

gp17 alone is sufficient to package linear DNA *in vitro*. gp17 exhibits a weak ATPase activity ($K_{cat} = \sim$1–2 ATPs hydrolyzed per gp17 molecule/min), which is stimulated by >50-fold by the small terminase protein gp16 (Baumann and Black, 2003; Leffers and Rao, 2000). Any mutation in the predicted catalytic residues of the N-terminal ATPase center results in a loss of stimulated ATPase and DNA packaging activities (Rao and Mitchell, 2001). Even subtle conservative substitutions such as aspartic acid to glutamic acid and vice versa in the Walker B motif resulted in complete loss of DNA packaging, suggesting that this ATPase provides energy for DNA translocation (Goetzinger and Rao, 2003; Mitchell and Rao, 2006).

The ATPase domain also exhibits DNA-binding activity, which may be involved in the DNA cutting and translocation functions of the packaging motor. Genetic evidence shows that gp17 may interact with gp32 (Franklin *et al.*, 1998; Mosig, 1998), but, without other phage components, highly purified preparations of gp17 do not show appreciable affinity for single-stranded DNA or dsDNA. In fact, nontag and full-length purified gp17 has little or no nuclease activity, although is able to cut and package circular plasmid DNAs together with other phage proteins (Baumann and Black, 2003; Black and Peng, 2006). Thus there seem to be complex interactions among the terminase proteins, the concatemeric DNA, and the DNA replication/recombination/repair and transcription proteins that transit DNA metabolism into the packaging phase (Black and Peng, 2006).

One of the ATPase mutants, the DE-ED mutant in which the sequence of Walker B and catalytic carboxylate was reversed, showed tighter binding to ATP than wild-type gp17 but failed to hydrolyze ATP (Mitchell and Rao, 2006). Unlike wild-type gp17 or the ATPase domain that failed to crystallize, the ATPase domain with the ED mutation crystallized readily, probably because it trapped the ATPase in an ATP-bound conformation. The X-ray structure of the ATPase domain was determined up to 1.8 Å resolution in different bound states: apo, ATP bound, and ADP bound (Sun *et al.*, 2007). It is a flat structure consisting of two subdomains: a large subdomain I (NsubI) and a smaller subdomain II (NsubII) forming a cleft in which ATP binds (Fig. 7A). The NsubI consists of the classic nucleotide-

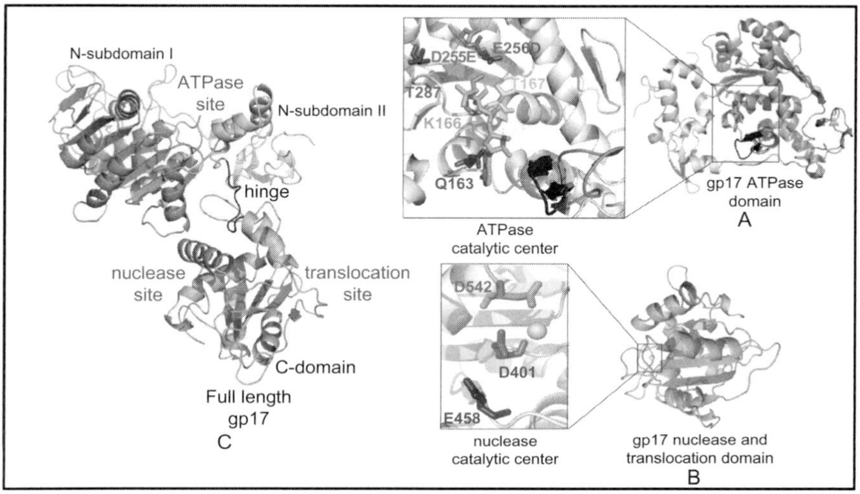

FIGURE 7 Structures of the T4 packaging motor protein gp17. Structures of the ATPase domain (A), nuclease/translocation domain (B), and full-length gp17 (C). Various functional sites and critical catalytic residues are labeled. For further details, see Sun *et al.* (2007, 2008). (See Page 8 in Color Section at the back of the book.)

binding fold (Rossmann fold), a parallel β sheet of six β strands interspersed with helices. The structure showed that the predicted catalytic residues are oriented into the ATP pocket, forming a network of interactions with bound ATP. These also include an arginine finger proposed to trigger βγ-phosphoanhydride bond cleavage. In addition, the structure showed movement of a loop near the adenine-binding motif in response to ATP hydrolysis, which may be important for the transduction of ATP energy into mechanical motion.

2. Nuclease

gp17 exhibits a sequence-nonspecific endonuclease activity *in vitro*, and *in vivo* upon overexpression in *E. coli*, apparently producing blunt ends (Bhattacharyya and Rao, 1993, 1994). Biochemical and structural studies suggest that this activity makes packaging initiation and headful termination cuts. In the infected cell, interaction of gp17 with gp16 and other phage components likely controls the frequency and specificity of this nuclease (see later; Alam *et al.*, 2008; Ghosh-Kumar *et al.*, 2011). In the T4-like phage IME08, sequence analysis of the mature DNA ends indicates that its terminase produces ends with a two base overhang at a preferred consensus sequence, TTGGAG (Jiang *et al.*, 2011).

Random mutagenesis of gene 17 and selection of mutants that lost nuclease activity identified a histidine-rich site in the C-terminal domain critical for DNA cleavage (Kuebler and Rao, 1998). Extensive site-directed mutagenesis of this region, combined with sequence alignments, identified a cluster of conserved aspartic acid and glutamic acid residues essential for DNA cleavage (Rentas and Rao, 2003). Unlike ATPase mutants, these mutants retained gp16-stimulated ATPase activity as well as DNA packaging activity as long as the substrate is a linear molecule. However, these mutants fail to package circular DNA as they are defective in cutting DNA that is required for packaging initiation.

The structure of the C-terminal nuclease domain from a T4 family phage, RB49, which has 72% sequence identity to the T4 C-domain, was determined to 1.16 Å resolution (Sun *et al.*, 2008)(Fig. 7B). It has a globular structure consisting mostly of antiparallel β strands forming an RNase H fold that is found in resolvases, RNase Hs, and integrases. As predicted from mutagenesis studies, structures showed that residues D401, E458, and D542 form a catalytic triad coordinating with the Mg^{2+} ion. In addition, the structure showed the presence of a DNA-binding groove lined with a number of basic residues. The acidic catalytic metal center is buried at one end of this groove. Together, these form the nuclease cleavage site of gp17.

The crystal structure of the full-length T4 gp17 (ED mutant) was determined to 2.8 Å resolution (Fig. 6C; Sun *et al.*, 2008). The N- and C-domain structures of the full-length gp17 superimpose with those solved using individually crystallized domains with only minor deviations. The full-length structure, however, has additional features relevant to the mechanism. A flexible "hinge" or "linker" connects the ATPase and nuclease domains. Previous biochemical studies showed that splitting gp17 into two domains at the linker retained the respective ATPase and nuclease functions but DNA translocation activity was completely lost (Kanamaru *et al.*, 2004). Second, the N- and C-domains have a >1000 square Å complementary surface area consisting of an array of five charged pairs and hydrophobic patches (Sun *et al.*, 2008). Third, the gp17 has a bound phosphate ion in the crystal structure. Docking of B-form DNA guided by shape and charge complementarity with one of the DNA phosphates superimposed on the bound phosphate aligns a number of basic residues, lining what appears to be a shallow translocation groove. Thus the C-domain appears to have two DNA grooves on different faces of the structure—one that aligns with the nuclease catalytic site and another that aligns with the translocating DNA (Fig. 7). Mutation of one of the groove residues (R406) showed a novel phenotype; loss of DNA translocation activity occurs but ATPase and nuclease activities are retained.

It is crucial that the nuclease activity of gp17 be controlled *in vivo* such that it is active at the initiation and termination steps of DNA packaging but inactive during translocation. Although it is clear that this must involve interactions with gp16, gp20 (portal), and other components, a basic mechanism by which the catalytic activity of the gp17's nuclease center can be controlled was hypothesized from structural and biochemical studies. Comparison of the X-ray structures of gp17 and cryoEM reconstruction of prohead-docked gp17 suggested that gp17 exists in two conformational states: tensed and relaxed (see later). Analysis of these states showed that the residues that line the nuclease groove come closer in the relaxed state, possibly "compressing" the DNA groove by ~4 Å. One of the mechanisms by which the headful nuclease is regulated might be by relaying conformational signals between the ATPase center to the nuclease center through a "communication track" consisting of residues from subdomain II, hinge, and β-hairpin (Ghosh-Kumar *et al.*, 2011). During active translocation, subunits would be in the nuclease-inactive relaxed state and unable to form the antiparallel dimer that is essential to make cuts in both strands. However, during the initiation (and termination) step, the free (or freed) gp17 subunits may form a holo-terminase complex with gp16 and other phage proteins to make a cut in the concatemeric viral genome. It is likely that communication with the portal is also essential for making the headful termination.

VI. PACKAGING MOTOR

A. Structure

A functional DNA packaging machine could be assembled by mixing proheads and purified gp17. The latter assembles into a packaging motor through specific interactions with the portal vertex (Lin *et al.*, 1999); such complexes, in a bulk *in vitro* assay, can package the 171-kb phage T4 DNA or any linear DNA (Black and Peng, 2006; Kondabagil *et al.*, 2006). If less than headful length DNA molecules are added as the DNA substrate, the motor shows discontinuous packaging, packaging multiple molecules one molecule after another (Leffers and Rao, 1996). This can lead to head filling when large plasmid DNA molecules are used (~30 kb)(Leffers and Rao, 1996) but with shorter DNAs, mostly partially filled heads with about six packaged DNA molecules are produced (Sabanayagam *et al.*, 2007; Zhang *et al.*, 2011). Although the unexpanded prohead is likely the true precursor for DNA packaging *in vivo*, in the *in vitro* assay, the expanded prohead, the partially full head, or even the full head can assemble the gp17 motor and drive efficient DNA translocation. In fact, packaged DNA of the virion can be emptied and refilled with DNA again (Zhang *et al.*, 2011). Packaging can also be studied in real time either by fluorescence correlation spectroscopy (Sabanayagam *et al.*, 2007) or at the single molecule level by optical tweezers (Fuller *et al.*, 2007). The translocation kinetics of rhodamine (R6G-labeled, 100-bp DNA) was measured by determining the decrease in the diffusion coefficient as the DNA gets confined inside the capsid. Fluorescence resonance energy transfer between green fluorescent protein-labeled proteins within the prohead interior and the translocated rhodamine-labeled DNA confirmed the ATP-powered movement of DNA into the capsid and the packaging of multiple segments per procapsid (Sabanayagam *et al.*, 2007). Analysis of Förster resonance energy transfer (FRET) dye pair end-labeled DNA substrates showed that upon packaging, the two ends of the packaged DNA were held 8–9 nm apart in the procapsid, likely fixed in the portal channel and crown, suggesting that a loop rather than an end of DNA is translocated following initiation at an end (Ray *et al.*, 2010a).

In the optical tweezers system, prohead–gp17 complexes are tethered to a microsphere coated with capsid protein antibody, and the biotinylated DNA is tethered to another microsphere coated with streptavidine. Microspheres are brought together into near contact, allowing the motor to capture the DNA. Single packaging events are monitored, and the dynamics of the T4 packaging process are quantified (Fuller *et al.*, 2007). The T4 motor, like the phi29 DNA packaging motor, generates forces as high as ~60 pN, which is ~20–25 times that of myosin ATPase, and a rate as high as ~2000 bp/sec, among the highest recorded to date. Slips and pauses occur but these are relatively short and rare and the motor recovers and recaptures DNA, continuing translocation. The high rate of translocation is in keeping with the need to package the 171-kb size T4 genome in about 5 minutes. The T4 motor generates enormous power; when an external load of 40 pN is applied, the T4 motor translocates at a speed of ~380 bp/sec. When scaled up to a macromotor, the T4 motor is approximately twice as powerful as a typical automobile engine.

CryoEM reconstruction of the packaging machine showed two rings of density at the portal vertex (Sun *et al.*, 2008) (Fig. 8). The upper ring is flat, resembling the ATPase domain structure, and the lower ring is spherical, resembling the C-domain structure. This was confirmed by docking of the X-ray structures of the domains into the cryoEM density. The motor has pentamer stoichiometry, with the ATP-binding surface facing the portal and interacting with it. It has an open central channel that is in line with the portal channel, and the translocation groove of the C-domain faces the channel. Minimal contacts exist between the adjacent subunits, suggesting that ATPases may fire relatively independently during translocation.

Unlike the cryoEM structure where the two lobes (domains) of the motor are separated ("relaxed" state), domains in the full-length gp17 are in close contact ("tensed" state) (Sun *et al.*, 2008). In the tensed state, the subdomain

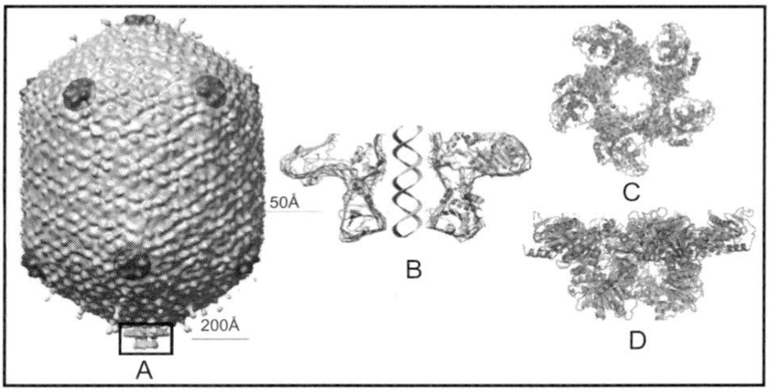

FIGURE 8 Structure of the T4 DNA packaging machine. (A) CryoEM reconstruction of the phage T4 DNA packaging machine showing the pentameric motor assembled at the special portal vertex. (B–D) Cross section, top, and side views of the pentameric motor, respectively, by fitting X-ray structures of the gp17 ATPase and nuclease/translocation domains into cryoEM density. (See Page 9 in Color Section at the back of the book.)

II of ATPase is rotated by 6° and the C-domain is pulled upward by 7 Å, equivalent to 2 bp. The "arginine finger" located between subI and NsubII is positioned toward the βγ-phosphates of ATP and the ion pairs are aligned.

B. Mechanism and dynamics

Of the many models proposed to explain the mechanism of viral DNA translocation, the portal rotation model (Hendrix, 1978) attracted the most attention and was often cited in other contexts in confirmation of the functional biological significance of symmetry mismatch. According to the original rotation model, the portal and DNA are locked like a nut and bolt. The symmetry mismatch between the 5-fold capsid and the 12-fold portal means that there are no reiterated specific portal–capsid subunit interactions, thereby enabling free rotation of the two multimeric interfaces; as proposed, this would allow the portal, or nut, to rotate, powered by ATP hydrolysis, causing the DNA, or bolt, to be translocated into the capsid. X-ray structures of conserved dodecameric Phi29 and SPP1 portals could be interpreted as consistent with the original portal rotation or newer, rotation-incorporating models such as the rotation–compression–relaxation (Simpson *et al.*, 2000), electrostatic gripping (Guasch *et al.*, 2002), and molecular lever (Lebedev *et al.*, 2007) models.

Protein fusions to either the N- or the C-terminal end of the T4 portal protein could be incorporated into up to approximately one-half of the dodecamer positions without loss of prohead function. As compared to the wild type, portals containing C-terminal GFP fusions but not N-terminal GFP fusions (Dixit *et al.*, 2011) lock the proheads into the unexpanded conformation unless terminase packages DNA, suggesting that the portal plays a central role in controlling prohead expansion. Expansion is required to protect the packaged DNA from nuclease but not for packaging itself as measured by fluorescence correlation spectroscopy (Ray *et al.*, 2009). Retention of the DNA packaging function of such portals is inconsistent with the portal rotation model, as rotation would require that bulky C-terminal GFP fusion proteins within the capsid rotate through the densely packaged DNA.

A more direct test tethered the portal to the capsid through Hoc interactions (Baumann *et al.*, 2006). Hoc-binding sites are not present in unexpanded proheads but are exposed following capsid expansion. Unexpanded proheads were first prepared with some of the 12 portal subunits replaced by the N-terminal Hoc–portal fusion proteins. The proheads were then expanded *in vitro* to expose Hoc-binding sites. The Hoc portion of the portal fusion would bind to the center of the nearest hexon, tethering one to five portal subunits to the capsid, thereby protecting Hoc from proteolysis. By this test and also by incorporation of Hoc–gp20 into active phage particles the Hoc was tethered but did not affect DNA packaging. Thus both N- and the C-terminal portal fusion protein portal results strongly suggested that portal rotation does not occur (Baumann *et al.*, 2006). This was supported more recently by single molecule fluorescence spectroscopy of actively packaging phi29 packaging complexes that apparently (to 99% certainty) failed to show rotation (Hugel *et al.*, 2007).

In a second class of models, the terminase not only provides the energy but also translocates DNA actively (Draper and Rao, 2007). Conformational changes in terminase domains cause changes in DNA-binding affinity, resulting in binding and releasing DNA, reminiscent of the inchworm-type translocation by helicases. gp17 and numerous large terminases possess an ATPase coupling motif that is commonly present in helicases and translocases (Draper and Rao, 2007). Mutations in the coupling motif present at the junction of NSubI and NSubII result in a loss of ATPase and DNA packaging activities.

The cryoEM and X-ray structures (Fig. 8), combined with the mutational analyses described earlier, led to the postulation of a terminase-driven packaging mechanism (Sun *et al.*, 2008). The pentameric T4 packaging motor can be considered to be analogous to a five-cylinder engine. It consists of an ATPase center in NsubI, which is the engine that provides energy. The C-domain has a translocation groove, which is the wheel that moves DNA. The smaller NsubII is the transmission domain, coupling the engine to the wheel via a flexible hinge. The arginine finger is a spark plug that fires ATPase when the motor is locked in the firing mode. Charged pairs generate electrostatic force by alternating between relaxed and tensed states (Fig. 9). The nuclease groove faces away from translocating DNA and is activated when packaging is completed.

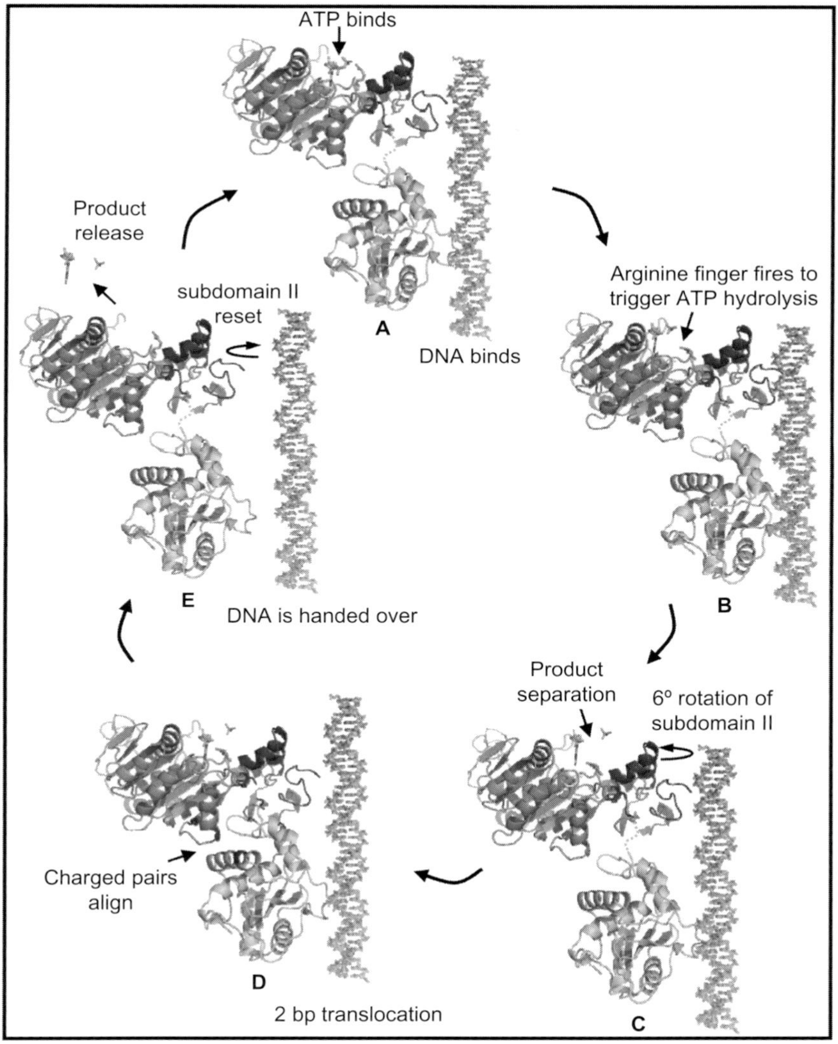

FIGURE 9 Model for the electrostatic force-driven DNA packaging mechanism. Schematic representation showing the sequence of events that occur in a single gp17 molecule to translocate 2 bp of DNA [for details, see text and Sun *et al.* (2008)]. (See Page 9 in Color Section at the back of the book.)

In the relaxed conformational state (cryoEM structure), the hinge is extended (Fig. 9). Binding of DNA to the translocation groove and of ATP to NsubI locks the motor in translocation mode (Fig. 9A) and brings the arginine finger into position, firing ATP hydrolysis (Fig. 9B). Repulsion between the negatively charged ADP(3-) and Pi(3-) drives them apart, causing NsubII to rotate by 6° (Fig. 9C), aligning the charge pairs between the N- and C-domains. This generates electrostatic force, attracting the C-domain–DNA complex and causing 7 Å upward movement, the tensed conformational state (X-ray structure)(Fig. 9D). Thus 2 bp of DNA is translocated into the capsid in one cycle. Product release and the loss of six negative charges causes NsubII to rotate back to the original position, misaligning the ion pairs and returning the C-domain to the relaxed state (Fig. 9E).

Translocation of 2 bp would bring the translocation groove of the adjacent subunit into alignment with the backbone phosphates. DNA is then handed over to the next subunit by the matching motor and DNA symmetries. Thus, ATPase catalysis causes conformational changes that generate electrostatic force, which is then converted to mechanical force. The pentameric motor translocates 10 bp (one turn of the helix) when all five gp17 subunits fire in succession, bringing the first gp17 subunit once again in alignment with the DNA phosphates. Synchronized orchestration movements of the motor translocate DNA up to ~2000 bp/sec.

DNA may not be translocated by a simple linear motion. *In vitro* packaging experiments with modified DNA substrates support a DNA torsional compression linear translocation mechanism in which the portal grips the DNA while a power stroke is applied by gp17 conformational changes (Oram *et al.*, 2008). This would transiently compress the DNA present in the translocation channel between the portal and the ATPase motor. Short (<200 bp) DNA substrate translocation by gp17 is blocked by nicks or other departures from the B form, suggesting that energy stored as DNA compression may be dissipated by a nick (Figs. 10A–10C). However, longer DNAs containing nicks or other abnormal structures are translocated apparently normally by both T4 and phi29 motors (Aathavan *et al.*, 2009; Oram *et al.*, 2008), presumably by multiple motor cycles. Use of a DNA leader of several kilobases joined to a 90-bp Y-DNA structure showed packaging of the leader segment while the dye containing Y-DNA was nuclease accessible; the Y-junction was arrested in proximity to a prohead portal containing GFP fusions, as evident by FRET transfer between the dye labeled Y-junction and the GFP-labeled portal (Ray *et al.*, 2010b)(Fig. 10D). Comparable stalled Y-DNA substrates containing FRET pair dyes in the Y-stem also suggested that the motor compresses or "crunches" the stem DNA held in the portal channel, as dyes separated by 10 or 14 bp undergo a comparable distance decrease measured by FRET at 22 and 24%, whether the dyes are on the same side or opposite side of the helix, supporting DNA compression rather than bending (Fig. 10E). Transient compression of 10–14 bp DNA in the portal translocation channel is thus proposed to be followed by release by the portal of B form DNA in 10 bp or longer steps. Other evidence in phages T4 and SPP1 also supports an active role for the portal as well as

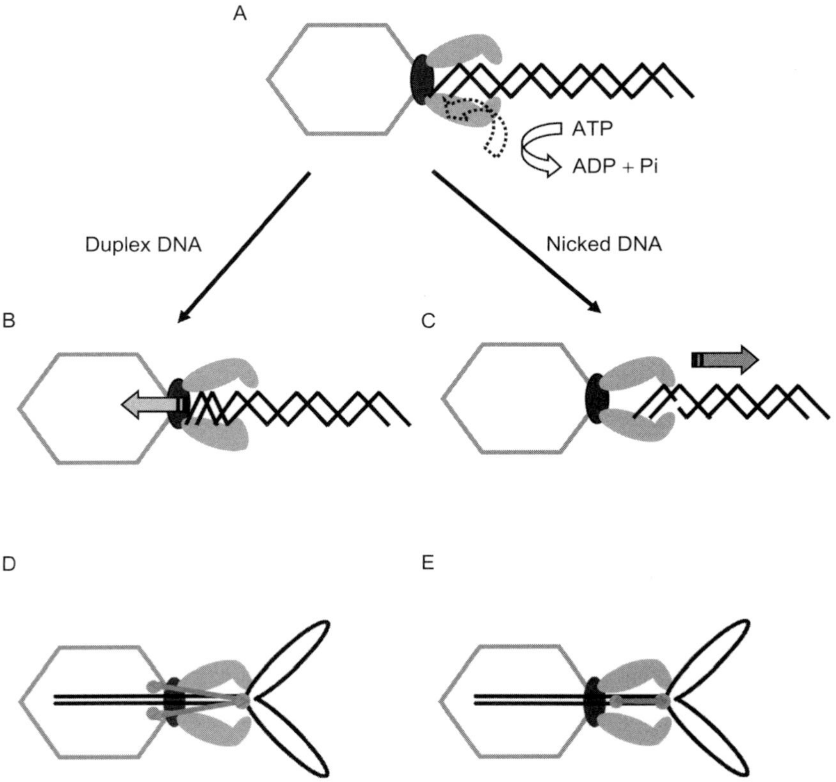

FIGURE 10 Model for the torsional compression portal-DNA-grip-and-release packaging mechanism (Oram *et al.*, 2008). An ATP-driven conformational change of the terminase to portal interaction is proposed to drive DNA into the prohead by a DNA compression motor stroke. (A–C) Short linear DNAs are packaged by the motor, whereas nicked or other abnormal structures containing DNA substrates are released; (D) kb DNA leader containing Y-DNA substrates is retained by the motor and the Y-junction (red dot) is anchored in the procapsid in proximity to portal GFP fusions; and (E) DNA compression of the Y-stem B two FRET pair dye-containing segment in the stalled complex is observed; fluorophore distances measured by FRET are given by narrow arrows (Ray *et al.*, 2010b). (See Page 10 in Color Section at the back of the book.)

terminase in DNA translocation (Cuervo *et al.*, 2007; Lin *et al.*, 1999; Dixit, Ray, and Black, in preparation).

A portal-bound T4 gp49 Holliday junction resolvase releases packaging-arrested Y- or X-branched structure-containing concatemers, its essential *in vivo* packaging function (Golz and Kemper, 1999). Consistent with this function and the torsional model is the release of DNA compression in arrested Y-DNA substrates by the addition of purified gp49

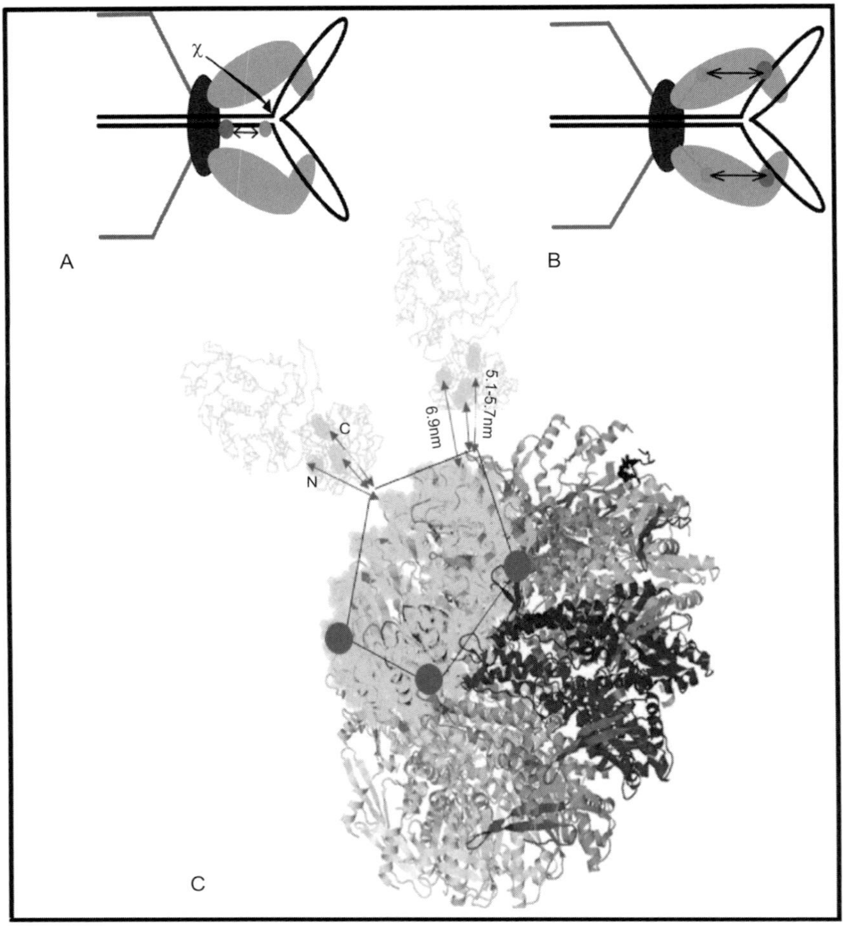

FIGURE 11 Both motor proteins and DNA undergo major conformational changes during DNA packaging. Schematic of FRET between different dyes within the DNA and site of action of T4 gp49 (χ). Portal-bound Holliday junction resolvase (gp49) action releases compression from the Y-DNA substrates (A). Resolvase release of trapped Y-DNAs is also correlated with increased distance (~0.6 nm, or 5.1 to 5.7 nm) between the terminase and the portal during packaging (B). Distances are measured from N and C termini of gp17 to the GFP–gp20 portal by FRET (C)(Dixit *et al.*, 2011). (See Page 11 in Color Section at the back of the book.)

enzyme *in vitro* (Fig. 11A) (Dixit *et al.*, 2011). Active terminases labeled at the N- and C-terminal ends with a single dye molecule showed that the FRET distance between the N-terminal GFP-labeled portal protein and gp17 is 6.9 nm for the N terminus and 5.7 nm for the C terminus (Figs. 11B and 11C). Packaging with a C-terminal fluorescent terminase on a

GFP-portal prohead shows a reduction in distance of 0.6 nm in the arrested Y-DNA as compared to linear DNA, and again the reduction is reversed by resolvase treatment, which is consistent with the tensed and relaxed terminase conformational changes (Fig. 9).

VII. CONCLUSIONS AND PROSPECTS

Although the aforementioned studies strongly suggest that phage T4 employs a linear motor mechanism to package DNA, some aspects of the mechanism are still speculative. More extensive mutagenesis and single molecule studies need to be performed to determine DNA step size, motor coordination, and specific roles of the structural parts of the terminase and portal proteins in translocation and DNA cutting. Other mechanistic questions that require deeper investigation include whether (i) the DNA structure is altered during normal linear translocation; (ii) to what extent and how the portal contributes to translocation; (iii) whether the DNA rotates during translocation to compensate for winding inside the capsid; and (iv) to what extent does a single conserved mechanism apply among tailed dsDNA phages and comparable portal-containing eukaryotic viruses; for example, herpes viruses.

Despite these remaining questions, it is clear from the discussion given previously that major advances have been made in understanding the phage T4 capsid structure and mechanism of DNA packaging. These advances, by combining genetics and biochemistry with structure and biophysics, set the stage to probe the packaging mechanism with even greater depth and precision. It is reasonable to hope that this would lead to the elucidation of catalytic cycle, mechanistic details, and motor dynamics to near atomic resolution. The accumulated and emerging basic knowledge should also lead to medical applications such as the development of vaccines and phage therapy.

ACKNOWLEDGMENTS

The authors thank Aparna Dixit, Bonnie Draper, Vishal Kottadiel, and Alice Kuaban for preparing the figures, references, proofreading, and helpful comments. Research in the authors' laboratories has been funded by the National Science Foundation (VBR: MCB-0923873) and National Institutes of Health (VBR: NIAID-AI081726; LWB: NIAID-AI011676). Special thanks to our present and former lab members for their contributions over the years.

REFERENCES

Aathavan, K., Politzer, A. T., Kaplan, A., Moffitt, J. R., Chemla, Y. R., Grimes, S., Jardine, P. J., Anderson, D. L., and Bustamante, C. (2009). Substrate interactions and promiscuity in a viral DNA packaging motor. *Nature* **461:**669–673.

Al-Zahrani, A. S., Kondabagil, K., Gao, S., Kelly, N., Ghosh-Kumar, M., and Rao, V. B. (2009). The small terminase, gp16, of bacteriophage T4 is a regulator of the DNA packaging motor. *J. Biol. Chem.* **284:**24490–24500.

Alam, T. I., Draper, B., Kondabagil, K., Rentas, F. J., Ghosh-Kumar, M., Sun, S., Rossmann, M. G., and Rao, V. B. (2008). The headful packaging nuclease of bacteriophage T4. *Mol. Microbiol.* **69:**1180–1190.

Alam, T. I., and Rao, V. B. (2008). The ATPase domain of the large terminase protein, gp17, from bacteriophage T4 binds DNA: Implications to the DNA packaging mechanism. *J. Mol. Biol.* **376:**1272–1281.

Andreadis, J. D., and Black, L. W. (1998). Substrate mutations that bypass a specific Cpn10 chaperonin requirement for protein folding. *J. Biol. Chem.* **273:**34075–34086.

Archer, M. J., and Liu, J. L. (2009). Bacteriophage T4 nanoparticles as materials in sensor applications: Variables that influence their organization and assembly on surfaces. *Sensors* **9:**6298–6311.

Bair, C. L., and Black, L. W. (2007). A type IV modification dependent restriction nuclease that targets glucosylated hydroxymethyl cytosine modified DNAs. *J. Mol. Biol.* **366:**768–778.

Bair, C. L., Rifat, D., and Black, L. W. (2007). Exclusion of glucosyl-hydroxymethylcytosine DNA containing bacteriophages is overcome by the injected protein inhibitor IPI*. *J. Mol. Biol.* **366:**779–789.

Bakkes, P. J., Faber, B. W., van Heerikhuizen, H., and van der Vies, S. M. (2005). The T4-encoded cochaperonin, gp31, has unique properties that explain its requirement for the folding of the T4 major capsid protein. *Proc. Natl. Acad. Sci. USA* **102:**8144–8149.

Baumann, R. G., and Black, L. W. (2003). Isolation and characterization of T4 bacteriophage gp17 terminase, a large subunit multimer with enhanced ATPase activity. *J. Biol. Chem.* **278:**4618–4627.

Baumann, R. G., Mullaney, J., and Black, L. W. (2006). Portal fusion protein constraints on function in DNA packaging of bacteriophage T4. *Mol. Microbiol.* **61:**16–32.

Bhattacharyya, S. P., and Rao, V. B. (1993). A novel terminase activity associated with the DNA packaging protein gp17 of bacteriophage T4. *Virology* **196:**34–44.

Bhattacharyya, S. P., and Rao, V. B. (1994). Structural analysis of DNA cleaved in vivo by bacteriophage T4 terminase. *Gene* **146:**67–72.

Black, L., and Thomas, J. A. (2012). Condensed Genome Structure in Viral Molecular Machines. (Rossmann and Rao, eds.).

Black, L. W. (1989). DNA packaging in dsDNA bacteriophages. *Annu. Rev. Microbiol.* **43:**267–292.

Black, L. W. (1995). DNA packaging and cutting by phage terminases: Control in phage T4 by a synaptic mechanism. *Bioessays* **17:**1025–1030.

Black, L. W., and Peng, G. (2006). Mechanistic coupling of bacteriophage T4 DNA packaging to components of the replication-dependent late transcription machinery. *J. Biol. Chem.* **281:**25635–25643.

Black, L. W., Showe, M. K., and Steven, A. C. (1994). Morphogenesis of the T4 head. *In* "Molecular Biology of Bacteriophage T4" (J. M. Karam, ed.), pp. 218–258. ASM Press, Washington, DC.

Calmat, S., Hendriks, J., van Heerikhuizen, H., Schmidt, C. F., van der Vies, S. M., and Peterman, E. J. (2009). Dissociation kinetics of the GroEL-gp31 chaperonin complex studied with Forster resonance energy transfer. *Biochemistry* **48:**11692–11698.

Cardarelli, L., Lam, R., Tuite, A., Baker, L. A., Sadowski, P. D., Radford, D. R., Rubinstein, J. L., Battaile, K. P., Chirgadze, N., Maxwell, K. L., and Davidson, A. R. (2010). The crystal structure of bacteriophage HK97 gp6: Defining a large family of head-tail connector proteins. *J. Mol. Biol.* **395:**754–768.

Cerritelli, M. E., Cheng, N., Rosenberg, A. H., McPherson, C. E., Booy, F. P., and Steven, A. C. (1997). Encapsidated conformation of bacteriophage T7 DNA. *Cell* **91:**271–280.

Cheng, H., Shen, N., Pei, J., and Grishin, N. V. (2004). Double-stranded DNA bacteriophage prohead protease is homologous to herpesvirus protease. *Protein Sci.* **13:**2260–2269.

Clare, D. K., Bakkes, P. J., van Heerikhuizen, H., van der Vies, S. M., and Saibil, H. R. (2006). An expanded protein folding cage in the GroEL-gp31 complex. *J. Mol. Biol.* **358:**905–911.

Clare, D. K., Bakkes, P. J., van Heerikhuizen, H., van der Vies, S. M., and Saibil, H. R. (2009). Chaperonin complex with a newly folded protein encapsulated in the folding chamber. *Nature* **457:**107–110.

Cuervo, A., Vaney, M. C., Antson, A. A., Tavares, P., and Oliveira, L. (2007). Structural rearrangements between portal protein subunits are essential for viral DNA translocation. *J. Biol. Chem.* **282:**18907–18913.

Desplats, C., and Krisch, H. M. (2003). The diversity and evolution of the T4-type bacteriophages. *Res. Microbiol.* **154:**259–267.

Dixit, A., Ray, K., Lakowicz, J. R., and Black, L. W. (2011). Dynamics of the T4 bacteriophage DNA packasome motor: Endo VII resolvase release of arrested Y-DNA substrates. *J. Biol. Chem.* **286:**18878–18889.

Draper, B., and Rao, V. B. (2007). An ATP hydrolysis sensor in the DNA packaging motor from bacteriophage T4 suggests an inchworm-type translocation mechanism. *J. Mol. Biol.* **369:**79–94.

Duijn, E. (2010). Current limitations in native mass spectrometry based structural biology. *J. Am. Soc. Mass Spect.* **21:**971–978.

Earnshaw, W. C., King, J., Harrison, S. C., and Eiserling, F. A. (1978). The structural organization of DNA packaged within the heads of T4 wild-type, isometric and giant bacteriophages. *Cell* **14:**559–568.

Fokine, A., Battisti, A. J., Kostyuchenko, V. A., Black, L. W., and Rossmann, M. G. (2006). Cryo-EM structure of a bacteriophage T4 gp24 bypass mutant: The evolution of pentameric vertex proteins in icosahedral viruses. *J. Struct. Biol.* **154:**255–259.

Fokine, A., Bowman, V. D., Battisti, A. J., Li, Q., Chipman, P. R., Rao, V. B., and Rossmann, M. G. (2007). Cryo-electron microscopy study of bacteriophage T4 displaying anthrax toxin proteins. *Virology* **367:**422–427.

Fokine, A., Chipman, P. R., Leiman, P. G., Mesyanzhinov, V. V., Rao, V. B., and Rossmann, M. G. (2004). Molecular architecture of the prolate head of bacteriophage T4. *Proc. Natl. Acad. Sci. USA* **101:**6003–6008.

Fokine, A., Islam, M. Z., Zhang, Z., Bowman, V. D., Rao, V. B., and Rossmann, M. G. (2011). Structure of the three N-terminal immunoglobulin domains of the highly immunogenic outer capsid protein from a T4-like bacteriophage. *J. Virol.* **85:**8141–8148.

Fokine, A., Leiman, P. G., Shneider, M. M., Ahvazi, B., Boeshans, K. M., Steven, A. C., Black, L. W., Mesyanzhinov, V. V., and Rossmann, M. G. (2005). Structural and functional similarities between the capsid proteins of bacteriophages T4 and HK97 point to a common ancestry. *Proc. Natl. Acad. Sci. USA* **102:**7163–7168.

Franklin, J. L., Haseltine, D., Davenport, L., and Mosig, G. (1998). The largest (70 kDa) product of the bacteriophage T4 DNA terminase gene 17 binds to single-stranded DNA segments and digests them towards junctions with double-stranded DNA. *J. Mol. Biol.* **277:**541–557.

Fuller, D. N., Raymer, D. M., Kottadiel, V. I., Rao, V. B., and Smith, D. E. (2007). Single phage T4 DNA packaging motors exhibit large force generation, high velocity, and dynamic variability. *Proc. Natl. Acad. Sci. USA* **104:**16868–16873.

Ghosh-Kumar, M., Alam, T. I., Draper, B., Stack, J. D., and Rao, V. B. (2011). Regulation by interdomain communication of a headful packaging nuclease from bacteriophage T4. *Nucleic Acids Res.* **39:**2742–2755.

Gao, S., and Rao, V. B. (2011). Specificity of interactions among the DNA packaging machine components of T4 related bacteriophages. *J. Biol. Chem.* **286,** 3944–3956.

Goetzinger, K. R., and Rao, V. B. (2003). Defining the ATPase center of bacteriophage T4 DNA packaging machine: Requirement for a catalytic glutamate residue in the large terminase protein gp17. *J. Mol. Biol.* **331:**139–154.

Golz, S., and Kemper, B. (1999). Association of holliday-structure resolving endonuclease VII with gp20 from the packaging machine of phage T4. *J. Mol. Biol.* **285:**1131–1144.

Guasch, A., Pous, J., Ibarra, B., Gomis-Ruth, F. X., Valpuesta, J. M., Sousa, N., Carrascosa, J. L., and Coll, M. (2002). Detailed architecture of a DNA translocating machine: The high-resolution structure of the bacteriophage phi29 connector particle. *J. Mol. Biol.* **315:**663–676.

Hendrix, R. W. (1978). Symmetry mismatch and DNA packaging in large bacteriophages. *Proc. Natl. Acad. Sci. USA* **75:**4779–4783.

Hugel, T., Michaelis, J., Hetherington, C. L., Jardine, P. J., Grimes, S., Walter, J. M., Falk, W., Anderson, D. L., and Bustamante, C. (2007). Experimental test of connector rotation during DNA packaging into bacteriophage phi29 capsids. *PLoS Biol.* **5:**e59.

Ishii, T., Yamaguchi, Y., and Yanagida, M. (1978). Binding of the structural protein soc to the head shell of bacteriophage T4. *J. Mol. Biol.* **120:**533–544.

Ishii, T., and Yanagida, M. (1977). The two dispensable structural proteins (soc and hoc) of the T4 phage capsid; their purification and properties, isolation and characterization of the defective mutants, and their binding with the defective heads in vitro. *J. Mol. Biol.* **109:**487–514.

Jiang, J., Abu-Shilbayeh, L., and Rao, V. B. (1997). Display of a PorA peptide from *Neisseria meningitidis* on the bacteriophage T4 capsid surface. *Infect. Immun.* **65:**4770–4777.

Jiang, X., Jiang, H., Li, C., Wang, S., Mi, Z., An, X., Chen, J., and Tong, Y. (2011). Sequence characteristics of T4-like bacteriophage IME08 genome termini revealed by high throughput sequencing. *Virol. J.* **8:**194.

Kanamaru, S., Kondabagil, K., Rossmann, M. G., and Rao, V. B. (2004). The functional domains of bacteriophage t4 terminase. *J. Biol. Chem.* **279:**40795–40801.

Keppel, F., Rychner, M., and Georgopoulos, C. (2002). Bacteriophage-encoded cochaperonins can substitute for Escherichia coli's essential GroES protein. *EMBO J.* **3:**893–898.

Kondabagil, K. R., and Rao, V. B. (2006). A critical coiled coil motif in the small terminase, gp16, from bacteriophage T4: Insights into DNA packaging initiation and assembly of packaging motor. *J. Mol. Biol.* **358:**67–82.

Kondabagil, K. R., Zhang, Z., and Rao, V. B. (2006). The DNA translocating ATPase of bacteriophage T4 packaging motor. *J. Mol. Biol.* **363:**786–799.

Krisch, H. M., and Comeau, A. M. (2008). The immense journey of bacteriophage T4: From d'Herelle to Delbruck and then to Darwin and beyond. *Res. Microbiol.* **159:**314–324.

Kuebler, D., and Rao, V. B. (1998). Functional analysis of the DNA-packaging/terminase protein gp17 from bacteriophage T4. *J. Mol. Biol.* **281:**803–814.

Lebedev, A. A., Krause, M. H., Isidro, A. L., Vagin, A. A., Orlova, E. V., Turner, J., Dodson, E. J., Tavares, P., and Antson, A. A. (2007). Structural framework for DNA translocation via the viral portal protein. *EMBO J.* **26:**1984–1994.

Leffers, G., and Rao, V. B. (1996). A discontinuous headful packaging model for packaging less than headful length DNA molecules by bacteriophage T4. *J. Mol. Biol.* **258:**839–850.

Leffers, G., and Rao, V. B. (2000). Biochemical characterization of an ATPase activity associated with the large packaging subunit gp17 from bacteriophage T4. *J. Biol. Chem.* **275:**37127–37136.

Li, Q., Shivachandra, S. B., Leppla, S. H., and Rao, V. B. (2006). Bacteriophage T4 capsid: a unique platform for efficient surface assembly of macromolecular complexes. *J. Mol. Biol.* **363:**577–588.

Li, Q., Shivachandra, S. B., Zhang, Z., and Rao, V. B. (2007). Assembly of the small outer capsid protein, Soc, on bacteriophage T4: A novel system for high density display of multiple large anthrax toxins and foreign proteins on phage capsid. *J. Mol. Biol.* **370:**1006–1019.

Lin, H., and Black, L. W. (1998). DNA requirements in vivo for phage T4 packaging. *Virology* **242:**118–127.

Lin, H., Rao, V. B., and Black, L. W. (1999). Analysis of capsid portal protein and terminase functional domains: Interaction sites required for DNA packaging in bacteriophage T4. *J. Mol. Biol.* **289:**249–260.

Lin, H., Simon, M. N., and Black, L. W. (1997). Purification and characterization of the small subunit of phage T4 terminase, gp16, required for DNA packaging. *J. Biol. Chem.* **272:**3495–3501.

Liu, J., and Mushegian, A. (2004). Displacements of prohead protease genes in the late operons of double-stranded-DNA bacteriophages. *J. Bacteriol.* **186:**4369–4375.

Lipińska, B., Rao, A. S., Bolten, B. M., Balakrishnan, R., and Goldberg, E. B. (1989). Cloning and identification of bacteriophage T4 gene 2 product gp2 and action of gp2 on infecting DNA *in vivo*. *J. Bacteriol.* **171:**488–497.

Malys, N., Chang, D. Y., Baumann, R. G., Xie, D., and Black, L. W. (2002). A bipartite bacteriophage T4 SOC and HOC randomized peptide display library: Detection and analysis of phage T4 terminase (gp17) and late sigma factor (gp55) interaction. *J. Mol. Biol.* **319:**289–304.

Miller, E. S., Kutter, E., Mosig, G., Arisaka, F., Kunisawa, T., and Ruger, W. (2003). Bacteriophage T4 genome. *Microbiol. Mol. Biol. Rev.* **67:**86–156.

Mitchell, M. S., Matsuzaki, S., Imai, S., and Rao, V. B. (2002). Sequence analysis of bacteriophage T4 DNA packaging/terminase genes 16 and 17 reveals a common ATPase center in the large subunit of viral terminases. *Nucleic Acids Res.* **30:**4009–4021.

Mitchell, M. S., and Rao, V. B. (2006). Functional analysis of the bacteriophage T4 DNA-packaging ATPase motor. *J. Biol. Chem.* **281:**518–527.

Mosig, G. (1998). Recombination and recombination-dependent DNA replication in bacteriophage T4. *Annu. Rev. Genet.* **32:**379–413.

Mullaney, J. M., and Black, L. W. (1998). Activity of foreign proteins targeted within the bacteriophage T4 head and prohead: Implications for packaged DNA structure. *J. Mol. Biol.* **283:**913–929.

Mullaney, J. M., Thompson, R. B., Gryczynski, Z., and Black, L. W. (2000). Green fluorescent protein as a probe of rotational mobility within bacteriophage T4. *J. Virol. Methods* **88:**35–40.

Oram, M., Sabanayagam, C., and Black, L. W. (2008). Modulation of the packaging reaction of bacteriophage T4 terminase by DNA structure. *J. Mol. Biol.* **381:**61–72.

Qin, L., Fokine, A., O'Donnell, E., Rao, V. B., and Rossmann, M. G. (2009). Structure of the small outer capsid protein, Soc: A clamp for stabilizing capsids of T4-like phages. *J. Mol. Biol.* **395:**728–741.

Rao, V. B., and Black, L. W. (1988). Cloning, overexpression and purification of the terminase proteins gp16 and gp17 of bacteriophage T4: Construction of a defined in-vitro DNA packaging system using purified terminase proteins. *J. Mol. Biol.* **200:**475–488.

Rao, V. B., and Black, L. W. (2005). "DNA Packaging in Bacteriophage T4 in Viral Genome Packaging Machines, Genetics, Structure, and Mechanism" (C. E. Catalano, ed.), pp. 40–58. Plenum Press.

Rao, V. B., and Feiss, M. (2008). The bacteriophage DNA packaging motor. *Annu. Rev. Genet.* **42:**647–681.

Rao, V. B., and Mitchell, M. S. (2001). The N-terminal ATPase site in the large terminase protein gp17 is critically required for DNA packaging in bacteriophage T4. *J. Mol. Biol.* **314:**401–411.

Rao, M., Peachman, K., Li, Q., Matyas, G., Shivachandra, S., Borschel, Morthole, V. I., Fernandez-Prada, C. R., Alving, C., and Rao, V. B. (2011). Highly effective generic adjuvant systems for orphan or poverty-related vaccines. *Vaccine* **29:**873–877.

Ray, K., Ma, J., Oram, M., Lakowicz, J. R., and Black, L. W. (2010a). Single molecule- and fluorescence correlation spectroscopy-FRET analysis of phage DNA packaging: Co-localization of the packaged phage T4 DNA ends within the capsid. *J. Mol. Biol.* **396:**1102–1113.

Ray, K., Oram, M., Ma, J., and Black, L. W. (2009). Portal control of viral prohead expansion and DNA packaging. *Virology* **391:**44–50.

Ray, K., Sabanayagam, C. R., Lakowicz, J. R., and Black, L. W. (2010b). DNA crunching by a viral packaging motor: Compression of a procapsid-portal stalled Y-DNA substrate. *Virology* **398:**224–232.

Ren, S. X., Ren, Z. J., Zhao, M. Y., Wang, X. B., Zuo, S. G., and Yu, F. (2009). Antitumor activity of endogenous mFlt4 displayed on a T4 phage nanoparticle surface. *Acta Pharmacol. Sin* **30:**637–645.

Ren, Z., and Black, L. W. (1998). Phage T4 SOC and HOC display of biologically active, full-length proteins on the viral capsid. *Gene* **215:**439–444.

Ren, Z. J., Baumann, R. G., and Black, L. W. (1997). Cloning of linear DNAs in vivo by overexpressed T4 DNA ligase: Construction of a T4 phage hoc gene display vector. *Gene* **195:**303–311.

Ren, Z. J., Lewis, G. K., Wingfield, P. T., Locke, E. G., Steven, A. C., and Black, L. W. (1996). Phage display of intact domains at high copy number: A system based on SOC, the small outer capsid protein of bacteriophage T4. *Protein Sci.* **5:**1833–1843.

Ren, Z. J., Tian, C. J., Zhu, Q. S., Zhao, M. Y., Xin, A. G., Nie, W. X., Ling, S. R., Zhu, M. W., Wu, J. Y., Lan, H. Y., Cao, Y. C., and Bi, Y. Z. (2008). Orally delivered foot-and-mouth disease virus capsid protomer vaccine displayed on T4 bacteriophage surface: 100% protection from potency challenge in mice. *Vaccine* **26:**1471–1481.

Rentas, F. J., and Rao, V. B. (2003). Defining the bacteriophage T4 DNA packaging machine: evidence for a C-terminal DNA cleavage domain in the large terminase/packaging protein gp17. *J. Mol. Biol.* **334:**37–52.

Repoila, F., Tetart, F., Bouet, J. Y., and Krisch, H. M. (1994). Genomic polymorphism in the T-even bacteriophages. *EMBO J.* **13:**4181–4192.

Rifat, D., Wright, N. T., Varney, K. M., Weber, D. J., and Black, L. W. (2008). Restriction endonuclease inhibitor IPI* of bacteriophage T4: A novel structure for a dedicated target. *J. Mol. Biol.* **375:**720–734.

Robertson, K. L., Soto, C. M., Archer, M. J., Odoemene, O., and Liu, J. L. (2011). Engineered T4 viral nanoparticles for cellular imaging and flow cytometry. *Bioconjug. Chem.* **22:**595–604.

Sabanayagam, C. R., Oram, M., Lakowicz, J. R., and Black, L. W. (2007). Viral DNA packaging studied by fluorescence correlation spectroscopy. *Biophys. J.* **93:**L17–L19.

Sathaliyawala, T., Islam, M. Z., Li, Q., Fokine, A., Rossmann, M. G., and Rao, V. B. (2010). Functional analysis of the highly antigenic outer capsid protein, Hoc, a virus decoration protein from T4-like bacteriophages. *Mol. Microbiol.* **77:**444–455.

Sathaliyawala, T., Rao, M., Maclean, D. M., Birx, D. L., Alving, C. R., and Rao, V. B. (2006). Assembly of human immunodeficiency virus (HIV) antigens on bacteriophage T4: A novel in vitro approach to construct multicomponent HIV vaccines. *J. Virol.* **80:**7688–7698.

Shivachandra, S. B., Li, Q., Peachman, K. K., Matyas, G. R., Leppla, S. H., Alving, C. R., Rao, M., and Rao, V. B. (2007). Multicomponent anthrax toxin display and delivery using bacteriophage T4. *Vaccine* **25:**1225–1235.

Shivachandra, S. B., Rao, M., Janosi, L., Sathaliyawala, T., Matyas, G. R., Alving, C. R., Leppla, S. H., and Rao, V. B. (2006). In vitro binding of anthrax protective antigen on bacteriophage T4 capsid surface through Hoc-capsid interactions: A strategy for efficient display of large full-length proteins. *Virology* **345:**190–198.

Simpson, A. A., Tao, Y., Leiman, P. G., Badasso, M. O., He, Y., Jardine, P. J., Olson, N. H., Morais, M. C., Grimes, S., Anderson, D. L., Baker, T. S., and Rossmann, M. G. (2000). Structure of the bacteriophage phi29 DNA packaging motor. *Nature* **408:**745–750.

Snyder, L., and Tarkowski, H. J. (2005). The N terminus of the head protein of T4 bacteriophage directs proteins to the GroEL chaperonin. *J. Mol. Biol.* **345:**375–386.

Stortelder, A., Hendriks, J., Buijs, J. B., Bulthuis, J., Gooijer, C., van der Vies, S. M., and van der Zwan, G. (2006). Hexamerization of the bacteriophage T4 capsid protein gp23 and its W13V mutant studied by time-resolved tryptophan fluorescence. *J. Phys. Chem. B* **110:**25050–25058.

Sun, S., Gao, S., Kondabagil, K., Xiang, Y., Rossmann, M. G., and Rao, V. B. (2012). Structure and function of the small terminase component of the DNA packaging machine from T4 like bacteriophages. *Proc. Natl. Acad. Sci. USA* **109**(3)**:**817–822.

Sun, S., Kondabagil, K., Draper, B., Alam, T. I., Bowman, V. D., Zhang, Z., Hegde, S., Fokine, A., Rossmann, M. G., and Rao, V. B. (2008). The structure of the phage T4 DNA packaging motor suggests a mechanism dependent on electrostatic forces. *Cell* **135:**1251–1262.

Sun, S., Kondabagil, K., Gentz, P. M., Rossmann, M. G., and Rao, V. B. (2007). The structure of the ATPase that powers DNA packaging into bacteriophage T4 procapsids. *Mol. Cell* **25:**943–949.

Tetart, F., Desplats, C., and Krisch, H. M. (1998). Genome plasticity in the distal tail fiber locus of the T-even bacteriophage: Recombination between conserved motifs swaps adhesin specificity. *J. Mol. Biol.* **282:**543–556.

Wang, G. R., Vianelli, A., and Goldberg, E. B. (2000). Bacteriophage T4 self-assembly: In vitro reconstitution of recombinant gp2 into infectious phage. *J. Bacteriol.* **182:**672–679.

Wikoff, W. R., Liljas, L., Duda, R. L., Tsuruta, H., Hendrix, R. W., and Johnson, J. E. (2000). Topologically linked protein rings in the bacteriophage HK97 capsid. *Science* **289:**2129–2133.

Wu, C. H., and Black, L. W. (1995). Mutational analysis of the sequence-specific recombination box for amplification of gene 17 of bacteriophage T4. *J. Mol. Biol.* **247:**604–617.

Wu, C. H., Lin, H., and Black, L. W. (1995). Bacteriophage T4 gene 17 amplification mutants: Evidence for initiation by the T4 terminase subunit gp16. *J. Mol. Biol.* **247:**523–528.

Wu, J., Tu, C., Yu, X., Zhang, M., Zhang, N., Zhao, M., Nie, W., and Ren, Z. (2007). Bacteriophage T4 nanoparticle capsid surface SOC and HOC bipartite display with enhanced classical swine fever virus immunogenicity: A powerful immunological approach. *J. Virol. Methods* **139:**50–60.

Yu, T. Y., and Schaefer, J. (2008). REDOR NMR characterization of DNA packaging in bacteriophage T4. *J. Mol. Biol.* **382:**1031–1042.

Zhang, Z., Kottadiel, V. I., Vafabakhsh, R., Dai, L., Chemla, Y. R., Ha, T., and Rao, V. B. (2011). A promiscuous DNA packaging machine from bacteriophage T4. *PLoS Biol.* **9:**e1000592.

CHAPTER 6

Phage λ—New Insights into Regulatory Circuits

Grzegorz Węgrzyn,* Katarzyna Licznerska,* and **Alicja Węgrzyn†**

Contents

* Department of Molecular Biology, University of Gdańsk, Gdańsk, Poland
† Laboratory of Molecular Biology (affiliated with the University of Gdańsk), Institute of Biochemistry and Biophysics, Polish Academy of Sciences, Gdańsk, Poland

Advances in Virus Research, Volume 82
ISSN 0065-3527, DOI: 10.1016/B978-0-12-394621-8.00016-9

Abstract

Bacteriophage λ, rediscovered in the early 1950s, has served as a model in molecular biology studies for decades. Although currently more complex organisms and more complicated biological systems can be studied, this phage is still an excellent model to investigate principles of biological processes occurring at the molecular level. In fact, very few other biological models provide possibilities to examine regulations of biological mechanisms as detailed as performed with λ. In this chapter, recent advances in our understanding of mechanisms of bacteriophage λ development are summarized and discussed. Particularly, studies on (i) phage DNA injection, (ii) molecular bases of the lysis-versus-lysogenization decision and the lysogenization process itself, (iii) prophage maintenance and induction, (iv), λ DNA replication, (v) phage-encoded recombination systems, (vi) transcription antitermination, (vii) formation of the virion structure, and (viii) lysis of the host cell, as published during several past years, will be presented.

I. INTRODUCTION: THE BACTERIOPHAGE λ PARADIGM

Bacteriophage λ has been used in studies on the molecular mechanism of biological processes for over half a century. In fact, it is now recognized as a paradigm and reference in many regulatory mechanisms occurring in both prokaryotic and eukaryotic systems. Therefore, this model virus, and processes crucial for its development, has been described in many review articles; this introductory section presents only a very concise overview on the phage life styles and basic mechanisms. For more detailed reviews on both λ-based advantages in basic research and phage-dependent biotechnological approaches, readers are advised to find various available articles, published in recent years, and exemplified by Gottesman and Weisberg (2004), Dodd *et al.* (2005), Ptashne (2005), Węgrzyn and Węgrzyn (2005), Van Duyne (2005), Garufi *et al.* (2005), Oppenheim *et al.* (2005), Court *et al.* (2007a), Thomason *et al.* (2007), Krupovic and Bamford (2008), and Valdez-Cruz *et al.* (2010). We do not refer here to 20th-century references describing the actual discovery of phage λ by Ester Lederberg; recognition of λ genome peculiarities by Al Hershey; genetic characterization of λ genes by Allan Campbell; characterization of the basic mechanism of phage-dependent cell lysis by Alina Taylor and Ry Young; first unraveling of the transcriptional patterns and discovery that both DNA strands can be transcribed by Karol Taylor and Waclaw Szybalski; and identification of crucial events in λ DNA replication by Ross Inman, Maria Schnos, Costa Georgopoulos, Roger McMacken, Karol Taylor, and Maciej Zylicz, to mention a few. We also do not refer here to the role of λ in the development of biotechnology, particularly genetic engineering, although we must remember that this bacteriophage, after suitable modifications, provided first cloning vectors and, employing λ genetic regulatory elements, allowed researchers to construct very useful tools for controlled expression of cloned genes (works by Waclaw Szybalski, Anna J, Podhajska, and Marian S. Sęktas must be acknowledged here). Furthermore, the first transcriptional and physical (by electron microscopy) maps of λ, including nomenclature and definition of regulatory elements (o_L, o_R, p_L, p_R, $p_{R'}$, p_I, p_M, *oop*, *ori*, *nutL*, *nutR*, *qut* and others) were developed by Waclaw Szybalski and his collaborators.

Phage λ virions adsorb on their host, *Escherichia coli*, cells employing the bacterial LamB receptor, located in external membrane, and playing the role of maltose porin in *E. coli*. Phage genome, the double-stranded DNA molecule composed of 48,502bp (with 12 nucleotide single-stranded, complementary ends), enters the host cell cytoplasm through the phage noncontractile tail, due to action of the *E. coli* enzymatic system, composed of PtsP and PtsM proteins, located in the internal membrane and used normally for the transportation of mannose. Following DNA circularization, the decision whether to lysogenize the host cell or to proceed to lytic development is made on the basis of the specific phage regulatory system, in which many phage- and host-encoded factors are employed. If the lysogenic pathway is chosen, phage DNA is incorporated into the host chromosome and persists in the form of a prophage. This state can be kept for many cell generations. Lytic development starts either after prophage induction (resulting in excision of the λ genome from the host chromosome) or when the lysis-versus-lysogenization decision is made in favor of the former option. At the early stage of lytic development, phage DNA is replicating predominantly according to the circle-to-circle (or θ) mode, which is switched later to the rolling circle (or σ) mode. The latter replication mechanism results in the appearance of long concatemeric DNA molecules encompassing several λ genomes. Simultaneously, general recombination of phage DNA takes place, and expression of phage genes occurs (which is controlled precisely by various mechanisms, including activation, repression, and antitermination of transcription, as well as RNA stability and efficiency of translation initiation), leading to the production of capsid proteins. During the process of the assembly of progeny virions, concatemeric λ DNA molecules are cut into monomers in the process called packaging, and maturation of capsids leads to the formation of progeny virions. Phage progeny is liberated from the host cell after its lysis, caused by products of phage genes. Importantly, all the stages of bacteriophage λ development are controlled precisely by mechanisms whose molecular details can be understood in more and more detail. These mechanisms also provide models for studies on biological processes in both bacterial cells and eukaryotic organisms.

This chapter does not refer further to basic information about particular processes. Rather, it focuses on novel findings, which provided new insights into the regulation of certain mechanisms operating during the development of bacteriophage λ.

II. EJECTION OF λ DNA FROM VIRION INTO THE HOST CELL

Findings on the mechanism of λ DNA ejection focused on physical and chemical factors that influence this process. Köster *et al.* (2009) demonstrated that an internal capsid pressure influences efficiency of the ejection process. A decrease in the capsid pressure correlated with a decrease in the efficiency of cell lysis caused by phage infection. Another study indicated that the ejection process is independent of ionic concentrations, which might seem quite surprising (Wu *et al.*, 2010). These studies provided important information about mechanistic processes occurring at the very first stage of phage λ development.

III. THE LYSIS-VERSUS-LYSOGENIZATION DECISION AND λ DNA INTEGRATION INTO HOST CHROMOSOME

A. The developmental decision

Shortly after phage λ DNA is ejected and enters the host cell, the major developmental decision must be done, namely whether to lysogenize the host or to perform lytic development. Several proteins and other factors play roles in the process of making the decision, and this process is complicated enough to make it impossible to predict precisely at the current state of knowledge. Although we can estimate what conditions support lysogenization and what is required for lytic development, it is impossible to calculate precisely what percentage of phages enters one or another pathway under certain conditions. Therefore, various methods of the simulation of the decision process are being developed. Building such models of the regulatory network is a usual step in attempts to understand the system in more detail (Avlund *et al.*, 2009a; Węgrzyn and Węgrzyn, 2005).

A very interesting quantitative kinetic analysis of such a network has been performed that suggested that the Cro protein may be the key player in sensing whether the host cell is infected by one or more phages (Kobiler *et al.*, 2005). In fact, a number of phages infecting a single host cell is one of the factors influencing the developmental decision. Recent simulation of the "counting" process suggested that apart from the already known importance of the ratio of Cro and cII proteins, additional regulatory mechanisms may exist (Avlund *et al.*, 2009b). Another study suggested that the presence of the restriction–modification system can influence phage λ development even if the phage has already been "adapted" to this particular system (Gregory *et al.*, 2010). The simulation used by the authors of that study showed that adaptation and readaptation to a particular restriction and modification system result in lower efficiency of phage propagation and delayed lysis of bacterial cells relative to non-restricting host bacteria.

Among proteins involved in the lysis-versus-lysogenization decision, Cro, cI, cII, and cIII appear to be crucial. Therefore, understanding the details of structures and functions of these proteins is necessary to build more solid models of phage developmental regulation. The Cro protein functions as a dimer, thus the efficiency of formation of such a structure determines its activity. Jia *et al.* (2005) demonstrated that native Cro monomers have compact structures, which are hard to unfold. However, unfolded or partially folded monomers appear to be preferred substrates for Cro dimer formation. This implicates that the assembly of Cro into a dimer is a slow process. Thus, *in vivo* Cro binding may be under kinetic rather than thermodynamic control.

Several papers on cII structure and function have been published. Datta *et al.* (2005b) demonstrated that 15C-terminal amino acid residues of cII are flexible and may act as a target for proteolysis in the host cell. In this 97 amino acid long protein, residues 70–82 were found to be necessary for tetramer formation, DNA binding, and transcription activation. The same research group (Datta *et al.*, 2005a) solved the three-dimensional structure of cII at 2.6 Å resolution. The tetrameric structure of cII is rather unusual, as it is a dimer of dimers, however, without a closed symmetry. Such a structure may be necessary for efficient binding of cII to major grooves of DNA, from one face of this molecule. Further studies by Parua *et al.* (2010a) indicated that residues 70, 74, and 78 are crucial for maintaining the tetrameric structure of cII. The model of binding of cII to one face of DNA was used to propose a mechanism for cII-mediated activation of transcription with involvement of the RNA polymerase α subunit (Kędzierska *et al.*, 2004). This can explain how cII stimulates promoter activity when it is bound at the site overlapping the −35 box and when it interacts with both σ^{70} and α subunits of RNA polymerase. Interestingly, one of the cII-activated promoters, p_{aQ}, has been found to be stimulated directly by guanosine tetraphosphate (ppGpp), the stringent control alarmone (Potrykus *et al.*, 2004).

The cII protein is degraded in *E. coli* cell by the host-encoded HflB protease (also called FtsH). Another host factor, the HflD protein, interacts with cII and facilitates its degradation. Parua *et al.* (2010b) demonstrated that HflD also negatively influences binding of cII to DNA, inhibiting transcription activation by this factor. Two other Hfl proteins,

HflC and HflK, make the HflB-mediated proteolysis of cII less efficient. Bandyopadhyay *et al.* (2010) have purified HflC and HflK separately (which was not achieved previously) and found that each of these proteins binds to HflB and inhibits cII proteolysis.

The best characterized inhibitor of HflB is λ cIII protein. Kobiler *et al.* (2007) showed that cIII functions as a competitive inhibitor of HflB, preventing binding of the cII protein by this protease. cIII–HflB interactions, but not those between cII and cIII, were found to be responsible for inhibition of the protease (Halder *et al.*, 2008). Kobiler *et al.* (2007) also suggested that oligomerization of cIII is required for its function as the HflB inhibitor. However, Halder *et al.* (2007) proposed that the cIII protein exists as a dimer under native conditions.

Apart from the regulatory factors influencing the lysis-versus-lysogenization decision known for many years, perhaps surprisingly, it is still possible to discover novel players in this game. One of them is the Ea8.5 protein, encoded by a λ gene located in the *b* (nonessential) region of the phage genome. Overproduction of this protein resulted in a decrease in the efficiency of lysogenization, most probably due to impairment in cII-mediated transcription activation (Łoś *et al.*, 2008).

Current knowledge about the regulatory network responsible for the control of phage λ development at the stage of the lysis-versus-lysogenization decision is presented schematically in Figure 1. This scheme is based on the previous picture by Węgrzyn and Węgrzyn (2005) but contains an update on regulations discovered during the last several years.

B. Antiphage response of the host cell

Recent discoveries informed us about the existence of the prokaryotic antiviral (antiphage) system. This system, abbreviated as CRISPR, consists of clusters of regularly interspaced short palindromic repeats in bacterial genomes (Brouns *et al.*, 2008). Because this topic is discussed in more detail in another chapter of this book, this chapter only signals its influence on lysogenization or lytic development. Namely, it appears that CRISPR affects lysogenization by phage λ, maintenance of the lysogenic state, and prophage induction (Edgar and Qimron, 2010). Interestingly, the function of CRISPR appears to be regulated by H-NS and LeuO proteins (Westra *et al.*, 2010).

C. Integration of λ DNA into host chromosome

If the lysogenic pathway of phage λ development is chosen, the crucial step to form the prophage is integration of the viral genome into the host chromosome. This reaction is catalyzed by the phage-encoded Int protein, also called integrase. Production of this enzyme is dependent on the

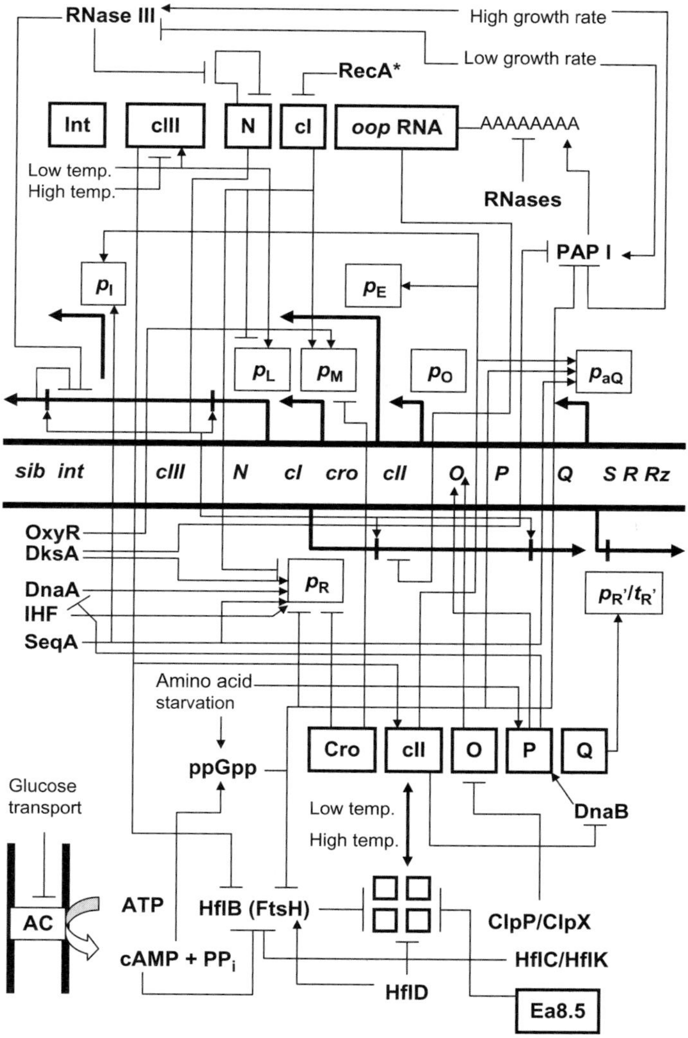

FIGURE 1 A regulatory network at the "lysis-versus-lysogenization" decision of bacteriophage λ. The crucial λ regulatory genes are presented between two thick horizontal lines that symbolize a fragment of the phage genome. Promoters are marked in thin boxes, and transcripts are shown as thick arrows, with arrowheads indicating directionality of transcription (*oop* RNA is an exception, see later). Phage λ gene products (proteins and one nontranslatable transcript, *oop* RNA) are marked in thick boxes. Host (*Escherichia coli*) proteins and specific conditions are presented without boxes. Regulatory processes are indicated as thin arrows and thin blunt-ended lines (positive regulations are represented by arrows and negative regulations are represented by blunt-ended lines). AC, adenylate cyclase; ATP, adenosine triphosphate; cAMP, cyclic AMP, High temp., high temperature; Low temp., low temperature; PAP I, poly(A) polymerase I (the *pcnB* gene product); ppGpp, guanosine tetraphosphate; PP_i, pyrophosphate. This figure is an updated version of the scheme published by Węgrzyn and Węgrzyn (2005).

function of the cII-dependent p_I promoter and the t_I terminator. Although transcription of the *int* gene is also possible from the p_L promoter, transcripts derived from this promoter do not terminate at t_I due to N-dependent antitermination, thus *int* mRNA is highly unstable due to the retroregulation mechanism (degradation of mRNA initiated at the specific secondary RNA structure located downstream of the coding sequence). Therefore, only transcripts that stop at t_I may give functional *int* mRNA. Detailed mapping of the t_I terminator has been performed only recently by Martinez-Trujillo *et al.* (2010), who identified sequences required for transcription termination at this site and mapped the alternative ends of the transcript. Another published paper focused on the integrase activity itself. In fact, novel integrase variants were also selected, which helped not only to understand biochemical function of this enzyme, but also provided interesting biotechnological tools (Tay *et al.*, 2010).

IV. PROPHAGE MAINTENANCE AND INDUCTION

Once the host cell is lysogenized by phage λ, the prophage state is stabilized by the action of the phage-encoded cI protein, which is a strong repressor of the main lytic promoters p_R and p_L and an activator of its own promoter, p_M. The prophage is maintained as long as cI is active. This repressor is sensitive to autoproteolysis, taking place upon interaction of cI with the activated form of the RecA protein, called RecA*, which changes its conformation after binding to single-stranded DNA fragments, appearing as a result of the damage of the genetic material (which is a signal triggering the specific cellular response, called the SOS response). Cleavage of the cI repressor results in derepression of p_R and p_L and subsequent excision of the prophage from the host chromosome, followed by phage lytic development. This drastic change in the bacteriophage life style is known as λ prophage induction and is a paradigm of the genetic switch—a process of the drastic changes in the pattern of expression of various genes leading to a major change in the life style.

A. Models of the genetic switch

A large body of data indicated that the prophage state is stable, showing that it is a specific type of cellular memory (Oppenheim *et al.*, 2005). In fact, studies indicated that this state is exceptionally stable, as the frequency of prophage induction events is lower than 10^{-8} per cell generation in absence of the SOS response (Cao *et al.*, 2008; Little and Michalowski, 2010; Morelli *et al.*, 2009; Zong *et al.*, 2010).

Despite this high stability, under natural conditions, prophage induction occurs relatively often due to frequent stress conditions met by lysogenic bacteria in their habitats. Therefore, modeling of the regulation of the switch is important in understanding this crucial process in phage life. Novel models of the λ genetic switch have been published in recent years. These include mathematical modeling, which described the switch as a piecewise, quadratic second-order differential equation (Laschov and Margaliot, 2009), and the experimental evolution approach, showing how lambdoid prophages can change their regulatory elements in response to selection (Refardt and Rainey, 2010). Other experimental and theoretical analyses suggested that some elements of the regulatory system may be dispensable, playing a role in modulating the response rather than determining the main event (Little, 2010), that a constant level of the cI repressor can be maintained over a wide range of its degradation rate, ensuring heritability of the lysogenic state (Cao *et al.*, 2010), and that this state is in a monostable regime rather than in a bistable regime (Lou *et al.*, 2007).

B. A role for Cro in λ prophage induction

An interesting question is the role of the Cro protein in λ prophage induction. Cro is known as a strong repressor of the p_M promoter (required for the *c*I gene expression) and a weak repressor of p_R and p_L, playing a role as a facilitator of lytic development. However, there were also reports indicating that Cro may have a role in the λ genetic switch. Recent works indicated that Cro is not crucial in the switch, but may moderate this process by a weak repression of early lytic promoters (Svenningsen *et al.*, 2005). Atsumi and Little (2006) replaced Cro with the Lac repressor and showed that a phage bearing the *lacI* gene and *lac* operators instead of *cro* is viable, but some details of the phage development are changed. This underlined once again the complexity of the regulatory system and its sensitivity to subtle changes in the regulatory processes (Ptashne, 2006). Further changes in the λ genome, namely replacement of the *c*I gene with the gene coding for Tet repressor, resulted in construction of a still viable phage, able to proceed either lytic or lysogenic development depending on levels of factors controlling particular promoters (Atsumi and Little, 2006). These experiments also suggested that we might be quite close to learning how to control phage λ development. However, Schubert *et al.* (2007) postulated that the role of Cro is critical, rather than modulatory, in prophage induction. They created mutants that strongly impaired Cro-mediated repression of p_M without significant changes in cI-mediated regulation of this promoter and demonstrated that inhibition of transcription from p_M is crucial for efficient prophage induction, as synthesis of new cI molecules can significantly impede phage lytic development.

C. Structure and function of the cI repressor

It is clear that prophage maintenance depends mainly on activity of the cI repressor. However, new insights into both structure and function of this protein have appeared recently. A role for the formation of cI oligomers bound to both o_R and o_L regions has been demonstrated previously, and their structures have been determined at high resolution by experiments employing an atomic force microscope (Wang *et al.*, 2009). The cI repressor binds to specific DNA sequences cooperatively; however, results by Babić and Little (2007) suggest that this cooperativity is not essential but rather is a refinement to a more basic circuit. It seems that the cooperative binding of cI may increase stability of the prophage under various environmental conditions. This could be supported by analyses based on a validated mathematical model (Gedeon *et al.*, 2008).

The cI-mediated activation of the p_M promoter has been studied by employing advanced biophysical (Bakk, 2005) and biochemical (Kędzierska *et al.*, 2007) methods. Results of such studies strongly suggest that cI contacts both σ^{70} and α subunits of the RNA polymerase at p_M (Kędzierska *et al.*, 2007) and that activation of this promoter can be two- to fourfold higher when DNA is looped (Anderson and Yang, 2008).

Determination of the crystal structure of the intact λ cI repressor dimer bound to DNA (Stayrook *et al.*, 2008) provided an excellent tool in understanding the function of this protein in more detail. In fact, this structure indicates that the two subunits of the cI dimer adopt different conformations (Hochschild and Lewis, 2009; Stayrook *et al.*, 2008).

D. Factors causing prophage induction

Because the cI repressor is sensitive to its autocleavage triggered by the RecA* protein, which also triggers the SOS response, λ prophage induction was usually studied in ultraviolet-irradiated cells. However, it is clear that other factors occurring in natural habitats of *E. coli* must also cause the genetic switch. Identification of such factors is important in understanding the physiology of λ and relative phages. Interestingly, it was demonstrated that significant differences exist in the mechanism of λ prophage induction and further lytic development between two experimental systems that are used commonly to study this phenomenon, namely the SOS-mediated induction of wild-type prophage and heat-mediated induction of a prophage bearing a mutation in the gene coding for a temperature-sensitive cI repressor (Rokney *et al.*, 2008). This finding strongly supports the prediction that our knowledge of prophage induction mechanisms, which is based on a very limited number of experimental systems, may be highly incomplete.

Shkilnyj and Koudelka (2007) found that a concentration of salt (NaCl) as high as 200 m*M* can cause efficient induction of a lambdoid prophage. This was an important discovery because it indicated that some conditions found in a natural environment may trigger the λ genetic switch. Another work demonstrated that λ prophage may be induced by acyl homoserine lactones (Ghosh *et al.*, 2009). These compounds are known as signaling molecules of quorum sensing in Gram-negative bacteria. Therefore, one may speculate that prophage induction may be a response to a high concentration of bacterial cells. Another factor triggering the induction of lambdoid prophages is hydrogen peroxide. This oxidative stress-mediating factor was shown to be an efficient inductor of not only λ but also other lambdoid phages, including Shiga toxin-converting phages (Łoś *et al.*, 2009, 2010). The mechanism of H_2O_2-caused prophage induction may involve action of the host-encoded OxyR protein, whose function in regulation of the activity of the λ p_M promoter has been described elsewhere (Glinkowska *et al.*, 2010).

E. Excision of the prophage

Following inactivation of the cI repressor, which is always the first step of λ prophage induction, excision of the phage DNA from the host chromosome is necessary to start lytic development. This excision is mediated by two phage-encoded proteins, Int and Xis, but is facilitated by at least four host proteins. The architecture of the nucleoprotein complex, containing six proteins and a λ DNA fragment encompassing the *att* site (the region of the site-specific recombination that occurs during prophage excision), has been studied using the **fluorescence resonance energy transfer** technique and employing a metric matrix distance-geometry algorithm (Sun *et al.*, 2006). These studies provided important data, which helped in understanding the actual structure of this complex in more detail. Another work underlined the important role of the Fis protein, a host DNA-binding protein, in λ DNA excision from the *E. coli* chromosome (Papagiannis *et al.*, 2007). In fact, upon prophage induction, a very high level of synthesis of the λ Xis protein has been observed; however, this protein was ineffective in the absence of Fis. It is also worth mentioning that some other lambdoid phages, such as the Shiga toxin-converting phage Φ24B, may encode additional factors controlling the excision events (Fogg *et al.*, 2011).

V. PHAGE λ DNA REPLICATION

Bacteriophage λ encodes only two proteins involved directly in replication of its genome, O and P proteins. The O protein binds to the *ori*λ region, and the P protein delivers the host-encoded DnaB helicase to this site. Then, a replication complex containing other replication proteins from the host cell (including DNA polymerase III holoenzyme, DnaG primase, and gyrase) is formed. However, contrary to host replication machinery, initiation of λ DNA bidirectional replication requires the activities of DnaJ, DnaK, and GrpE chaperones. Interestingly, this initiation is not effective in the absence of transcription proceeding near the *ori*λ region. This transcription, called transcriptional activation of the *origin*, somehow activates the replication complex to start bidirectional DNA replication. After several rounds of such a replication, a switch from circle-to-circle (or θ) to rolling-circle (or σ) replication mode occurs, which leads to the production of long concatemeric λ DNA molecules, used for packaging of the phage genomes into newly assembled capsids.

During recent years, work on the regulation of λ DNA replication focused on the process of transcriptional activation of *ori*λ. This was because this process, rather than the assembly of the replication complex per se, appears to trigger the replication event. Moreover, because only a unidirectional replication can start from *ori*λ in the absence of transcriptional activation, it was proposed that the switch from θ to σ replication mode is due to an inhibition of transcription; after one round of unidirectional replication, the 5′ end of the replicating DNA strand can be displaced by the growing 3′ end, which can result in initiation of rolling-circle replication (Węgrzyn and Węgrzyn, 2005). As the transcription that activates DNA replication from *ori*λ starts from the p_R promoter, any factors that influence its strength may also influence λ genome replication. Thus, studies on the function and regulation of this promoter have a high impact on our understanding of the regulation of λ DNA replication.

A novel technique enabling calibration of biochemical parameters *in vivo* was tested to characterize the p_R promoter (Rosenfeld *et al.*, 2005). This provided important details on the function of this promoter, particularly suggesting that its activity fluctuates during the cell cycle. Then, the isomerization step of transcription initiation from p_R was characterized in more detail (Kontur *et al.*, 2006). It is worth mentioning that this promoter is controlled by many factors, including negative (cI, Cro, ppGpp) and positive (DnaA, SeqA) regulators. Interestingly, both DnaA and SeqA proteins stimulate the transcription from p_R and bind downstream of the transcription start point, which is very rare in prokaryotic systems (Łyżeń *et al.*, 2006). SeqA was found to activate p_R at the stage of promoter clearance (Łyżeń *et al.*, 2006). This protein influences DNA topology, and it was demonstrated that another such protein, IHF, can also stimulate p_R-initiated transcription (Łyżeń *et al.*, 2008). Interestingly, the SeqA protein may also control λ DNA replication indirectly by influencing the stability of the replication complex (this complex, under standard laboratory conditions of cultivation of bacteria, is stable and can be inherited by one of two daughter DNA molecules for many cell generations if bound to a plasmid derived from bacteriophage λ)(Narajczyk *et al.*, 2007a). Another stimulator of the p_R promoter has been identified, namely the DksA protein (Łyżeń *et al.*, 2009). Interestingly, this protein often acts together and synergistically with guanosine tetraphosphate (ppGpp), the stringent control alarmone, whereas their actions are independent and antagonistic at p_R (ppGpp inhibits and DksA stimulates p_R activity)(Łyżeń *et al.*, 2009). Interestingly, the phage-encoded P protein was found to inhibit biochemical activities of DnaA, which normally functions as a stimulator of p_R-initiated transcription (Datta *et al.*, 2005c, d). All these new discoveries were bases for the updated proposal of the mechanism of the switch from θ to σ replication mode during the lytic development of bacteriophage λ, published by Narajczyk *et al.* (2007b) and presented schematically in Figure 2.

Szambowska *et al.* (2011) provided another input into the puzzle of the regulatory mechanism of λ DNA replication regulation. It has been demonstrated that RNA polymerase interacts directly with the λ O protein. This may indicate that RNA polymerase can play an additional role in λ DNA replication initiation apart from transcribing the *ori*λ region. Moreover, it was found that binding of the λ O protein to tandem copies of specific DNA sequences affects DNA topology more significantly than suggested earlier (Chen *et al.*, 2010). Perhaps a large nucleoprotein structure is built whose activity is controlled in a more complex manner than predicted previously. This speculation may be supported by other recent discoveries that some lambdoid bacteriophages bear six O-binding sequences at the *ori* region instead of four (like in λ)(Nejman *et al.*, 2009) and that single amino acid substitutions in O and P proteins may drastically change requirements for factors influencing the transcriptional activation of *ori*λ (such as DnaA, ppGpp, and DksA)(Nejman *et al.*, 2011). In addition, activity of the CbpA protein, which can replace DnaJ in λ DNA replication, is controlled at multiple levels (Chenoweth and Wickner, 2008). Perhaps newly developed methods, such as allowing either direct observation of enzymes replicating DNA molecules (Kulczyk *et al.*, 2010) or dynamic analyses of phage–host interactions at the protein level (Maynard *et al.*, 2010), should help in understanding the complicated regulation of λ DNA replication initiation in more detail.

At the late stage of λ DNA replication, when long concatemers of the phage genome are produced as a result of rolling-circle replication, it is necessary to protect these linear "tails" from host nucleases. The RecBCD nuclease is one of them, and λ encodes an inhibitor of this enzyme, called Gam. The crystal structure of Gam has been resolved, which helped to show that this protein inhibits RecBCD by preventing it from binding DNA (Court *et al.*, 2007b). Interestingly, it seems that λ encodes its own inhibitor of DNA replication. It was known that the cII protein is toxic

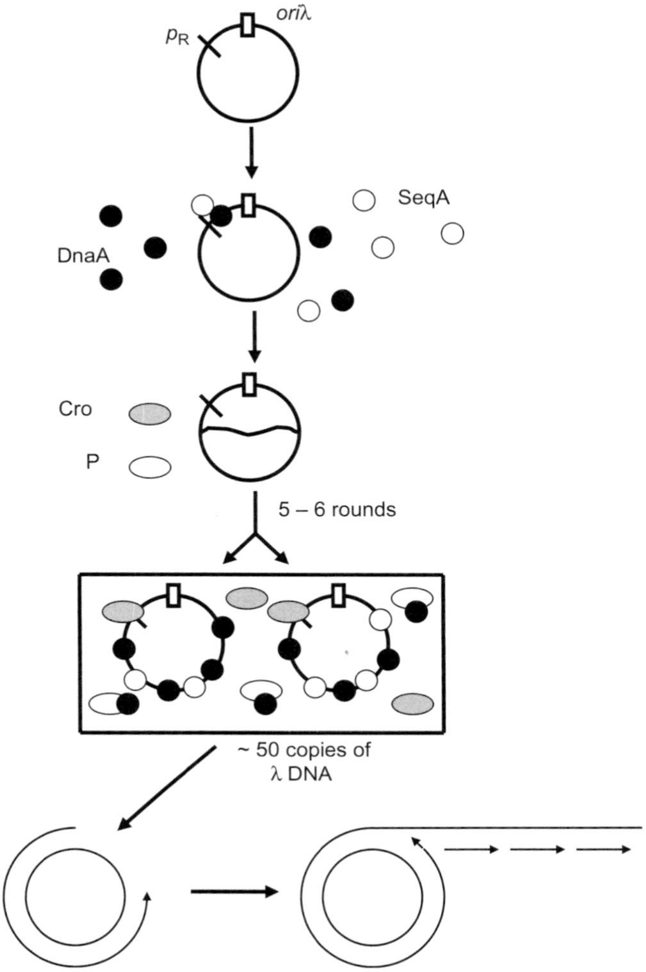

FIGURE 2 Model for regulation of the switch from circle-to-circle (θ) to rolling-circle (σ) replication of bacteriophage λ DNA. Shortly after infection, there are a few copies (or even just one copy) of phage DNA (thick circles, representing double-stranded DNA molecules) and a certain number of free DnaA (small filled circles) and SeqA (small open circles) molecules, which can potentially bind to weak DnaA or GATC boxes, respectively, at the p_R promoter (a thick slash) region. This binding is required to stimulate transcription from p_R, which activates the *origin* region (rectangle). Such an activation facilitates formation of the preprimosomal complex, consisting of the λ O protein (the replication initiator protein that binds to four iterone sequences located in *ori*λ), the λ P protein (a protein that delivers the host-encoded DNA helicase to *ori*λ), and the DnaB protein (the DNA helicase). DnaA- and SeqA-mediated stimulation of transcriptional activation of *ori*λ allows for initiation of bidirectional θ replication of λ DNA. After five to six rounds of such a replication, about 50 copies of λ genome appear. Due to the presence of many

when overproduced in *E. coli* cells; however, the mechanism of its toxicity remained unknown. Kędzierska *et al.* (2003) found that DNA replication is impaired in cells overexpressing the *c*II gene efficiently, which suggests that the replication machinery may be a target for cII toxic activity.

VI. GENERAL RECOMBINATION SYSTEM ENCODED BY λ

As an alternative to the hypothesis that the switch from θ to σ replication mode is the result of a change from bidirectional to unidirectional replication (Narajczyk *et al.*, 2007b), it has been suggested that this switch depends on formation of the recombination intermediate (reviewed by Weigel and Seitz, 2006). This could be possible, as bacteriophage λ encodes its own system for general genetic recombination, called Red, and is composed of two proteins: Exo and Bet. This system is known as one of the most efficient recombination pathways identified to date. Interestingly, recent works indicated another possible link between replication and recombination.

Although mechanisms of Red recombination based on strand annealing or strand invasion have been proposed previously, neither of them could explain the results of experiments in which crosses between a nonreplicating linear λ genome and a replicating plasmid bearing a cloned fragment of λ DNA were studied (Poteete, 2008). Therefore, another mechanism, in which the replisome invasion is considered, has been proposed. This mechanism implies that replication is directly involved in Red recombination (Poteete, 2008). Further work by Maresca *et al.* (2010) demonstrated that Red recombination indeed requires replication of the target molecule. They proposed a model in which the formation of single-stranded DNA heteroduplexes at the replication fork is required. The third work implicating the requirement of

DnaA- and SeqA-binding sequences in λ DNA, cellular DnaA and SeqA proteins are titrated out. Moreover, continuous production of the λ P protein (small open oval) results in its accumulation and interaction with DnaA, causing inhibition of activities of the latter protein. In addition, the high amount of the Cro repressor (small gray oval), which blocks transcription from p_R when occurring at elevated concentrations in cell, is also produced. Therefore, transcriptional activation of *ori*λ becomes impaired. This leads to the initiation of unidirectional θ replication, which, after one round, switches to the σ mode because a round of unidirectional θ replication initiated at *ori*λ is followed by displacement of the 5′ end of the newly synthesized leading strand by its growing 3′ end (arrow). For simplification of the diagram, at the switch from θ to σ replication, only the fate of the parental strand (thin lines) directing the synthesis of the leading strand is shown. This figure is an updated version of the scheme published by Węgrzyn and Węgrzyn (2005).

DNA replication in Red recombination was published by Mosberg *et al.* (2010). They suggested that the Exo protein degrades one λ DNA strand and leaves the other strand intact as single-stranded DNA, which is annealed at the replication fork in the reaction catalyzed by the Bet protein. Altogether, these studies indicate that DNA replication is required for efficient recombination by the Red system.

VII. TRANSCRIPTION ANTITERMINATION

Transcription antitermination is a specific mechanism of regulation of the efficiency of RNA production. It is based on the formation of complexes between RNA polymerase and regulatory proteins, which prevent transcription termination at otherwise functional terminator regions. Bacteriophage λ encodes two antitermination proteins, N and Q. In N-dependent antitermination, a large nucleoprotein complex is formed, which contains RNA polymerase, the λ N protein, and a set of host-encoded regulatory proteins, including NusA, NusB, NusE, and NusG. The N antitermination complex, assembled on the RNA strand formed just after transcription of the corresponding DNA fragment, is one of the largest prokaryotic transcription complexes. Therefore, recent works have focused on more detailed description of the structure of this complex and interactions between its elements to better understand its function.

Conant *et al.* (2005) described quantitatively the binding states of the N protein, while Horiya *et al.* (2009) concluded about similar issues on the basis of replacement of *boxB* RNA (a part of the transcript that, together with *boxA* RNA, forms a structure recognized by the antitermination complex) and N with the heterologous system composed of interacting RNA and peptide molecules. That work led to a determination of spatial requirements for RNA–protein interactions. A role for RNA looping was tested by Conant *et al.* (2008), who proposed that such a looping facilitates interaction of RNA with other elements of the N antitermination complex. Interestingly, it appears that p_R promoter activity can influence assembly of the N antitermination complex significantly (Zhou *et al.*, 2006). Finally, interactions of the NusA protein with the λ N protein and the specific RNA sequence were studied by Prasch *et al.* (2006 and 2009, respectively), and affinity of the NusB–NusE complex to *boxA* RNA was determined by Burmann *et al.* (2010). All these studies provided very important information, which now can be used for building more detailed models on the function of the large nucleoprotein complex enabling N-dependent transcription antitermination. It appears significant, that one of the first examples of the new field of ***Synthetic Biology*** was the organic chemical synthesis of the antitermination *nut* sites and its mutants (Brown and Szybalski, 1985).

In the Q-dependent antitermination system, a significantly less complicated complex is formed, which includes only RNA polymerase, the Q protein, and NusA. Contrary to the N protein, the *Q* gene product interacts with DNA rather than with RNA. However, Nickels *et al.* (2006) demonstrated that RNA-mediated destabilization of the interaction between σ^{70} and β subunits is required for binding of the Q protein to the RNA polymerase holoenzyme. In fact, it was subsequently demonstrated that Q is a stable component of the transcription elongation complex (Deighan and Hochschild, 2007) and that this antitermination protein contacts the β-flap domain of RNA polymerase (Deighan *et al.*, 2008).

VIII. FORMATION OF MATURE PROGENY VIRIONS

Assembly of mature virions is a complicated process in which many kinds of proteins, which build phage head and tail, are involved. Although all players of this reaction are perhaps already identified, details of particular steps of this process and detailed roles of particular proteins remained poorly understood. For example, it was known that protein cross-linking and proteolytic maturation events are necessary for phage head formation, but the specific protease remained unknown. Medina *et al.* (2010) demonstrated that the phage-encoded C protein is the specific protease responsible for λ procapsid maturation. Interestingly, it was found that major head and tail proteins, E and V, respectively, as well as a capsid decoration protein D, are associated with transition metals (Zhang *et al.*, 2011).

λ DNA packaging into preformed heads, together with addition of the preformed tail, finalizes the virion assembly. The packaging reaction is catalyzed by the complex of the phage-encoded A and Nu1 proteins, called the terminase complex, and the specificities of terminases from different lambdoid phages have been studied, including phylogenetic analyses of these enzymes (Feiss *et al.*, 2010). An interesting work has been published by Yang *et al.* (2009), who determined that phage λ DNA is packaged at the rate of about 120bp per second at 4°C and at 1m*M* ATP. Although the packaging reaction usually occurs at higher temperatures in a natural environment, these results provide important information facilitating estimation of the actual rate of this reaction proceeding inside the host cell.

IX. HOST CELL LYSIS

The final step of bacteriophage λ development is lysis of the host cell. This reaction is performed by two major phage-encoded proteins, S holin and R transglycosylase, and two accessory proteins, Rz and Rz1. All these proteins are products of phage genes transcribed from the λ late promoter

$p_{R'}$. Work by Shao and Wang (2009) led to a description of the model for prediction of the effects of $p_{R'}$ activity on lysis time.

The specific roles of Rz and Rz1 proteins were poorly understood for a relatively long time. It is worth mentioning here that the existence and functionality of *Rz* and *Rz1* genes have been discovered by Alina Taylor and co-workers (Hanych *et al.*, 1993; Taylor *et al.*, 1983) relatively late, well after determining the whole sequence of the phage λ genome. Now we know that Rz and Rz1, classified as spanins, are involved in disrupting the outer membrane of the bacterial cell. Structures of these proteins were determined in more detail recently (Berry *et al.*, 2010), which was the basis for speculations on the model of Rz-Rz1-dependent disruption of the outer membrane.

The protein making holes in the cytoplasmic membrane is encoded by the phage λ *S* gene. Using cryoelectron microscopy, Dewey *et al.* (2010) demonstrated that holes made by this protein are surprisingly large, having an average diameter of 340 nm, while some holes were even as large as 1 μm in diameter. Interestingly, most cells had only one large hole.

An old and intriguing question was why cell lysis occurs at the very end of phage development, while genes coding for S, R, Rz, and Rz1 proteins are located at the beginning of the $p_{R'}$ operon. It was known that the *S* cistron encodes two proteins, the holin (called S105) and the antiholin (S107), due to the presence of two alternative translation start sites. However, it remained unclear what actually triggers lysis, initiated by formation of the hole in the cytoplamic membrane. White *et al.* (2011) suggest that lysis occurs when the holin reaches a critical concentration and nucleates to form rafts. Therefore, a significant amount of time is needed for the production of sufficient amounts of the S105 protein, and the presence of S107 prevents premature formation of S105 rafts.

X. CONCLUDING REMARKS

In 2001, David Friedman and Don Court published an article entitled "Bacteriophage λ: Alive and Well and Still Doing Its Thing" to indicate that this classical model, used for years in biological research, could still serve as an excellent tool in studies on molecular mechanisms of basic cellular processes. It appears obvious that after 10 years, the statement by Friedman and Court (2001) is still valid and that the use of bacteriophage λ (as well as other lambdoid phages) in basic and applied research may be required to solve some still difficult problems in understanding molecular mechanisms of biological processes. This statement is supported by the number of excellent papers describing studies on λ, published in previous several years and cited in this article, that concerned various aspects of phage biology, particularly phage DNA injection, molecular bases of the lysis-versus-lysogenization decision and the lysogenization process itself, prophage maintenance and induction, DNA replication, genetic recombination, transcription antitermination, formation of the virion structure, and lysis of the host cell.

ACKNOWLEDGMENT

This work was supported by Ministry of Science and Higher Education, Poland (Grant N N301 192439 to AW).

REFERENCES

Anderson, L. M., and Yang, H. (2008). DNA looping can enhance lysogenic CI transcription in phage lambda. *Proc. Natl. Acad. Sci. USA* **105:**5827–5832.

Atsumi, S., and Little, J. W. (2006). A synthetic phage lambda regulatory circuit. *Proc. Natl. Acad. Sci. USA* **103:**19045–19050.

Avlund, M., Dodd, I. B., Semsey, S., Sneppen, K., and Krishna, S. (2009a). Why do phage play dice? *J. Virol.* **83:**11416–11420.

Avlund, M., Dodd, I. B., Sneppen, K., and Krishna, S. (2009b). Minimal gene regulatory circuits that can count like bacteriophage lambda. *J. Mol. Biol.* **394:**681–693.

Babić, A. C., and Little, J. W. (2007). Cooperative DNA binding by CI repressor is dispensable in a phage lambda variant. *Proc. Natl. Acad. Sci. USA* **104:**17741–17746.

Bakk, A. (2005). Transcriptional activation mechanisms of the PRM promoter of lambda phage. *Biophys. Chem.* **114:**229–234.

Bandyopadhyay, K., Parua, P. K., Datta, A. B., and Parrack, P. (2010). *Escherichia coli* HflK and HflC can individually inhibit the HflB (FtsH)-mediated proteolysis of lambdaCII in vitro. *Arch. Biochem. Biophys.* **501:**239–243.

Berry, J., Savva, C., Holzenburg, A., and Young, R. (2010). The lambda spanin components Rz and Rz1 undergo tertiary and quaternary rearrangements upon complex formation. *Protein Sci.* **19:**1967–1977.

Brown, A. L., and Szybalski, W. (1985). Transcriptional antitermination activity of the synthetic *nut* elements of coliphage lambda. I. Assembly of the nutR recognition site from boxA and nut core elements. *Gene* **39:**121–127.

Brouns, S. J., Jore, M. M., Lundgren, M., Westra, E. R., Slijkhuis, R. J., Snijders, A. P., Dickman, M. J., Makarova, K. S., Koonin, E. V., and van der Oost, J. (2008). Small CRISPR RNAs guide antiviral defense in prokaryotes. *Science* **321:**960–964.

Burmann, B. M., Luo, X., Rösch, P., Wahl, M. C., and Gottesman, M. E. (2010). Fine tuning of the *E. coli* NusB:NusE complex affinity to BoxA RNA is required for processive antitermination. *Nucleic Acids Res* **38:**314–326.

Cao, Y., Lu, H. M., and Liang, J. (2008). Stochastic probability landscape model for switching efficiency, robustness, and differential threshold for induction of genetic circuit in phage. *Conf. Proc. IEEE Eng. Med. Biol. Soc.* pp. 611–614.

Cao, Y., Lu, H. M., and Liang, J. (2010). Probability landscape of heritable and robust epigenetic state of lysogeny in phage lambda. *Proc. Natal. Acad. Sci. USA* **107:**18445–18450.

Chen, B., Xiao, Y., Liu, C., Li, C., and Leng, F. (2010). DNA linking number change induced by sequence-specific DNA-binding proteins. *Nucleic Acids Res.* **38:**3643–3654.

Chenoweth, M. R., and Wickner, S. (2008). Complex regulation of the DnaJ homolog CbpA by the global regulators sigmaS and Lrp, by the specific inhibitor CbpM, and by the proteolytic degradation of CbpM. *J. Bacteriol.* **190:**5153–5161.

Conant, C. R., Goodarzi, J. P., Weitzel, S. E., and von Hippel, P. H. (2008). The antitermination activity of bacteriophage lambda N protein is controlled by the kinetics of an RNA-looping-facilitated interaction with the transcription complex. *J. Mol. Biol.* **384:**87–108.
Conant, C. R., Van Gilst, M. R., Weitzel, S. E., Rees, W. A., and von Hippel, P. H. (2005). A quantitative description of the binding states and in vitro function of antitermination protein N of bacteriophage lambda. *J. Mol. Biol.* **348:**1039–1057.
Court, D. L., Oppenheim, A. B., and Adhya, S. L. (2007a). A new look at bacteriophage lambda genetic networks. *J. Bacteriol.* **189:**298–304.
Court, R., Cook, N., Saikrishnan, K., and Wigley, D. (2007b). The crystal structure of lambda-Gam protein suggests a model for RecBCD inhibition. *J. Mol. Biol.* **371:**25–33.
Datta, A. B., Panjikar, S., Weiss, M. S., Chakrabarti, P., and Parrack, P. (2005a). Structure of lambda CII: Implications for recognition of direct-repeat DNA by an unusual tetrameric organization. *Proc. Natl. Acad. Sci. USA* **102:**11242–11247.
Datta, A. B., Roy, S., and Parrack, P. (2005b). Role of C-terminal residues in oligomerization and stability of lambda CII: Implications for lysis-lysogeny decision of the phage. *J. Mol. Biol.* **345:**315–324.
Datta, I., Banik-Maiti, S., Adhakari, L., Sau, S., Das, N., and Mandal, N. C. (2005c). The mutation that makes *Escherichia coli* resistant to lambda P gene-mediated host lethality is located within the DNA initiator Gene *dnaA* of the bacterium. *J. Biochem. Mol. Biol.* **38:**89–96.
Datta, I., Sau, S., Sil, A. K., and Mandal, N. C. (2005d). The bacteriophage lambda DNA replication protein P inhibits the *oriC* DNA- and ATP-binding functions of the DNA replication initiator protein DnaA of *Eshcerichia coli*. *J. Biochem. Mol. Biol.* **38:**97–103.
Deighan, P., and Hochschild, A. (2007). The bacteriophage lambdaQ anti-terminator protein regulates late gene expression as a stable component of the transcription elongation complex. *Mol. Microbiol.* **63:**911–920.
Deighan, P., Diez, C. M., Leibman, M., Hochschild, A., and Nickels, B. E. (2008). The bacteriophage lambda Q antiterminator protein contacts the beta-flap domain of RNA polymerase. *Proc. Natl. Acad. Sci. USA* **105:**15305–15310.
Dewey, J. S., Savva, C. G., White, R. L., Vitha, S., Holzenburg, A., and Young, R. (2010). Micron-scale holes terminate the phage infection cycle. *Proc. Natl. Acad. Sci. USA* **107:**2219–2223.
Dodd, I. B., Shearwin, K. E., and Egan, J. B. (2005). Revisited gene regulation in bacteriophage lambda. *Curr. Opin. Genet. Dev.* **15:**145–152.
Edgar, R., and Qimron, U. (2010). The *Escherichia coli* CRISPR system protects from λ lysogenization, lysogens, and prophage induction. *J. Bacteriol.* **192:**6291–6294.
Feiss, M., Reynolds, E., Schrock, M., and Sippy, J. (2010). DNA packaging by lambda-like bacteriophages: Mutations broadening the packaging specificity of terminase, the lambda-packaging enzyme. *Genetics* **184:**43–52.
Fogg, P. C., Rigden, D. J., Saunders, J. R., McCarthy, A. J., and Allison, H. E. (2011). Characterization of the relationship between integrase, excisionase and antirepressor activities associated with a superinfecting Shiga toxin encoding bacteriophage. *Nucleic Acids Res.* **39:**2116–2129.
Friedman, D. I., and Court, D. L. (2001). Bacteriophage lambda: Alive and well and still doing its thing. *Curr. Opin. Microbiol.* **4:**201–207.
Garufi, G., Minenkova, O., Lo Passo, C., Pernice, I., and Felici, F. (2005). Display libraries on bacteriophage lambda capsid. *Biotechnol. Annu. Rev.* **11:**153–190.
Gedeon, T., Mischaikow, K., Patterson, K., and Traldi, E. (2008). Binding cooperativity in phage lambda is not sufficient to produce an effective switch. *Biophys. J.* **94:**3384–3392.
Ghosh, D., Roy, K., Williamson, K. E., Srinivasiah, S., Wommack, K. E., and Radosevich, M. (2009). Acyl-homoserine lactones can induce virus production in lysogenic bacteria: An alternative paradigm for prophage induction. *Appl. Environ. Microbiol.* **75:**7142–7152.

Glinkowska, M., Łoś, J. M., Szambowska, A., Czyż, A., Całkiewicz, J., Herman-Antosiewicz, A., Wróbel, B., Węgrzyn, G., Węgrzyn, A., and Łoś, M. (2010). Influence of the *Escherichia coli oxyR* gene function on lambda prophage maintenance. *Arch. Microbiol.* **192:**673–683.

Gottesman, M. E., and Weisberg, R. A. (2004). Little lambda, who made thee? *Microbiol. Mol. Biol. Rev.* **68:**796–813.

Gregory, R., Saunders, V. A., and Saunders, J. R. (2010). Rule-based simulation of temperate bacteriophage infection: Restriction-modification as a limiter to infection in bacterial populations. *Biosystems* **100:**166–177.

Halder, S., Banerjee, S., and Parrack, P. (2008). Direct CIII-HflB interaction is responsible for the inhibition of the HflB (FtsH)-mediated proteolysis of *Escherichia coli* sigma(32) by lambda CIII. *FEBS J.* **275:**4767–4772.

Halder, S., Datta, A. B., and Parrack, P. (2007). Probing the antiprotease activity of lambda-CIII, an inhibitor of the *Escherichia coli* metalloprotease HflB (FtsH). *J. Bacteriol.* **189:**8130–8138.

Hanych, B., Kędzierska, S., Walderich, B., Uznański, B., and Taylor, A. (1993). Expression of the *Rz* gene and the overlapping *Rz1* reading frame present at the right end of the bacteriophage lambda genome. *Gene* **129:**1–8.

Hochschild, A., and Lewis, M. (2009). The bacteriophage lambda CI protein finds an asymmetric solution. *Curr. Opin. Struct. Biol.* **19:**79–86.

Horiya, S., Inaba, M., Koh, C. S., Uehara, H., Masui, N., Ishibashi, M., Matsufuji, S., and Harada, K. (2009). Analysis of the spacial requirements for RNA-protein interactions within the N antitermination complex of bacteriophage lambda. *Nucleic Acids Symp. Ser.* **53:**91–92.

Jia, H., Satumba, W. J., Bidwell, G. L., 3rd, and Mossing, M. C. (2005). Slow assembly and disassembly of lambda Cro repressor dimers. *J. Mol. Biol.* **350:**919–929.

Kędzierska, B., Glinkowska, M., Iwanicki, A., Obuchowski, M., Sojka, P., Thomas, M. S., and Węgrzyn, G. (2003). Toxicity of the bacteriophage lambda cII gene product to *Escherichia coli* arises from inhibition of host cell DNA replication. *Virology* **313:**622–628.

Kędzierska, B., Lee, D. J., Węgrzyn, G., Busby, S. J., and Thomas, M. S. (2004). Role of the RNA polymerase alpha subunits in CII-dependent activation of the bacteriophage lambda pE promoter: Identification of important residues and positioning of the alpha C-terminal domains. *Nucleic Acids Res.* **32:**834–841.

Kędzierska, B., Szambowska, A., Herman-Antosiewicz, A., Lee, D. J., Busby, S. J., Węgrzyn, G., and Thomas, M. S. (2007). The C-terminal domain of the *Escherichia coli* RNA polymerase alpha subunit plays a role in the CI-dependent activation of the bacteriophage lambda pM promoter. *Nucleic Acids Res.* **35:**2311–2320.

Kobiler, O., Rokney, A., Friedman, N., Court, D. L., Stavans, J., and Oppenheim, A. B. (2005). Quantitative kinetic analysis of the bacteriophage lambda genetic network. *Proc. Natl. Acad. Sci. USA* **102:**4470–4775.

Kobiler, O., Rokney, A., and Oppenheim, A. B. (2007). Phage lambda CIII: A protease inhibitor regulating the lysis-lysogeny decision. *PLoS One* **2:**e363.

Kontur, W. S., Saecker, R. M., Davis, C. A., Capp, M. W., and Record, M. T., Jr. (2006). Solute probes of conformational changes in open complex (RPo) formation by *Escherichia coli* RNA polymerase at the lambdaPR promoter: Evidence for unmasking of the active site in the isomerization step and for large-scale coupled folding in the subsequent conversion to RPo. *Biochemistry* **45:**2161–2177.

Köster, S., Evilevitch, A., Jeembaeva, M., and Weitz, D. A. (2009). Influence of internal capsid pressure on viral infection by phage lambda. *Biophys. J.* **97:**1525–1529.

Krupovic, M., and Bamford, D. H. (2008). Holin of bacteriophage lambda: Structural insights into a membrane lesion. *Mol. Microbiol.* **69:**781–783.

Kulczyk, A. W., Tanner, N. A., Loparo, J. J., Richardson, C. C., and van Oijen, A. M. (2010). Direct observation of enzymes replicating DNA using a single-molecule DNA stretching assay. *J. Vis. Exp.* **37:**1689.

Little, J. W. (2010). Evolution of complex gene regulatory circuits by addition of refinements. *Curr. Biol.* **20:**R724–R734.

Little, J. W., and Michalowski, C. B. (2010). Stability and instability in the lysogenic state of phage lambda. *J. Bacteriol.* **192:**6064–6076.

Lou, C., Yang, X., Liu, X., He, B., and Ouyang, Q. (2007). A quantitative study of lambda-phage SWITCH and its components. *Biophys. J.* **92:**2685–2693.

Łoś, J. M., Łoś, M., Węgrzyn, A., and Węgrzyn, G. (2008). Role of the bacteriophage lambda *exo-xis* region in the virus development. *Folia Microbiol.* **53:**443–450.

Łoś, J. M., Łoś, M., Węgrzyn, G., and Węgrzyn, A. (2009). Differential efficiency of induction of various lambdoid prophages responsible for production of Shiga toxins in response to different induction agents. *Microb. Pathog.* **47:**289–298.

Łoś, J. M., Łoś, M., Węgrzyn, A., and Węgrzyn, G. (2010). Hydrogen peroxide-mediated induction of the Shiga toxin-converting lambdoid prophage ST2-8624 in *Escherichia coli* O157:H7. *FEMS Immunol. Med. Microbiol.* **58:**322–329.

Łyżeń, R., Kochanowska, M., Węgrzyn, G., and Szalewska-Pałasz, A. (2008). IHF- and SeqA-binding sites, present in plasmid cloning vectors, may significantly influence activities of promoters. *Plasmid* **60:**125–130.

Łyżeń, R., Kochanowska, M., Węgrzyn, G., and Szalewska-Palasz, A. (2009). Transcription from bacteriophage lambda pR promoter is regulated independently and antagonistically by DksA and ppGpp. *Nucleic Acids Res.* **37:**6655–6664.

Łyżeń, R., Węgrzyn, G., Węgrzyn, A., and Szalewska-Pałasz, A. (2006). Stimulation of the lambda pR promoter by *Escherichia coli* SeqA protein requires downstream GATC sequences and involves late stages of transcription initiation. *Microbiology* **152:**2985–2992.

Maresca, M., Erler, A., Fu, J., Friedrich, A., Zhang, Y., and Stewart, A. F. (2010). Single-stranded heteroduplex intermediates in lambda Red homologous recombination. *BMC Mol. Biol.* **11:**54.

Martínez-Trujillo, M., Sánchez-Trujillo, A., Ceja, V., Avila-Moreno, F., Bermúdez-Cruz, R. M., Court, D., and Montañez, C. (2010). Sequences required for transcription termination at the intrinsic lambda t1 terminator. *Can. J. Microbiol.* **56:**168–177.

Maynard, N. D., Birch, E. W., Sanghvi, J. C., Chen, L., Gutschow, M. V., and Covert, M. W. (2010). A forward-genetic screen and dynamic analysis of lambda phage host-dependencies reveals an extensive interaction network and a new anti-viral strategy. *PLoS Genet.* **6:** e1001017.

Medina, E., Wieczorek, D., Medina, E. M., Yang, Q., Feiss, M., and Catalano, C. E. (2010). Assembly and maturation of the bacteriophage lambda procapsid: gpC is the viral protease. *J. Mol. Biol.* **401:**813–830.

Morelli, M. J., Ten Wolde, P. R., and Allen, R. J. (2009). DNA looping provides stability and robustness to the bacteriophage lambda switch. *Proc. Natl. Acad. Sci. USA* **106:**8101–8106.

Mosberg, J. A., Lajoie, M. J., and Church, G. M. (2010). Lambda Red recombineering in *Escherichia coli* occurs through a fully single-stranded intermediate. *Genetics* **186:**791–799.

Narajczyk, M., Barańska, S., Szambowska, A., Glinkowska, M., Węgrzyn, A., and Węgrzyn, G. (2007a). Modulation of lambda plasmid and phage DNA replication by *Escherichia coli* SeqA protein. *Microbiology* **153:**1653–1663.

Narajczyk, M., Barańska, S., Węgrzyn, A., and Węgrzyn, G. (2007b). Switch from theta to sigma replication of bacteriophage lambda DNA: Factors involved in process and a model for its regulation. *Mol. Genet. Genomics* **278:**65–74.

Nejman, B., Łoś, J. M., Łoś, M., Węgrzyn, G., and Węgrzyn, A. (2009). Plasmids derived from lambdoid bacteriophages as models for studying replication of mobile genetic elements responsible for the production of Shiga toxins by pathogenic *Escherichia coli* strains. *J. Mol. Microbiol. Biotechnol.* **17:**211–220.

Nejman, B., Nadratowska-Wesołowska, B., Szalewska-Pałasz, A., Węgrzyn, A., and Węgrzyn, G. (2011). Replication of plasmids derived from Shiga toxin-converting bacteriophages in starved *Escherichia coli. Microbiology* **157:**220–233.

Nickels, B. E., Roberts, C. W., Roberts, J. W., and Hochschild, A. (2006). RNA-mediated destabilization of the sigma(70) region 4/beta flap interaction facilitates engagement of RNA polymerase by the Q antiterminator. *Mol. Cell* **24:**457–468.

Oppenheim, A. B., Kobiler, O., Stavans, J., Court, D. L., and Adhya, S. (2005). Switches in bacteriophage lambda development. *Annu. Rev. Genet.* **39:**409–429.

Papagiannis, C. V., Sam, M. D., Abbani, M. A., Yoo, D., Cascio, D., Clubb, R. T., and Johnson, R. C. (2007). Fis targets assembly of the Xis nucleoprotein filament to promote excisive recombination by phage lambda. *J. Mol. Biol.* **367:**328–343.

Parua, P. K., Datta, A. B., and Parrack, P. (2010a). Specific hydrophobic residues in the alpha4 helix of lambda CII are crucial for maintaining its tetrameric structure and directing the lysogenic choice. *J. Gen. Virol.* **91:**306–312.

Parua, P. K., Mondal, A., and Parrack, P. (2010b). HflD, an *Escherichia coli* protein involved in the lambda lysis-lysogeny switch, impairs transcription activation by lambda CII. *Arch. Biochem. Biophys.* **493:**175–183.

Poteete, A. R. (2008). Involvement of DNA replication in phage lambda Red-mediated homologous recombination. *Mol. Microbiol.* **68:**66–74.

Potrykus, K., Węgrzyn, G., and Hernandez, V. J. (2004). Direct stimulation of the lambdapaQ promoter by the transcription effector guanosine-3′,5′-(bis)pyrophosphate in a defined *in vitro* system. *J. Biol. Chem.* **279:**19860–19866.

Prasch, S., Jurk, M., Washburn, R. S., Gottesman, M. E., Wöhrl, B. M., and Rösch, P. (2009). RNA-binding specificity of *E. coli* NusA. *Nucleic Acids Res* **37:**4736–4742.

Prasch, S., Schwarz, S., Eisenmann, A., Wöhrl, B. M., Schweimer, K., and Rösch, P. (2006). Interaction of the intrinsically unstructured phage lambda N protein with *Escherichia coli* NusA. *Biochemistry* **45:**4542–4549.

Ptashne, M. (2005). Regulation of transcription: From lambda to eukaryotes. *Trends Biochem. Sci.* **30:**275–279.

Ptashne, M. (2006). Lambda's switch: Lessons from a module swap. *Curr. Biol.* **16:**R459–R462.

Refardt, D., and Rainey, P. B. (2010). Tuning a genetic switch: Experimental evolution and natural variation of prophage induction. *Evolution* **64:**1086–1097.

Rokney, A., Kobiler, O., Amir, A., Court, D. L., Stavans, J., Adhya, S., and Oppenheim, A. B. (2008). Host responses influence on the induction of lambda prophage. *Mol. Microbiol.* **68:**29–36.

Rosenfeld, N., Young, J. W., Alon, U., Swain, P. S., and Elowitz, M. B. (2005). Gene regulation at the single-cell level. *Science* **307:**1962–1965.

Schubert, R. A., Dodd, I. B., Egan, J. B., and Shearwin, K. E. (2007). Cro's role in the CI Cro bistable switch is critical for λ's transition from lysogeny to lytic development. *Genes Dev.* **21:**2461–2472.

Shao, Y., and Wang, I. N. (2009). Effect of late promoter activity on bacteriophage lambda fitness. *Genetics* **181:**1467–7145.

Shkilnyj, P., and Koudelka, G. B. (2007). Effect of salt shock on stability of λ^{imm434} lysogens. *J. Bacteriol.* **189:**3115–3123.

Stayrook, S., Jaru-Ampornpan, P., Ni, J., Hochschild, A., and Lewis, M. (2008). Crystal structure of the lambda repressor and a model for pairwise cooperative operator binding. *Nature* **452:**1022–1025.

Sun, X., Mierke, D. F., Biswas, T., Lee, S. Y., Landy, A., and Radman-Livaja, M. (2006). Architecture of the 99 bp DNA-six-protein regulatory complex of the lambda *att* site. *Mol. Cell.* **24:**569–580.

Svenningsen, S. L., Costantino, N., Court, D. L., and Adhya, S. (2005). On the role of Cro in lambda prophage induction. *Proc. Natl. Acad. Sci. USA* **102:**4465–4469.

Szambowska, A., Pierechod, M., Węgrzyn, G., and Glinkowska, M. (2011). Coupling of transcription and replication machineries in λ DNA replication initiation: Evidence for direct interaction of *Escherichia coli* RNA polymerase and λO protein. *Nucleic Acids Res.* **39:**168–177.

Tay, Y., Ho, C., Droge, P., and Ghadessy, F. J. (2010). Selection of bacteriophage lambda integrases with altered recombination specificity by *in vitro* compartmentalization. *Nucleic Acids Res.* **38:**e25.

Taylor, A., Benedik, M., and Campbell, A. (1983). Location of the *Rz* gene in bacteriophage lambda. *Gene* **26:**159–163.

Thomason, L., Court, D. L., Bubunenko, M., Costantino, N., Wilson, H., Datta, S., and Oppenheim, A. (2007). Recombineering: Genetic engineering in bacteria using homologous recombination. *Curr. Protoc. Mol. Biol.* Chapt. 1, 1.16.

Valdez-Cruz, N. A., Caspeta, L., Pérez, N. O., Ramirez, O. T., and Trujillo-Roldán, M. A. (2010). Production of recombinant proteins in *E. coli* by the heat inducible expression system based on the phage lambda pL and/or pR promoters. *Microb. Cell Fact* **9:**18.

Van Duyne, G. D. (2005). Lambda integrase: Armed for recombination. *Curr. Biol.* **15:** R658–R660.

Wang, H., Finzi, L., Lewis, D. E., and Dunlap, D. (2009). AFM studies of lambda repressor oligomers securing DNA loops. *Curr. Pharm. Biotechnol.* **10:**494–501.

Weigel, C., and Seitz, H. (2006). Bacteriophage replication modules. *FEMS Microbiol. Rev.* **30:**321–381.

Węgrzyn, G., and Węgrzyn, A. (2005). Genetic switches during bacteriophage lambda development. *Prog. Nucleic Acid Res. Mol. Biol.* **79:**1–48.

Westra, E. R., Pul, U., Heidrich, N., Jore, M. M., Lundgren, M., Stratmann, T., Wurm, R., Raine, A., Mescher, M., Van Heereveld, L., Mastop, M., Wagner, E. G., Schnetz, K., Van Der Oost, J., Wagner, R., and Brouns, S. J. (2010). H-NS-mediated repression of CRISPR-based immunity in *Escherichia coli* K12 can be relieved by the transcription activator LeuO. *Mol. Microbiol.* **77:**1380–1393.

White, R., Chiba, S., Pang, T., Dewey, J. S., Savva, C. G., Holzenburg, A., Pogliano, K., and Young, R. (2011). Holin triggering in real time. *Proc. Natl. Acad. Sci. USA* **108:**798–803.

Wu, D., Van Valen, D., Hu, Q., and Phillips, R. (2010). Ion-dependent dynamics of DNA ejections for bacteriophage lambda. *Biophys. J.* **99:**1101–1109.

Yang, Q., Catalano, C. E., and Maluf, N. K. (2009). Kinetic analysis of the genome packaging reaction in bacteriophage lambda. *Biochemistry* **48:**10705–10715.

Zhang, Y., Thompson, R., and Caruso, J. (2011). Probing the viral metallome: Searching for metalloproteins in bacteriophage λ: The hunt begins. *Metallomics* **3:**472–481.

Zhou, Y., Shi, T., Mozola, M. A., Olson, E. R., Henthorn, K., Brown, S., Gussin, G. N., and Friedman, D. I. (2006). Evidence that the promoter can influence assembly of antitermination complexes at downstream RNA sites. *J. Bacteriol.* **188:**2222–2232.

Zong, C., So, L. H., Sepúlveda, L. A., Skinner, S. O., and Golding, I. (2010). Lysogen stability is determined by the frequency of activity bursts from the fate-determining gene. *Mol. Syst. Biol.* **6:**440.

CHAPTER 7

The Secret Lives of Mycobacteriophages

Graham F. Hatfull

Contents

Department of Biological Sciences, Pittsburgh Bacteriophage Institute, University of Pittsburgh, Pittsburgh, Pennslyvania, USA

Advances in Virus Research, Volume 82
ISSN 0065-3527, DOI: 10.1016/B978-0-12-394621-8.00015-7

Abstract

The study of mycobacteriophages provides insights into viral diversity and evolution, as well as the genetics and physiology of their pathogenic hosts. Genomic characterization of 80 mycobacteriophages reveals a high degree of genetic diversity and an especially rich reservoir of interesting genes. These include a vast number of genes of unknown function that do not match known database entries and many genes whose functions can be predicted but which are not typically found as components of phage genomes. Thus many mysteries surround these genomes, such as why the genes are there, what do they do, how are they expressed and regulated, how do they influence the physiology of the host bacterium, and what forces of evolution directed them to their genomic homes? Although the genetic diversity and novelty of these phages is full of intrigue, it is a godsend for the mycobacterial geneticist, presenting an abundantly rich toolbox that can be exploited to devise new and effective ways for understanding the genetics and physiology of human tuberculosis. As the number of sequenced genomes continues to grow, their mysteries continue to thicken, and the time has come to learn more about the secret lives of mycobacteriophages.

I. INTRODUCTION

Mycobacteriophages are viruses that infect mycobacterial hosts. Interest in these viruses first arose in the late 1940s, and more than 300 publications followed in the 1950s, 1960s, and 1970s. Many of these studies focused on descriptions of new mycobacteriophages and their characteristics and utility in phage typing of clinical specimens. There was a significant decline in the next two decades with fewer than 100 papers published, followed by a resurgence in the early 1990s, and over 250 publications in the following two decades. This resurgence was fueled by the pioneering work of Dr. Jacobs and colleagues in using mycobacteriophages to deliver foreign DNA into mycobacteria (Jacobs, 2000; Jacobs *et al.*, 1987) and by the advent of the genomics era.

The utility of exploiting mycobacteriophages to understand their pathogenic hosts—such as *Mycobacterium tuberculosis* and *Mycobacterium leprae*, the causative agents of human tuberculosis and leprosy, respectively—is enhanced by complications in growth and manipulation of their bacterial hosts (Jacobs, 1992). *M. tuberculosis* can be propagated in the laboratory with relative ease, except that it grows extremely slowly, with a doubling time of 24h, and virulent strains require biosafety level III containment. *M. leprae* cannot be grown readily under defined laboratory conditions and no simple genetic tools are available (Scollard *et al.*, 2006). Mycobacteriophages multiply relatively quickly (plaques appear on a lawn of *M. tuberculosis* in 3–4days, whereas colonies take 3–4weeks to grow) and can be grown easily to high titers (Jacobs, 2000). However, isolation of new mycobacteriophages on slow-growing strains such as *M. tuberculosis* is complicated because contamination becomes a serious problem—everything else grows faster than *M. tuberculosis*. Ever since the late 1940s it has been commonplace to use relatively fast-growing saprophytic nonpathogenic strains such as *Mycobacterium smegmatis* (doubling time ~3 hours) to isolate and propagate mycobacteriophages (Mizuguchi, 1984). Some of these phages also infect *M. tuberculosis*, although many do not. However, these host preferences may be derived from host surface differences rather than metabolic restrictions on gene expression, DNA replication, packaging, or lysis (Hatfull, 2010; Hatfull *et al.*, 2010).

The application of more sophisticated molecular genetic approaches has made mycobacteriophages important tools in mycobacterial genetics and has been taken advantage of in numerous ways. However, the genomic characterization of mycobacteriophages has also shown them to be enormously diverse, rendering them as fruitful subjects for addressing broader questions in viral diversity and elucidating evolutionary mechanisms (Hatfull, 2010). These dual approaches - exploration and exploitation - work well together such that key questions about mycobacteriophage biology and how they can be utilized are expanding faster than answers can be obtained. Mycobacteriophage genomics hint at a vast array of genetic and molecular secrets that await discovery, and it would seem that mycobacteriophage investigations have a very promising future—that the best is still to come.

Finally, the enormous diversity of mycobacteriophages lends them for use in an integrated research–education platform in viral discovery and genomics (Hanauer *et al.*, 2006; Hatfull *et al.*, 2006). The Science Education Alliance program of the Howard Hughes Medical Institute has facilitated implementation of mycobacteriophage discovery for freshman undergraduate students in 44 institutions in the United States since 2008, with more than 800 students engaged, hundreds of new mycobacteriophages isolated, and many dozens of genomes sequenced and analyzed (Caruso *et al.*, 2009; Pope *et al.*, 2011). This platform could be readily extended to the use of alternative bacterial hosts with the potential to have a substantial impact on the broader field of bacteriophage diversity, relieving the major limitations in the area, which are no longer in DNA sequence acquisition technologies but in obtaining individual isolates for further characterization.

This chapter discusses the current state of mycobacteriophage genomics, our current understanding of mycobacteriophage molecular biology, and the variety of ways in which mycobacteriophages have been exploited for both genetic and clinical applications. A number of other reviews on various aspects of mycobacteriophages may be useful to the reader (Hatfull, 1994, 1999, 2000, 2004, 2006, 2008, 2010; Hatfull *et al.*, 1994, 2008; Hatfull and Jacobs, 1994, 2000; McNerney, 1999; McNerney and Traore, 2005; Stella *et al.*, 2009).

II. THE MYCOBACTERIOPHAGE GENOMIC LANDSCAPE

A. Overview of 80 sequenced mycobacteriophage genomes

Consideration of mycobacteriophage diversity as revealed by their genomic characterization is a suitable starting point for this review, and a genome-based taxonomy—albeit one that is intentionally barely hierarchical—imposes a degree of order that is useful in discussing their biology. Currently, a total of 80 different phage genome sequences have been described and compared (Pope *et al.*, 2011), and as of the time of writing (January 2011) another 80 unpublished sequenced genomes are available (http://www.phagesdb.org). The discussion here is restricted primarily to the 80 published genomes listed in Table I. Mycobacteriophage genomes vary in length from 42 to 164 kbp with an average of 69.2 kbp (Table I). Genome sizes are distributed across this spectrum, although with a notable absence of phages with genomes between 110 and 150 kbp. All of the virions contain linear double-stranded DNA (dsDNA) molecules, but two different types of genome termini are observed. Approximately 60% of the phage genomes have defined ends with short single-stranded DNA (ssDNA) termini (4–14 bases), all of which have 3′ extensions. The other 40% are terminally redundant and circularly

TABLE I Genometrics of 80 sequenced mycobacteriophage genomes

Cluster	Phage	Size (bp)	GC%	#ORFs	tRNA #	tmRNA #	Ends[a]	Accession #	Origins[b]	Reference
A1	Bethlehem	52250	63.3	87	0	0	10-base 3′	AY500153	Bethlehem, PA	Hatfull *et al.*, 2006
A1	Bxb1	50550	63.7	86	0	0	9-base 3′	AF271693	Bronx, NY	Mediavilla *et al.*, 2001
A1	DD5	51621	63.4	87	0	0	10-base 3′	EU744252	Upp. St. Clair, PA	Hatfull *et al.*, 2010
A1	Jasper	50968	63.7	94	0	0	10-base 3′	EU744251	Lexington, MA	Hatfull *et al.*, 2010
A1	KBG	53572	63.6	89	0	0	10-base 3′	EU744248	Kentucky	Hatfull *et al.*, 2010
A1	Lockley	51478	63.4	90	0	0	10-base 3′	EU744249	Pittsburgh, PA	Hatfull *et al.*, 2010
A1	Skipole	53137	62.7	102	0	0	10-base 3′	GU247132	Champlin Park, MN	Pope *et al.*, 2011
A1	Solon	49487	63.8	86	0	0	10-base 3′	EU826470	Solon, IA	Hatfull *et al.*, 2010
A1	U2	51277	63.7	81	0	0	10-base 3′	AY500152	Bethlehem, PA	Hatfull *et al.*, 2006
A2	Che12	52047	62.9	98	3	0	10-base 3′	DQ398043	Chennai, India	Hatfull *et al.*, 2006
A2	D29	49136	63.5	77	5	0	9-base 3′	AF022214	California	Ford *et al.*, 1998
A2	L5	52297	62.3	85	3	0	9-base 3′	Z18946	Japan	Hatfull *et al.*, 1993
A2	Pukovnik	52892	63.3	88	1	0	10-base 3′	EU744250	Ft. Bragg, NC	Hatfull *et al.*, 2010
A2	RedRock	53332	64.5	95	1	0	10-base 3′	GU339467	Sedona, AZ	Pope *et al.*, 2011
A3	Bxz2	50913	64.2	86	3	0	10-base 3′	AY129332	Bronx, NY	Pedulla *et al.*, 2003
A4	Eagle	51436	63.4	87	0	0	10-base 3′	HM152766	Fredericksburg, VA	Pope *et al.*, 2011
A4	Peaches	51376	63.9	86	0	0	10-base 3′	GQ303263.1	Monroe, LA	Pope *et al.*, 2011
B1	Chah	68450	66.5	104	0	0	Circ Perm	FJ174694	Ruffsdale, PA	Hatfull *et al.*, 2010
B1	Colbert	67774	66.5	100	0	0	Circ Perm	GQ303259.1	Corvallis, OR	Pope *et al.*, 2011
B1	Fang	68569	66.5	102	0	0	Circ Perm	GU247133	O'Hara Twp, PA	Pope *et al.*, 2011
B1	Orion	68427	66.5	100	0	0	Circ Perm	DQ398046	Pittsburgh, PA	Hatfull *et al.*, 2006

(*continued*)

TABLE I *(continued)*

Cluster	Phage	Size (bp)	GC%	#ORFs	tRNA #	tmRNA #	Ends[a]	Accession #	Origins[b]	Reference
B1	PG1	68999	66.5	100	0	0	Circ Perm	AF547430	Pittsburgh, PA	Hatfull *et al.*, 2006
B1	Puhltonio	68323	66.4	97	0	0	Circ Perm	GQ303264.1	Baltimore, MD	Pope *et al.*, 2011
B1	Scoot17C	68432	66.5	102	0	0	Circ Perm	GU247134	Pittsburgh, PA	Pope *et al.*, 2011
B1	UncleHowie	68016	66.5	98	0	0	Circ Perm	GQ303266.1	St. Louis, MO	Pope *et al.*, 2011
B2	Qyrzula	67188	69.0	81	0	0	Circ Perm	DQ398048	Pittsburgh, PA	Hatfull *et al.*, 2006
B2	Rosebush	67480	69.0	90	0	0	Circ Perm	AY129334	Latrobe, PA	Pedulla *et al.*, 2003
B3	Phaedrus	68090	67.6	98	0	0	Circ Perm	EU816589	Pittsburgh, PA	Hatfull *et al.*, 2010
B3	Phlyer	69378	67.5	103	0	0	Circ Perm	FJ641182.1	Pitsburgh, PA	Pope *et al.*, 2011
B3	Pipefish	69059	67.3	102	0	0	Circ Perm	DQ398049	Pittsburgh, PA	Hatfull *et al.*, 2006
B4	Cooper	70654	69.1	99	0	0	Circ Perm	DQ398044	Pittsburgh, PA	Hatfull *et al.*, 2006
B4	Nigel	69904	68.3	94	1	0	Circ Perm	EU770221	Pittsburgh, PA	Hatfull *et al.*, 2010
C1	Bxz1	156102	64.8	225	35	1	Circ Perm	AY129337	Bronx, NY	Pedulla *et al.*, 2003
C1	Cali	155372	64.7	222	35	1	Circ Perm	EU826471	Santa Clara, CA	Hatfull *et al.*, 2010
C1	Catera	153766	64.7	218	35	1	Circ Perm	DQ398053	Pittsburgh, PA	Hatfull *et al.*, 2006
C1	ET08	155445	64.6	218	30	1	Circ Perm	GQ303260.1	San Diego, CA	Pope *et al.*, 2011
C1	LRRHood	154349	64.7	224	30	1	Circ Perm	GQ303262.1	Santa Cruz, CA	Pope *et al.*, 2011
C1	Rizal	153894	64.7	220	35	1	Circ Perm	EU826467	Pittsburgh, PA	Hatfull *et al.*, 2010
C1	Scott McG	154017	64.8	221	35	1	Circ Perm	EU826469	Pittsburgh, PA	Hatfull *et al.*, 2010
C1	Spud	154906	64.8	222	35	1	Circ Perm	EU826468	Pittsburgh, PA	Hatfull *et al.*, 2010
C2	Myrna	164602	65.4	229	41	0	Circ Perm	EU826466	Upp. St. Clair, PA	Hatfull *et al.*, 2010
D	Adjutor	64511	59.7	86	0	0	Circ Perm	EU676000	Pittsburgh, PA	Hatfull *et al.*, 2010
D	Butterscotch	64562	59.7	86	0	0	Circ Perm	FJ168660	Pittsburgh, PA	Hatfull *et al.*, 2010
D	Gumball	64807	59.6	88	0	0	Circ Perm	FJ168661	Pittsburgh, PA	Hatfull *et al.*, 2010
D	P-lot	64787	59.7	89	0	0	Circ Perm	DQ398051	Pittsburgh, PA	Hatfull *et al.*, 2006

D	PBI1	64494	59.7	81	0	0	Circ Perm	DQ398047	Pittsburgh, PA	Hatfull *et al.*, 2006
D	Troll4	64618	59.6	88	0	0	Circ Perm	FJ168662	Silver Springs, MD	Hatfull *et al.*, 2010
E	244	74483	62.9	142	2	0	9-base 3′	DQ398041	Pittsburgh, PA	Hatfull *et al.*, 2006
E	Cjw1	75931	63.1	141	2	0	9-base 3′	AY129331	Pittsburgh, PA	Pedulla *et al.*, 2003
E	Kostya	75811	62.9	143	2	0	9-base 3′	EU816591	Washington, DC	Hatfull *et al.*, 2010
E	Porky	76312	62.8	147	2	0	9-base 3′	EU816588	Concord, MA	Hatfull *et al.*, 2010
E	Pumpkin	74491	63.0	143	2	0	9-base 3′	GQ303265.1	Holland, MI	Pope *et al.*, 2011
F1	Ardmore	52141	61.5	87	0	0	?	GU060500	C'nty Waterford, Ireland	Henry *et al.*, 2010
F1	Boomer	58037	61.1	105	0	0	10-base 3′	EU816590	Pittsburgh, PA	Hatfull *et al.*, 2010
F1	Che8	59471	61.3	112	0	0	10-base 3′	AY129330	Chennai, India	Pedulla *et al.*, 2003
F1	Fruitloop	58471	61.8	102	0	0	10-base 3′	FJ174690	Latrobe, PA	Hatfull *et al.*, 2010
F1	Llij	56852	61.5	100	0	0	10-base 3′	DQ398045	Pittsburgh, PA	Hatfull *et al.*, 2006
F1	Pacc40	58554	61.3	101	0	0	10-base 3′	FJ174692	Pittsburgh, PA	Hatfull *et al.*, 2010
F1	PMC	56692	61.4	104	0	0	10-base 3′	DQ398050	Pittsburgh, PA	Hatfull *et al.*, 2006
F1	Ramsey	58578	61.2	108	0	0	10-base 3′	FJ174693	White Bear, MN	Hatfull *et al.*, 2010
F1	Tweety	58692	61.7	109	0	0	10-base 3′	EF536069	Pittsburgh, PA	Pham *et al.*, 2007
F2	Che9d	56276	60.9	111	0	0	10-base 3′	AY129336	Chennai, India	Pedulla *et al.*, 2003
G	Angel	41441	66.7	61	0	0	11-base 3′	EU568876.1	O'Hara Twp, PA	Sampson *et al.*, 2009
G	BPs	41901	66.6	63	0	0	11-base 3′	EU568876	Pittsburgh, PA	Sampson *et al.*, 2009
G	Halo	42289	66.7	64	0	0	11-base 3′	DQ398042	Pittsburgh, PA	Hatfull *et al.*, 2006
G	Hope	41901	66.6	63	0	0	11-base 3′	GQ303261.1	Atlanta, GA	Pope *et al.*, 2011
H1	Konstantine	68952	57.3	95	0	0	Circ Perm	FJ174691	Pittsburgh, PA	Hatfull *et al.*, 2010
H1	Predator	70110	56.3	92	0	0	Circ Perm	EU770222	Donegal, PA	Hatfull *et al.*, 2010
H2	Barnyard	70797	57.3	109	0	0	Circ Perm	AY129339	Latrobe, PA	Pedulla *et al.*, 2003
I1	Brujita	47057	66.8	74	0	0	11-base 3′	FJ168659	Virginia	Hatfull *et al.*, 2010
I1	Island3	47287	66.8	76	0	0	11-base 3′	HM152765	Pittsburgh, PA	Pope *et al.*, 2011

(continued)

TABLE I *(continued)*

Cluster	Phage	Size (bp)	GC%	#ORFs	tRNA #	tmRNA #	Ends[a]	Accession #	Origins[b]	Reference
I2	Che9c	57050	65.4	84	0	0	10-base 3′	AY129333	Chennai, India	Pedulla *et al.*, 2003
J	Omega	110865	61.4	237	2	0	4-base 3′	AY129338	Upp. St. Clair, PA	Pedulla *et al.*, 2003
K1	Angelica	59598	66.4	94	1	0	11-base 3′	HM152764	Clayton, MO	Pope *et al.*, 2011
K1	CrimD	59798	66.5	95	1	0	11-base 3′	HM152767	Williamsburg, VA	Pope *et al.*, 2011
K2	TM4	52797	68.1	89	0	0	10-base 3′	AF068845	Colorado	Ford *et al.*, 1998b
L	LeBron	73453	58.8	123	9	0	10-base 3′	HM152763	Allensville, NC	Pope *et al.*, 2011
Sin	Corndog	69777	65.4	122	0	0	4-base 3′	AY129335	Pittsburgh, PA	Pedulla *et al.*, 2003
Sin	Giles	53746	67.5	78	0	0	14-base 3′	EU203571	Pittsburgh, PA	Morris *et al.*, 2008
Sin	Wildcat	78296	56.9	148	24	1	11-base 3′	DQ398052	Latrobe, PA	Hatfull *et al.*, 2006
	TOTAL	5,734,561		9,013	375					
	AVERAGE	71,683	63.83	112.66	4.69					

[a] Indicates whether the genome termini are circularly permuted or if they have defined ends with the length and polarity of the ssDNA extension.
[b] The geographic location from where the phage was isolated is shown.

permuted, although the extent of the redundancy has not been determined for any of these phages (Table I). Although the average GC% content is similar to their common host *M. smegmatis*, there is substantial variation in GC% content, ranging from 56 to 69%. The implications of this are discussed further later (see Section II.D).

B. Grouping of mycobacteriophages into clusters and subclusters

Nucleotide sequence comparisons using dot plots clearly show that some mycobacteriophages are more closely related than others (Fig. 1). Although a seemingly crude approach, grouping phages according to this relatedness offers a useful and pragmatic approach that recognizes this basic level of diversity. Seventy-seven of the 80 sequenced phages can be placed in a total of 12 different clusters (A–L) with the remaining 3 considered as singletons, of which no closely related phages have yet been identified (Table I). Two of the 80 phages, Omega and LeBron, have been assigned to clusters (J and L, respectively) because they have close relatives among the sequenced but yet to be published mycobacteriophage genomes. In the case of Cluster J there are two phages in addition to Omega that form this cluster, whereas for Cluster L there are six additional phages related to LeBron. The detailed discussions that follow are constrained to just those 80 published genomes shown in Table I.

Cluster assignment is performed primarily according to recognizable nucleotide sequence similarity that spans more than 50% of the genome length with one or more other genomes (Fig. 1) (Hatfull, 2010). The advantage of using dot plot analyses for this is that it provides a method for resolving two of the most difficult scenarios that emerge: (1) when two genomes appear to have diverged substantially such that they share DNA sequence similarity over a substantial portion of their genomes, but the degree of similarity is relatively low, and (2) when two genomes share segments of DNA sequence similarity that are very similar to each other, but extend only over a relatively small portion of the genomes (i.e., <50%). In practice, relatively few such scenarios arise, and in most cases cluster assignment is straightforward. Dot plot analyses and average nucleotide identity (ANI) parameters suggest that some clusters can be further divided into subdivisions referred to as subclusters (Fig. 1) (Hatfull, 2010). Phages of different subclusters within the same cluster often share similar genome organizations and many genes are clearly orthologues as revealed by amino acid sequence comparisons of their products, but with relatively low degrees of nucleotide similarity (Fig. 1, Table I; see also Fig. 3B).

A hallmark of all or most phage genomic architectures is that they are mosaic, built from segments that have distinct evolutionary histories and

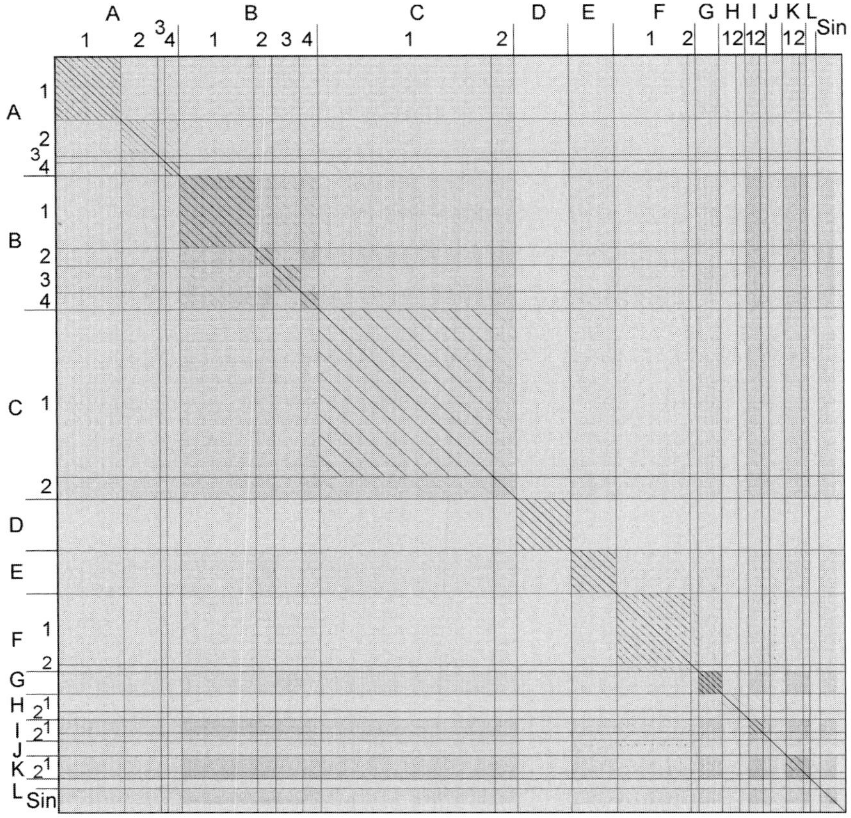

FIGURE 1 Dot plot nucleotide comparison of 80 mycobacteriophage genomes. A single FASTA-formatted file was generated containing the nucleotide sequences of all 80 sequenced and published mycobacteriophages, joined in the same order as presented in Table I. This 5.7 Mbp file was then compared to itself using the dot matrix program GEPARD (Krumsiek *et al*., 2007). The assignment of clusters and subclusters is shown at the top and on the left.

that have been exchanged horizontally over an extended period of time; this is certainly true of mycobacteriophages (Hendrix, 2002; Hendrix *et al.*, 1999, 2000; Pedulla *et al.*, 2003). This makes any form of hierarchical classification of whole genomes difficult, and reticulate systems provide fuller and likely more accurate portrayals of their evolution (Lawrence *et al.*, 2002; Lima-Mendez *et al.*, 2008). Organization into clusters and subclusters should not be interpreted as representing any well-defined boundaries between different types of viruses, but rather a reflection of incomplete sampling of a large and diverse population of viruses occupying positions on a broad spectrum of multidimensional relationships.

As such, cluster and subcluster structures are not likely to be stable, and as more mycobacteriophages are discovered and sequenced, clusters are expected to undergo further subdivision, and differences between members of particular clusters could become reduced to that seen between subclusters. Clusters thus do not represent lineages per se, but do provide a convenient means of representing the heterogeneity of the currently sequenced phages. For example, differences between genomes within subclusters are likely to represent relatively recent evolutionary events, and a good example is the discovery of the novel MPME transposons from comparison of Cluster G phage genomes (Pope *et al.*, 2011; Sampson *et al.*, 2009).

C. Relationships between viral morphologies and cluster types

All mycobacteriophages discovered to date are tailed phages and contain dsDNA genomes (i.e., of the order Caudovirales). However, only two of the three major families of the Caudovirales are represented, and no Podoviridae—with short stubby tails—have been reported. Seventy-one of the mycobacteriophages are Siphoviridae with long flexible noncontractile tails and 9 are Myoviridae with contractile tails. All Myoviridae are in Cluster C, containing capsids approximately 80 nm in diameter and genomes 153.7 to 164.6 kbp long. Most Siphoviridae contain isometric heads—ranging from 48 to 75 nm in diameter—but several contain prolate heads, including all three members of Cluster I and the singleton Corndog (Table I)(Hatfull *et al.*, 2010; Pope *et al.*, 2011). Tail lengths of Siphoviridae vary by nearly threefold, from 110 nm to 300 (Hatfull *et al.*, 2010).

D. Relationships between GC% and cluster types

Different clusters—and in some cases subclusters—have distinctive GC% contents (Fig. 2). For example, all Cluster A genomes range between 62.3% GC% (L5) and 64.5% GC% (RedRock) (Table I), and the only other genomes that lie within this range are those in cluster E (Fig. 2). Cluster D phages are all 59.6–59.7% GC% and no other phages lie within this range. In Cluster B, the four subclusters differ somewhat in GC% with little overlap between their ranges of values, and Subclusters B1 and B4 contain genomes with the highest GC% content of any of the mycobacteriophages. At the other extreme, Cluster H and the singleton Wildcat have the lowest GC% content (56.3–57.3%). The reason why GC% should vary by cluster is not known, but an intriguing idea is that the different clusters (and perhaps in some cases subclusters) have distinct host ranges, notwithstanding that they are all capable of infecting *M. smegmatis*, a requirement of their isolation procedure; codon usage analyses are consistent

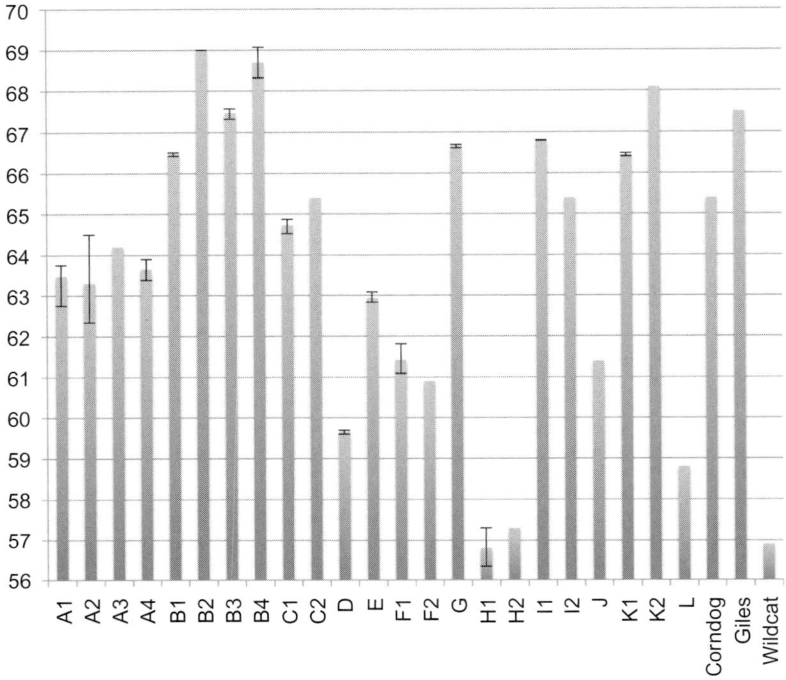

FIGURE 2 Relationships between genome GC% and cluster/subcluster types. The average GC% is shown for each cluster or subcluster of the mycobacteriophages, with variants showing extreme values within each group.

with this (Hassan *et al.*, 2009). As such, distinctions between clusters may arise from partial genetic isolation, with either host range or host availability imposing constraints on the exchange of genetic information between the genomes over a short—but evolutionary significant—time frame. Limited host range data are available for some of the phages (Rybniker *et al.*, 2006), but detailed host preferences for the larger collection of phages have yet to be determined.

E. Mycobacteriophage phamilies

Although nucleotide sequence comparisons are useful for clustering closely related phages, identification of homologues that diverged longer ago can usually only be identified by comparison of the predicted amino acid sequences. The computer program Phamerator performs automated assembly of genes into phamilies (phams) based on pairwise comparisons using both Clustal and BlastP searches using current threshold levels of 32.5% amino acid sequence identity and 10^{-50} E values, respectively

(Cresawn *et al.*, manuscript in preparation; Hatfull *et al.*, 2006; Pope *et al.*, 2011). The 80 published genomes encode a total of 9015 predicted genes, which assemble into 2343 phams of which 1106 (47.2%) are orphams (phams containing only a single gene member)(Pope *et al.*, 2011). Phamerator enables two helpful types of representation of these data. The first is the display of genome maps illustrating both regions of DNA similarity and representing individual genes according to the phamily to which they belong. The second is the use of phamily circles to display which of the component genomes contain members of any particular phamily. Phamily circles are especially useful for examining the phylogenies of adjacent genes in a genome and identifying where recombination events have adjoined genes or gene segments, each of which have distinct evolutionary histories (Hatfull *et al.*, 2006).

Database searches show that about 80% of mycobacteriophages phamilies have no identifiable homologues outside of mycobacteriophages; this high proportion of novel sequences reflects a common finding in phage genomics (Abedon, 2009; Casas and Rohwer, 2007; Comeau *et al.*, 2008; Hatfull, 2010; Hatfull *et al.*, 2006, 2010; Krisch and Comeau, 2008). As a consequence, functions of the vast majority of mycobacteriophage gene functions remain unknown. Exceptions are the approximately 10% of phamilies that are homologues of proteins with known functions and those that constitute operons of virion structural and assembly genes, that at least in phages with a siphoviral morphology have a well-conserved synteny (Hatfull *et al.*, 2010; Pope *et al.*, 2011).

F. Genome organizations

Mycobacteriophage genome organizations are well conserved among phages within clusters, and there are therefore 15 types to be considered, clusters A through L, and three singleton genomes. With the exception of Cluster C genomes, all have *siphoviral* morphologies and contain a predicted long operon of the virion structure and assembly genes, which are typically represented in the left parts of the genomes and transcribed rightward (Hatfull, 2010). In the smallest genomes (Cluster G), there are about 25 genes in this operon spanning 24 kbp, or 57% of the total genome length. In contrast, in the Cluster J phage Omega, there are 48 genes in this presumed late operon, although likely only approximately half of these have roles in virion structure and assembly genes, and the functions of most of the others are unknown (see Section III.J). However, they include a putative glycosyl transferase (gp16), a putative *O*-methyl transferase (gp17), a putative kinase (gp2), and a putative enoyl-CoA hydratase (gp4). It is not known if these are required for phage propagation or what their specific roles are.

The genomes of Clusters A, E, F, G, I, J, K, L, and all three singletons have defined ends with short (4–14 base) 3′ ssDNA extensions (Table I).

Phages in Cluster B, C, D, and H have genomes lacking defined ends, and genome sequencing data suggest that these all have circularly permuted, terminally redundant ends. In Clusters F, G, I, and K genomes, genes encoding large terminase subunits and, in some cases, genes encoding small terminase subunits can be identified close to the physical left end of the genomes, their presumed sites of action in DNA packaging. However, in other genomes, these terminase genes are displaced from the physical genome ends, with additional genes (mostly of unknown function) in the intervening space. The largest distance is in Corndog, where the large terminase subunit gene (*32*) is over 13 kbp from the left end.

Genes encoding integrases can be identified in most genomes of Clusters A, E, F, G, I, J, K, L, and singleton Giles, and are located near the center of their genomes, regardless of a span of a nearly threefold difference in genome size (Hatfull, 2006). The furthest from the midpoint is in the Cluster I phage, Brujita (39% of genome length from the left end). Putative lysis genes can be identified in all of the siphoviral mycobacteriophage genomes, although they may be located either to the left of terminase genes (as in Cluster A) or to the right of the virion structure and assembly genes (Hatfull, 2010). In Cluster C genomes, linkage of the virion structure and assembly genes is much less obvious, and the identities of relatively few have been determined. Further details of each of the genome types are discussed.

III. PHAGES OF INDIVIDUAL CLUSTERS, SUBCLUSTERS, AND SINGLETONS

A. Cluster A

Cluster A is one of the largest clusters of mycobacteriophages, containing 17 of the 80 (21%) sequenced genomes. It is also highly diverse and currently contains four subclusters (A1–A4), although many of the unpublished genomes also belong to this cluster and are expected to expand the number of subclusters to at least six. Subclusters A1 and A2 predominate, with nine current members of A1 and five of A2 (Table I). There is no obvious geographical preference for the Cluster A phages having been isolated from two countries outside the United States and from 11 states within the United States (Table I)(Pope *et al.*, 2011). Cluster A genome lengths lie within a relatively narrow range (49,136–53,572 bp) and also occupy a narrow range of GC% (62.3–64.5%), slightly lower than that of their *M. smegmatis* host (67.4%)(Table I). Subclusters A2 and A3 phages all contain at least one tRNA gene, whereas Subclusters A1 and A4 do not. At least four of the Subcluster A2 phages infect *M. tuberculosis* efficiently [L5, D29, Che12, and Pukovnik (Fullner and Hatfull, 1997;

Gomathi *et al.*, 2007; Hatfull *et al.*, 2010; Kumar *et al.*, 2008)] but none of the phages in Subclusters A1 or A3 do so. It is not yet known if either of the Subcluster A4 phages infect *M. tuberculosis* or do so efficiently.

All Cluster A phages are either temperate or recent derivatives of temperate parents. Perhaps the best-studied member of the cluster is L5 (Subcluster A2), which was isolated in Japan in 1960 (Doke, 1960) and was the first sequenced mycobacteriophage (Hatfull and Sarkis, 1993). L5 forms evidently turbid plaques from which lysogens can be recovered readily and which are both immune to superinfection by L5 and release phage particles into culture supernatants during liquid growth. Phage L1 is a closely related temperate phage with the same restriction pattern as L5, but is naturally temperature sensitive (Doke, 1960; Lee *et al.*, 1991); its genome sequence has not yet been reported. Lysogens have also been generated for phages Bxb1, DD5, Jasper, Skipole, Solon, RedRock, Eagle, Peaches, Pukovnik, and Che12 (Kumar *et al.*, 2008; Pope *et al.*, 2011). D29, which was isolated more than 50years ago (Froman *et al.*, 1954), forms clear plaques and kills a high proportion of infected cells. Genomic characterization suggests that it has lost a segment of approximately 3 kbp from the right end corresponding to the position of the repressor gene of L5, and D29 remains subject to L5 superinfection immunity (Ford *et al.*, 1998a). This deletion event could have occurred relatively recently, and it was noted previously that the current isolate of D29 is just one of several plaque morphotypes in the original isolate (Bowman, 1958; Hatfull, 2010). Bethlehem, KBG, Lockley, and Bxz2 also form clear plaques and fail to form lysogens (Pope *et al.*, 2011), and we speculate that the temperate nature of Cluster A phages tends to either be selected against during plaque isolation or be lost during subsequent laboratory propagation. Cluster A phages fall into three major immunity groups, which correspond closely to subclusters A1, A2, and A4 (see Section V.A.1). Lysogens of any of the temperate Subcluster A1 phages confer superinfection immunity to other A1 phages (i.e., they are homoimmune) but not to phages of other A subclusters. Similarly, phages within each of Subclusters A2 and A4 are homoimmune but heteroimmune with other Cluster A phages (Pope *et al.*, 2011). Bxz2 is currently the sole member of Subcluster A3 and does not form lysogens. However, the genome contains a putatively defective repressor gene, and because it is not subject to immunity by any of the other Cluster A phages, it likely corresponds to a derivative of a fourth distinct immunity group (Pope *et al.*, 2011). A map of the L5 genome, as a representative of Cluster A phages, is shown in Figure 3A. There are several notable features. First, the virion structure and assembly genes—several of which were identified through N-terminal sequencing of virion proteins (Hatfull and Sarkis, 1993)—are arranged in canonical order, encoding terminase, portal, protease, scaffold, capsid, major tail subunit tail assembly

A

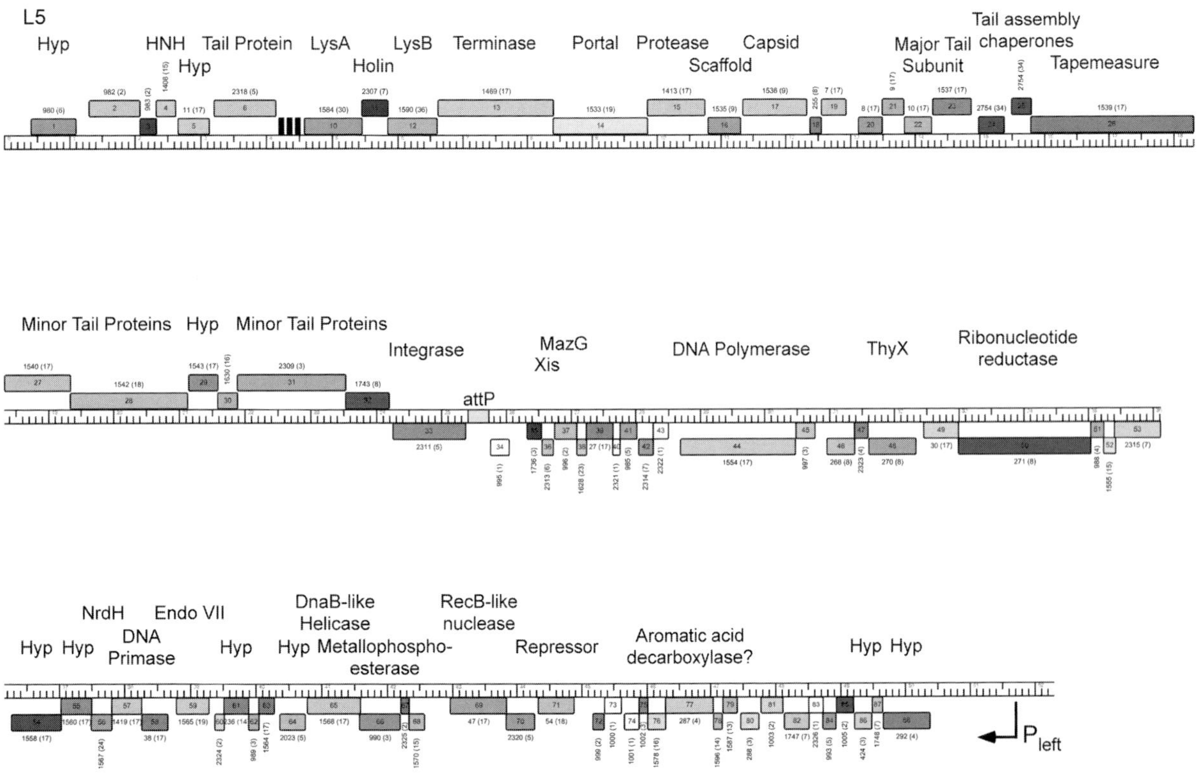

L5
Hyp
HNH
Hyp
Tail Protein
LysA
Holin
LysB
Terminase
Portal
Protease
Scaffold
Capsid
Major Tail Subunit
Tail assembly chaperones
Tapemeasure
Minor Tail Proteins
Hyp
Minor Tail Proteins
Integrase
attP
MazG
Xis
DNA Polymerase
ThyX
Ribonucleotide reductase
Hyp
Hyp
NrdH
DNA Primase
Endo VII
Hyp
Hyp
DnaB-like Helicase
Metallophospho-esterase
RecB-like nuclease
Repressor
Aromatic acid decarboxylase?
Hyp
Hyp
P left

B

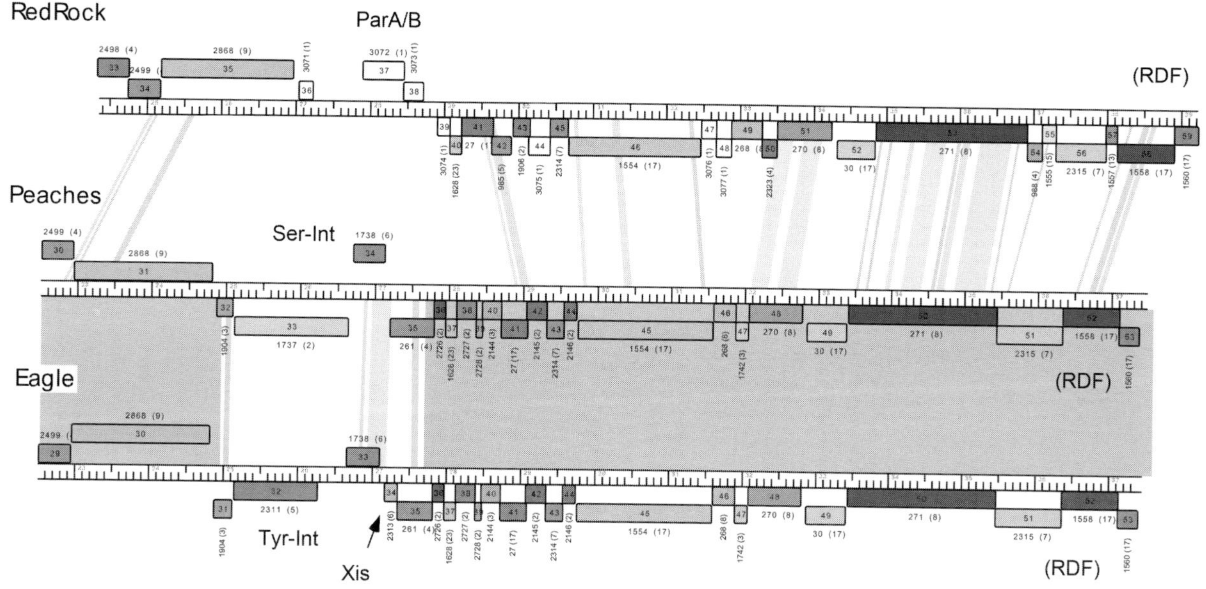

FIGURE 3 Organizational feature of Cluster A genomes. (A) Map of the L5 genome. The L5 genome (a member of Subcluster A2) is represented as a horizontal bar with markers, and the predicted ORFs are shown as colored boxes either above (rightward transcribed) or below (leftward transcribed) the genome. Gene names are shown inside the boxes, and phams to which they belong are indicated above, with the total number of pham members shown in parentheses. ORFs are color coded according to their pham memberships (i.e., all members of the same pham are the same color) and those shown in white are orphans, phams that contain only a single gene member. tRNA genes are shown as short black bars. Putative gene functions identified either experimentally or as predicted bioinformatically are shown above the genes; genes whose products match a substantial number of conserved hypothetical proteins are designated Hyp (for conserved hypothetical). Bioinformatically

chaperones, tapemeasure, and minor tail proteins (Fig. 3A), although an additional tail protein (gp6) is encoded between the terminase and the left end of the genome (Hatfull and Sarkis, 1993); no gene encoding a putative small terminase subunit has been identified. The ~6.5-kbp region between the large terminase and the left end contains three closely linked tRNA genes, as well as the lysis system. Second, the integration system is encoded in the middle of the genome, and all genes to its left are transcribed rightward and all genes to its right are transcribed leftward; the genome can therefore be split conveniently into left and right arms (Fig. 3A). Third, L5 encodes its own DNA polymerase, a Pol I-like enzyme that lacks the 5′–3′ exonuclease domain (Hatfull and Sarkis, 1993), although it is not known if it required phage DNA replication. L5 does

---

designated functional assignments were determined using BLASTP against the nonredundant protein database at GenBank and HHPRed (Soding, 2005). The map was generated using the program Phamerator (S. Cresawn and Graham F. Hatfull, manuscript in preparation). (B) Cassettes for genome stabilization encoding tyrosine-integrases, serine-integrases, or ParA/B partitioning functions. The figure shows representations of the central parts of three Cluster A genomes: RedRock (Subcluster A2), Peaches (Subcluster A4), and Eagle (Subcluster A4). Each genome is represented as horizontal bars with markers, and genes are represented as described (A). Maps were generated in Phamerator, and nucleotide sequence similarity between adjacent genomes is shown between them; the strength of similarity is shown according to the color spectrum, with red being the weakest and violet the strongest. Thus Peaches and Eagle are very closely related at the nucleotide level (justifying their grouping into the same subcluster, A4), except for the ~2.6-kbp central segment. RedRock is more distantly related to Peaches (and therefore to Eagle too) as shown by a sporadic, shorter, weaker segment of homology. Note that even in the absence of extensive nucleotide similarity the synteny of this region (other than in the central segment) is largely conserved, and the gene order in RedRock is similar to that in Peaches and Eagle. This can be seen by gene/pham color coding and by pham designations above the genes. Within the central region, Peaches and Eagle are different, and Peaches encodes a serine-integrase and Eagle encodes a tyrosine-integrase (gp32). The segment in Eagle that differs between the two phages also encodes the Xis protein [gp34; a member of a large highly diverse group of proteins acting as recombination directionality factors, or RDFs (Lewis and Hatfull, 2001)] that controls the directionality of integrase-mediated recombination. The central segment in Eagle contains a serine-integrase gene (*33*), but does not contain a putative RDF gene that acts with the integrase. However, the Peaches RDF is likely gp52, which is related to the known RDF of the Bxb1 serine-integrase system (Ghosh *et al.*, 2006). Related proteins are encoded in all Cluster A members, even those with tyrosine-integrases such as Eagle (shown as RDF in parentheses in the figure) and likely perform additional functions in DNA replication. The Peaches serine-integrase could thus have been acquired from an Eagle-like ancestor, without concomitant RDF acquisition. In RedRock, the central segment has no integrase gene, but instead has two genes (*37* and *38*) encoding ParA and ParB functions, respectively. The RedRock prophage presumably replicates extrachromosomally and is stably maintained by these partitioning functions. (See Page 12 in Color Section at the back of the book.)

not encode its own RNA polymerase and uses the host RNA polymerase for all its transcription, being sensitive to the addition of rifampicin (Hatfull and Sarkis, 1993). Genes *48* and *50* are expressed early in lytic growth and encode flavin-dependent thymidylate synthase (ThyX) and ribonucleotide reductase functions, respectively, and the two proteins form a complex during lytic growth (Bhattacharya *et al.*, 2008). Mutants affecting two genes in L1 are implicated in the regulation of late gene expression, although locations of the mutations relative to the genome map are not known (Datta *et al.*, 2007).

There is considerable variation in the tRNA genes present in Cluster A genomes. Subclusters A1 and A4 have none, the Subcluster A3 phage Bxz2 has three ($tRNA^{asn}$, $tRNA^{trp}$, and $tRNA^{leu}$), and Subcluster A2 phages differ between one and five (Table I). Of the five tRNA genes in D29 ($tRNA^{asn}$, $tRNA^{trp}$, $tRNA^{gln}$, $tRNA^{glu}$, and $tRNA^{tyr}$), the first three are also present in L5, and the $tRNA^{glu}$ and $tRNA^{tyr}$ genes could have been lost by a simple deletion (Ford *et al.*, 1998a; Hatfull and Sarkis, 1993). Interestingly, Che12 has three similar tRNA genes, but the order of the latter two is reversed. Pukovnik and RedRock each have a single tRNA gene ($tRNA^{gln}$ and $tRNA^{trp}$, respectively), and although the RedRock $tRNA^{trp}$ is a close relative of that in L5, D29, and Bxz2, the $tRNA^{gln}$ is not closely related to the $tRNA^{gln}$ genes in L5 and D29. It has been noted that the frequencies of usage of the five amino acids corresponding to the tRNA specificities are high in D29 relative to *M. tuberculosis*, but that the specific roles of their tRNAs are uncertain because they are not well conserved among the related genomes (Kunisawa, 2000).

A leftward promoter (P_{left}) located at the right end of the genome is responsible for early expression of right arm genes and is directly under repressor control (Brown *et al.*, 1997; Nesbit *et al.*, 1995); a promoter for expression of the virion structure and assembly genes has yet to be identified. Three additional promoters are located upstream of the repressor, but it is unclear what specific roles these play (Nesbit *et al.*, 1995). Because the *attP* site is located at the 5′ side of the integrase gene, a promoter is presumably located between the *attP* crossover site and the start of the *int* gene. A detailed description of transcription patterns and a full constitution of promoter and terminator signals have yet to be elucidated for L5 or any mycobacteriophage.

A particularly intriguing feature of L5 and the other Cluster A genomes is that they contain multiple repressor binding sites—referred to as stoperators—in addition to operator sites at P_{left} (Brown *et al.*, 1997) (see Section V.A.1). In L5, there are a total of 24 sites corresponding to the 13-bp asymmetric consensus sequence 5′-GGTGGMTGTCAAG (where M is A or C) to which the repressor binds; these are located predominantly within short intergenic regions and in one orientation relative to the direction of transcription (Brown *et al.*, 1997). When one or more of these sites is positioned between a promoter and a reporter gene, there is a repressor-dependent reduction of reporter gene activity; activity is dependent on orientation of the site relative to the direction of transcription and is amplified by multiple site insertions (Brown *et al.*, 1997). It is proposed that the repressor mediates termination of transcription rather than initiation and perhaps plays a role in ensuring transcriptional silence of phage genes in a lysogen that might otherwise be deleterious to lysogenic growth (Brown *et al.*, 1997). Several L5 genes have been implicated in cytotoxicity (Donnelly-Wu *et al.*, 1993), with strong evidence for genes *64* (Chattoraj *et al.*, 2008), *77*, *78*, and *79* (Rybniker *et al.*, 2008). All other Cluster A phages contain multiple stoperator sites with as many as 36 predicted in Jasper (Pope *et al.*, 2011). The consensus sequence is similar within each of the subclusters, but differs from subcluster to subcluster, consistent with these playing an important role—along with their cognate repressors—in determining immunity specificities.

Although there is considerable sequence diversity in Cluster A genomes, overall organizations are similar (Hatfull *et al.*, 2010). An interesting point of departure though is the use of different types of integration systems. All of the genomes in Cluster A2 encode tyrosine-integrases, whereas all A1 and A3 genomes encode serine-integrases. Interestingly, although the two phages in Cluster A4, Eagle and Peaches, are otherwise extremely similar to each other (97.5% average nucleotide identity), Eagle encodes a tyrosine-integrase and Peaches encodes a serine-integrase (Fig. 3B). The segment of DNA differing between the two genomes includes the tyrosine integrase, *attP* and the excise gene in Eagle, and the serine-integrase gene in Peaches (Pope *et al.*, 2011). We note that this genetic swap does not include the Peaches recombination directionality factor (RDF), which is presumably encoded by gene *52*, a close relative of the known RDF of Bxb1 gp47 (Ghosh *et al.*, 2006), but which is more than 9 kbp away (Fig. 3B) (see Section V.B). However, homologues of the Bxb1 RDF are present in all Cluster A genomes, regardless of whether they utilize a tyrosine-integrase or a serine-integrase, and they presumably perform additional functions, perhaps in DNA replication (Fig. 3B). Thus Peaches could conceivably have acquired the serine-integrase from an Eagle-like parent without the necessity to also acquire the RDF function (Fig. 3B). Curiously, RedRock encodes neither a tyrosine- nor a serine-integrase, but in the middle of the genome (where the integrase is located in related genomes) codes for ParA and ParB proteins, suggesting that RedRock lysogens are maintained extrachromosomally and that the ParAB system acts to provide maintenance of the prophage (Fig. 3B). This raises the question as to how an extrachromosomal RedRock prophage is replicated because its close relatives in Subcluster A2 clearly do integrate into the host chromosome and replication functions are presumably switched off. There are few clues from genome comparisons as to how these presumed differences in replication requirements are accomplished and warrant a detailed experimental investigation. There are at least two examples of intein insertions in Cluster A phages evident from comparative genomic analyses. One of these is the intein in the Cluster A1 phage Bethlehem located within the terminase large subunit gene (*10*), which is absent from all other Cluster A terminases. Related copies of this intein are present in terminases of the Cluster L phage Omega (gp11) and Cluster E phages Cjw1 and Kostya (gp8 and gp9, respectively) and in a phamily of genes of unknown function in Cluster C phages (e.g., ET08 gp202); it is also commonly associated with most mycobacterial DnaB-like helicases. Interestingly, the Bethlehem gp10 intein utilizes a novel mechanism of protein splicing (Tori *et al.*, 2009). The second type of intein is present in Cluster A1 phages Bethlehem gp51, KBG gp53, and U2 gp50, which are homologues of the RDF protein that controls the directionality of Bxb1 serine-integrase-mediated site-specific recombination (see Section V.B.2)(Ghosh *et al.*, 2006). Relatives of this intein are also present in Cluster C phages Cali gp3, ET08 gp3, and LRRHood gp3 that encode putative nucleotidyltranferases (see Section III.C).

Cluster A virions are all siphoviral in their morphology and contain isometric heads attached to long flexible tails. The lengths of their tails are relatively short (~115 nm) compared to other mycobacteriophages (Hatfull *et al.*, 2010), and a close correlation exists between tail length and length of the gene encoding the tapemeasure protein (Pedulla *et al.*, 2003). Between the major tail subunit gene (*23*) and the tapemeasure protein gene (*26*) are two open reading frames (*24* and *25*) that are functionally analogous to the *G* and *T* genes of phage Lambda (Levin *et al.*, 1993) and are expressed via a −1 programmed translational frameshift (Xu *et al.*, 2004); these are thought to act as tail assembly chaperones. Interestingly, although the capsid subunits are not closely related to the well-studied phage HK97 capsid subunit, L5, D29, and Bxb1 (and presumably all other Cluster A phages) share the property of wholesale covalent cross-linking of capsid subunits (Ford *et al.*, 1998a; Hatfull and Sarkis, 1993; Mediavilla *et al.*, 2000; Popa *et al.*, 1991).

Genome comparison of Cluster A phages shows that Bethlehem contains a segment including genes *71* and *72* that is absent from other Subcluster A1 phage genomes. The gene products are distantly related to genes found in Omega (*21*, *22*) postulated to be part of an IS110-like transposon (see Section III.J). Bethlehem therefore likely carries a distantly related member of this poorly characterized transposon family.

B. Cluster B

Cluster B contains 15 phages whose genomes range from 67,118 to 70,654 bp. There are four subclusters, B1–B4, and there are a number of notable differences among them. Cluster B virions contain a linear genome with terminally redundant and circularly permuted ends (Table I). The left ends of B1 genomes are arbitrarily designated as the first base of the putative small terminase gene, but in Subclusters B2, B3, and B4, additional genes are closely linked with intergenic spaces to the left of putative terminase large subunit genes, and thus the left end of the genome is arbitrarily designated at the first noncoding intergenic gap encountered to the left of the terminase gene. Cluster B genomes could use specific packaging sites near the terminase gene to initiate headful packaging, but none have been identified. Cluster B phages typically form plaques that are neither clear nor evidently turbid, but have a somewhat hazy appearance. However, stable lysogens have not been reported for any Cluster B phage, and they behave as lytic rather than temperate phages. The genomes provide few clues as to their life styles, and none encode identifiable integrases, transposases, or partitioning functions. Also, none encode recognizable repressors, although these are generally diverse at the sequence level, and it is noteworthy that the L5 repressor (gp71) has no close relatives outside of the mycobacteriophages; repressors can thus be overlooked easily. Like Cluster A phages, Cluster B phages encode their own DNA polymerase—also a Pol I-like enzyme—as well as a putative primase/helicase protein. The genome organization of the Cluster B representative, Rosebush, is shown in Figure 4. The virion structure and assembly genes are shown in the left part of the genome and transcribed rightward; the long tapemeasure protein gene is a striking feature because of its length (5.6 kbp), reflecting the long tail (235 nm) in Rosebush and other Cluster B virions (Hatfull *et al.*, 2010). However, there are several notable features of this presumed operon. First, it is interrupted by a number of genes of unknown function, transcribed in both forward and reverse directions (Fig. 4). The leftward-transcribed genes *8* and *23* both have homologues outside of Cluster B, raising confidence in their annotation and identification. There are also four rightward-transcribed genes (*9–12*) between the terminase (*7*) and portal (*13*) genes; gp8 is related to Holliday Junction resolving RuvC-like proteins, but the others are of unknown function. Second, there are five predicted rightward-transcribed genes between the major tail subunit gene (*21*) and the tapemeasure protein gene (*29*; and two transcribed leftward) instead of the more typical pair of genes as seen in L5 (Fig. 3A). Moreover, the two genes in L5 (*24* and *25*) encoding tail assembly chaperones are expressed via a programmed translational frameshift, a highly conserved feature in virtually all dsDNA-tailed viruses, especially those with siphoviral morphotypes (Xu *et al.*, 2004). Cluster B phages appear to be notable exceptions, and although one or more of genes *24–28* could perhaps act as tail assembly chaperones, there is no evidence supporting expression via frameshifting. We note that none of these have mycobacteriophage-related proteins outside of Cluster B phages. Two of the Cluster B3 phages, Pipefish and Phlyer, contain intein insertions in their terminase large subunit genes (gp6).

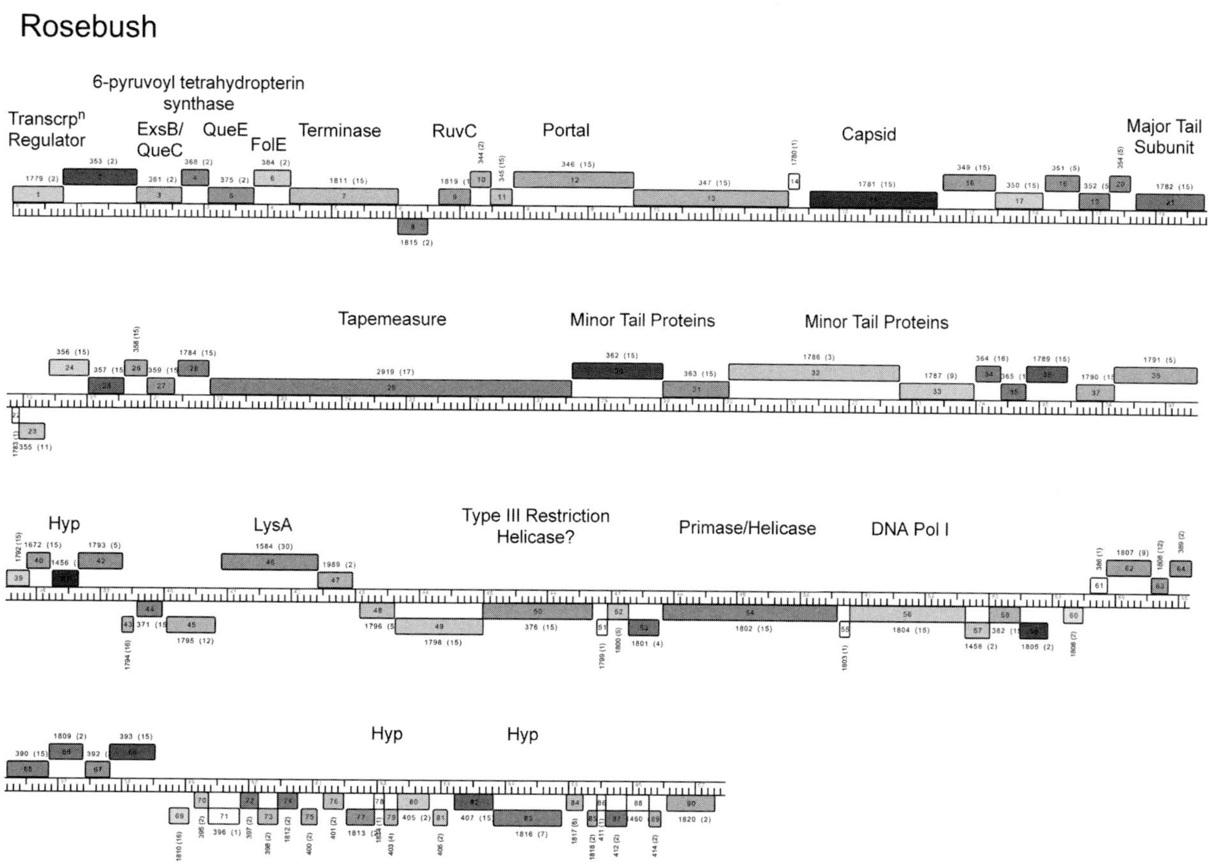

FIGURE 4 Map of the phage Rosebush genome, a member of Subcluster B2. See Figure 3A for further details on genome map presentation. (See Page 14 in Color Section at the back of the book.)

The extent of the Rosebush virion structure and assembly operon is unclear, although closely linked open reading frames continue after the tapemeasure protein gene through to gene *42* (there is a small noncoding gap between genes *36* and *37*), all or many of which are likely to encode minor tail proteins (Fig. 4). Genes to the right of gene *42* are arranged in five putative operons containing genes *43–45*, *46–47*, *48–60*, *61–68*, and *69–90*. The vast majority of these 46 genes are of unknown function, although the *48–60* operon includes genes encoding DNA replication functions. Rosebush gene *46* corresponds to lysin A, and it is plausible that *47* encodes a holin required for the delivery of Lysin A to its peptidoglycan substrate. Rosebush and its Subcluster B2 associate Qyrzula are notable in that they are among the few mycobacteriophages that do not encode a Lysin B protein (Payne *et al.*, 2009); Subcluster B1, B3, and B4 phages all encode a Lysin B protein. Subcluster B2 phages (Rosebush and Qyrzula) have an intriguing set of six genes (*1–6*) located between the terminase gene (*7*) and the arbitrarily designated genome left end (Figs. 4 and 5). Genes *3–6* are predicted to encode QueC, QueD, QueE, and QueF proteins, respectively, strongly implicating them in the biosynthesis of queuosine from GTP; HHPred analysis (Soding, 2005) predicts that gp2 is a queuine-tRNA-ribosyltransferase (Fig. 5). Queuosine is a modified base found commonly in bacterial tRNAs in the first anticodon position, although known queuosine biosynthetic genes are absent from mycobacterial genomes and the presence of queuosine as a tRNA modification has not been reported. Neither Rosebush nor any other Cluster B phage encodes its own tRNAs, and we therefore predict that in Rosebush (and Qyrzula) infections, that host tRNAs are modified with Queuosine; whether this alters the specificity of the translational apparatus or simply enhances the efficiency of translation is not known. Rosebush gene *1* is predicted to encode a transcription regulator but HHPred predicts relationships to ParB, KorB, and SopB proteins involved in chromosome partitioning; the proximity to the Queuosine biosynthetic operon suggests that gp1 could be involved in regulating this process.

The architectures of Subcluster B1, B3, and B4 genomes are similar to those of Rosebush and Qyrzula (Subcluster B2), including the long putative operon containing the virion structure and assembly genes and the five predicted operons to its right. There is an interesting difference within the structural gene organization in that—like Subcluster B2—Subclusters B1, B2, and B4 also encode homologues of RuvC (Rosebush *9*), but is transcribed in the opposite direction (Fig. 5). Maintenance of the RuvC gene in this location—notwithstanding the genetic rearrangements—is consistent with this playing an important role in virion assembly, perhaps in resolving any residual Holliday Junctions in replicated DNA molecules that otherwise would not be packaged. We note that HJ resolvases are also present within the virion structure gene operons of some other mycobacteriophages (e.g., Cluster E phages 244, Pumpkin and Porky).

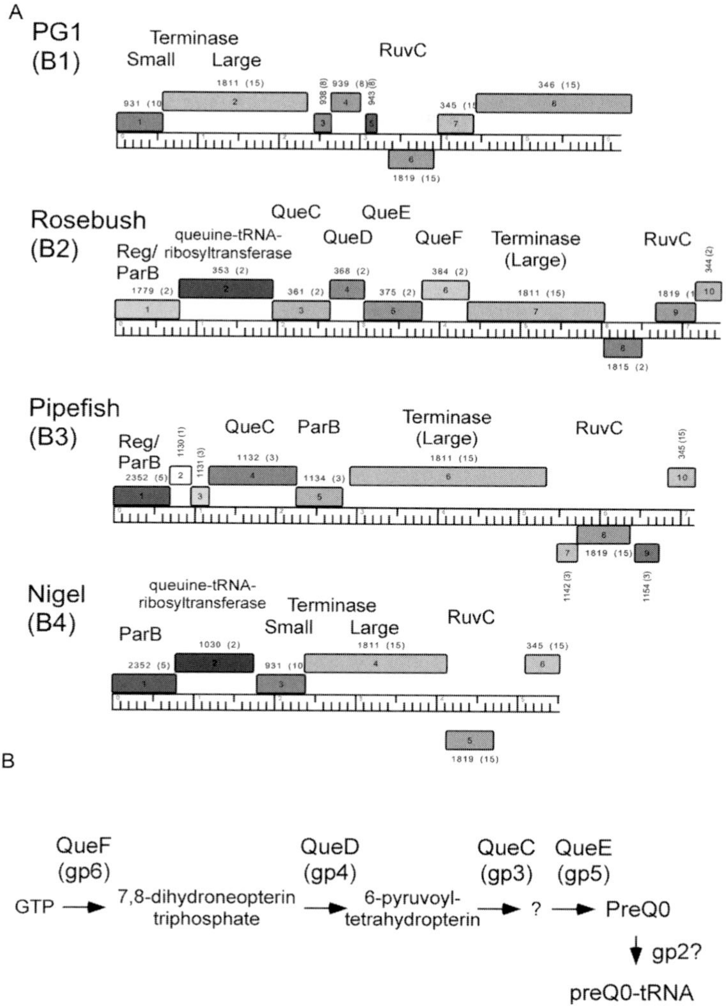

FIGURE 5 An unusual group of genes in Cluster B genomes. (A) Genome maps of segments of phages PBI1, Rosebush, Pipefish, and Nigel, representing Subclusters B1, B2, B3, and B4, respectively, are shown. Segments correspond to the extreme left ends of the genomes as they are standardly represented, although these genomes are circularly permuted and terminally redundant, the left end is arbitrarily designated and is not a physically defined left end. However, genes to the left (i.e., represented at the extreme right end of the genomes) are not evidently related to any of the functions described here. Five of the six genes to the left of the terminase large subunit gene of Rosebush (*7*) are predicted to be involved in queuosine biosynthesis, which presumably is used for tRNA modification (see text), as well as a predicted regulator with ParB-like features (gene *1*). PG1 contains none of these but has a putative terminase small subunit gene (*1*). Pipefish and Nigel have a subset of the Rosebush functions, including some but not all of

Notable differences exist in the region to the left of terminase genes, where putative queuosine biosynthetic genes are located in Subcluster B2 phages (Fig. 5). Subcluster B1 and B4 phages encode a terminase small subunit gene in this location, but Subcluster B2 and B3 phages appear to lack this function. Subcluster B4 phages (e.g., Nigel), however, encode two additional rightward-transcribed proteins (gp1 and gp2), and gp1 has similarities to ParB proteins, mirroring the putative function of Rosebush gp1 (although the two proteins are not obviously homologues). Interestingly, gp2 is predicted by HHPred to encode a queuine-tRNA-ribosyltransferase, although the protein is not obviously related to Rosebush gp2, and the Subcluster B4 phages do not encode other queuosine biosynthetic genes (Fig 5). The subcluster B3 phage Pipefish has five genes upstream of the terminase large subunit gene (5) (Phaedrus has only four of these), and it is noteworthy that Pipefish gp1 is a distant relative of Rosebush gp1, and Pipefish gp4 is likely a QueC-like protein, but without significant sequence similarity to Rosebush gp3 (Fig. 5). Pipefish gp5 is related to ParB-like proteins, although not related with significant sequence similarity to Rosebush gp1. The roles of the gene encoded to the left of the terminase genes in the Cluster B phages is therefore a substantial mystery, especially as there is great sequence divergence but with conservation of some common functions.

C. Cluster C

Cluster C currently contains nine phages divided into two subclusters, C1 and C2; C2 contains just a single phage, Myrna. All form plaques that are not completely clear, but also do not form stable lysogens, and their genomes do not encode any recognizable features of temperate phages. In general, the eight Cluster C1 phages are very similar to each other, while Myrna differs in a variety of ways. The C1 genomes are all relatively long but similarly sized (153.7–156.1 kbp) and Myrna is substantially larger (165.6 kbp), the largest of all the sequenced mycobacteriophage genomes. All of the Cluster C phages have myoviral morphologies with 80-nm-diameter isometric heads and modest length (85 nm) contractile tails. Cluster C genomes are circularly permuted and terminally

---

the genes. (B) Biosynthetic pathways for queuosine biosynthesis. QueF (Rosebush gp6) is a member of the family of GTP cyclohydrolases (which also includes FolE) and converts GTP to 7,8-dihydroneopterin triphosphate. This is converted to 6-pyruvoyltetrahydropterin by QueD (Rosebush gp4), which is then converted to PreQ0 by QueC and QueE (Rosebush gp3 and gp5, respectively). PreQ0 is transferred to a tRNA substrate by a queuine-tRNA-ribosyltransferase (Rosebush gp2). None of these phages appears to encode enzymes that would further process the PreQ0-tRNA to Q-tRNA. (See Page 15 in Color Section at the back of the book.)

redundant, although some (such as Bxz1) are unusual in that there is a long run of G-C residues in the genome through which sequencing reactions typically fail (Pedulla *et al.*, 2003). The left end of Bxz1 is arbitrarily designated as being the first unique base position after the G-string, and the other Cluster C phage genomes are represented and numbered accordingly.

The genome organization of the Cluster C representative Bxz1 is shown in Figure 6. It differs from all other mycobacteriophages in that few of the virion structure and assembly genes have been defined and it is not obvious that they are organized syntenically with respect to the genomes of Siphoviridae. However, Bxz1 gp124 and gp125 are similar to major structural proteins of mycobacteriophage I3 (Ramesh and Gopinathan, 1994), and gp129 is a putative tapemeasure protein, located immediately downstream of two genes (*127* and *128*) predicted to be expressed via a programmed translational frameshift (Fig. 6). Bxz1 gp135 and gp143 have similarity to other mycobacteriophage minor tail proteins, gp133 has features suggesting it is a plausible minor tail protein, and gp137 is related to Baseplate J proteins. Thus the region encoding genes *124–143* likely corresponds to a set of genes involved in tail structure and assembly; we note that Bxz1 gp114 has similarities to some other mycobacteriophage tail proteins (Fig. 6). The location of the putative major capsid subunit is unclear, although gp112 has similarity to phage head decoration proteins.

Bxz1 and its fellow Cluster C phages have a variety of genes whose putative functions suggest enticing aspects of its biology. For example, Bxz1 gp29 is a Band-7 family protein with two strongly predicted membrane-spanning segments at its N terminus (Fig. 6). Is this protein associated with host membranes during phage replication and, if so, what is its function? Is it possible that there are membranes associated with the virion itself and that gp29 is a virion protein? Answers to these questions remain unresolved. Bxz1 also encodes two putative glycosyltransferases and a galactosyltransferase protein, and gp230 is related to Ro proteins (Fig. 6). These phages also encode a large number of tRNA genes (>30; Table I)(Sahu *et al.*, 2004) as well as a tmRNA gene, a putative initiation factor (gp200), and a peptidyl-tRNA hydrolase (gp164), suggesting substantial modifications to the host translational machinery (Fig. 6). They encode a large DNA Pol III α subunit and DnaB/C proteins implicated in replication initiation and synthesis. In addition to RecA (gp220), Bxz1 also encodes a RusA-like Holliday junction resolvase (gp205). Many Cluster C genomes contain one or more intein insertions. Phage ET08 has a total of five inteins, more than any other mycobacteriophage genome. One of these is located in the gp3 putative nucleotidyltranferase gene, and Cali gp3 and LRRHood share this intein, as do the RDF proteins of some Cluster A1 phages (see Section III.A). A second is in ET08 gene *79* and

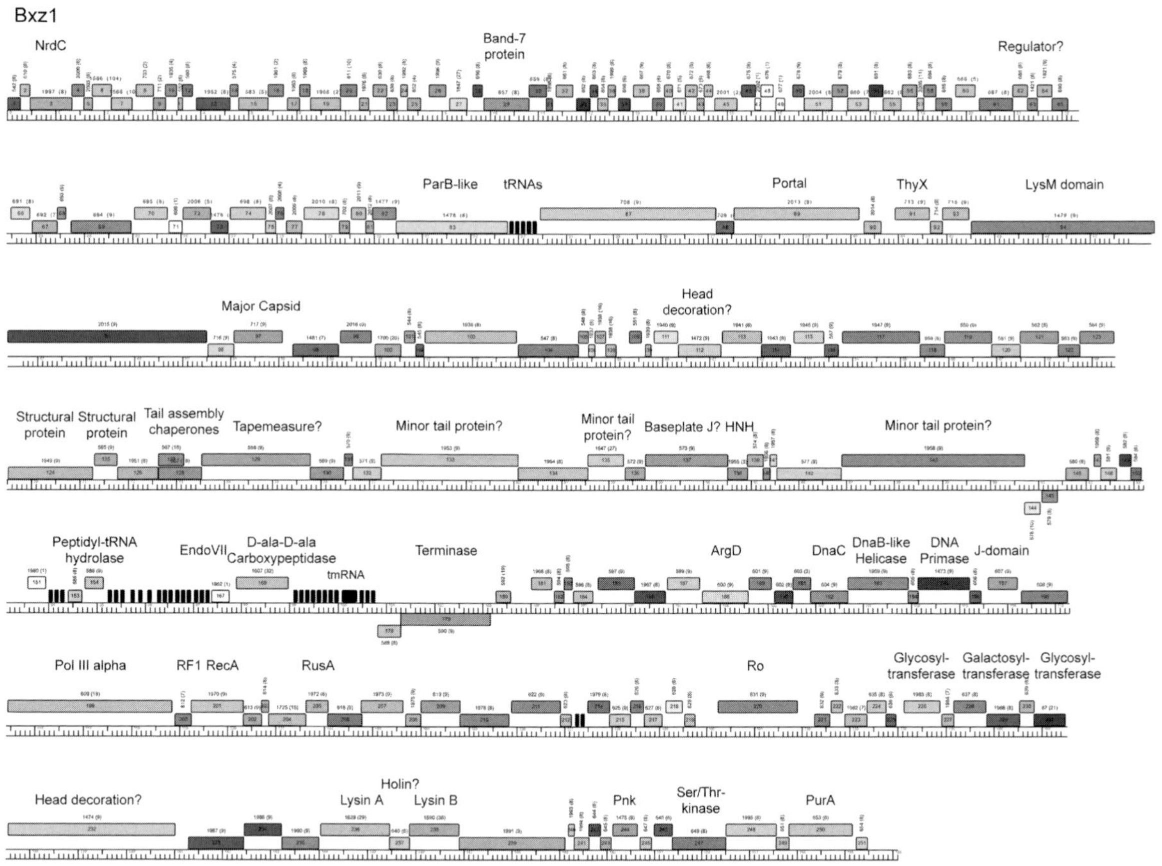

FIGURE 6 Map of the phage Bxz1 genome, a member of Subcluster C1. See Figure 3A for further details on genome map presentation. (See Page 16 in Color Section at the back of the book.)

none of the Cluster C homologues share this intein, although relatives are present in some terminase genes of nonmycobacteriophage phages, as well as host DnaB-like helicases. The third ET08 intein is in gp202, which is also present in homologues in ScottMcG, Spud, Catera, and Rizal, as well as in some Cluster A terminases (see Section III.A). ET08 gp239 contains an intein present in all Cluster C homologues except for Bxz1 gp239 and Myrna 246, but is related (28% amino acid identity) to the fifth ET08 intein, which is in gp248; none of the Cluster C homologues of ET08 gp248 contain this intein. Because Cluster C phages are not apparently temperate, it is no surprise that they do not encode integrase or partitioning functions. However, it is striking that one of these—phage LRRHood—carries a close relative of the Cluster A repressors (Pope *et al.*, 2011). Specifically, LRRHood gp44 is near identical to the known Bxb1 repressor (gp69) and differs in only one amino acid substitution (see Section III.A, Section V.A.1, and Fig. 20). However, the LRRHood genome does not contain even a single copy of the 13-bp stoperator sequence, which is the known binding site for this repressor; this, therefore, does not appear to be an immunity system for LRRHood. Presumably LRRHood has "stolen" the repressor from a Cluster A1-like phage, and because there are only between 5- and 11-bp differences between LRRHood gene *44* and Cluster A1 repressor genes, this presumably occurred relatively recently in evolutionary time. A plausible role for LRRHood gp44 might be to exclude other phages from superinfecting cells that are undergoing lytic growth of the phage. We note that another example of apparent repressor theft is seen in Cluster F phages (see Sections III.F and V.A.1).

Although the Subcluster C2 phage Myrna shares its myoviral morphology and genome size with Subcluster C1 phages, it is substantially different at the genomic level. There are segments where synteny is maintained, including regions corresponding to Bxz1 genes *117–143* and *187–211*, but other regions, including the leftmost ~31 kbp, have very few genes in common. These Cluster C phages therefore represent a really intriguing collection of viruses whose genomes suggest many secrets waiting to be discovered.

D. Cluster D

There are six closely related phages in Cluster D, and no subcluster divisions (Table I). These phages form plaques that are not completely clear, and although not evidently turbid as phages such as L5 or Bxb1, it is possible that they are temperate and form lysogens at low frequency. However, the genomes reveal no features associated with temperate phages such as integrases or repressors. The genomes are presumed to be circularly permuted and terminally redundant, and the left end is arbitrarily designated at a noncoding gap 6-7 genes to the left of the terminase large subunit gene; a gene map of the Cluster D representative, PBI1, is shown in Figure 7. All of the PBI1 genes, except for *41* and *42*, are transcribed rightward (Fig. 7), and the virion structure and assembly genes are located in the left part of the genome (*8-31*); although confident assignments can be ascribed to the terminase large subunit gene (*8*), portal (*9*), tapemeasure (*27*), and minor tail proteins (*30, 31*), assignments of capsid subunit (*17*) and major tail subunit genes (*24*) are less confident. None of the virion proteins have been characterized experimentally. The lysis cassette, including lysin A and lysin B genes (*35* and *39*, respectively), is located immediately downstream of the virion structure and assembly operon (Fig. 7). PBI1 genes *36-38* all encode proteins with putative transmembrane domains (two, four, and one, respectively) and at least one may act as a holin. None of the Cluster D phages infect *M. tuberculosis*. Genes in the right half of the genome contain several that are likely involved in DNA replication, including a helicase (*50*), a DNA polymerase III α subunit (*63*), and a primase/polymerase gene (*81*). Several other genes in this region are implicated in nucleotide metabolism, including a putative nucleotide-binding protein (*56*), a putative deoxyribonucleotidase (*59*), and ThyX (*65*). Gene *57* encodes a 309 residue protein containing a highly acidic segment in its C-terminal half (76 of 79 resides are aspartic or glutamic acids) and has weak similarity to CDC45-related proteins, implicating it in a possible role in initiation of DNA replication.

A particularly intriguing aspect of Cluster D phages is genes immediately to the left of the terminase large subunit gene (*8*). Five of the six Cluster D phages (PBI1, Adjutor, P-lot, Butterscotch, and Troll4) each encode a 256 residue protein (PBI gp7; Fig. 7) with similarity to vegetative insecticidal protein 2 (VIP2) family proteins encoding actin-ADP-ribosylating toxins (Han *et al.*, 1999). In phage Gumball, the gene encoding VIP2 is part of a single gene (*6*) that also contains sequences corresponding to PBI1 gene *6*, as a result of an additional A-residue immediately upstream of the VIP2 moiety. Although this mutation could have arisen during propagation of the phage in the laboratory and perhaps renders it nonfunctional, an alternative possibility is that PBI1 gene *6* encodes a function related to VIP2-like activity and that the two units can function either as independent proteins (as in PBI gp6 and gp7) or as two domains of a single protein (as in Gumball gp6). None of these proteins contain putative signal sequences and it is plausible that they are expressed late in lytic growth (together with their closely linked virion structure and assembly genes) and are released upon cell lysis, a parallel scenario to the expression and release of toxin from *E. coli* phage 933W (Tyler *et al.*, 2004). In light of the similarity to insecticidal toxins, it is puzzling as to what role is played by PBI1 gp7 and its related proteins. Do Cluster D phages confer insecticidal properties to infected cells and, if so, what is the natural bacterial host and what insects are affected? These questions remain unresolved.

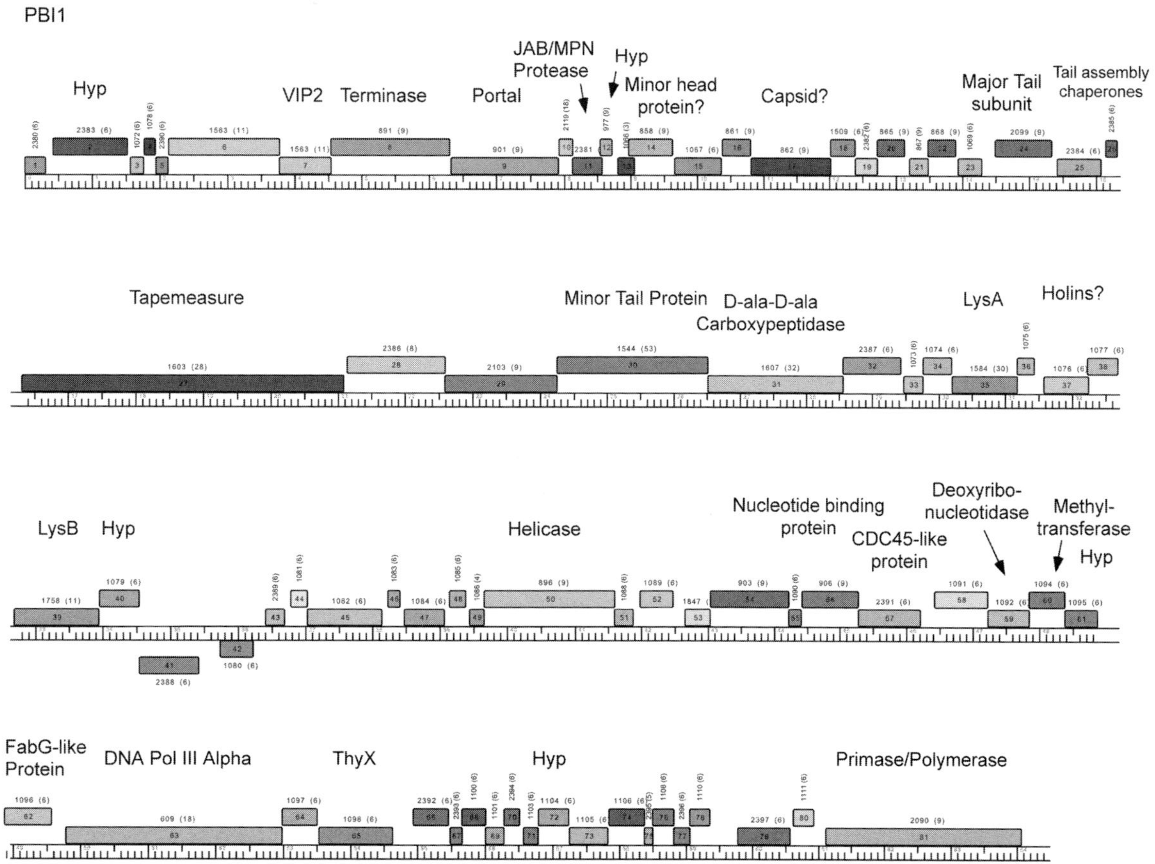

FIGURE 7 Map of the phage PBI1 genome, a member of Cluster D. See Figure 3A for further details on genome map presentation. (See Page 17 in Color Section at the back of the book.)

E. Cluster E

There are five Cluster E phages, all are very similar to each other at the nucleotide sequence level—albeit with a variety of replacements, insertions, and deletions—and there are no subcluster divisions (Table I). They form plaques with some turbidity and likely form lysogens, although stable lysogens have not been reported for any Cluster E phages. Preliminary data from our laboratory suggest that Cjw1 may form stable lysogens at 42 °C, but not at 37 °C. None of these phages infect *M. tuberculosis*. A map of the genome organization of the Cluster E representative phage, Cjw1, is shown in Figure 8. It encodes both small and large subunits of terminase, and the small subunit gene is approximately 1 kbp away from the physical left end of the genome (Fig. 9). The terminase large subunit genes of Cjw1 and Kostya (gp8 and gp9 respectively) both contain inteins related to those in some of the Cluster A1 terminases (see Section III.A). The virion structure and assembly operon spans from genes *5* to *28* and follows the canonical synteny, but with two notable interruptions. First, in Pumpkin, Porky, and 244, but not in Cjw1 or Kostya, an EndoVII Holiday junction resolvase gene lies between the frameshifting tail assembly chaperones and the tapemeasure protein. Although this is an unusual position for a gene insertion, presence of a HJ resolvase in the virion structural operon is reminiscent of the organization of Cluster B genomes (see Section III.B). Presumably it is not essential for growth because Cjw1 and Kostya lack this gene and no obvious candidates exist for providing this function elsewhere in their genomes. Second, in Cjw1 (and Pumpkin and 244), two small genes are located between the terminase small and large subunit genes, transcribed in the opposite direction and of unknown function (Fig. 8); one of these (Cjw1 *17*) is absent from Kostya and Porky. Genes related to Cjw1 *8* are also found in Cluster B4 genomes (e.g., Cooper *98*), elevating confidence in their assignment. The lysis cassette lies to the right of the virion structure and assembly genes and includes both Lysin A and Lysin B genes (Cjw1 *32* and *35*, respectively); Cjw1 gene *33* is a strong candidate for encoding a holin with two membrane-spanning domains. Genes to the right of this appear to be organized into six putative operons: (1) leftward-transcribed genes *38–52*, (2) genes *53* (integrase) and *54* that are transcribed rightward, (3) leftward-transcribed genes *55* and *56*, (4) genes *57–128* transcribed rightward, (5) leftward-transcribed genes *129–142*, and (6) genes *143* and *144* that are transcribed rightward. In general, this organization is conserved in all Cluster E phages. Although the vast majority of genes in Cluster E genomes are of unknown function (>80%), several of those that do have functional assignments are rare among the mycobacteriophages and are of considerable interest. First, Cjw1 gene *39* encodes a relative of Lsr2, a regulatory protein present in the mycobacterial host that coordinates expression of a

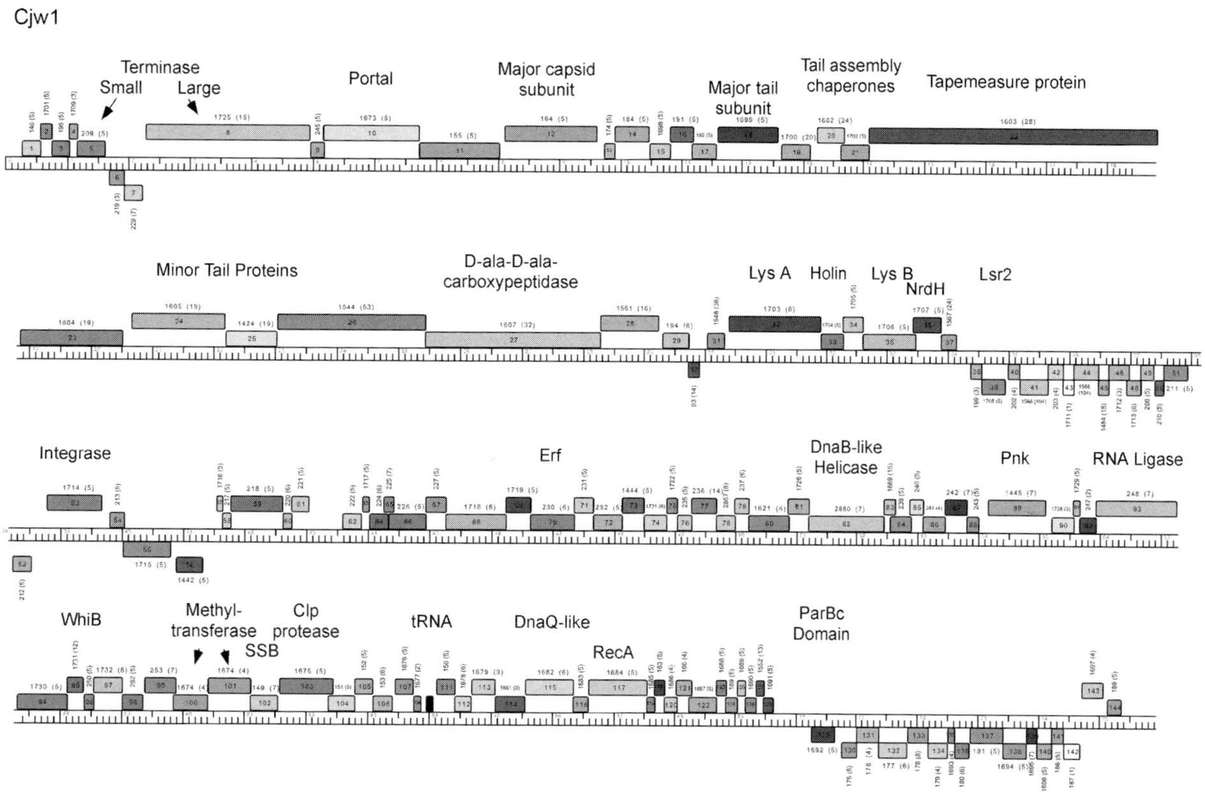

FIGURE 8 Map of the phage Cjw1 genome, a member of Cluster E. See Figure 3A for further details on genome map presentation. (See Page 18 in Color Section at the back of the book.)

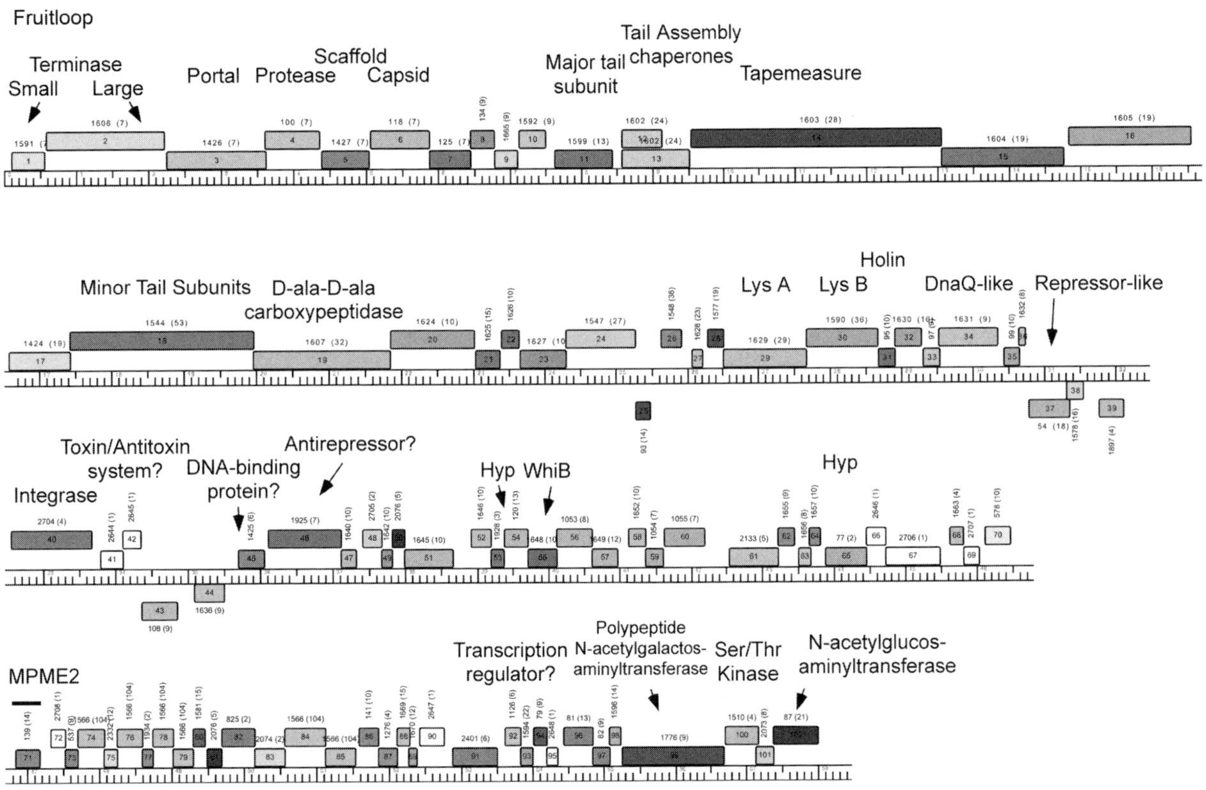

FIGURE 9 Map of the phage Fruitloop genome, a member of Subcluster F1. See Figure 3A for further details on genome map presentation. (See Page 19 in Color Section at the back of the book.)

large number of host genes. Its role in Cjw1 is not clear, although we note that Cluster J genomes also encode related proteins (see Sections III.J). It is plausible that this confers the repressor function in Cluster E phages, although this would be a very unusual form of phage regulation.

Second, the long operon *57–128* includes several genes encoding putative functions in DNA replication, recombination, RNA metabolism, and nucleotide metabolism. Cjw1 gene *70* is related to Erf-family recombinases and presumably mediates general recombination functions. Although several other mycobacteriophages encode RecA-like and RecET-like recombination functions, Cluster E phages—along with Omega (Cluster L) and Wildcat—are the only ones with an Erf-like protein; somewhat surprisingly, Cluster E phages also encode RecA homologues (e.g., Cjw1 gp117). Cjw1 encodes a $tRNA^{gly}$ gene (*109*) in this region, as well as a tRNA-like gene in the small intergenic gap between genes *108* and *109*, although the noncanonical tRNA structure and four-base anticodon suggest that this is either nonfunctional or perhaps plays a role in translational frameshifting. The RNA Ligase encoded by Cjw1 gene *93* is also unusual but related proteins are also encoded in Cluster L and J phages. Likewise, Cjw1 gene *102* encoding a single-stranded binding protein (SSB) is only found elsewhere in Cluster L phages and the singleton Wildcat. Cjw1 gp89 has been shown to be a bifunctional polynucleotide kinase (Pnk) with both kinase and phosphatase domains, and it was noted that because Cjw1—like Omega (see Section III.J)—also encodes an RNA Ligase (gp93), that these might act to evade an RNA-damaging antiviral host response (Zhu *et al.*, 2004). Interestingly, we note that the Cluster L phage LeBron also encodes similar Pnk and RNA Ligase proteins (see Section III.L; Fig. 15). Moreover, this highly diverse set of phages also encodes one or more tRNA genes (Table I). Finally, Cjw1 gene *115* encodes a protein with similarity to DnaQ-like proteins, suggesting a possible role in DNA repair or perhaps in phage replication itself. Roles for these interesting proteins, their biochemical activities, and their expression patterns await elucidation.

F. Cluster F

Cluster F is one of the more diverse groups of phages at the nucleotide sequence level. There are a total of 10 members, with all but 1 (Che9d) constituting Subcluster F1, and Che9d forming Subcluster F2. They form somewhat turbid plaques from which stable lysogens can be recovered (Pham *et al.*, 2007). Genomes vary somewhat in length, ranging from 52.1 kbp [Ardmore, (Henry *et al.*, 2010b)] to 59.5 kbp [Che8 (Pedulla *et al.*, 2003)](Table I), but all have defined cohesive termini. None of the Cluster F phages infect *M. tuberculosis*. The complete sequence of mycobacteriophage Ms6 is not yet available, but from sequenced segments of the genome it seems probable that it belongs to the F cluster. The genome map of the Cluster F representative phage, Fruitloop, is shown in Figure 9. Fruitloop encodes both terminase small and large subunits, and the small subunit gene is very close to the physical left end of the genome (Fig. 9). The virion structure and assembly operon extends from gene *1* to gene *24*, transcribed rightward, and is fairly canonical with regard to the common syntenic organization (Fig. 9). The block of genes corresponding to the region from Fruitloop *11* (major ail subunit) to gene *23* (putative minor tail protein) is the most highly conversed segment among Cluster F1 phages at the nucleotide level. The region to the left of Fruitloop gene *11* is substantially different in Ramsey and Boomer, although the genes are likely to confer similar functions in DNA packaging and head assembly. The lysis cassette lies to the right of the virion structure and assembly genes and includes lysin A (*29*) and lysin B (*30*) genes; gp31 is a likely Holin and contains a single predicted transmembrane domain (Fig 9). Immediately to the right of the lysis cassette is a DnaQ-like gene (*34*) implicated either in DNA repair, or perhaps DNA replication itself, as in Cluster E phages.

Genes in the right part of the Fruitloop genome are organized into four possible operons: (1) genes *37–39*, (2) genes *40–42*, (3) genes *43* and *44*, and (4) genes *45–102*. The first of these is of particular interest, as Fruitloop genes *37–39* do not have closely related counterparts in other Cluster F phages, but are homologues to genes in Cluster A phages; gp37 and gp38 are close relatives of Bxb1 gp69 and gp70, respectively; and gp39 is most closely related to Jasper gp92. Bxb1 gp69 is a well-characterized repressor related to the L5 repressor (Jain and Hatfull, 2000) and its presence in Fruitloop is somewhat surprising (see Sections III.A and V.A.1). Two lines of evidence suggest that Fruitloop gp37 is not involved directly in the immunity regulation of Fruitloop itself, in that it is absent from all other Cluster F genomes, and there is not an abundant array of stoperator sites throughout the Fruitloop genome as there are in Bxb1 and its relatives (see Section III.A). There is, however, a single putative 13-bp repressor-binding site located upstream of gene *39* near a strongly predicted putative leftward promoter, which thus may be involved in autoregulation of its expression. We also note that the nucleotide sequences of Fruitloop *37* and *38* are ~98% identical with their Bxb1 homologues, suggesting that these were acquired very recently in evolutionary times. A plausible scenario is that Fruitloop has stolen these genes from a Cluster A-like phage for purposes of conferring a rogue immunity status, providing protection to Fruitloop lysogens from superinfection by Cluster A-type phages that have a Bxb1 type of immunity. A similar example of apparent repressor theft occurs in the Subcluster C1 phage LRRHood (see Section III.C).

The rightward operon encompassing genes *40–42* contains the integrase gene (*40*) plus two genes (*41*, *42*) that are candidates for forming a toxin–antitoxin (TA) system; Fruitloop gp41 is the putative toxin and gp42 is the putative antitoxin. There are no identifiable relatives of these in other Cluster F phages or indeed in any other mycobacteriophages. TA systems generally are not common in phage genomes, although the well-studied plasmid addiction system of phage P1 is within this general class (Lehnherr *et al.*, 1993). However, it seems unlikely that genes *41* and *42* are involved in plasmid maintenance of Fruitloop similar to P1 because Fruitloop encodes an integrase (gp40) and presumably provides prophage maintenance through stable integration. Because it has been reported that TA systems can provide protection to bacterial cultures by conferring abortive infection (Fineran *et al.*, 2009), an intriguing hypothesis is that this Fruitloop TA system has been acquired to provide protection to Fruitloop lysogens by infection from other phages. In this model, addition of the Bxb1 repressor and the putative TA system has been selected for by the same core property of providing survival of the host to subsequent viral attack.

The vast majority of genes in the Fruitloop rightward operon containing genes *45–102* are of unknown function, but several are of interest. First, gene *45* encodes a helix-turn-helix DNA-binding protein, which could either provide repressor activity or possibly a Cro-like function. Second, Fruitloop gene *55* encodes a WhiB-family transcriptional regulator protein, and although WhiB-related proteins are encoded by several mycobacteriophages, their roles remain unclear (Rybniker *et al.*, 2010). Third, gene *100* encodes a putative serine-threonine kinase of unknown function, and it is unclear whether it is phosphorylating host or phage proteins. Fourth, there are two putative genes encoding glycosyltransferase enzymes, although the roles and the targets of these are also unknown. Finally, gene *71* is part of a MycobacterioPhage Mobile Element (MPME2)(see Sections III.G and IV) that is prevalent throughout Cluster F phages and was first identified through genome comparison of Cluster G phages (Sampson *et al.*, 2009).

G. Cluster G

Four Cluster G phages are extremely closely related to each other at the nucleotide sequence level and there are no subcluster divisions. These phages have among the smallest mycobacteriophage genomes, ranging from 41.1 to 42.3 kbp in length. As discussed later, the primary cause for length differences is the presence/absence of a novel small putative mobile genetic element (MPME), which is absent from Angel and present as a single copy but in a different location in each of the other three genomes. Cluster G phages form lightly turbid plaques from which stable lysogens can be recovered (Sampson *et al.*, 2009). They do not infect *M. tuberculosis* at high efficiency, but mutants arise at a frequency of $\sim 10^{-5}$ that have acquired the ability to infect *M. tuberculosis* at equal efficiency to

M. smegmatis (Sampson *et al.*, 2009). A genome map of the Cluster G representative phage, BPs, is shown in Figure 10. Cluster G genomes have defined cohesive ends, and a putative small terminase subunit gene (*1*) is located near the left physical end of the genome (Fig. 10). The virion structure and assembly operon encompasses genes *1* through *26* and is organized with canonical synteny. The lysis cassette follows immediately after and contains both lysin A and lysin B genes (*27* and *28*, respectively; Fig.10); gene *29* encodes a putative holin, based on the presence of two putative transmembrane domains. The only leftward-transcribed genes in the genome are *32* and *33*, encoding the integrase and repressor proteins, respectively, and genes to their right are all transcribed rightward. Most of the genes in this rightward operon are of unknown function, although it also includes a RecET recombination system (gp42 and gp43) and a RuvC-like Holliday Junction resolvase (gp51).

A rather striking feature of repressor/integrase gene organization is that the crossover site for integrase-mediated, site-specific recombination within the phage attachment site (*attP*) is located within the coding region for the repressor (Sampson *et al.*, 2009). As a consequence, two different types of gene product are expressed from gene *33*: a 130 residue product from the viral genome and a 97 residue product from an integrated prophage. The 97 residue protein confers immunity and provides the repressor function, whereas the virally expressed 130 residue protein does not. Integration and excision would therefore seem to play critical roles in the decision between lysogenic and lytic growth, having a direct impact on whether an active or an inactive repressor is expressed (Sampson *et al.*, 2009). A particularly interesting question arises as to how the genetic switch operates and whether phage-encoded cII and/or cIII analogues modulate the frequencies of lysogeny or whether this is accomplished solely by the gene *32*–*33* cassette (see Section V.A.2).

The close nucleotide sequence similarity between Cluster G genomes proved crucial in identification of a new class of ultrasmall mobile genetic elements present in mycobacteriophages (Sampson *et al.*, 2009). For example, when BP is compared with the other three genomes, it is apparent that the small open reading frames BPs *57* and *59* form a single gene in Angel, Hope, and Halo (Pope *et al.*, 2011; Sampson *et al.*, 2009). Alignment of DNA sequences shows that there is a precise insertion of 445 bp, including the open reading frame for gene *58*, in BPs relative to the other phages (Figs. 10B and 10C). Such alignments also show similar relationships reflecting an insertion in Halo that has occurred at a target within the homologue of BP gene *54* and in Hope at a target within the homologue of BP gene *56* (Fig. 10C). Alignment of the inserted sequences shows that there are two types of these MPME elements, MPME1 (in Hope and BPs) and MPME2 (in Halo), that share 78% nucleotide sequence

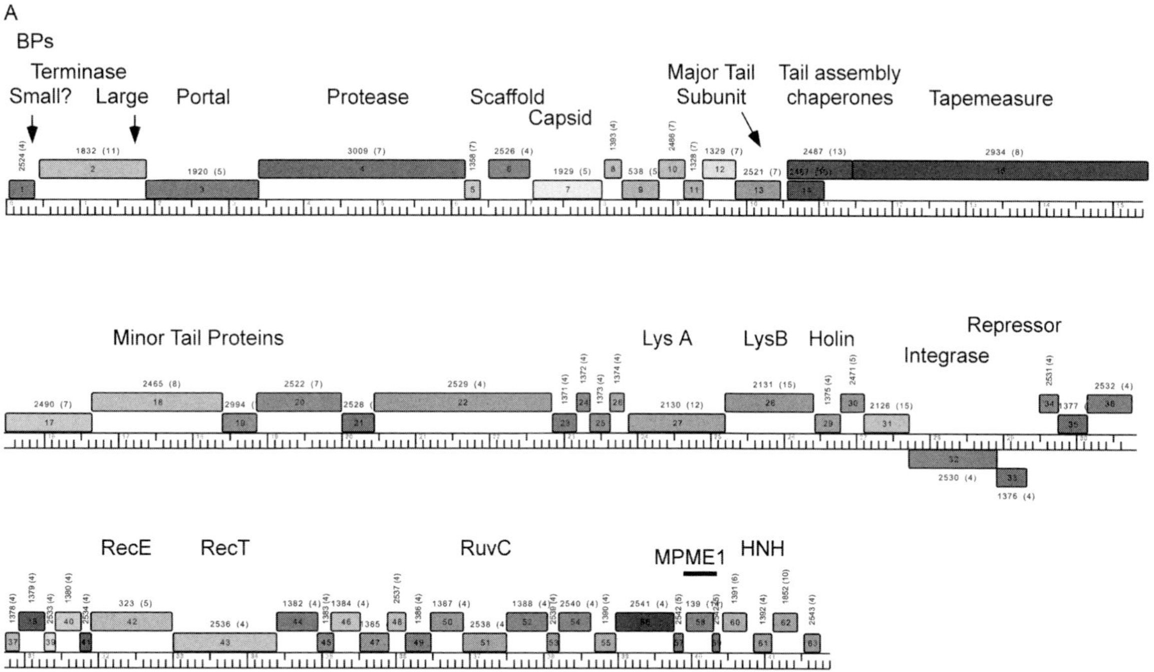

FIGURE 10 (Continued)

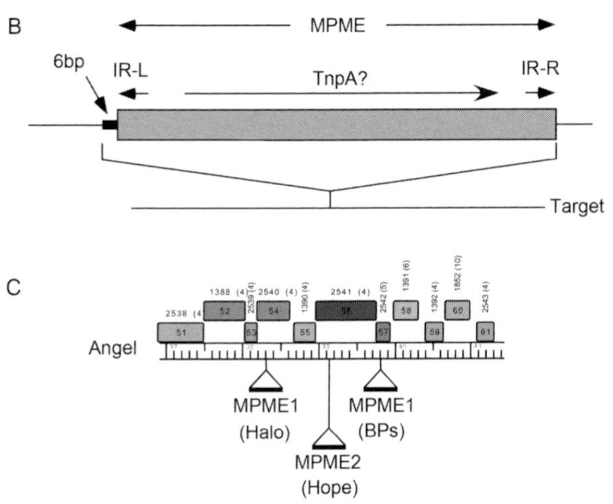

FIGURE 10 Features of Cluster G phages. (A) Map of the phage BPs genome, a member of Cluster G. See Figure 3A for further details on genome map presentation. (B) MPME elements are 439–440 bp in length and are flanked by 11-bp imperfect inverted repeats, IR-L and IR-R. Insertion into the target is associated by a 6-bp insertion between IR-L and the target, and its origin is unknown. (C) A 5.5-kbp segment at the extreme right end of the Angel genome is shown, illustrating the positions of the insertion of MPME1 elements in BPs and Halo, and an MPME2 insertion in phage Hope. (See Page 20 in Color Section at the back of the book.)

similarity (Sampson *et al.*, 2009); Angel is devoid of these elements (Sampson *et al.*, 2009). Comparison against other mycobacteriophages shows that there is a single copy of MPME1 in Cluster F phages Fruitloop, PMC, Llij, Boomer, Che8, Tweety, Ardmore and Pacc40; in Cluster I phages Brujita and Island 3; and a partial copy in Corndog. There are no related copies in any of the sequenced mycobacterial genomes or elsewhere (Sampson *et al.*, 2009).

Although the size of each of the MPME1 insertions is 445 bp, the mobile element itself appears to be 439 bp long, with two imperfect 11-bp inserted repeats (IR) at the extreme ends (Sampson *et al.*, 2009) (Fig. 10B). At the right end, IR-R is joined to the target sequence without addition or duplication of any target sequences (Fig.10B). However, at the left junction, there is an insertion of 6 bp between IR-L and the target DNA. This 6-bp segment is different in many of the insertions and does not correspond to target duplication. It therefore remains a mystery as to where this 6 bp originates from and what mechanism of transposition could be involved in generating these types of products. Transposition is presumably mediated by the 123 residue product encoded within the MPME element (Fig. 10B), although this is both remarkably small and shows no motifs to structural elements common to other transposases.

The MPME elements show that the three genes containing insertions (corresponding to BPs genes *54*, *56*, and the gene *57–59* interruption; Fig. 10) are presumably nonessential for phage growth. However, because there are no obvious differences in growth of the four Cluster G phages, this provides little information as to what the genes actually do and why they may have been acquired by the phages—they are simply nonessential. Cluster G genomes are suitable substrates for BRED manipulation (see Section VII.A.6), and four additional genes (BPs genes *44*, *49*, *50*, and *52*) have also been shown to be nonessential because viable deletion mutants can be constructed readily. This raises the question as to whether it is generally true that a high proportion of genes constituting the nonvirion structure and assembly genes are nonessential in the mycobacteriophages and, if so, what forces drive the evolutionary of this large number of mysterious genes.

H. Cluster H

There are three phages assigned to Cluster H: two (Predator and Konstantine) in Subcluster H1 and one (Barnyard) in Subcluster H2 (Hatfull *et al.*, 2010; Pope *et al.*, 2011). The cluster is quite diverse and many differences exist among the three constituting genomes. These phages form plaques that are not evidently turbid, but also not completely clear, although stable lysogens have not been recovered; the genomes also do not possess features of temperate phages such as integrase or repressor genes. They all have termini consistent with the viral chromosomes being circularly permuted and terminally redundant. They have among the lowest of the GC% content of the mycobacteriophages (Fig. 2), and none of them infect *M. tuberculosis*.

The genomic organization of a Cluster H representative, Barnyard, is shown in Figure 11. The left end of Barnyard is designated arbitrarily at the first noncoding interval to the left of the terminase large subunit gene (*6*), and the functions of the five intervening genes are unknown; none of these is an obvious candidate for a terminase small subunit gene (Fig. 11). All of the predicted genes are transcribed in the rightward direction as shown in Figure 11, and an obvious genomic feature is the presence of the large number of orphams (genes belonging to a phamily that has only a single member), nearly 60% of all 109 predicted Barnyard genes. Such a high proportion of orphams is not unexpected in singleton phage genomes where other closely related phages have yet to be isolated, but for Barnyard this reflects the degree to which it—as a Subcluster H2 phage—differs from Subcluster H1 phages. Subcluster H1 phages Predator and Konstantine have 18 and 25% orphams, respectively, again reflecting the generally high diversity of this cluster. This is in marked contrast to, for example, Cluster G phages, which differ mostly by just a relatively modest number of nucleotide differences (see Section III.G).

The virion structure and assembly genes span from gene *6* through to gene *36*, and genes encoding the terminase large subunit (*6*), portal (*7*), a putative protease (*17*), capsid subunit (*21*), major tail subunit (*30*), tail assembly chaperones expressed by a putative programmed translational frameshift (*31* and *32*), tapemeasure protein, and putative minor tail proteins (*34-36*) are predicted (Fig. 11). Although these genes are in canonical order, several small additional genes are present between the putative portal and protease genes, and also between the capsid and major tail subunit genes. The tapemeasure gene is notable due to its impressive length (6.1 kbp), corresponding to the very long tails of the Cluster H phages (~300 nm). In Predator—although not the other Cluster H phages—there is a putative Endo VII Holliday Junction resolvase (*10*) located between the portal and protease genes, reminiscent of the location of functionally related genes in some Cluster B and E genomes. The lysis cassette, containing lysin A (*39*) and lysin B (*40*) genes, as well as a putative holin (*41*), lies immediately to the right of the virion structure and assembly operon (Fig. 11). Of the 67 genes in the Barnyard genome to the right of the lysis cassette, only 13 have homologues in other mycobacteriophages; 8 of these are found only in Subcluster H1 phages. Five genes in this region can be assigned putative functions, including a Helicase (*65*), a putative nucleotide-binding protein (*75*), an α subunit of DNA polymerase III (*80*), a peptidase (*94*), and a large primase/polymerase gene (*108*).

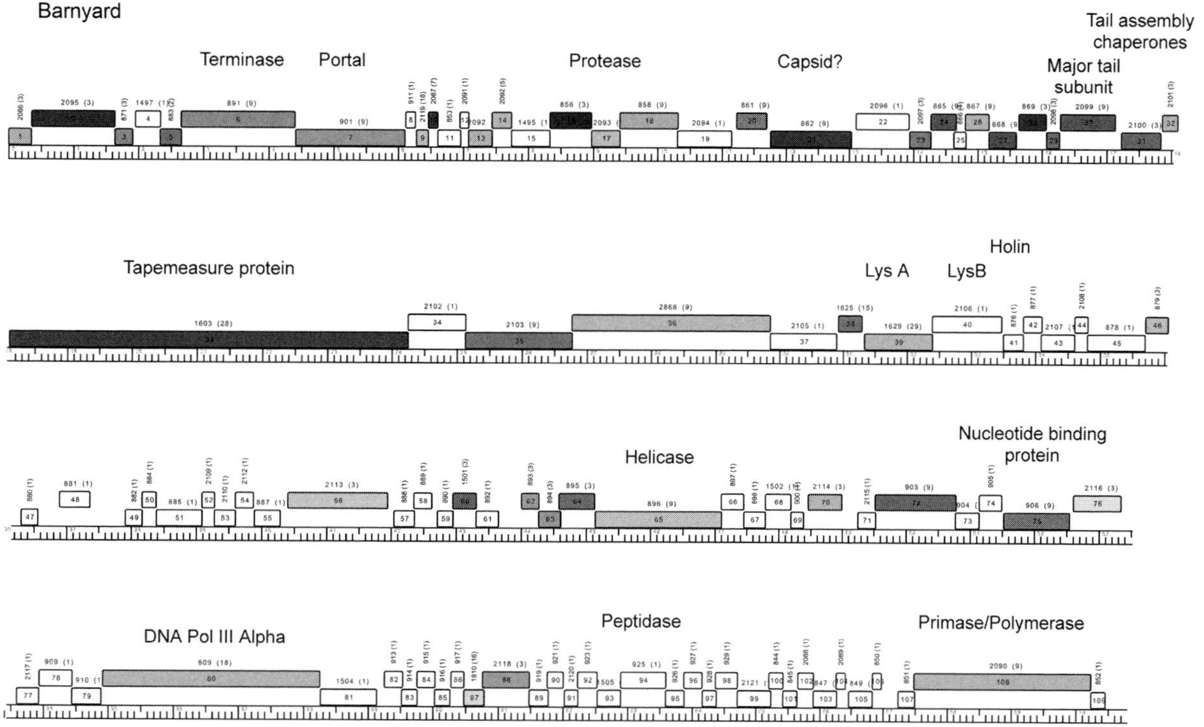

FIGURE 11 Map of the phage Barnyard genome, a member of Subcluster H1. See Figure 3A for further details on genome map presentation. (See Page 22 in Color Section at the back of the book.)

The GC% of Cluster H phages is among the lowest of all the mycobacteriophages (56.3–57.3%, Fig. 2), and only the singleton Wildcat shares a GC% content lower than 60% (56.9%). This may reflect a preference of the Cluster H phages for hosts that are more distantly related to *M. smegmatis*, and although the majority of members of the Actinomycetales have GC% contents that are above 60%, some—such as *M. leprae* (57.8%)—do have a substantially lower GC% content. Such a different host preference may account for notable differences of Cluster H phages from other mycobacteriophages and the high proportions of orphams (Fig. 11). Cluster H phages thus remain largely unexplored and would seem to warrant extensive further analysis both in regards to the determination of gene function and expression and in elucidation of their host ranges.

I. Cluster I

Cluster I contains three phage members: two (Brujita and Island3) in Subcluster I1 and one (Che9c) in Subcluster I2. They are quite diverse at the sequence level, although the two Subcluster I1 phages share nucleotide sequence similarity across most of their genomes. Che9c is both more distantly related and has a substantially larger genome (57 kbp) than Subcluster I1 genomes (~47 kbp). Cluster I phages form somewhat turbid plaques, although lysogens have not been well characterized. They do, however, encode genes common to temperate phages such as an integration system; no repressor genes have been described. All Cluster I phages have defined cohesive termini, and gene *1* is a reasonable candidate for encoding a terminase small subunit (Fig. 12). None of the Cluster I phages infect *M. tuberculosis*. A notable morphological feature of Cluster I phages is that they contain prolate heads, with a length to width ratio of approximately 2.5:1 (Hatfull *et al.*, 2010). The genome organization of a Cluster I representative, Che9c, is shown in Figure 12. The virion structure and assembly operon (gene *1–22*) is syntenically canonical, and genes encoding terminase large subunit (*2*), portal (*4*), protease (*5*), capsid (*6*), major tail subunit (*12*), tail assembly chaperones (*13* and *14*), tapemeasure (*15*), and minor tail proteins (*16–19*) can be predicted confidently (Fig. 12). The lysis cassette lies to the right of the virion structure and assembly operon and includes genes encoding lysin A (*25*), lysin B (*26*), and a putative holin (27) containing two predicted membrane-spanning domains. To the right of the lysis genes are perhaps four operons: (1) gene *30* and *31* transcribed leftward, (2) genes *32–36* transcribed rightward, (3) genes *37–46* transcribed leftward (although two large intergenic regions exist between genes *39* and *40* and between *45* and *56* so there could be multiple operons; Fig. 12), and (4) genes *47–84* transcribed rightward.

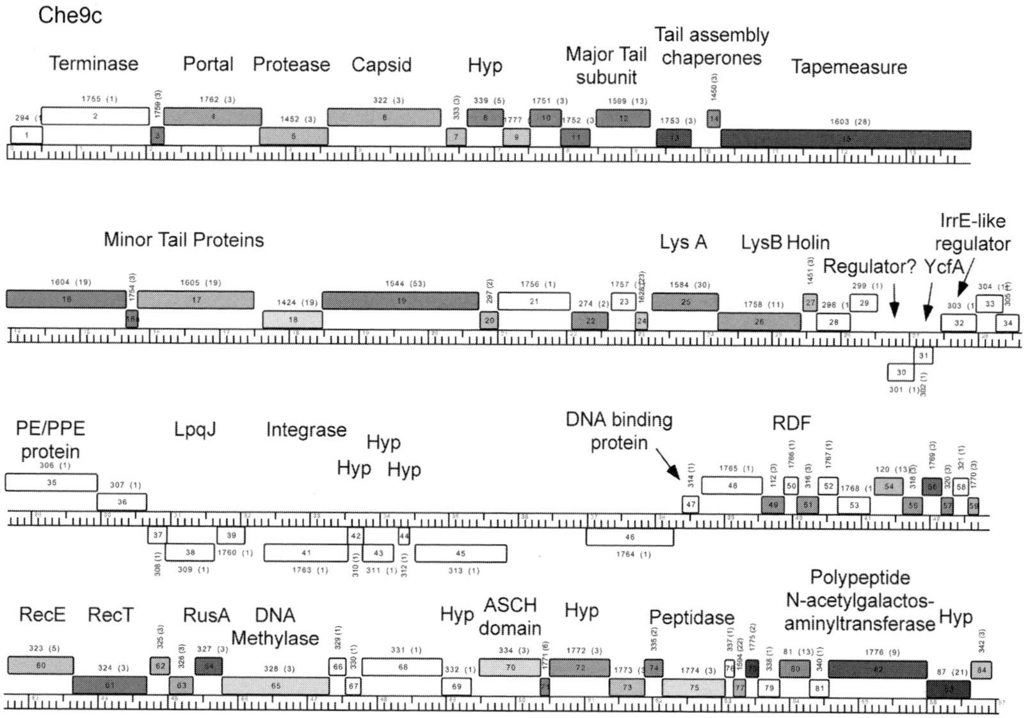

FIGURE 12 Map of the phage Che9c genome, a member of Subcluster I2. See Figure 3A for further details on genome map presentation. (See Page 23 in Color Section at the back of the book.)

Several of the genes in these operons encode putative transcriptional regulators, including genes *30, 32, 46,* and *47*. Che9c gp32 is unusual in that it is related to the IrrE regulator of *Deinococcus radiodurans* and there are no similar genes elsewhere in the mycobacteriophages; it contains both putative DNA recognition and protease motifs. Gene *46* encodes a large protein with a putative helix-turn-helix motif near its N terminus, and gene *47* encodes a smaller helix-turn-helix containing a predicted DNA-binding protein. Any of these could plausibly play the role of the phage repressor, but it is curious that there are such a variety of putative regulatory proteins. Che9c gene *41* encodes a tyrosine-integrase and a putative Xis encoded by gene *50*, displaced almost 7 kbp from the *int* gene. A putative *attP* common core can be identified immediately adjacent to the integrase gene, and Che9c is predicted to integrate at an *attB* site overlapping a host $tRNA^{tyr}$ gene (see Section V.B.1 and Table II). Cluster I1 phages have a different integration specificity and are predicted to integrate into a $tRNA^{thr}$ gene (see Section V.B.1 and Table II). The *attP* site of Che9c is notable because whereas other phages that integrate into tRNA genes carry the 3′ end of the host tRNA, Che9c is predicted to encode a complete $tRNA^{tyr}$ gene at this position. However, the predicted tRNA has a number of nonstandard features that bring into question whether this is either expressed or functional, and it is possible that it is just a bioinformatic quirk.

The segment to the left of the Che9c integrase contains several genes of interest, including one encoding a putative PE/PPE-like protein (*35*), and an LpqJ-like predicted lipoprotein containing a single transmembrane domain; gp38 is also a predicted membrane protein with three membrane-spanning domains. Although nothing is known about the expression patterns of Che9c or any other Cluster I phages, it is tempting to suggest that these genes were acquired relatively recently from a bacterial chromosome through an errant excision process and are expressed from an integrated prophage, perhaps conferring new properties to lysogenic strains. Che9c gp38 is related more closely to the LpqJ protein of *M. smegmatis* (Msmeg_0704 product) than to other bacteria, but these share only 42% amino acid sequence identity, making it unlikely that it was a recent acquisition from this host specifically. The region to the right of integrase contains five leftward-transcribed genes, all of which are orphams, and genes *42, 43,* and *44* are all related to large families of hypothetical bacterial proteins of unknown function; there is an unusually large noncoding region (~1.1 kbp) between genes *45* and *46*. The GC % content of the 4.7-kbp gene *42–46* region is substantially different (59.9%) from the overall GC% of the Che9c genome (65.4%), consistent with the interpretation that it has been acquired relatively recently by horizontal genetic exchange, most likely from a bacterial host. Furthermore, the genome organization of Che9c is similar to Subcluster I1 phages to the left of Che9c gene *19*, and the similarity—although still rather weak—does not pick up again until to the right of Che9c gene *51*. Differences in lengths of the intervening regions account for the 10-kbp differences in overall genome lengths. The rightward-transcribed operon containing genes *47–64* encodes a number of genes of interest. These include RecET-like genes (*60* and *61*), which are of note because they have been exploited to develop a system for recombineering in mycobacteria (van Kessel and Hatfull, 2007, 2008a,b; van Kessel *et al.*, 2008) and of the mycobacteriophages themselves (Marinelli *et al.*, 2008)(see Section VII. A.6). Gene *64* encodes a RusA-like Holliday Junction resolvase, gene *75* encodes a putative peptidase, and gene *82* encodes a protein predicted to encode polypeptide *N*-acetylgalactosaminyltransferase activity; Subcluster F1 and the Subcluster C2 phage Myrna encode similar enzymes, and it is of considerable interest to identify which proteins—either phage or perhaps host encoded—are targets of glycosylation. Interestingly, Subcluster I1 phages Brujita and Island3 lack homologues of Che9c gene *82*, but at the right ends of their genomes encode a protein with a different sequence but which is predicted to have the same activity as Che9c gp82. Cluster I genomes clearly are rich in features of interest and warrant substantial further investigation.

J. Cluster J

Cluster J contains the published genome Omega and unpublished phages LittleE and Baka; the genome organization of Omega is shown in Figure 13. Omega forms slightly turbid plaques from which stable lysogens can be recovered (G. Broussard and Graham F. Hatfull, unpublished results) and does not infect *M. tuberculosis*. The genome is 110 kbp long and contains defined cohesive termini, although with unusually short 4-base single-stranded DNA extensions (Pedulla *et al.*, 2003)(see Section VI.A). The left end is ~1.5 kbp from the putative terminase small subunit gene, and the virion structure and assembly genes extend to approximately gene *44* (Fig. 13). Genes encoding terminase small (*3*) and large subunits (*11*), portal (*13*), protease (*14*), capsid (*15*), major tail subunit (*31*), tail assembly chaperones expressed via a programmed translational frameshift (*32* and *33*), tapemeasure (*34*), and minor tail proteins (*35–40*) can be predicted with reasonable confidence (Fig. 13); the terminase large subunit contains an intein similar to that in some Cluster A and E terminases (see Sections III.A and III.E). However, this operon contains many interruptions with insertions of genes transcribed both in forward and reverse directions (Fig. 13). For example, there are seven small open reading frames (*4–10*) of unknown function between terminase small and large subunit genes, and immediately to the right of capsid genes are open reading frames encoding putative glycosyltransferase and *O*-methyltransferase activities

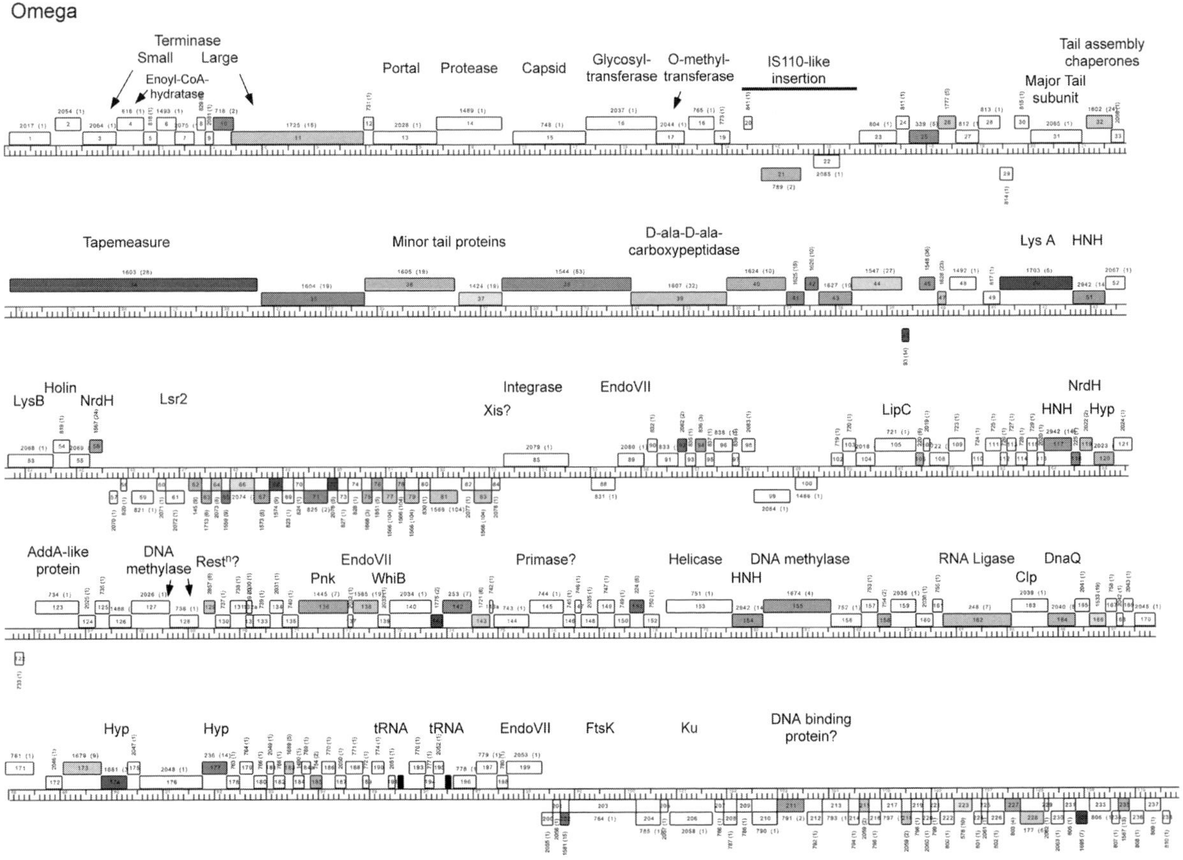

FIGURE 13 Map of the phage Omega genome, a member of Cluster J. See Figure 3A for further details on genome map presentation. (See Page 24 in Color Section at the back of the book.)

(Fig. 13). It is unclear what the specific functions of these genes are, although their location within the virion structure and assembly operon suggests the intriguing possibility that they are modifying virion proteins. Leftward-transcribed genes *21* and *22* are of unknown function, although there are homologues of both in the Subcluster A1 phage, Bethlehem (gp71 and gp72)(see Section III.A). Omega gp21 also has weak sequence similarity to IS110 family transposases, and thus both Omega genes *21* and *22* could conceivably belong to an uncharacterized IS110-like transposon. The absence of this segment in the LittleE genome—as well as the related insertion in Bethlehem—strongly supports this possibility.

The Omega lysis cassette lies to the right of the virion structure and assembly genes, and includes lysin A (*50*) and lysin B (*53*) genes—separated by an HNH domain gene (*51*) and a gene of unknown function (*52*)—as well as a putative holin (*54*). To the right of this is a leftward-transcribed operon containing 28 small open reading frames. Accurate annotation of the many small genes in phage genomes is an ongoing challenge, but this operon presents a good example of the utility of comparative genomic analyses because more than half of these are related to genes in mycobacteriophages in other clusters and subclusters (Fig. 13); Omega genes *77*, *79*, *81*, and *83* all belong to a relatively large phamily with representatives of the total of 104 members in virtually all clusters of phages. Gene *61* is of interest because it encodes a homologue of the host *lsr2* gene, which has been shown to be a global regulator of gene expression in *M. tuberculosis* (Colangeli *et al.*, 2007, 2009)(see Section III.E). It is unclear what functional role there could be for Omega gp16, perhaps acting as a regulator of phage gene expression, but more enticingly as a possible regulator that reprograms host gene expression of lysogenic strains.

Although the Omega genome is replete with orphams and genes of unknown function, several genes with predicted functions make them odd denizens of a phage genome. For example, gene *206* encodes a homologue of bacterial Ku-like proteins involved in mediating nonhomologous end joining (NHEJ)(Pitcher *et al.*, 2007). Ku-like proteins typically act together with a dedicated DNA ligase (Lig IV), which is absent from the Omega genome (Pitcher *et al.*, 2006). Interestingly, the NHEJ system seems to be required for Omega infection and presumably is required for genome recircularization upon infection (Pitcher *et al.*, 2006). Further details are provided in Section VI.B. We note that the only other mycobacteriophage to encode a Ku-like protein is the singleton mycobacteriophage Corndog (gp87)(see Section III.M.1) and it too has 4-base ssDNA extensions (Pitcher *et al.*, 2006).

Just to the left of the Ku-like protein gene, gene *203* encodes an FtsK-like protein. Bacterial FtsK proteins, including that of *M. smegmatis*, contain three domains: an N-terminal domain involved in membrane association; a central domain containing motor functions, including a

AAA-ATPase motif; and a short C-terminal domain (gamma) that confers the specificity of DNA binding (Sivanathan *et al.*, 2006). Its primary function in bacteria is to facilitate proper segregation of daughter chromosomes at cell division. Omega gp203 is a 443 residue protein that lacks the N-terminal domain of bacterial FtsK proteins and contains just the core domain and a putative C-terminal gamma domain. However, although the core domain is quite closely related to that of *M. smegmatis* FtsK (60% amino acid identity), the gamma domain is distinctly different and is not closely related to the gamma domains of any other known FtsK proteins. The presence of this FtsK-like gene in a phage genome is highly unusual, and its role is unknown (Pedulla *et al.*, 2003). It is possible that it acts on a host chromosome that contains different 8-bp asymmetric FtsK Orienting Polar Sequences (KOPS) targeting sequences for gp203 recognition, and although such a host has apparently yet to be described genomically, it is not obviously *M. smegmatis*. Alternatively, it could be acting on the Omega genome itself, and it would be of interest to determine bioinformatically or experimentally if Omega contains KOPS-like gp203-binding sites. We note, however, that gp203 is unlikely to simply facilitate partitioning of extrachromosomally replicating prophage molecules because Omega encodes an integrase (gp85), as well as a putative excise (gp84), and Omega lysogens contain an integrated prophage (G. Broussard and Graham F. Hatfull, unpublished results)(Fig. 13). Omega also encodes a number of proteins predicted to be involved in DNA metabolism, including a DnaQ-like protein (*183*), DNA methylases (*127, 128, 165*), and an AddA-like protein. Curiously, it encodes three proteins with sequence similarity to EndoVII Holliday Junction resolvases (*89, 138*, and *199*). It is unclear why any phage genome would need to encode HJ resolvase activity in three separate genes. Omega—with its curious collection of genes with predicted functions and its vast array of hundreds of genes of unknown function—clearly warrants much more detailed investigations to understand gene expression and gene function and how these contribute to the overall biology of this phage and its Cluster J relatives. Omega also encodes a bifunctional polynucleotide kinase (gp136, Pnk) similar to that of Cjw1 and is proposed to act with the RNA Ligase (gp162) to evade an RNA-damaging host response (Zhu *et al.*, 2004). Omega encodes two putative tRNAs: a $tRNA^{gly}$ (gene *192*) closely related to the one in Cjw1 and a noncanonical putative tRNA with a 4-base anticodon also similar to that encoded by Cjw1 (see Section III.E).

K. Cluster K

Cluster K contains three genomes divided into two subclusters: K1 and K2. Subcluster K1 contains Angelica and CrimD, and Subcluster K2 contains TM4; three additional unpublished phages also belong to this cluster

(Pope *et al.*, 2011). TM4 and its derivatives are perhaps the most widely utilized in mycobacterial genetics, and a map of TM4 genome organization is shown in Figure 14. All of the Cluster K phages infect *M. tuberculosis* as well as *M. smegmatis*, and TM4 was originally isolated by recovery from a putative lysogenic strain of *Mycobacterium avium* (Timme and Brennan, 1984). TM4 forms clear plaques on mycobacterial lawns, whereas the other Cluster K phages form turbid plaques from which stable lysogens can be recovered.

Cluster K phages contain defined cohesive termini (Table I), and the terminase large subunit gene is located near the physical left end (Fig. 14). All of the genes are transcribed in the rightward direction, with the exception of genes *39–41*. Genes *4* through *25* encode the virion structure and assembly functions, and genes encoding terminase large subunit (*4*), portal (*5*), protease (*6*), scaffold (*8*), capsid (*9*), major tail subunit (*14*), tail assembly chaperones expressed via a programmed translational frameshift (*15, 16*), tapemeasure protein (*17*), and minor tail proteins (*18–25*) can be predicted with reasonable confidence (Fig. 14). To the right of the structural genes lies the lysis cassette, and this included both lysin A (*29*) and lysin B (*30*) genes, as well as a gene (*31*) encoding a putative Holin that has four predicted transmembrane domains. The remainder of the genome encodes several genes with predicted functions that are of interest. TM4 gene *49* encodes a putative WhiB-like protein, and it is noteworthy that WhiB family proteins are found in a variety of the mycobacteriophages, including phages of Clusters E, F, and J (see Sections III.E, III.F, and III.J). Ryniker and colleagues (2010) showed that TM4 gp49 is highly expressed soon after infection and functions as a dominant negative regulator of the host WhiB2 protein, and when TM4 gp49 is expressed in *M. smegmatis* it induces septation inhibition. TM4 gene *69* is not an essential gene for viral propagation (Rybniker *et al.*, 2010) but is implicated in mediating superinfection exclusion.

TM4 also encodes a large putative Primase/Helicase enzyme (gp70) and a RusA-like Holliday Junction resolvase (gp71). TM4 gene *79* encodes an SprT-like protease of unknown function, although it is noteworthy that a number of mycobacteriophages encode proteases of a variety of types in the nonstructural parts of their genomes; phages in Clusters A, C, and K also encode SprT-like proteases, Cluster E phages encode a Clp-like protease, and Cluster I phages encode a predicted peptidase. These are distinct from proteases encoded as part of the virion structural gene operon where they play a role in capsid assembly, although a variety of different types of enzymes appear to perform that function. Presumably there are protease-required processing events involved outside of capsid assembly, but these remain poorly understood. At least 3 to 4 kbp of the TM4 genome must be nonessential for growth because a variety of shuttle phasmids have been constructed in which parts of the genome are

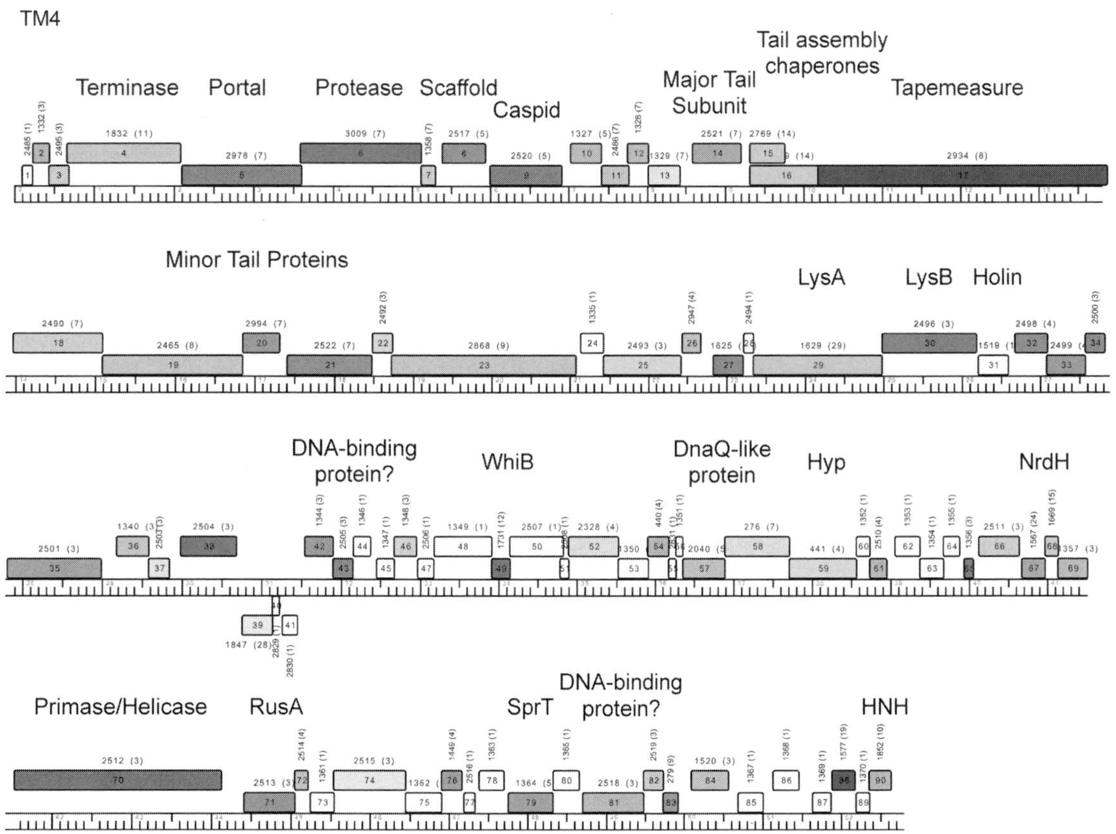

FIGURE 14 Map of the phage TM4 genome, a member of Subcluster K2. See Figure 3A for further details on genome map presentation. (See Page 25 in Color Section at the back of the book.)

replaced by a cosmid vector (Bardarov *et al.*, 1997; Jacobs *et al.*, 1987, 1989). The extent of the deleted regions is not yet clear but is within the right half of the genome.

The two Subcluster K1 phages, Angelica and CrimD, are quite similar to each other and both are rather different from TM4 at the nucleotide sequence level (Pope *et al.*, 2011). Nonetheless, many of the genes are homologues when compared at the amino acid level, especially in the virion structure and assembly operons. A notable difference though is that Cluster K1 genomes are approximately 7 kbp larger than TM4 (Table I); this difference is largely accounted for by a large insertion in the middle of K1 genomes relative to TM4. It thus seems likely that TM4 acquired a large central deletion, possibly during its time of isolation, a scenario reminiscent of the properties of some of the Cluster A phages, such as D29 (see Section III.A). Unfortunately, K1 and K2 genomes are insufficiently similar at the nucleotide level to determine precisely how such a deletion might have occurred. The central segment of K1 phage genomes encodes an integrase and a plausible transcriptional regulator, consistent with the idea that the nontemperate behavior of TM4 arises as a consequence of this deletion. Putative *attP* sites are located adjacent to their integrase genes, and Angelica and CrimD are predicted to integrate at an *attB* site overlapping the host tmRNA gene (see Section V.B.1 and Table II). These are the only mycobacteriophages known to use this integration site (*attB-9*, Table II) and all others that encode a tyrosine-integrase integrate into known host tRNA genes (see Section V.B.1 and Table II). We also note that both Angelica and CrimD encode a $tRNA^{trp}$ gene (gene *5*) located between the terminase large subunit gene (*8*) and the left physical end. This tRNA gene is similar to $tRNA^{trp}$ genes encoded by L5 and D29 (95% identity across 59 of the 75 bp; see Section III.A) as well as the *M. smegmatis* mc^2155 host tRNA gene (Msmeg_1343, 90% over 73 bp) and presumably could have been acquired from either a phage or a host genome. Derivatives of TM4 are perhaps the most widely used mycobacteriophages in mycobacterial genetics because of their ability to infect both fast- and slow-growing mycobacteria and the availability of TM4 shuttle phasmids that can be manipulated readily (Jacobs *et al.*, 1987). Shuttle phasmids are chimeras containing a mycobacteriophage moiety and an *E. coli* cosmid moiety, such that they can be propagated as large plasmids in *E. coli* and as phages in mycobacteria (Jacobs *et al.*, 1987, 1991). Construction of shuttle phasmids involves a step in which these chimeras are packaged into phage λ heads *in vitro* (Jacobs *et al.*, 1991) and thus it is not surprising that TM4 shuttle phasmids contain deletions of phage DNA such as to accommodate the cosmid vector insertion. TM4 shuttle phasmids can be manipulated readily using standard genetic and molecular biology approaches in *E. coli* and have been exploited for the delivery of transposons (Bardarov *et al.*, 1997), reporter genes (Jacobs *et al.*, 1993), and allelic exchange substrates (Bardarov *et al.*, 2002)(see Section VII.A).

L. Cluster L

Cluster L contains just a single published phage genome, LeBron; however, six additional unpublished phages also fall within Cluster L. LeBron is anticipated to be competent to form lysogens and encodes its own tyrosine-integrase (Pope *et al.*, 2011). It is relatively recently isolated and little is known about its general biological properties. A map of the LeBron genome organization is shown in Figure 15.

A LeBron gene encoding a terminase large subunit (*4*) is located approximately 1.0 kbp from the physical left end of the genome, and there are three small open reading frames predicted in the intervening region. One of these (*2*) is a candidate for encoding a terminase small subunit, and has sequence similarity to a gene immediately upstream of the terminase large subunit gene in Omega (Fig. 13). Within the putative virion structure and assembly operon (*2–24*) genes encoding a terminase large subunit (*4*), portal (*5*), protease (*6*), capsid (*7*), major tail subunit (*13*), tail assembly chaperones (*14* and *15*), tapemeasure protein, and minor tail proteins (*17–24*) are predicted. The lysis cassette follows this and includes both lysin A (*25*) and lysin B (*26*) genes and the putative holin gene *27*. To the right there are short leftward-transcribed operons, including a tyrosine-integrase gene, followed by a longer rightward-transcribed operon that includes several genes implicated in regulation and nucleotide or DNA metabolism. These include a putative WhiB regulator (gp78), a kinase (gp74), a ssDNA-binding protein (gp81), an RNA ligase (gp65), a putative DNA Pol II (gp93), EndoVII (gp96) and RusA (gp76) Holliday Junction resolvases, a DnaB-like helicase, a ribosephosphate kinase (gp71), and an Erf-like general recombinase (gp60). The kinase may act with the RNA Ligase to evade an RNA-damaging host response as proposed to Cjw1 and Omega (Zhu *et al.*, 2004). Two additional genes encode a putative esterase (gp46) of unknown specificity and a nicotinate phosphoribosyltransferase (gp72). The rightmost 10 kbp of the LeBron genome contains mostly leftward-transcribed genes of unknown function, with the exception of gene *128* that encodes an AAA-ATPase protein. About 80% of LeBron genes remain of unknown function. LeBron also encodes nine tRNA genes ($tRNA^{leu}$, $tRNA^{thr}$, $tRNA^{lys}$, $tRNA^{tyr}$, $tRNA^{trp}$, $tRNA^{leu}$, $tRNA^{his}$, $tRNA^{cys}$, and $tRNA^{lys}$) of unknown function, but we note that while most of these correspond to codons used highly in the LeBron genome (see Section III.A), one of them ($tRNA^{leu}$) has an anticodon corresponding to the rare codon 5′-CUA. Overall, LeBron is an interesting genome with numerous genes, suggesting an intriguing but poorly understood biology.

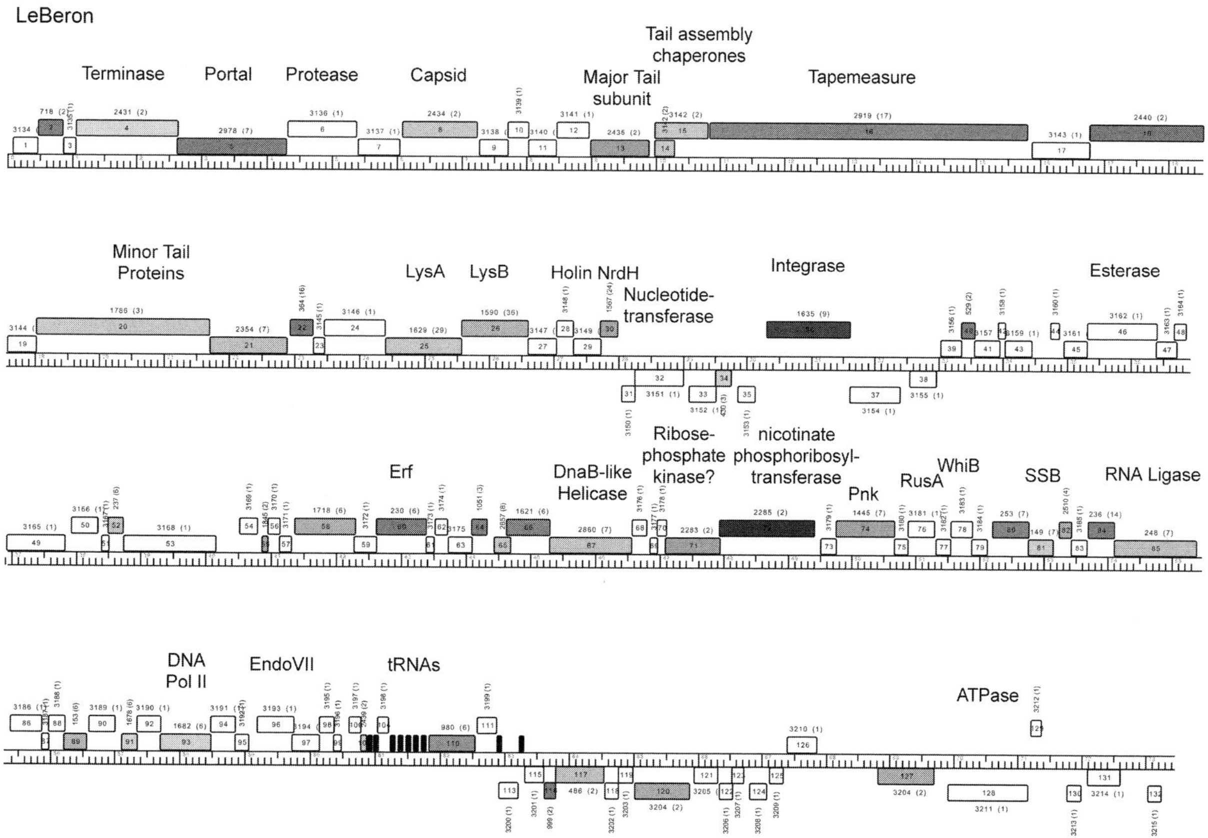

FIGURE 15 Map of the phage LeBron genome, a member of Cluster L. See Figure 3A for further details on genome map presentation. (See Page 26 in Color Section at the back of the book.)

M. Singletons

1. Corndog

The singleton phage Corndog has an unusual morphology with a prolate head having approximately a 4:1 length-to-width ratio. It looks like a corndog! Corndog forms plaques on *M. smegmatis* that are neither completely clear nor turbid, and stable lysogens have not been reported. Corndog does not infect *M. tuberculosis*. A map of the Corndog genome organization is shown in Figure 16.

Corndog contains a 69.8-kbp genome with defined cohesive ends, having 4-base ssDNA extensions as described earlier for Omega (see Section III.J). However, the putative terminase large subunit gene (*32*) is located ~13.5 kbp away and there are 31 predicted genes between it and the left physical end. Most of these 31 genes are of unknown function and the majority are orphams. However, genes *6* and *7* have regions associated with DNA methylases, gene *11* encodes an Endo VII Holiday Junction resolvase, gene *22* encodes a primase/polymerase, and gene *29* has an HNH domain. Gene *25* is part of a truncated copy of an MPME1 mobile element (Sampson *et al.*, 2009)(see Fig. 10B). Genes *1–12* are organized into an apparent leftward-transcribed operon, whereas genes *13–31* are transcribed rightward and could be part of the viral structure and assembly operon that continues to the right.

The Corndog virion structure and assembly genes (*32–67*) containing genes encoding the terminase large subunit (*32*), portal (*34*), protease (*39*), capsid (*41*), major tail subunit (*49*), tail assembly chaperones expressed via a programmed translational frameshift (*54* and *55*), tapemeasure protein (*57*), and minor tail proteins (*58-67*) can be predicted confidently. Although these genes appear in the canonical order, their synteny is disrupted in at least five locations. Two of these involve HNH insertions (genes *33* and *56*) and one (*40*) is a single small gene inserted between the putative protease and capsid genes that does not appear similar to scaffold proteins (Fig. 16). A fourth is between the putative major tail subunit gene and the tail assembly chaperones and contains four small open reading frames, one of which (*51*) has homologues in some Cluster C phages. Another of these (*53*) is a predicted transcriptional regulator. The fifth syntenic interruption is between the putative portal protein and the protease (Fig. 16). Four open reading frames are present (genes *35–38*), and three of them (*35, 37,* and *38*) are predicted to encode an *O*-methytransferase, a polypeptide *N*-acetylgalactose aminyltransferase, and a glycosyltransferase, respectively. These are similar functions to genes located within the virion structure and assembly operon of Omega (Fig. 13), except that in Omega they are inserted between the capsid and major tail subunit genes. It is not known if these enzymes are responsible for modification of virions, although this is an intriguing possibility that warrants investigation.

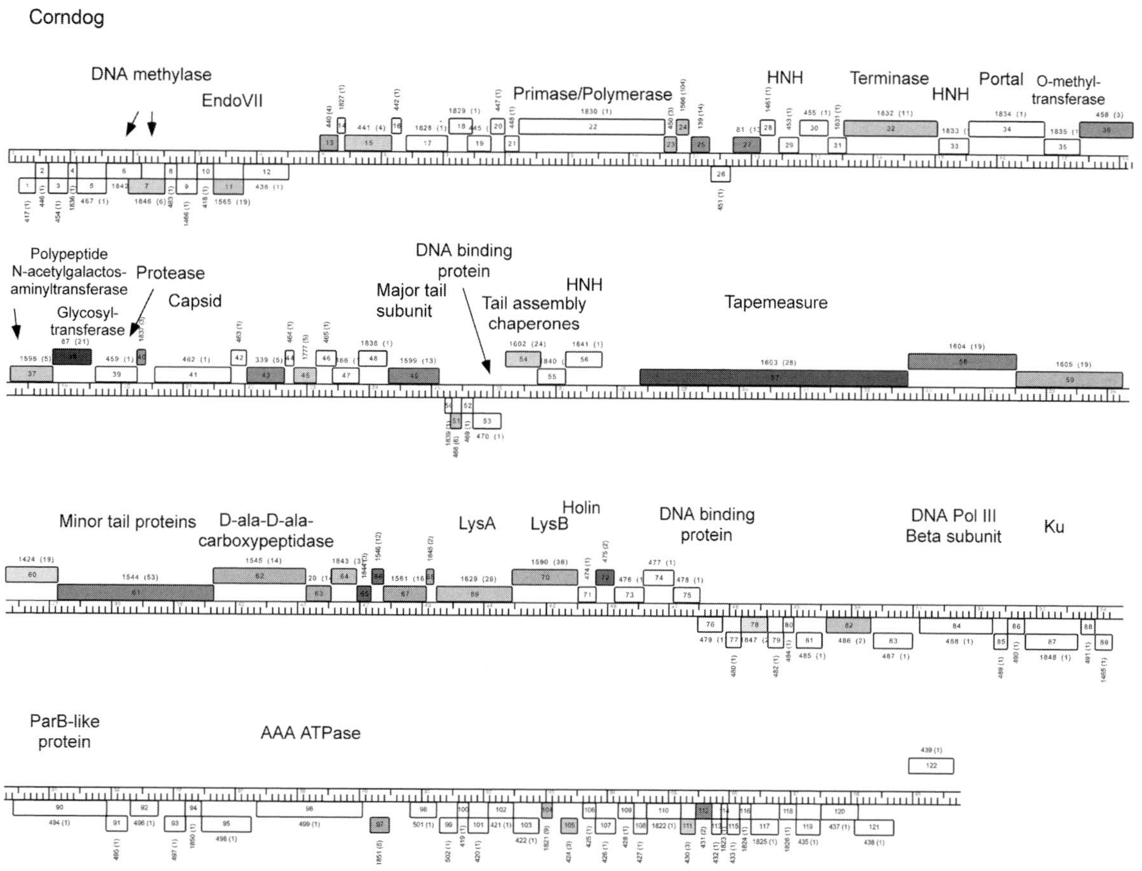

FIGURE 16 Map of the singleton phage Corndog genome. See Figure 3A for further details on genome map presentation. (See Page 27 in Color Section at the back of the book.)

Sequences of the virion structure genes provide few clues to the unusual Corndog prolate head morphology. The putative capsid subunit (gp41) is not closely related to any other mycobacteriophage-encoded proteins, and its closest homologue is a gene within the *Bifidobacterium dentium* genome, although the two proteins are only 26% identical. It does, however, contain a predicted capsid domain and HHPred reports similarity to the HK97 capsid structure. We note that the Corndog capsid subunit sequence has no evident sequence similarity with the capsid subunits of Cluster I phages, which also have prolate heads but with a different length:width ratio (see Section III.I).

The lysis cassette lies to the right of the virion structure and assembly operon and contains lysin A (*69*) and lysin B (*70*) genes, as well as a putative holin (*71*). To the right of that is a long leftward-transcribed operon (*76–121*) amazingly enriched for orphams (only 7 of the 57 genes have evidently related genes in other mycobacteriophages), although several genes in this operon have interesting predicted functions (Fig. 16). First, Corndog gene *87* encodes a Ku-like protein distantly related to both the *M. smegmatis* homologue (Msmeg_5580; 35% amino acid sequence identity) and the Ku-like gp206 protein encoded in phage Omega (32% amino acid sequence identity) (Pitcher *et al.*, 2006). Both Corndog and Omega share the features of having cohesive genome termini with 4-base extensions and encoding Ku-like proteins, supporting the idea that Ku-like proteins play a role in NHEJ-mediated genome circularization upon infection (see Sections III.J and VI.B). Second, Corndog gene *82* encodes a putative DNA polymerase III β subunit implicated in acting as a loading clamp in DNA replication. It is unclear whether this gene is required for Corndog replication and what role it could play. However, this is the only occurrence of this particular function in any of the mycobacteriophage genomes. It is not obvious that it was acquired recently from a bacterial host, in that its closest relative is the clamp loader of *Saccharopolyspora erythraea* and the proteins share only 23% amino acid sequence identity. There are no other closely related phage-encoded homologues. A third gene of interest is Corndog *96* encoding an AAA-ATPase, although it is not closely related to other AAA-ATPases encoded by other mycobacteriophages (such as LeBron gp128, Myrna gp262, or Che8 gp69). Its origins are also unclear, and the closest homologue is a protein encoded by *Haliangium ochraceum*, which shares 33% amino acid sequence identity. There are no closely related homologues in other phage genomes.

Finally, Corndog gp90 is a ParB-like protein implicated in chromosome partitioning. A plausible role for such a function could be to provide stability to an extrachromosomally replicating Corndog prophage, although because lysogens have not yet been recovered, this is unclear. An alternative possibility is that this protein plays a different role such as a regulatory function rather than partitioning per se. We note in this regard that there is no putative ParA protein, a striking difference from the putative partitioning functions in RedRock (see Fig. 3B and Section III.A). There is no obvious homologue of Corndog gp90 in *M. smegmatis*, but there are related genes in other mycobacterial strains and other Actinomycetales genomes, with the closest homologue being *Mycobacterium kansasii* Spo0J (46% amino acid sequence identity).

2. Giles

The singleton Giles forms lightly turbid plaques on *M. smegmatis* from which stable lysogens can be recovered (Morris *et al.*, 2008). Giles does not infect *M. tuberculosis*. A map showing Giles genome organization is shown in Figure 17. The genome has defined cohesive termini with long (14-base) single-stranded DNA extensions. Like other singleton genomes, Giles contains a high proportion of orphams and only 14 of the predicted 78 genes have readily identifiable homologues in other mycobacteriophages (Fig. 17). Giles gene *1* corresponds to a possible terminase small subunit gene, but is separated by three short open reading frames on the opposite strand from the terminase large subunit gene (Fig. 17). The putative virion structure and assembly operon extends from gene *1* to gene *36* and has several striking features. First, most of these are orphams, reflecting the considerable sequence divergence from other mycobacteriophages. Genes encoding the terminase large subunit (*5*), portal (*6*), protease (*7*), capsid (*9*), tail assembly chaperones expressed via a programmed frameshift (*17* and *18*), tapemeasure (*19*), and minor tail proteins (*21–28*, *36*) can be identified readily (Fig. 17). Gene *8* may encode a scaffold-like protein, based on its position in the operon, but no major tail subunit gene can be predicted confidently. Second, the lysis cassette lies upstream of gene *36*, which has been shown experimentally to be a virion-associated protein, and thus the lysis cassette—including lysin A (*31*), lysin B (*32*), and putative holin genes (*33*)—appears to lie within this operon (Morris *et al.*, 2008). In most other mycobacteriophages genomes (with the notable exception of Cluster A phages), it is noteworthy that the location of the lysis cassette is immediately downstream of the virion structure and assembly operon, and because the virion proteins have been characterized experimentally for few of these phages, it is plausible that they may also have genes downstream of the lysis cassette that encode virion proteins; it is also plausible that Giles gene *36* was acquired relatively recently, and we note that other mycobacteriophage genomes contain tail genes in noncanonical positions [e.g., L5 gene *6*; (Hatfull and Sarkis, 1993)]. A more curious feature of this part of the Giles genome is the presence of the integration cassette—including *integrase* and *xis* genes, as well as *attP*—between the minor tail subunit genes and the lysis cassette (Morris *et al.*, 2008). It is plausible that this cassette relocated to

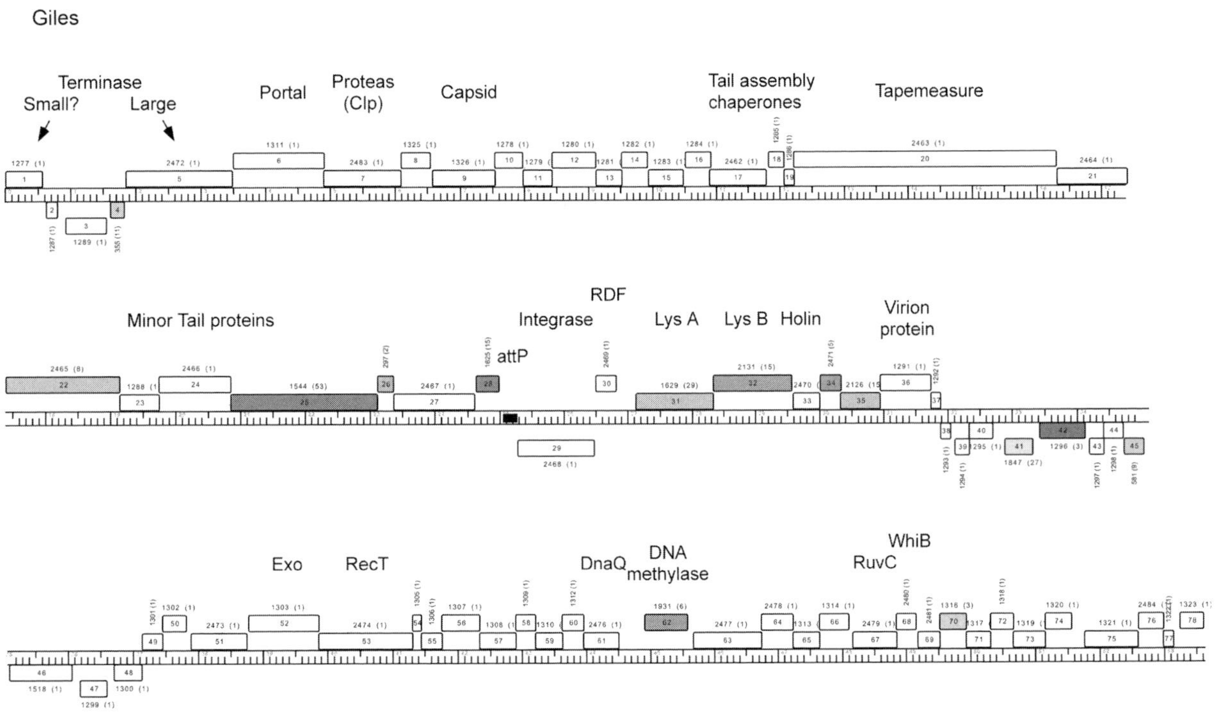

FIGURE 17 Map of the singleton phage Giles genome. See Figure 3A for further details on genome map presentation. (See Page 28 in Color Section at the back of the book.)

this position by an errant recombination event encoded by the integrase protein (Morris *et al.*, 2008). To the right of the virion structure and assembly operon there is one leftward-transcribed operon and one to the right. The leftward operon contains genes *38–48*, all of which are of unknown function. The rightward operon contains genes *49–78* and although the vast majority of these are orphams and have no known function, several do have potential functions of interest. For example, genes *52* and *53* encode putative exonuclease and RecT proteins, respectively, and presumably mediate homologous recombination events; this is supported experimentally (van Kessel and Hatfull, 2008a). Gene *61* encodes a DnaQ-like enzyme, a common function among many mycobacteriophage genomes, although the diversity of the encoded proteins is very high, and the closest homologue of Giles gp61 is a related gene encoded by *Kineococcus radiotolerans* (30% amino acid sequence identity). Gene *62* encodes a putative DNA methylase, although its function is unclear; DNA methylases can be components of restriction–modification systems, although if this were the case in Giles, it is unclear which gene might encode the putative restriction function. Gene *67* encodes a RuvC-like Holliday Junction resolvase, extending this as one of the most common functions encoded by mycobacteriophage genomes, albeit through the use of different classes of genes (i.e., RuvA, RusA, EndoVII); gene *68* encodes a WhiB-like gene regulator.

Although it was reported initially that the Giles genome was 54,512 bp in length (Morris *et al.*, 2008), this includes a segment at the extreme right end that was included due to an assembly error. Correction of the sequence generates a 53,746 bp genome, and the putative *metE*-like gene reported as Giles gene *79* is not actually part of the Giles genome (Hatfull *et al.*, 2010).

3. Wildcat

The singleton Wildcat forms plaques on *M. smegmatis* that are not evidently turbid, but not completely clear, and stable lysogens have not been reported. There is also no evidence for prophage stabilization functions in the genome, such as integrase or partitioning functions, nor are there any obvious candidates for a phage repressor. It does not infect *M. tuberculosis*. The genome is 78.3 kbp in length and contains defined cohesive termini with 11-base 3′ ssDNA extensions (Table I). A map of the Wildcat genome organization is shown in Figure 18. As with other singleton phages (and Clusters J and L for which only a single published genome is discussed here), there is a very high proportion (84%) of orphams.

Wildcat gene *26* encodes the putative terminase large subunit and is located over 8 kbp away from the physical left end of the genome. Immediately to its left are two other rightward-transcribed genes (*24* and *25*) of unknown function, although one of these could plausibly

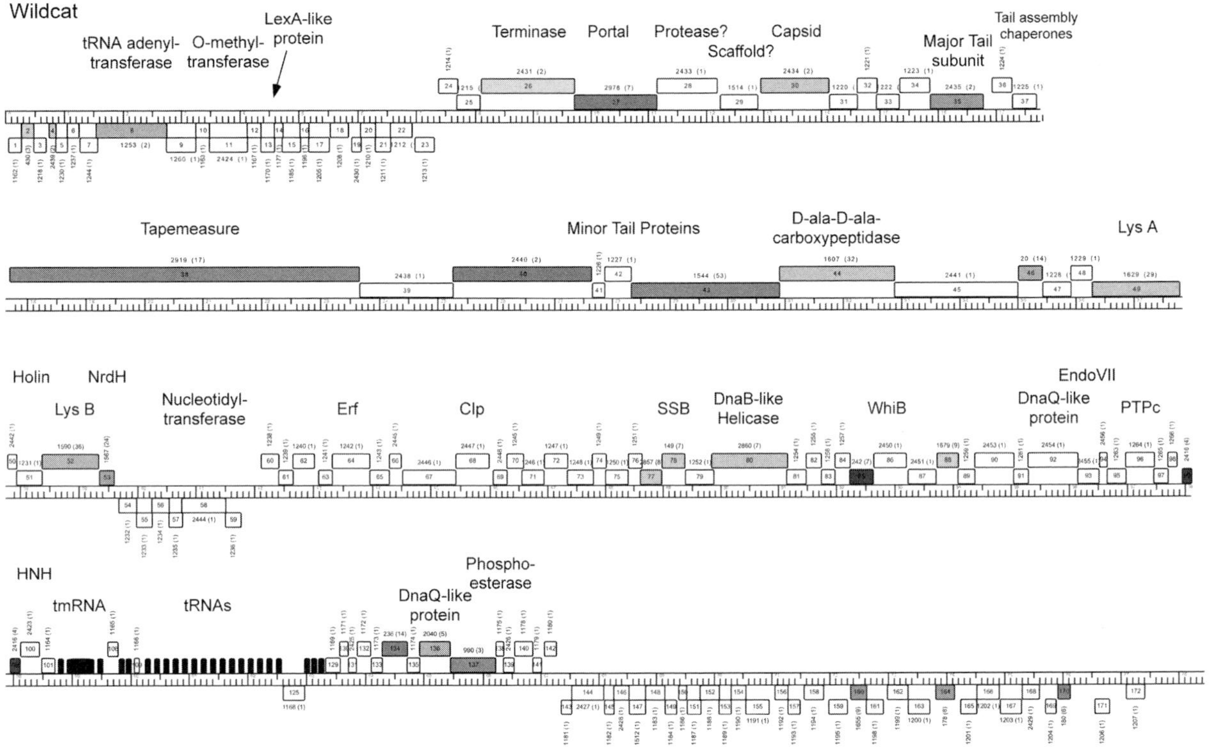

FIGURE 18 Map of the singleton phage Wildcat genome. See Figure 3A for further details on genome map presentation. (See Page 29 in Color Section at the back of the book.)

encode a terminase small subunit. Between the left end and gene *24* is a leftward-transcribed operon (genes *1–23*) containing mostly genes of unknown function. The presence of "additional" genes in this part of the genome (it is more typical for terminase genes to be close to their sites of action, i.e., near the genome end) extends a theme observed in the genomes of Clusters A, B, D, and Corndog. Three of these genes can be assigned putative functions. Gene *13* encodes a putative LexA-like transcriptional regulator, and gene *11* encodes a putative *O*-methyltransferase, a function seen in other genomes, including Omega and Corndog, although Wildcat gene *11* is quite different in sequence to these. Wildcat gene *8* encodes putative tRNA adenyltransferase activity, presumably involved in CCA addition to tRNAs (Fig. 18). The only other mycobacteriophage with a similar function is Myrna (gp28), although it shares no more similarity to Myrna gp28 (~35% amino acid identity) than it does to host PncA proteins. Both Wildcat and Myrna encode a large number of tRNAs, which may belie the requirement to encode this function, although we note that Cluster C1 phages also encode a large number of tRNAs but appear to lack such an activity.

The virion structure and assembly operon (genes *26–45*) is fairly canonical with uninterrupted synteny, and genes encoding putative terminase large subunit (*26*), portal (*27*), capsid (*30*), major tail subunit (*35*), tail assembly chaperone expressed via a programmed translational frameshift (*36* and *37*), tapemeasure (*38*), and minor tail proteins (*39–45*) can be predicted with confidence; gene *28* is a distant relative of LeBron gene *6* and likely encodes a protease, and gene *29*—a distant relative of LeBron gene *7*—is a strong candidate for encoding a scaffold protein in light of its location within the operon (Fig. 18). Wildcat gp44 has putative D-alanyl-D-alanine carboxypeptidase activity common to that of β-lactamase enzymes, which is commonly encoded by genes located among other tail genes, as in Subcluster A1, Clusters C, D, E, J, and singleton Corndog. The role of these putative proteins is unknown, but they presumably are involved in either cell wall binding or facilitating receptor recognition (see Section VI.A). The lysis cassette lies to the right of the virion structure and assembly operon and includes lysin A (*49*) and lysin B (*52*) genes and a putative holin gene (*51*) located between them. Immediately to the right of the lysis cassette is a small gene encoding a putative NrdH-like glutaredoxin. The role for such a redoxin is unknown, but genes with related functions are found in a number of other mycobacteriophages and there are many examples of them located in a similar position, just downstream of the lysis cassette. Examples are found in Cjw1, Omega, and LeBron, but in Cluster A and K phages it is encoded elsewhere in the genome. To the right of the lysis cassette are three apparent operons: two transcribed leftward (genes *54–59*; genes *143–172*) flanking a rightward operon (genes *60–142*). The two leftward-transcribed operons are virtually devoid of any genes with predicted functions, the exception being gene *58* encoding a putative nucleotyltransferase. Only three of this entire repertoire of genes (*160*, *164*, and *170*) have homologues in other mycobacteriophages. The rightward operon contains several genes of interest, including the putative recombinase Erf (gene *64*), a Clp protease (gene *68*), SSB (gene *78*), a DnaB-like helicase (gene *80*), a WhiB-like regulator (gene *86*), two DnaQ-like but distantly related proteins (encoded by genes *92* and *136*), a putative PTPc-like phosphatase (gene *96*), and a putative phosphoesterase of unknown specificity (gene 137).

This operon also contains an impressive array of tRNA genes (23 in total), as well as a tmRNA gene. Although several proposals have been presented to explain the potential roles for mycobacteriophage-encoded tRNAs (Hassan *et al.*, 2009; Kunisawa, 2000; Sahu *et al.*, 2004), the variety of their numbers and types in mycobacteriophages is amazing (Table I). In addition to Wildcat, all Cluster C phages also have a large number of tRNA genes, Cluster E phages have two, Subcluster K1 phages have one, just one of the Cluster B phages has one (Nigel), and Subclusters A2 and A3 have between one and five. Thus for some phages it appears advantageous to have virtually a complete coding set of tRNA genes, whereas others appear to require no tRNA genes at all. Wildcat is the only phage outside of Cluster C that also encodes a tmRNA. It seems plausible that the phage-encoded tmRNA may serve to increase the efficiency of release of ribosomes from broken or otherwise damaged mRNAs, optimize translation efficiencies by maximizing the size of the pool of available ribosomes, or monitor protein folding (Hayes and Keiler, 2010). tRNA genes may also play a general role in enhancing the frequency of translation, although we cannot rule out that at least some of the tRNAs may be involved in the introduction of noncanonical amino acids into proteins.

In summary, Wildcat is certainly quite a wild phage with numerous features of interest that deserve a more detailed investigation. As with other singletons, annotation and interpretation of the genome lacks from the insights provided by the availability of more closely related phages.

IV. MYCOBACTERIOPHAGE EVOLUTION: HOW DID THEY GET TO BE THE WAY THEY ARE?

The collection of sequenced mycobacteriophage genomes —with its clusters of closely related phages, as well as those that are distantly related—provides abundant insights into their evolution, and viral evolution in general. The most prominent feature of their genomic architectures is that they are mosaic, that is, that the structure of each genome can be explained as being constructed from a set of modules that are being exchanged among the phage population (Hatfull, 2010; Hatfull *et al.*, 2006, 2008;

Pedulla *et al.*, 2003). These modules—composed of either single genes or larger groups of genes—are thus located in different phage genomes that are otherwise not closely related. This mosaicism can be seen at two different levels of comparative genomic analyses. When genomes are compared at the nucleotide sequence level, relatively recent exchange events can be seen, and several examples have been described previously (Hatfull *et al.*, 2010; Pope *et al.*, 2011). However, the pervasive nature of the genomic mosaicism is manifested by looking at protein sequence comparisons because shared gene ancestries can be detected even when they have diverged sufficiently long ago in evolutionary time that nucleotide sequence commonality is no longer evident (Pedulla *et al.*, 2003). Representing genome mosaicism through a comparison of standard approaches such as phylogenetic trees is made complicated by the fact that the component genes may be resident in entirely different genomes (Hatfull *et al.*, 2006). An alternative method of representation uses phamily circles in which each of the genomes in the analysis is placed around the circumference of a circle and an arc is drawn between those genomes that share a gene member of a particular phamily of related genes (Hatfull *et al.*, 2006). Examples in Figure 19 show phamily circles for five of their eight consecutive genes (*21–28*) in the Che9c genome; the remaining three genes are orphams and thus no related mycobacteriophage genes have yet been identified. One of these genes (Che9c *22*) has only a single related gene (forming Pham274), which is located in the Subcluster A3 genome, Bxz2 (Fig. 19). Che9c gene *24* is related to 22 other genes in Pham1628 and is found in a variety of genomes in Clusters A, F, I, J, and K, but not in Bxz2 (Fig. 19). Che9c genes *25, 26*, and *27* (members of Phams 1584, Pham1758, and Pham 1451, respectively) have relationships that are different yet again. Thus all the individual genes in this region appear to have distinct evolutionary histories and arrived at their genomic locations through different evolutionary journeys (Fig. 19). This mosaicism complicates greatly the task of constructing whole genome phylogenies, as the genome relationships are fundamentally reticulate in nature (Lawrence *et al.*, 2002; Lima-Mendez *et al.*, 2007).

A hallmark feature of phage genome mosaicism is that in cases where recombination events can be inferred, they occurred at gene boundaries, or occasionally at domain boundaries. Usually this is observed where closely related phage genomes have undergone relatively recent recombination events and genome discontinuities can be seen at the nucleotide level (Hatfull, 2008; Pope *et al.*, 2011). In such cases, it is not uncommon for sequence discontinuity to appear precisely at or very close to the start and/or stop codons of genes (Hendrix, 2002). This could occur either by a process of targeted recombination events or as a consequence of functional selection from a large number of possible exchange events, most of which generate nonviable progeny and were subsequently lost from the

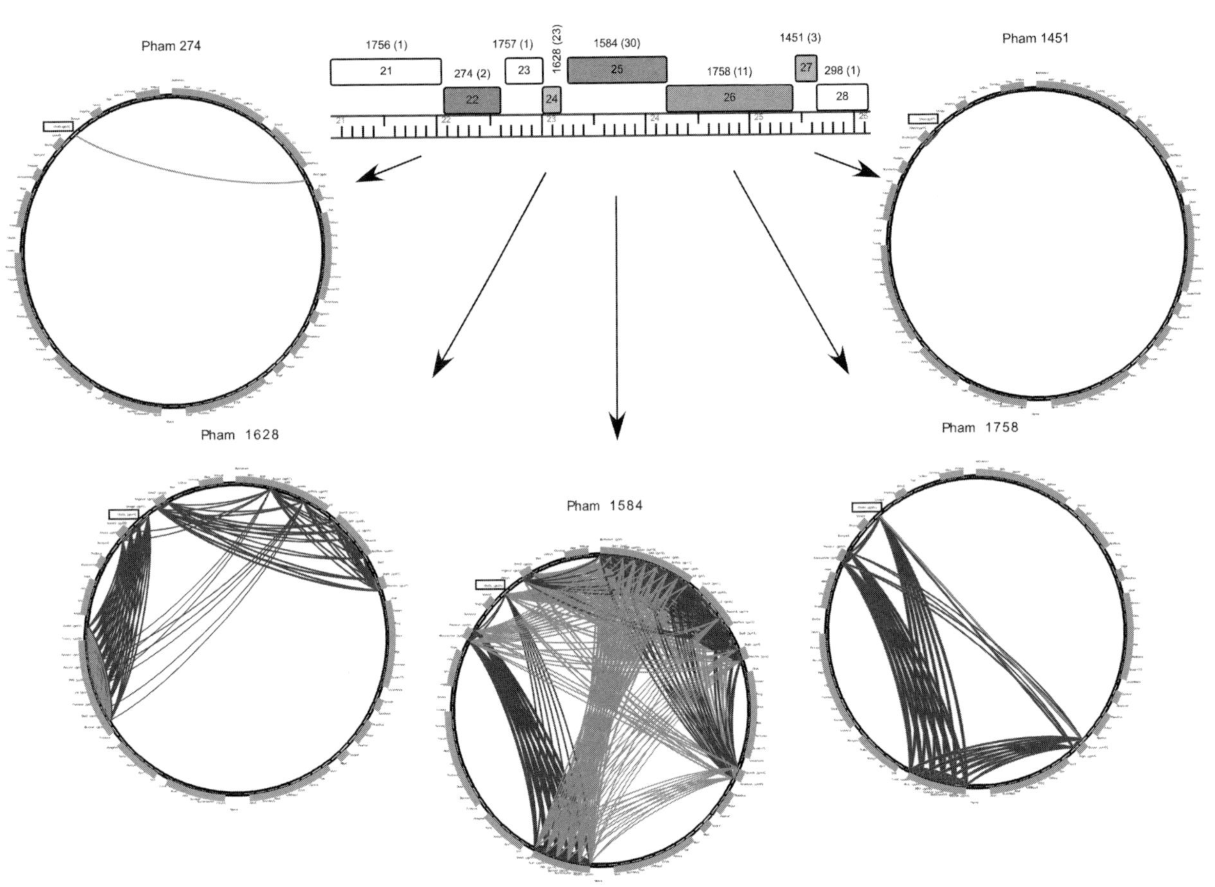

FIGURE 19 Genome mosaicism in the Che9c genome. A segment of the Subcluster I2 Che9c genome encoding genes *21–28* is shown, as described for Fig. 12. Genes *21*, *23*, and *28* are single members of orphams and thus are shown as white boxes. Genes *22* and *24–27* each have relatives in other mycobacteriophage genomes; these are represented as phamily circles for the five respective phams. In each phamily circle, all

population (Hendrix *et al.*, 1999). It should be noted that interpreting where recombination events occur is complicated by the fact that subsequent rearrangement events could have occurred between the genomes being compared (Pedulla *et al.*, 2003). In addition, it is impossible to know what specific recombination events gave rise to the numerous mosaic relationships revealed only through amino acid sequence comparisons.

How does genome mosaicism arise? First it is helpful to note that while mutational changes involving nucleotide substitutions clearly occur and are an important component of phage evolution, this does not contribute directly to genome mosaicism, and acquisition of genome segments from other contexts—either phage or host—by horizontal genetic exchange offers a more general explanation (Hendrix *et al.*, 1999, 2000). Homologous recombination between genome segments with extensive sequence similarity also plays an important role in genome evolution in that it can generate new combinations of gene content, but does not—with the exception of the process described later—create new gene boundaries that are the key to juxtaposing one module next to another.

Four known mechanisms are likely to make substantial contributions to the creation of new module boundaries, although their relative importance is ill-defined. The first is the process of homologous recombination events occurring at short conserved sequences at gene boundaries (Susskind and Botstein, 1978). This process has been proposed in other phages (Clark *et al.*, 2001), and there are a few examples in which this could have played a role in mycobacteriophage mosaicism (Pope *et al.*, 2011). We also note that the 13-bp stoperator sites present in Cluster A genomes, which are predominantly located near gene boundaries, could be ideal targets for such targeted recombination events (see Sections III.A and V.A.1). For the most part, however, short conserved boundary sequences are not obvious at most of the mosaic boundaries that can be identified (Pedulla *et al.*, 2003). However, most of these are revealed through amino acid sequence comparisons and occurred long ago in evolutionary time such that any conservation at the boundaries would

80 genomes are represented (in the same order and grouped according to cluster/subcluster) around the circumference of the circle, and an arc is drawn between those members of the genomes containing a gene that is a member of that pham. Red and blue arcs show BlastP and ClustalW comparisons, respectively, and the thickness of the arc reflects strengths of the relationships. The position of Che9c is boxed in each circle. In Pham 1451 there are only two relatives present in the Subcluster I1 genomes. In Pham 274, there is only a single relative that is in the unrelated Subcluster A3 genome, Bxz2. Phams 1628, 1584, and 1758 each have multiple members but distributed among different clusters and subclusters. This suggests that each of the eight Che9c genes, *21–28*, have arrived in Che9c through distinct evolutionary journeys. This mosaicism is a hallmark of bacteriophage genomic architectures. (See Page 30 in Color Section at the back of the book.)

have been long lost. The second process is site-specific recombination events in which secondary sites have been used by a site-specific recombinase to give rise to insertions in atypical locations. Although this is unlikely to be a predominant process, phages often encode site-specific recombinases, including both tyrosine- and serine-family integrases. One notable example is observed in phage Giles (see Section III.M.2), in which the integration cassette is located among the tail genes, and could have moved there through integrase acting at a secondary site within the virion structure and assembly operon (Morris *et al.*, 2008). The third process is by movement of mobile elements such as transposons and other mobile elements such as inteins, homing endonucleases, and introns, generating both insertions into new genomic locations, and by transposase-mediated rearrangements such as adjacent deletions and inversions. Transposons are not common in mycobacteriophages but several have been recognized. The strongest evidence is for MPME elements found in Cluster G and many Cluster F genomes (see Sections III.G and III.F); these are clearly involved in interrupting what are otherwise conserved gene syntenies (Sampson *et al.*, 2009). Another example is the putative IS110 family insertion sequence present in Omega and its distant relative in Bethlehem (see Sections III.A and III.J). Numerous examples of inteins and HNH-like homing endonucleases exist throughout genomes, but no mycobacteriophage introns have been described. The fourth—and probably the most important contributor—is illegitimate or nontargeted recombination processes that occur without requirement for extensive sequence identity (Hendrix, 2003; Hendrix *et al.*, 1999; Pedulla *et al.*, 2003). It is unclear what mediates such events, although it is noteworthy that bacteriophages commonly encode their own general recombinases, such as phage λ Red systems, RecET-like systems, and P22 Erf-like systems. Mycobacteriophages are no exception, and there are now many examples of RecT-like recombinases (associated with several different types of exonuclease, some of which are related to RecE and some which are not), Erf-like functions, and RecA-like proteins. Examples of some of these phage-encoded recombinases are known to mediate recombination over shorter segments of sequence identity than is typically favored by host recombination systems; they can also tolerate substantial differences between recombining partners (Martinsohn *et al.*, 2008). Although the efficiency of recombination at ultrashort sequence commonalities (such as codons or ribosome-binding sites) is expected to occur at very low frequencies, and multiple events may be required to generate viable progeny, with a potentially long evolutionary history (2 to 3 billion years?) and a high incidence of infection (estimated to be about 10^{23} infections per second globally), inefficiency is unlikely to be an impediment to generating the extent of mosaicism seen in the phage population today. It should also be noted that such illegitimate recombination events are likely to occur more frequently between phages and their host genomes that are often 100 times larger, consistent with the common finding of host genes in phage genomes (Pedulla *et al.*, 2003). There are numerous examples of genes present within mycobacteriophages that are not typically present in phage genomes, with the queuosine biosynthesis genes in Rosebush (see Figs. 4 and 5) being a good example (Pedulla *et al.*, 2003). Finally, we note that generating new gene boundaries either by transposition or by illegitimate recombination is a highly creative process in that DNA sequence elements can be placed together in combinations that did not exist previously in nature. Although most illegitimate recombination events are expected to make genomic trash, the process is one of very few that can create entirely new types of genes.

Comparative genomic analysis of SPO1-like phages led to the suggestions that newly acquired genes are, on average, relatively small (Stewart *et al.*, 2009), and a similar conclusion arises from a comparison of mycobacteriophage genomes (Hatfull *et al.*, 2010). This is consistent with the predominant role of illegitimate recombination because most events are likely to occur within reading frames, and thus selection for function is expected to drive toward functional domains rather than multidomain proteins. This could also account for the reason that phage genes are, on average, only about two-thirds the average size of host genes (Hatfull *et al.*, 2010).

V. ESTABLISHMENT AND MAINTENANCE OF LYSOGENY

Temperate phages are of particular interest for a variety of reasons. For example, they typically employ gene regulatory circuits that can provide insights into novel systems for gene expression and control, as well as being potentially useful for genetic manipulation of the host. Similarly, phage integration provide insights into mechanisms of site-specific recombination and how directionality is controlled, as well as providing the basis for novel plasmid vectors for host genetics (Hatfull, 2010). Temperate phages also often carry genes expressed from the prophage state and contribute to lysogenic conversion of the physiological state of the host. All of these aspects are applicable to mycobacteriophages, and the intimacy of phage–host relationships inherent in temperate phages is particularly intriguing.

A. Repressors and immunity functions

1. Cluster A immunity systems

Genes encoding phage repressors have been identified in remarkably few mycobacteriophages, and there is no complete understanding of life cycle regulation in any of them. Perhaps the best studied are the immunity systems of mycobacteriophage L5—and its unsequenced but closely related phage L1 (Subcluster A2)(Lee *et al.*, 1991)—where the repressor has been identified and characterized (Bandhu *et al.*, 2009, 2010; Brown *et al.*, 1997; Donnelly-Wu *et al.*, 1993; Ganguly *et al.*, 2004, 2006, 2007; Nesbit *et al.*, 1995; Sau *et al.*, 2004)(see Section III.A). A number of other phages encode related repressors, including other Cluster A members, and the Cluster C phage, LRRHood, and the Cluster F phage, Fruitloop, although they are diverse at the sequence level, and pairwise relationships between repressors from different Subclusters in Custer A can be below 30% amino acid sequence identity. The Subcluster A1 phage Bxb1 repressor is the only other one that has been analyzed in any detail (Jain and Hatfull, 2000).

The L5 repressor (gp71) is a 183 residue protein containing a strongly predicted helix-turn-helix DNA-binding motif and was identified through two key observations. First, when the repressor gene is expressed in the absence of any other phage-encoded functions it confers immunity to superinfection by L5. Second, mutations in the repressor confer a clear plaque phenotype; point mutations in the repressor gene can lead to a temperature-sensitive clear-plaque phenotype and lysogens that are thermoinducible (Donnelly-Wu *et al.*, 1993). Unlike most other well-studied repressors, it is predominantly a monomer in solution and recognizes an asymmetric sequence in DNA (Bandhu *et al.*, 2010; Brown *et al.*, 1997). A primary target of regulation is the early lytic promoter P_{left}, which is situated at the right end of the genome and transcribed leftward (Fig. 3A). The L5 P_{left} promoter is highly active and contains −10 and −35 sequences corresponding closely to the consensus sequences for *E. coli* sigma-70 promoters (Nesbit *et al.*, 1995). There are two 13-bp repressor-binding sites at P_{left}, one of which (site 1) overlaps the −35 sequence and the other (site 2) is located ~100 bp downstream within the transcribed region. L5 gp71 binds to these two sites independently, and binding to site 2 does not substantially influence repression of P_{left}; when gene *71* is provided on an extrachromosomal plasmid, P_{left} is downregulated about 50-fold (Brown *et al.*, 1997) through repressor binding to site 1. The binding affinity of gp71 for site 1 is modest, with a K_d of about 5×10^{-8} *M*, and binding to site 2 is about 5- to 10-fold weaker (Brown *et al.*, 1997). Interestingly, repression by binding at site 1 may not be mediated by promoter occlusion, but rather by RNA polymerase retention at the promoter. This is indicated by the observation that phage mutants can be isolated [designated as class III (Donnelly-Wu *et al.*, 1993)] that have mutations within the repressor gene but have a dominant-negative phenotype, being competent to infect a repressor-expressing strain. Such gp71 variants could thus bind to site I without retaining RNA polymerase and prevent that action of wild-type gp71. A surprising observation was that the L5 genome contains a large number of potential repressor-binding sites located throughout the genome (see Section III.A). Initially, a total of 30 putative sites were identified, 24 of which (including sites 1 and 2) were shown biochemically to be bound by gp71 (Brown *et al.*, 1997). These sites conform to the asymmetric consensus sequence 5′-GGTGGMTGTCAAG (M is either A or C), where eight of the positions are absolutely conserved and three others contain only a single departure from the consensus (Brown *et al.*, 1997); roles for the six nonbinding sites cannot be ruled out, as weaker gp71 association may be biologically relevant. Sites are not positioned randomly in the genome but have two important features in common. First, they are oriented in predominantly just one direction relative to the direction of transcription. Thus of the five sites located within the left arm (between the physical left end and the integrase gene; Fig. 3), four (sites 20–23) within the predicted rightward-transcribed region (genes *1–32*) are oriented in the "−" direction; the other (site 24, located between the physical left ends and gene 1) is in the "+" orientation (Brown *et al.*, 1997). It is not known if this segment is transcribed or not. In the right arm (between the integrase gene and the physical right end, Fig. 3A), all of the sites in the leftward-transcribed region (genes *23–88*) are oriented in the "+" orientation. Second, sites are typically located within short intergenic intervals, often overlapping the putative start and stop codons of adjacent genes. When one or more of these binding sites is inserted between a heterologous promoter (hsp60) and a reporter gene (*FFLux*), binding of gp71 has a polar effect on gene expression. This effect is repressor dependent, is strongly influenced by orientation of the site relative to transcription, and is amplified by the presence of multiple sites (Brown *et al.*, 1997). Because repressor binding appears to prevent transcription elongation rather than initiation, these sites (other than site 1) are referred to as "stoperators" (Brown *et al.*, 1997). The mechanism by which this occurs is unknown, but an attractive model is that the repressor interacts directly with RNA polymerase and perhaps retains it at the stoperator site, consistent with the model for action as a repressor by RNA polymerase retention at site 1. It is postulated that these sites play a role in silencing the L5 prophage, ensuring that phage genes potentially deleterious to growth of a lysogen are not expressed from errant transcription events during lysogeny (Brown *et al.*, 1997). However, there is not yet any formal demonstration that additional promoters are not overlapping all or some of these sites or that removal of any of these sites influences either prophage stability or fitness of L5 lysogens.

The regulation of L5 gene *71* is poorly understood, although there is a set of three putative promoters located upstream in the gene *71–72* intergenic region that are presumably responsible for gp71 synthesis from a prophage. The reason for three promoters is unclear. Curiously, even though these three promoters are downregulated during lytic growth, evidence shows that the repressor gene is transcribed during early lytic growth from transcripts arising from P_{left} (Fig. 3)(Nesbit *et al.*, 1995). Presumably, other phage-encoded functions prevent gp71 from acting during lytic growth, although none have been identified. Another conundrum arises because the three promoters upstream of gene *71* are active in a nonlysogen, such that although it is simple to model how lysogeny is established, it is less easy to imagine how lytic growth ensues after infection. Presumably, either the action of gp71 itself is modulated — perhaps either by post-translational modification or by degradation—or a second regulator prevents expression during the establishment of lytic growth. Although no additional L5 genes have been specifically identified as playing a role in the L5 lytic–lysogenic decision, clear plaque mutants have been identified with reduced frequencies of lysogeny (similar to cIII mutants of phage λ), and genes located within the region to the right of gene *71* are implicated (Donnelly-Wu *et al.*, 1993; Sarkis *et al.*, 1995). Finally, we note that L5 lysogens are not strongly inducible by DNA-damaging agents, even though this is a common feature of many other temperate phages. Lysogens do undergo spontaneous induction to release particles into the supernatant of a liquid culture, but the nature of repressor loss-of-function is not known.

Mycobacteriophage Bxb1 encodes a related repressor (gp69), although it shares only 41% amino acid identity with L5 gp71 and the two phages are heteroimmune (Jain and Hatfull, 2000; Mediavilla *et al.*, 2000). However, there are many common features of the two immunity systems, including multiple promoters upstream of the repressor gene [two in Bxb1 (Jain and Hatfull, 2000)], a repressor-regulated early lytic promoter, and multiple stoperator sites located throughout the genomes. In Bxb1 there are 34 putative 13-bp asymmetric stoperator sites, corresponding to the consensus 5′-GTTACGWDTCAAG (W is A or T), with notable differences from the L5 consensus at positions 1, 4, and 5. Most of these share the same features of the L5 stoperators in being located within short intergenic regions and oriented in one direction relative to the direction of transcription. Bxb1 gp69 binds with a similar affinity to its binding site as L5 gp71 does to its sites, but recognition of each other's sites occurs only at a much lower affinity (~1000-fold lower), accounting for their heteroimmune phenotype.

Prior to its genomic characterization, mycobacteriophage D29 was thought to be a substantially different phage than L5 and others, partly because it forms completely clear plaques and partly because it infects *M. tuberculosis* readily [L5 also infects *M. tuberculosis*, but has specific requirements for high calcium concentrations that D29 does not (Fullner and Hatfull, 1997)]. Genomic analysis showed that it is a derivative of a temperate parent that has suffered a 3.1-kbp deletion removing the repressor and several closely linked genes (Ford *et al.*, 1998a). The deletion event likely occurred relatively recently—perhaps at the time of its isolation (see Section III.A)—and D29 is subject to gp71-mediated L5 immunity (Ford *et al.*, 1998a). Most of the stoperator sites identified in L5 are present at similar positions in L5, and the 13-bp consensus sequence is the same as that of L5.

More recently, the bioinformatic analysis of immunity specificities has been extended to all of the known Cluster A phages and concludes that these specificities closely mirror the subcluster divisions. That is, all of the phages within a subcluster form a homoimmune group, but none offers immunity to phages from other subclusters (Pope *et al.*, 2011). Although several other Cluster A phages appear to contain defective repressors such that stable lysogens cannot be recovered, all contain predicted stoperator sites varying in number from 23 in Che12 and Bxz2 to 36 in Jasper (Pope *et al.*, 2011). From all 17 Cluster A phages, a total of 453 potential sites have been identified, and although only those in L5 and Bxb1 have been shown to be true binding sites, some general features are evident. In particular, positions 1 and 13 are absolutely conserved (G in both positions) and positions 9 and 12 are highly conserved (T and A, with nine and two departures, respectively). Positions 2 through 6 maybe the primary determinants of specificity and can likely be discriminated by differences in the second helix of the HTH motifs of the repressors (Pope *et al.*, 2011), although this awaits detailed experimental analysis. DNA protection and mutational analysis of L1 repressor binding are consistent with the bioinformatic findings (Bandhu *et al.*, 2010).

Mycobacteriophage Fruitloop and LRRHood—members of Clusters F and C, respectively—both contain genes related to the Cluster A repressors, even though stable lysogens have not been reported for either phage. Comparisons of repressor genes show that these are very closely related to the repressor of Cluster A1 phages (Fig. 20), and LRRHood gp44 has only a single amino acid departure from Bxb1 gp69 (Pope *et al.*, 2011). However, neither LRRHood nor Fruitloop contains multiple binding sites related to the Bxb1 stoperators. There is not a single potential repressor-binding site in the LRRHood genome, and in Fruitloop there is just a single site located upstream of gene *39* that could play a role in autoregulation. As discussed previously, the presence of repressor genes in these phages could have been selected to confer protection to either lysogens or infected cells from superinfection with Bxb1-like Cluster A1 phages (see Section III.F).

2. Cluster G immunity systems

Putative repressor genes have also been identified in Cluster G phages, such as BPs and its closely related relatives Halo, Angel, and Hope (Sampson *et al.*, 2009). These phages are temperate, and stable lysogens can be recovered from infected cells. The BP repressor (gp33) is not closely related to Cluster A-encoded or any other phage repressors but does

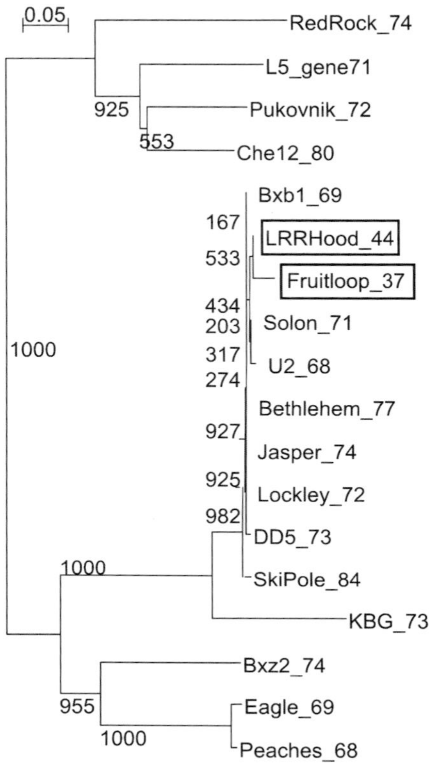

FIGURE 20 Phylogenetic tree of mycobacteriophage repressor proteins. The neighbor-joining phylogenetic tree of mycobacteriophage repressor-like proteins was generated from an alignment created in Cluster X and drawn using NJPlot. All of the repressors shown are encoded by Cluster A phages, with the exception of LRRHood and Fruitloop (boxed), which are members of Subclusters C1 and F1, respectively. THe LRRHood and Fruitloop repressors are related more closely to those in Cluster A1 (i.e., Bxb1 and its relatives) than to others.

contain a putative helix-turn-helix DNA-binding motif, and expression of gp33 confers immunity to superinfection by all of the Cluster G phages (see Section III.G). The repressor gene is located immediately upstream of the integrase gene (32) and the two genes are predicted to overlap. A notably unusual feature of genome organization is that the crossover site for integrative recombination at *attP* is located within the repressor gene itself, such that the gene product expressed from the prophage is 33 residues shorter than the virally encoded form. This suggests the possibility that integration plays a central regulatory role in the lytic–lysogenic decision.

B. Integration systems

Phages within Clusters A, E, F, G, I, and K and singletons Giles, Omega, and LeBron all encode integrases, mostly of the tyrosine-recombinase family. However, several phages—all within the Cluster A—encode serine-integrases, including all of Subcluster A1, Subcluster A3, and Peaches of Subcluster A4. Although most of the Subcluster A2 phages encode tyrosine-integrases, an interesting exception is RedRock, which encodes putative ParA and ParB proteins at the same genomic location as its Subcluster A2 relative encode tyrosine-integrases (Fig. 3B).

1. Tyrosine-integrase systems

The best studied of mycobacteriophage tyrosine-integrases is that encoded by L5. L5 integrase is a distant relative of the phage λ prototype, but shares many central features. For example, L5 gpInt (gp33) contains two DNA-binding specificities: one encoded in a small (65 residue) N-terminal domain that binds to arm-type sites in *attP* and a second within the larger C-terminal domain that recognizes core-type sequences in *attP* and *attB* (Peña *et al.*, 1997). Amino acid residues critical for the chemistry of strand exchange, including catalytic tyrosine, are all well conserved. In the L5 genome, the *attP* site is located to the 5′ side of the integrase gene and is ~250 bp long, containing core-type integrase-binding sites flanked by arm-type integrase-binding sites (Peña *et al.*, 1997); the *attB* site overlaps a $tRNA^{gly}$ gene in the *M. smegmatis* genome (Lee and Hatfull, 1993; Lee *et al.*, 1991). L5 integrase-mediated integrative recombination requires L5 gpInt, a host-encoded mycobacterial integration host factor (mIHF), and *attP* and *attB* DNAs (Lee and Hatfull, 1993; Pedulla *et al.*, 1996). DNA supercoiling stimulates recombination *in vitro*, but this is observed if either of the DNA molecules is supercoiled (Peña *et al.*, 1998). The host factor mIHF is quite distinct from other IHF-like proteins, and its name reflects its function rather than any sequence or structural similarity (Pedulla *et al.*, 1996). It contains a single subunit with DNA -binding properties, is an essential gene in *M. smegmatis* (Pedulla and Hatfull, 1998), but does not appear to bind either *attP* or *attB* with any specificity. Nonetheless, it strongly promotes the formation of stable tertiary complexes containing gpInt, mIHF, and *attP* DNA (Pedulla *et al.*, 1996). Interestingly, there appear to be alternative pathways for the assembly of synaptic complexes (Peña *et al.*, 2000) containing *attB*, and within which strand exchange occurs. Cleavage occurs seven bases apart within the core region to generate 5′ extensions, and cleavage is associated with covalent linkage of gpInt to the 3′ ends of the DNA (Peña *et al.*, 1996); the 7-bp overlap region corresponds to the anticodon loop of the $tRNA^{gly}$ gene at *attB*. The directionality of L5 integrase-mediated recombination is determined by recombination directionality factor gp36 (gpXis)(Lewis and Hatfull, 2000). L5 gp36 is small (56 residues) and binds to four putative-binding sites in *attP* and *attR* (Lewis and Hatfull, 2003). L5 gp36 is proposed to impart a substantial DNA bend at these sites and thus dictates the ability of integrase to form recombinagenic protein–DNA complexes (Lewis and Hatfull, 2003). When bound to *attR* it promotes formation of a complex in which gpInt is bound simultaneously to the core and arm-type sites in *attR* to form an intasome that can synapse with at *attL*–intasome (Lewis and Hatfull, 2003). L5 gp36 thus strongly stimulates excisive recombination. In contrast, when L5 gp36 is bound to *attP* DNA, it discourages formation of an intasome-like structure that can synapse with *attB* DNA, which inhibits integrative recombination (Lewis and Hatfull, 2003).

2. Serine-integrase systems

Serine-integrases are unrelated to tyrosine-integrases and typically contain an N-terminal domain of 140–150 residues related to the catalytic domain of transposon resolvases such as Tn*3* and $\gamma\delta$ and a large C-terminal domain with DNA-binding activity (Smith and Thorpe, 2002). The best-studied of the mycobacteriophage-encoded systems is that of Bxb1, although the related system encoded by the prophage-like element, ϕRv1, has also been investigated. Bxb1 gpInt (gp35) is a 500 residue two-domain protein that catalyzes site-specific recombination between an *attP* site located to the 5′ side of the gene *35* and an *attB* site located with the *M. smegmatis groEL1* gene (Kim *et al.*, 2003). Both *attP* and *attB* are small, and the minimally required sites contain 48 and 38 bp, respectively (Ghosh *et al.*, 2003). Strand exchange occurs at the centers of these sites, and Bxb1 gpInt cleaves to generate two-base 3′ extensions; strand exchange involves the formation of gpInt–DNA covalent linkages with the serine at position 10 linked to the 5′ ends of the DNA (Ghosh *et al.*, 2003). Bxb1 gp35 efficiently mediates site-specific recombination between *attP* and *attB in vitro* to generate *attL* and *attR*, and no additional proteins are required (Ghosh *et al.*, 2003; Kim *et al.*, 2003). The reaction is not stimulated significantly by DNA supercoiling and does not require the addition of either metal ions or high-energy cofactors (Ghosh *et al.*, 2003). This reaction is strongly directional, and Bxb1 gpInt alone does not catalyze recombination between *attL* and *attR*; it also fails to catalyze recombination between any pair of sites other than between *attP* and *attB*. The simplicity of this reaction greatly facilitates biochemical dissection of the reaction, with the origins of the site specificity and the control of directionality as central questions of interest.

Both *attP* and *attB* are quasi-symmetric in nature, being composed of imperfectly inverted repeats flanking the 5′-GT central dinucleotide, and gpInt binds to each site as a dimer (Ghosh *et al.*, 2005). However, the P and P′ half sites in *attP* are distinctly different from the B and B′ *attB* half-sites, although all four half-sites contain a 5′ ACNAC motif in symmetrically related positions (Ghosh *et al.*, 2003). These structures raise several interesting questions. First is the issue as to whether gpInt contains a single DNA recognition motif that somehow adapts to interact with the two different types of half-sites or whether there are two separate structural motifs, each capable of recognizing either B-type or B-type half-sites. Thus far there is no evidence for more than one type of DNA recognition motif, and the only mutants that discriminate between binding to *attP* and *attB* have substitutions in the putative linker region that joins the two domains (Ghosh *et al.*, 2005). A second issue is in regard to the relative orientation of synapsis, as each site is quasi-symmetrical, and presumably synapsis is mediated by protein–protein interactions between gpInt dimers bound to *attP* and *attB*. Interestingly, synapsis does indeed appear to occur in an orientation-independent manner, and it is only the asymmetric 5′-GT central dinucleotide that determines the orientation of integration (Ghosh *et al.*, 2003). Thus wild-type *attP* and *attB* sites can synapse in both possible orientations (these are referred to as parallel and antiparallel alignments, although the actual configurations are not known) with equal probabilities. In the productive orientation, one helix can rotate 180° around the other to generate a recombinant configuration within which religation to the partner DNA can proceed. In the nonproductive configuration, after 180° rotation of the helices, bases at the central dinucleotide are noncomplementary and ligation does not occur (Ghosh *et al.*, 2003). However, rotation can proceed for one or more subsequent rounds to realign the central nucleotide bases such that they are in the parental—and thus ligatable—position. Changing a single base in the central dinucleotide of both *attP* and *attB* such that the central nucleotides are palindromic thus leads to complete loss of orientation specificity, with approximately equal efficiencies of ligation of the P half-site with B and B′; likewise for P′ (Ghosh *et al.*, 2003). Site specificity for integrative recombination likely results from the specificity for synapsis. That is, even though integrase binds as a dimer to all four possible sites, *attP*, *attB*, *attL*, and *attR*, synapsis only occurs between gpInt-bound *attB* and *attP* complexes. The molecular basis for this is not known, but presumably gpInt adopts different conformations when bound to the four different sites, such that noncognate combinations are excluded conformationally. This raises the question as to how excision occurs, where gpInt bound to *attL* and *attR* must somehow presumably adopt conformations that are productive (Ghosh *et al.*, 2006). Genetic analysis identified a second phage-encoded protein, gp47, acting as an RDF in that it is required to enable integrase-mediated excisive recombination between *attL* and *attR*. Bxb1 gp47 also inhibits integrative recombination (Ghosh *et al.*, 2006), a common property of RDF proteins (Lewis and Hatfull, 2001). The molecular mechanism by which Bxb1 gp47 switches site specificity for gpInt is not known; it does not bind DNA, but rather associates with gpInt–DNA complexes and seems to do so differently depending on which type of site is bound

(Ghosh *et al.*, 2006). This is at least consistent with a model in which gp47 modulates the conformation of gpInt, enabling productive configurations when it is bound to *attL* and *attR*, but not when it is bound to *attP* and *attB*. A notable consequence of the finding that *attP* and *attB* are essentially symmetrical for the purposes of synapsis is that *attL* and *attR*, both of which contain one B-type and one P-type half-site, are essentially identical (Ghosh *et al.*, 2006, 2008). This predicts that asymmetry of the central dinucleotide again plays a critical role in determining productive recombination for excision, as gpInt bound to *attL* is expected to promote synapsis just as efficiently with itself as with gpInt bound to *attR*. This is confirmed experimentally, because switching the central dinucleotide of *attL* to make it palindromic is sufficient to generate an efficient three-component system requiring just gpInt, gp47, and the mutant *attL* site (Ghosh *et al.*, 2008). It is also noteworthy that the asymmetry of *attL* and *attR* (each containing one B-type and one P-type half-site) is also reflected in the orientation of synapsis. Thus in each of the synaptic interactions observed, an gpInt protomer bound to a B-type half-site must interact with one bound to a P-type half-site (Ghosh *et al.*, 2008).

Bxb1 255 residue gp47 is an unusual RDF and has no sequence similarity to other RDF proteins. It is not closely linked to the integrase gene as is often observed for RDF's, but is located approximately 5 kbp to its right, among genes predicted to be involved in DNA replication, including DNA polymerase and DNA primase genes (Ghosh *et al.*, 2006). Strangely, there are relatives of Bxb1 gp47 in all 17 of the Cluster A phages, including all those that encode tyrosine-integrases, and indeed in L5 where all the phage-encoded genes required for efficient site-specific recombination—both integrative and excisive—are known (see Sections III.A and V.B.1 and Fig. 3B). The simplest interpretation is that Bxb1 gp47 is a dual function protein, fulfilling a common role of Cluster A phages—most likely in DNA replication—but also co-opted for use as an RDF in Bxb1. This raises the question as to whether homologues of Bxb1 gp47 also perform the RDF function in those phages that encode more distantly related serine-integrases, such as Bxz2 and Peaches (all of the serine-integrase encoded by Subcluster A1 phages are very similar to each other), or whether alternative proteins have been adopted (see Section III.A).

The serine-integrase system encoded by the *M. tuberculosis* prophage-like element ϕRv1 sheds some light on at least some of the questions raised by the Bxb1 system (Bibb *et al.*, 2005; Bibb and Hatfull, 2002). Like Bxb1, requirements for integration *in vitro* are simple, requiring *attP* and *attB* partner DNAs, and ϕRv1 gpInt (Bibb and Hatfull, 2002). However, the reaction is somewhat slow and inefficient relative to the Bxb1 reaction. Interestingly, the ϕRv1 element integrates into a repetitive element in *M. tuberculosis* and is therefore found in several different chromosomal locations. The putative *attB* sites differ for each of the repeated sequences, although four of them are active as sites for recombination (Bibb and

Hatfull, 2002). The RDF protein for the ϕRv1 system has been identified, and the 73 residue protein (Rv1584c) is completely unrelated to Bxb1 gp47 (Bibb and Hatfull, 2002). Yet more surprising, the ϕRv1 RDF is related to Xis-like proteins associated with tyrosine-integrases, including L5 gp36 (Bibb and Hatfull, 2002). While the ϕRv1 RDF may have DNA-binding activity, this does not appear to be required for excision, as only the same minimal sequences are required for excision as they are for integration, all of which are apparently involved in close interactions with gpInt (Bibb *et al.*, 2005). It is thus likely that the mechanism of action of the Bxb1 RDF, both in stimulating excision and in inhibiting integration, is mediated by direct interactions in ϕRv1 gpInt or gpInt–DNA complexes (Bibb *et al.*, 2005).

3. Integration specificities of mycobacteriophage integrases

For most phages that encode a tyrosine-integrase, a putative *attP* core site can be identified bioinformatically. The basis for this is the observation that most of these utilize a host tRNA gene for integration, with strand exchange occurring somewhere within the gene, and the phage genome carries the 3′ part of the tRNA gene such that a functional gene is reconstructed following integration. Although recombination itself likely only requires identity between *attP* and *attB* at the 7–8 bp constituting the overlap region between sites of strand cleavage, the requirement for tRNA reconstruction usually extends the sequence identity (or near-identity) to as much as 45 bp, which can be identified readily in a BLASTN search. Furthermore, the *attP* site is typically located near the integrase gene and is usually in an intergenic noncoding interval. As a result, these regions can be used to search sequence databases, followed by determination of whether any matching sequences overlap host tRNA genes. Within the phage genome, it is often possible to identify pairs of short (10–11 bp) sequences flanking the *attP* core that correspond to putative arm-type integrase-binding sites (Morris *et al.*, 2008). Using this strategy, putative *attB* sites can be predicted for most of the mycobacteriophages that encode tyrosine-integrases (Table II). For some of these, including L5, Tweety, BPs, Ms6, and Giles (Freitas-Vieira *et al.*, 1998; Lee *et al.*, 1991; Morris *et al.*, 2008; Pham *et al.*, 2007; Sampson *et al.*, 2009), good experimental evidence supports *attB* site usage. Others await experimental verification. However, for Cluster E phages, as well as LeBron, bioinformatic identification has proven difficult, and *attB* site identification will likely require experimental approaches. One plausible explanation for this is if they either do not use tRNAs for integration or the positions of strand exchange are so close to the 3′ end of a tRNA gene that they are carrying only a minimal segment of homology to the host genome. An alternative explanation is that these phages do not normally infect *M. smegmatis* or any closely related strains, and the *attB* site is simply not present in *M. smegmatis*. This would seem unlikely for Cluster E phages because at least for some, lysogens have been recovered.

TABLE II Integration specificities of mycobacteriophage integration sites in *M. smegmatis* mc^2155 and *M. tuberculosis* H37Rv

attB	tRNA	M. smeg	M. tb H37Rv	Phages	Cluster	Int
attB-1	tRNA-gly	Msmeg_4676 (4764493–4764563)	NT02MT2675 (2765539–2765609)	L5_33, D29_33, Che12_36, Pukovnik_35	A2	Tyr
				Eagle_32,	A4	Tyr
attB-2	tRNA-Lys	Msmeg_4746 (484790–4847983)	NT02MT2737 (2835492–2835564)	Che8_46, Boomer_46, Llij_40, PMC_38, Tweety_43, Pacc40_40, Ramsey_44	F1	Tyr
attB-3		Msmeg_5156		Bxz2_34	A3	Ser
attB-4	tRNA-Lys	Msmeg_5758 (5834573–5834645)	NT02MT0910 (92387–923798)	Unpublished phages[a]	N	Tyr
attB-5	tRNA-Thr	Msmeg_6152 (6221063–6220991)	NT02MT3969 (4081434–4081359)	Brujita_33, Island3_33	I1	Tyr
attB-6	tRNA-Arg	Msmeg_6349 (6410438–6410366)	NT02MT4110 (4216934–4216862)	BPs_32, Halo_32, Angel_32, Hope_32	G	Tyr
attB-7	GroEL1	MSMEG_0880		Bxb1_35, U2_36, Bethlehem_36 DD5_38, Jasper_38, KBG_38. Lockley_38, Solon_37, SkiPole_40,	A1	Ser
attB-8	tRNA-Tyr	Msmeg_1166 (1228393–1228478)	No	Che9c_41	I2	Tyr
attB-9	tmRNA	Msmeg_2093 (2169257– 2169625)	No	Angelica_41, CrimD_41	K1	Tyr

attB-10	tRNA-Ala	MSMEG_2138 (2213142–2213214)	NT02MT3342 (3431909–3431837)	Fruitloop_40, Ardmore_36, Ms6_int	F1	Tyr
attB-11	tRNA-Leu	Msmeg_3245 (3328766–3328690)	NT02MT1769 (1828086–1828010)	Omega_85	J	Tyr
attB-12	tRNA-Pro	Msmeg_3734 (3800622–3800546)	NT02MT1869 (1946611–1946684)	Giles_29	Sin	Tyr
attB-13	tRNA-Met	Msmeg_4452 (4532894–4532821)	NT02MT2502 (2581835–2581762)	Che9d_50	F2	Tyr
Unassigned[b]				Cjw1_53, 244_53, Kostya_53, Porky_51, Pumpkin_54,	E	Tyr
				Peaches_33	A4	Ser
				LeBron_36	L1	Tyr

[a] Two phages have been identified that utilize this site but the phage sequences are as yet incomplete and are not yet published.
[b] *attB* sites have yet to be identified for these phages.

Confident bioinformatic identification of *attB* sites for phages encoding serine-integrases is currently not possible. These generally do not integrate into tRNA genes, and the segment of *attP* homology to the host chromosome can be as small as 3 bp (Smith and Thorpe, 2002). The *attB* sites for both Bxb1 (which is likely used for all Subcluster A1 phages because their integrases are extremely similar) and Bxz2 have been identified (Kim *et al.*, 2003; Pham *et al.*, 2007). The Bxb1 *attB* site is located within the host *groEL1* gene, and integration results in inactivation of the gene with interesting physiological consequences (Kim *et al.*, 2003). Specifically, Bxb1 lysogens are defective in the formation of mature biofilms, revealing the novel function of GroEL1, which acts as a dedicated chaperone for mycolic acid biosynthesis (Ojha *et al.*, 2005). Bxz2 integrates into the extreme 5′ end of the *M. smegmatis* gene Msmeg_5156 (Pham *et al.*, 2007), although no physiological consequences have been examined. The *attB* site for the more distantly related Peaches has yet to be identified. The propensity for phages encoding serine-integrases to integrate within host protein-coding genes with opportunities to influence their physiology makes this class of mycobacteriophages of particular interest.

In total, 13 distinct *attB* sites have been identified or predicted (Table II). To facilitate discussion of these *attB* sites, we have designated them *attB1*–*attB13*, with *attB1* denoting the L5 site. The others are ordered according to their location in the *M. smegmatis* genome, proceeding in a clockwise direction (Fig. 21). Related sites for 9 of these are also present in *M. tuberculosis*; these are numbered according to the same designations (Table II). The placement of these on a circular representation of the *M. tuberculosis* genome illustrates the lack of synteny between these two strains of mycobacteria (Fig. 21). Distribution of *attB* sites is of interest in part because of the utility of using phage integrase-based integration-proficient plasmid vectors, which have the advantage of constructing single-copy recombinants that are genetically stable in the absence of selection (see Section VII.A.1). While additional *attB* specificities would likely be welcome, the current distribution of sites enables the potential use of integration-proficient vectors to introduce genetic elements of choice at different locations relative to the chromosomal origin of DNA replication.

VI. MYCOBACTERIOPHAGE FUNCTIONS ASSOCIATED WITH LYTIC GROWTH

A. Adsorption and DNA injection

Unfortunately, rather little is known about the repertoire of bacterial surface molecules used by mycobacteriophages to specifically recognize their hosts. A *M. smegmatis* peptidoglycolipid, mycoside C(sm), has been

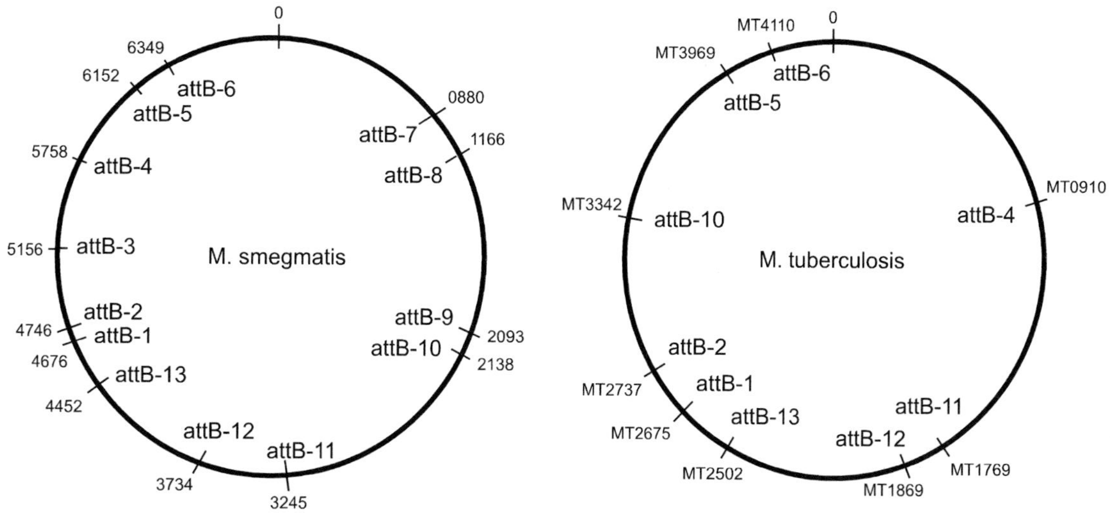

FIGURE 21 Locations of predicted mycobacteriophage *attB* sites in *M. smegmatis* and *M. tuberculosis* genomes. The predicted *attB* sites for all mycobacteriophage genomes containing integrase genes and for which *attP* and *attB* sites can be predicted or are identified experimentally are shown on a circular representation of the *M. smegmatis* genome. Those *attB* sites that are also present in *M. tuberculosis* H37Rv are shown on a circular representation of the *M. tuberculosis* H37Rv genome on the right. The *attB* designation is conserved between the two strains, such that for example *attB-4* has the same sequence in both strains, notwithstanding the different chromosomal position. All *attB* sites and their specific locations are listed in Table II.

purified and proposed to play a role in binding of the uncharacterized phage D4 (Furuchi and Tokunaga, 1972), and a set of lyxose-containing molecules have been proposed as receptors for the uncharacterized phage Phlei (Bisso *et al.*, 1976; Khoo *et al.*, 1996). In addition, a single methylated rhamnose residue on the cell wall-associated glycopeptidolipid has been implicated in the adsorption of phage I3 to *M. smegmatis* (Chen *et al.*, 2009). No protein-based receptors for mycobacteriophages have been reported.

Overexpression of a single *M. smegmatis* protein, Mpr, is sufficient to confer high levels of resistance to phage D29 (Barsom and Hatfull, 1996), which may occur through placement of the gene on an extrachromosomal plasmid (Barsom and Hatfull, 1996), expressing it from a strong promoter (Barsom and Hatfull, 1996), or through activation by adjacent transposon insertion (Rubin *et al.*, 1999). It is plausible that spontaneous D29 resistance could occur by localized genome amplification of the *mpr* locus that is genetically unstable and recombines back to a single copy in the absence of selection. The cellular role of Mpr is not known and is not present in *M. tuberculosis*. Interestingly, the 215 residue Mpr protein contains a 125 reside domain at its extreme C-terminus belonging to the Telomeric repeat-binding factor 2 (TRF2) superfamily implicated in recognition and binding to TTAGGG-like telomeric repeats. This is an unexpected function for a bacterium that does not contain a linear genome, although it is of interest because of the observation that overexpression of Mpr appears to specifically inhibit injection of D29 DNA. Perhaps this gene has been acquired specifically to prevent phage infection under certain circumstances.

Presumably, specific recognition of the bacterial host is accomplished through structures encoded at the tips of phage tails. Genomic analysis shows that those phages with a siphoviral morphotype encode five to eight putative minor tail protein genes downstream of the tapemeasure protein gene, and for a few phages these have been confirmed experimentally (Ford *et al.*, 1998b; Hatfull and Sarkis, 1993; Morris *et al.*, 2008). Many mycobacteriophages also encode three to four relatively small genes at the end of the virion structural and assembly operon that may also be involved in tail assembly. Which of these tail proteins is specifically involved in host recognition is unclear. Interestingly, a number of genomes (e.g., in Clusters/Subclusters A1, C1, D, E, F, H1, I1, and J and singletons Corndog and Wildcat) encode a putative β-lactamase-like D-alanyl-D-alanine carboxypeptidase activity that is presumably involved in modification of the cell wall and perhaps facilitates productive association of the tail tip with the cell wall or membrane. However, none of these have been characterized. The lengths of mycobacteriophage tails, especially those with siphoviral morphologies, vary considerably, and the lengths of tapemeasure protein genes vary correspondingly (Pedulla *et al.*, 2003). These proteins are of particular interest because many of them encode short sequence motifs associated with peptidoglycan hydrolysis, suggesting functionalities in addition to a role in phage tail assembly (Lai *et al.*, 2006; Pedulla *et al.*, 2003). Initially, three distinct motifs were discovered (Pedulla *et al.*, 2003), although the expanded genomic set suggests that there are at least seven different motifs (L. Marinelli and Graham F. Hatfull, manuscript in preparation). The motif present in TM4 (motif 3) has been shown to be nonessential for viability, although mutants in whom it has been deleted or inactivated have reduced abilities to infect stationary phase cells (Piuri and Hatfull, 2006). Because removal of the motif results in a corresponding reduction in tail length, the tapemeasure protein must be able to adopt two different conformations, an extended rod-like structure involved in tail assembly and a component of the complete tail structure, and a folded structure that has enzymatic activity (Piuri and Hatfull, 2006). Even though peptidoglycan-hydrolyzing motifs cannot be identified in all mycobacteriophages, it seems probable that all can form this proposed alternative folded state. Because most of the mycobacteriophage tapemeasure proteins also contain putative transmembrane-spanning domains—as many as nine in the Cluster G genomes—an intriguing role for all these tapemeasure proteins is as a membrane-located pore through which DNA traverses to gain entry into the cell (L. Marinelli and Graham F. Hatfull, manuscript in preparation).

B. Genome recircularization

Virion DNA is expected to be linear in all mycobacteriophages, and in Clusters A, E, F, G, I, J, K, and L and in singletons Corndog, Giles, and Wildcat, the genomes have defined ends with short ssDNA extension varying from 4 to 14 bases in length; all have 3′ extensions (Table I). Following DNA injection, all are expected to be circularized at an early stage, prior to either DNA replication or integration. The specific requirement for circularization has not been examined, but it is expected that for most phages it is dependent on the action of the host DNA ligase. The kinetics of recirularization are not known.

Two phages with defined ends appear to use a different and an unusual mechanism involving nonhomologous end joining (NJEJ). Phages Omega and Corndog both have 4-base ssDNA extensions, substantially shorter than all the others (9–14 bases), and are different from all other mycobacteriophages in that they also encode a Ku-like protein that facilitates DNA end association in bacterial NHEJ systems (see Sections III.J and III.M.1)(Pitcher *et al.*, 2006). Both *M. smegmatis* and *M. tuberculosis* encode NHEJ systems, including a Ku-like protein and an associated DNA ligase (Lig IV)(Aniukwu *et al.*, 2008; Pitcher *et al.*, 2005). Interestingly, efficient infection of *M. smegmatis* by Corndog and Omega is dependent on the host Lig IV, but not on the host Ku-70 protein (Pitcher *et al.*, 2006). Presumably, the phage-encoded Ku-70 is required for infection, although this has yet to be demonstrated. A conundrum in the implication of NHEJ in genome circularization is that either the phage Ku-70 gene would need to be expressed immediately upon DNA injection or the protein would need to be encapsulated in the phage capsids. Unfortunately, mutants of Omega or Corndog lacking Ku-like genes have yet to be constructed.

It is likely that mycobacteriophages with terminally redundant ends are circularized by homologous recombination. However, it is unclear whether this is dependent on phage-encoded functions or whether host recombinases are utilized. In phage P22 it is proposed that the Erf recombinase, which is essential for phage growth, promotes genome circularization (Botstein and Matz, 1970). We note that mycobacteriophages in Clusters E and L and the singleton Wildcat all encode Erf-family proteins, but their genomes have defined ends, not terminally redundant ends; presumably the Erf-like proteins they encode perform alternative functions. Of those phages that do have terminally redundant ends, only Cluster C phages encode an obvious recombinase, a RecA-like protein (gp201). Cluster B, D, and H phages do not encode a recognizable recombinase at all, thus presumably either exclusively use host recombination enzymes for genome circularization or encode novel recombinases currently uncharacterized.

C. DNA replication

Mycobacteriophage DNA replication represents another understudied but interesting aspect of their biology. Presumably, replication involves a combination of phage- and host-encoded functions and is initiated at one or more origins of replication in the phage genome, although none have been identified. Many of the genomes do not encode their own DNA polymerase and presumably use one or more of the resident polymerases. Others do encode their own DNA polymerase, although both DNA Pol I and Pol III subunits are well represented; LeBron unusually encodes a DNA Pol II-like protein. It has not been shown, however, that any of these are essential for replication or whether host enzymes can be utilized if phage polymerases are inactivated. For the most part, other components of the replication machinery presumably are provided by the host, although we note that Corndog unusually encodes a Pol III clamp loader-like protein (see Section III.M.1). Many genomes also encode predicted DNA primases, although there is great diversity among the types of proteins encoded. For example, in some phages (e.g., in Cluster A), primase functions are associated with two adjacent open reading frames, raising the possibility that a functional enzyme is generated by an unusual translation event (such as a programmed frameshift or a ribosome hop) or by processing at the RNA level. In other phages, the primase function is associated with proteins that also provide either predicted helicase activities (as in Cluster B and K phages) or polymerase functions (as in Clusters D and H and Corndog). Many of the phages also encode apparent stand-alone canonical DNA helicases, frequently of the DnaB family. The gp65 protein of D29 has been characterized and shown to be a structure-specific nuclease with a preference for forked DNA structures (Giri *et al.*, 2009).

Most mycobacteriophages encode a Holliday Junction resolvase, although many different types are represented, including those related to Endo VII, RuvC, and RusA, and they are present in a multitude of genomic locations. Notable exceptions are phages in Clusters D, E, F, and H, raising the possibilities that either these employ different strategies in replication and recombination or encode one or more novel classes of HJ resolvases that have not been recognized previously. Because Holliday junctions are strongly deleterious to DNA packaging and many of the phages encode recombination-promoting proteins, we favor the second of these explanations.

D. Virion assembly

Identification of genes involved in virion structure and assembly is facilitated by their conserved gene order (at least in Siphoviridae), even though the sequences are highly diverse. For example, capsid subunits can be identified in most of the genomes—with perhaps the greatest ambiguity in Cluster H phages (Fig. 11)—although they represent many different sequence phamilies (Hatfull *et al.*, 2010). Nonetheless, it is plausible that they contain similar protein folds to that of the HK97 capsid subunit that is also present in other viruses that lack substantial sequence similarity with it (Fokine *et al.*, 2005; Hendrix, 2005; Johnson, 2010; Wikoff *et al.*, 2000). In some of the genomes there is an identifiable scaffold protein encoded immediately upstream of the capsid gene, and in L5, gp16 has been shown to be a component of head-like particles but absent from intact virions (Hatfull and Sarkis, 1993). Putative scaffold genes are present in many other mycobacteriophage genomes but are divergent at the sequence level and their assignments remain tentative until there is further experimental support. In some mycobacteriophages, the scaffold function may be provided by an N-terminal domain of the capsid protein itself, as in HK97 (Duda *et al.*, 1995). In L5, D29, and TM4 there is strong evidence that the major capsid subunit is covalently cross-linked (Ford *et al.*, 1998a,b; Hatfull and Sarkis, 1993), as described for HK97, and it may be a common feature among mycobacteriophages (Hatfull and Sarkis,

1993). But not all do so, and there is evidence against it in Giles (Morris *et al.*, 2008).

As noted earlier, all three phages in Cluster I and the singleton Corndog have prolate heads in contrast to all other mycobacteriophages, which have isometric heads. The dimensions are somewhat different, with Cluster I phages having a length:width ratio of 2.5:1 and Corndog a ratio of 4:1. However, no evident sequence similarity exists between their capsid subunits, although we note that a closely related protein to Cluster I capsid subunits is encoded within the genome of *Streptomyces scabies* (55% amino acid identity), suggesting the presence of a prophage capable of generating prolate-headed particles. It is unclear how the length of these prolate capsids is determined, and at least in Cluster I phages, there is no evidence from genome analysis of genes encoding additional capsid-associated proteins (see Section III.I and Fig. 12). In Corndog, this is less clear because of the greater complexity of the virion structure and assembly operon (Fig. 16).

Like many other phages with siphoviral morphologies, most mycobacteriophages contain a set of four to eight genes located between the major capsid and the major tail subunit that are likely involved in the head–tail joining process. For the most part, these genes are shared among genomes within a subcluster, but only in a few instances are relatives observed in other mycobacteriophages. When they do, they are typically (although not always) as groups of genes that appear to be traveling together; one example is PBI1 genes *20–22* (Fig. 7), which have homologues in similar genomic locations in Cluster H phages (Fig. 11). These observations are consistent with the idea that these head–tail connector proteins have to interact with each other physically.

One of the most highly conserved features of tail assembly genes of Siphoviridae is the expression of two genes between the major tail subunit and the tapemeasure protein genes that are expressed via a programmed translational frameshift to produce tail assembly chaperones (Xu *et al.*, 2004). A programmed frameshift can be identified in nearly all mycobacteriophage genomes, and the majority (Cluster A, C, D, E, and G and Wildcat) use a canonical -1 frameshift as described for phage λ proteins G and G-T (Levin *et al.*, 1993), whereas others (Clusters F and I, Corndog, and Omega) use a $+1$ (or possibly -2). Somewhat surprising given the strong conservation of this feature among phages of diverse bacterial hosts, no similar frameshifting events have been identified in Cluster B phages.

Some mycobacteriophages have the unusual feature of sharing short but related C-terminal extensions on the ends of their capsid and major tail subunits (Hatfull, 2006). This was first evident from sequencing of the Bxb1 genome (Mediavilla *et al.*, 2000), as the predicted capsid and major tail subunits are both ~85 residues longer than their counterparts in L5 as a consequence of C-terminal extensions. Moreover, Bxb1 extensions are related to each other (47% amino acid identity). Related sequences are present in all other Subcluster A1, A4, B2, and B3 phages situated similarly at the C-terminal ends of their capsids and major tail subunits. HHPred analysis shows predicted structural similarity of these to the C-terminal part of the phage λ major tail subunit (gpV), which is part of the large Big-2 family of Ig-like domains (Fraser *et al.*, 2006; Pell *et al.*, 2009, 2010). Related sequences are found in some other mycobacteriophage proteins, including several copies in a putative minor tail protein in Bxb1 (gp23) and in putative structural proteins in Cluster C1 phages (e.g., Bxz1 gp24). Wildcat and LeBron capsid and major tail subunits have similar types of extensions, and although the sequences are distinct from the others, they are related to other Ig-like domains. The presence of these Ig-like domains is relatively common in phage structural proteins, although their functional roles are unclear. Removal of the C-terminal domain from Lambda gpV results in a 100-fold reduction in viability and a possible defect in tail assembly (Pell *et al.*, 2010), although the relationships between the closely related mycobacteriophage capsids and major tail subunits containing these extensions (e.g., Bxb1) and those that do not (e.g. L5) suggest that these are likely not essential in these mycobacteriophage contexts.

E. Lysis

All mycobacteriophage genomes sequenced to date contain an identifiable lysin A (endolysin) gene encoding a peptidoglycan-hydrolyzing enzyme. However, sequence comparisons show that they are highly modular in nature and encompass a broad span of predicted enzyme specificities; there is no single amino acid sequence motif in common to all. Despite their very different sequences, the Lysin A proteins of D29, Ms6, and TM4 have all been shown to have peptidoglycan-hydrolyzing activity (Garcia *et al.*, 2002; Henry *et al.*, 2010a; Payne *et al.*, 2009). Inactivation of the lysin A gene in Giles (*31*) results in the loss of phage release without interruption of particle assembly (Marinelli *et al.*, 2008; Payne *et al.*, 2009). Delivery of the peptidoglycan hydrolase to its target is likely facilitated by a holin protein; in most mycobacteriophages, a putative holin gene can be identified closely linked to lysin A and encoding a small protein with one or more strongly predicted membrane-spanning domains. However, these are highly diverse at the sequence level and none have been examined experimentally. Interestingly, in phage Ms6, gene *1* product (a close relative of Fruitloop gp28, 97% amino acid identity; Fig. 9) has chaperone-like features and interacts directly with the endolysin to facilitate delivery to its peptidoglycan target in a holin-independent manner (Catalao *et al.*, 2010). Relatives of this protein are also encoded in Subcluster A1 genomes, where it is also closely linked to the lysis genes—and in the Subcluster K1 phage TM4 (gp90), where it is not. Mycobacteriophages are unusual in encoding a second lysis protein, Lysin B, that promotes efficient lysis of the host (Gil *et al.*, 2008; Payne *et al.*, 2009). Deletion of the Giles lysin B gene (*32*) does not lead to loss of viability but gives a reduction of plaque size and the number of particles contained therein (Payne *et al.*, 2009). A few phages do not encode Lysin B, including Che12, Subcluster B2 phages, and the C2 phage Myrna, although not all of these form noticeably small plaques. In Myrna, there is an additional unrelated gene (*244*) implicated in lysis, although it is not known if it substitutes for lysin B activity (Payne *et al.*, 2009). Lysin B has been shown to be a lipolytic enzyme (Gil *et al.*, 2008; Payne *et al.*, 2009), and the crystal structure of D29 Lysin B reveals structural similarity to cutinase family enzymes (Payne *et al.*, 2009). D29 Lysin B has activity as a mycolylarabinogalactan esterase and is proposed to separate the mycolic acid-rich mycobacterial outer membrane from its arabinogalactan anchor (Payne *et al.*, 2009). Ms6 Lysin B acts similarily (Gil *et al.*, 2010). Lysin B can thus be thought of as providing a function analogous to the Rz/Rz1 or spanning proteins encoded by phages of Gram-negative bacteria, which play a role in compromising the integrity of the outer membrane through fusion to the cytoplasmic membrane and facilitating complete lysis (Berry *et al.*, 2008).

VII. GENETIC AND CLINICAL APPLICATIONS OF MYCOBACTERIOPHAGES

Mycobacteriophages have played a central role in the development of tuberculosis genetic systems (Jacobs, 2000), and the large set of sequenced mycobacteriophage genomes provides a rich source of materials for applications both in mycobacterial genetics and in potential clinical applications. Some of these take advantage of the use of whole phage particles, and in these applications the host range is likely to be especially important. Because only Subcluster A2 and Cluster K phages infect *M. tuberculosis* efficiently, these have proven the most useful for these utilities. For development of genetic tools, host range is of less concern because most, if not all of, the genetic functionalities are likely to function equally well in both *M. smegmatis* and *M. tuberculosis*. In at least one example, the reason for lack of phage infection of *M. tuberculosis* can be ascribed to a failure of either adsorption or DNA injection, not phage metabolism per se (R. Dedrick and Graham F. Hatfull, unpublished observations).

A. Genetic tools

1. Integration-proficient vectors

Integration-proficient vectors are those that carry the integration apparatus of a temperate phage and have no other means of DNA replication. The first to be constructed were derived from L5 (Lee *et al.*, 1991), although others with different chromosomal targets have since been reported (Freitas-Vieira *et al.*, 1998; Morris *et al.*, 2008; Murry *et al.*, 2005; Pham *et al.*, 2007). The only phage requirements are the integrase gene and a functional *attP* site; because the *attP* site is typically closely linked to the integrase gene, simple versions of these vectors often can be constructed by inserting a single DNA fragment into a nonreplicating plasmid vector. It should be noted though that although integrase genes can usually be identified readily easily, identification of a functional *attP* is more error-prone. As discussed previously, the core region within which recombination occurs can usually be identified readily, but *attP* function usually requires flanking sequences containing arm-type integrase-binding sites (see Section IV.B.1). In some phages, these too can be predicted bioinformatically (Morris *et al.*, 2008), but in other genomes, this is more difficult. Nonetheless, the functional requirements for *attP* are usually encompassed with a region no larger than about 250–300 bp. Plasmid derivatives can also be used in which the *attP* site and the integrase gene are introduced on separate fragments (Huff *et al.*, 2010).

Of the potential 13 different *attB* sites that can be used for vector construction (Table II, Fig. 21), vectors have been described for at least six of them: *attB-1* (Lee *et al.*, 1991), *attB-2* (Pham *et al.*, 2007), *attB-6* (Sampson *et al.*, 2009), *attB-7* (Kim *et al.*, 2003), *attB-10* (Freitas-Vieira *et al.*, 1998), and *attB-12* (Morris *et al.*, 2008). Two additional site specificities have been described that use integration systems derived from Streptomyces phages or plasmids. Vectors derived from plasmid pSAM2 integrate site specifically into at *attB* overlapping $tRNA^{pro}$ gene Msmeg_6204 and the $tRNA^{pro}$ gene located between Rv3684 and Rv3685c in *M. tuberculosis* H37Rv (Martin *et al.*, 1991; Seoane *et al.*, 1997). Phage phiC31-derived vectors (using a serine-integrase) integrate into an *attB* site located within the putative glutamyl-tRNA(Gln) amidotransferase gene Msmeg_3400 and presumably inactivate it; there are three potential *attB* sites in *M. tuberculosis* (Murry *et al.*, 2005). L5 integration-proficient plasmids have also been manipulated such as to carry an additional *attB* site that will accept secondary integration events (Saviola and Bishai, 2004), and the Ms6 system has been manipulated so as to use alternative $tRNA^{ala}$ genes as integration sites (Vultos *et al.*, 2006). Integrated sequences can also be switched efficiently by introduction of a second plasmid with the same integration specificity and a second selectable marker (Pashley and Parish, 2003).

Integration-proficient plasmid vectors have several advantages over extrachromosomally replicating vectors. For example, introduction of genes in single copy typically avoids the overexpression seen with extrachromosomal vectors and thus avoids complications that can be encountered in complementation experiments. Second, they can have greater stability in the absence of selection relative to extrachromosomal vectors, provided that the phage-encoded excise functions are not also present. Even in the absence of excise, integrase-mediated excise-independent excisive recombination can lead to plasmid loss, which may be exacerbated in recombinants that are at a selective disadvantage (Springer *et al.*, 2001). Improved stability can be provided if the integrase gene is absent from the recombinant, which can be accomplished either by introducing the integrase gene on a second, nonreplicating plasmid that is subsequently lost (Peña *et al.*, 1997) or by site specifically removing the integrase gene from the recombinant (Huff *et al.*, 2010). Phage-encoded, site-specific recombination systems are also useful for efficient modification of recombinants, and excisive recombination by L5 integrase has been used to demonstrate gene essentiality (Parish *et al.*, 2001).

2. Selection by immunity

Temperate phages are immune to phage superinfection. If the superinfecting phage is defective in lysogeny and thus efficiently kills the bacterial cells, then this provides an effective means for using phage immunity functions—and repressor genes specifically—as selectable markers. This has been demonstrated using the L5 repressor gene (*71*) and using either D29 or a clear-plaque mutant of L5 (Donnelly-Wu *et al.*, 1993). Phage particles can be spread readily onto solid media prior to plating of cells, and relatively large numbers of cells can be plated and still get efficient killing of nontransformed cells. The obvious advantage of such systems is that they avoid the use of antibiotics and are thus useful for constructing complex recombinants where relatively few markers are available, for manipulation of strains that are extensively drug resistant even without manipulation, and to minimize biosafety concerns of generating highly drug-resistant forms of pathogenic strains. Because there are a substantial number of distinct mycobacteriophage immune specificities (see Section V.A), there is the potential to construct a large collection of compatible markers, at least for *M. smegmatis*. Although there are a large number of phages that encode identifiable integrases, only in Cluster A and Cluster G phages have repressor genes been identified. Phage repressors appear to be highly diverse and, in many cases, will need to be identified experimentally. We also note that immune selection requires the isolation of a clear-plaque derivative of the phage, and we note that because many of the temperate mycobacteriophages often form lysogens at relatively low frequencies, this is not always a simple task.

3. Generalized transduction

Generalized transduction is one of the most useful tools broadly implemented in bacterial genetics because it provides a simple means of moving genetic markers and mutations into different strain backgrounds. As such, it becomes easy to construct isogenic strains and to thus draw confident conclusions about the correlation between genotype and phenotype. Generalized transducing phages are typically those that package their DNA by headful-packaging systems and thus have genomes that are terminally redundant and circularly permuted. As described earlier, there are many mycobacteriophages in this class, including those in Clusters B, C, D, and H.

The first mycobacteriophage demonstrated to mediate generalized transduction of *M. smegmatis* was I3 (Raj and Ramakrishnan, 1970), a myovirus whose complete genome has not been sequenced, but is likely to be a Cluster C-like phage. It has been shown subsequently that Bxz1 is also a generalized transducing phage and can be used to efficiently exchange genetic markers between strains of *M. smegmatis* (Lee *et al.*, 2004). It is highly likely that other members of Cluster C behave similarly. Transduction by phages of Clusters A, D, and H has yet to be demonstrated.

No phages capable of generalized transduction of *M. tuberculosis* have been identified. This is unfortunate because there is a particular need for such phages to construct isogenic strains, especially for the analysis of mutations that occur in clinical isolates and that may be suspected of contributing to drug resistance or pathogenicity phenotypes. We note that none of the known mycobacteriophages with circularly permuted terminally redundant genomes infect *M. tuberculosis*. Although such phages may well exist in nature and await isolation, there is also the possibility that the idiosyncrasies of homologous recombination systems in *M. tuberculosis*—especially their proclivity for illegitimate recombination of linear DNA substrates introduced by electroporation (Kalpana *et al.*, 1991)—could have thwarted the successful evolution of such phages.

4. Transposon delivery

Several transposons have been described that can be used for insertional mutagenesis in *M. tuberculosis* and *M. smegmatis* (Cirillo *et al.*, 1991; Fomukong and Dale, 1993; Guilhot *et al.*, 1992; Rubin *et al.*, 1999). Mycobacteriophages offer attractive systems for transposon delivery because of the high efficiency of infection and the ability to deliver transposon DNA to nearly every cell in a liquid culture. This is especially important given the relatively low frequencies of movement of most transposons in bacteria. Phage delivery of transposons has the additional advantage over plasmid delivery systems in that the mutants recovered result from independent transposition events, providing the opportunity to generate mutant libraries of a large number of different insertions (Lamichhane *et al.*, 2003; Sassetti *et al.*, 2001), which is critical for applications such as transposon site hybridization (Sassetti *et al.*, 2003).

For efficient phage-mediated transposon delivery, it is important that lytic growth of the phage does not lead to cell death (Kleckner *et al.*, 1991).

Conditionally replicating mutants of both D29 and TM4 have been described that grow normally at 30°C but fail to replicate at higher temperatures (37°C for TM4, 38.5°C for D29)(Bardarov *et al.*, 1997). These mutants were isolated to ensure low frequencies of reversion to wild-type replication patterns, a potential concern when seeking selection of relatively low frequency transposition events. Coupling of these mutant phages with shuttle phasmids enables introduction of a variety of transposons of choice and has created a facile system for mutagenesis of a variety of mycobacterial strains. The phasmids can be prepared and grown in *M. smegmatis* at 30°C, but then used to infect *M. smegmatis* or *M. tuberculosis* at the nonpermissive temperature and selection for transposon mutants on solid media (Bardarov *et al.*, 1997).

5. Specialized transducing phages

Conditionally replicating mycobacteriophages also provide a powerful approach to the delivery of allelic exchange substrates for constructing mycobacterial mutants, including gene knockout and gene replacement mutants (Bardarov *et al.*, 2002). The approach is similar to that for transposon delivery, and construction of a phasmid carrying a DNA substrate in which an antibiotic resistance marker is flanked by 500–1000 bp corresponding to the flanking sequences of the gene to be replaced. Following infection, gene replacement mutants can be selected by antibiotic resistance. Because effectively every cell can be infected by the phage, the number of recombinants should be very high, even if gene replacement occurs in only a relatively small proportion of cells. In practice, only perhaps 10^{-6} or fewer cells generate recombinants, although a very high proportion of these result from homologous recombination at the intended site (Bardarov *et al.*, 1997), in contrast to the high proportion of illegitimate events observed when introducing linear DNA fragments by electroporation (Kalpana *et al.*, 1991). The reason why the recovery of recombinants is relatively inefficient is not known, although it suggests that there is a substantial opportunity to increase the recovery of the number of recombinants.

6. Mycobacterial recombineering

Recombineering [genetic engineering using recombination (Court *et al.*, 2002)] offers a general approach to constructing mutant bacterial derivatives by taking advantage of the high frequencies of homologous recombination that can be accomplished by the expression of phage-encoded recombination systems. Perhaps the most widely used system in *E. coli* is the λ-encoded Red system in which three proteins, Exo, Beta, and Gam, contribute to recombination proficiency. Exo is an exonuclease that degrades one strand of dsDNA substrates, Beta is a protein that promotes pairing of complementary DNA strands, and Gam is an inhibitor of RecBCD (Court *et al.*, 2002). When either dsDNA or short ssDNA substrates are introduced into *E. coli* by electroporation, recombination with a chromosomal or plasmid target occurs efficiently; in some configurations, desired recombinants can be identified even without selection. Similar systems have been described that utilize the RecET system encoded by the *E. coli rac* prophage (Murphy, 1998; Zhang *et al.*, 1998).

The *E. coli* recombineering systems do not function well in mycobacteria, especially when using dsDNA substrates (van Kessel and Hatfull, 2007, 2008a). Mycobacterial-specific recombineering systems have been developed using mycobacteriophage-encoded recombinases, especially those related to the RecET systems (van Kessel and Hatfull, 2007, 2008a, b; van Kessel *et al.*, 2008), such as genes *60* and *61* of phage Che9c (Fig. 12). When both Che9c gp60 and gp61 are expressed from an inducible expression system in *M. smegmatis* or *M. tuberculosis*, recombination frequencies are elevated substantially. Introduction of a dsDNA allelic exchange substrate in which 500–1000 bp of chromosomal homology flank an antibiotic resistance marker, followed by selection, generates recombinants efficiently (van Kessel and Hatfull, 2007). dsDNA recombineering works well and reproducibly in *M. smegmatis*, but anecdotal reports suggest that it may be somewhat more erratic in *M. tuberculosis*, perhaps due to irreproducibility of efficient expression of the recombinases.

Recombineering using ssDNA substrate requires only short synthetic oligonucleotide-derived substrates, provided that mutations are introduced that confer a selectable phenotype (van Kessel and Hatfull, 2008a). Interestingly, in both *M. smegmatis* and *M. tuberculosis* there is a very substantial strand bias, such that oligonucleotides with complementary sequences can yield recombinants at frequencies differing by more than 10^4-fold (van Kessel and Hatfull, 2008a). For engineering purposes it is therefore important that the most efficient of the two possible oligonucleotides is used, which is usually that corresponding to the leading strand of chromosomal DNA replication (i.e., can base pair with the template for lagging strand synthesis). ssDNA recombineering can be used to generate recombinants in the absence of direct selection using coelectroporation of two oligonucleotides: one designed to introduce the desired mutation and one that can be used for selection. A high proportion of selected recombinants also carry the unselected mutation and can be detected by physical screening (van Kessel and Hatfull, 2008a).

Recombineering provides an especially powerful tool for genetic manipulation of the mycobacteriophages themselves (Marinelli *et al.*, 2008; van Kessel *et al.*, 2008). The Bacteriophage Recombineering of Electroporated DNA (BRED) system involves coelectroporation of a phage genomic DNA substrate and a short (~200 bp) dsDNA substrate in a strain in which recombineering functions have been induced. Plaques can then be recovered on solid media in an infectious center configuration in which each electroporated cell that has taken up phage DNA gives rise to a plaque. When individual plaques are screened for the presence of either wild-type or mutant alleles at the targeted site, all contain the wild-type allele, but 10% or more also contain the mutant allele (Marinelli *et al.*, 2008). The desired phage mutant can then be recovered from this mixed primary plaque by replating and testing individual secondary plaques. In this way, two rounds of polymerase chain reaction analysis of 12–18 plaques typically generates the desired mutant, provided that the mutant is viable. In at least some cases, nonviable plaques can be recovered by complementation (Marinelli *et al.*, 2008; Payne *et al.*, 2009). BRED can be used to introduce insertions, deletions, and point mutations into mycobacteriophage genomes (Marinelli *et al.*, 2008).

B. Clinical tools

1. Phage-based diagnosis of *M. tuberculosis*

The ability of mycobacteriophages to infect mycobacterial hosts specifically and efficiently has led to three types of systems for the diagnosis of *M. tuberculosis* infections. There is a particular need for such systems because the diagnosis of human tuberculosis is complicated by the slow growth of the bacteria, the need to determine drug susceptibility profiles, and the fact that the demographic and geographic areas of greatest need often have only minimal resources to devote to this issue. An inexpensive, rapid, simple diagnostic system for drug susceptibility testing of *M. tuberculosis* is therefore highly desirable.

The first phage-based diagnostic developed was the phage-typing approach in which a substantial number of mycobacteriophages were isolated whose host ranges were informative about the identity of any unknown host (Engel, 1975; Redmond and Ward, 1966). In this way, an unknown clinical isolate could be tested for susceptibility to a set of phages and preliminary identification was obtained within a few days. Of particular note in this regard is the use of phage DS6A, whose host range is restricted to bacteria of the *M. tuberculosis* complex, including *Mycobacterium bovis*, *Mycobacterium africanum*, *Mycobacterium canetti*, and *Mycobacterium microti* (Bowman, 1969; Jones, 1975). DS6A has not yet been characterized genomically. Although phage typing is useful for strain identification, it does not readily provide information about drug susceptibility profiles.

A second phage-based diagnostic system is the phage amplification biological assay (PhaB), which is based on the ability of mycobacteriophages to infect and amplify in *M. tuberculosis* if present in a clinical sample, followed by enumeration of particles using *M. smegmatis* as a host (Eltringham *et al.*, 1999; Watterson *et al.*, 1998; Wilson *et al.*, 1997).

Phage D29 has been the primary focus for this approach because it infects both *M. tuberculosis* and *M. smegmatis* and produces large, clear, easily identifiable plaques. The system has been evaluated with clinical specimens in several studies and has been used to discern rifampicin-resistant and rifampicin-sensitive hosts (Albert *et al.*, 2001, 2002a,b, 2004; McNerney *et al.*, 2000; Pai *et al.*, 2005). The third approach is the use of reporter mycobacteriophages in which recombinant phages carrying a reporter gene, such as firefly luciferase (FFlux) or GPF (or related fluorescence genes), can be used to detect the physiological status of the cell rapidly, thus reporting on drug susceptibilities (Jacobs *et al.*, 1993; Piuri *et al.*, 2009). These phages can be constructed readily using either shuttle phasmid technology or BRED recombineering (Marinelli *et al.*, 2008) and can be used in several configurations depending on the reporter gene used and the detection technology available. Fluoromycobacteriophages have some notable advantages in that it is possible to detect single cells following infection and the signal is retained after fixation, providing additional biosafety and assay flexibility (Piuri *et al.*, 2009). The assay is rapid, and the use of light-emitting, diode-based microscopes provides a potentially simple clinical configuration. Establishment of efficient phage infection conditions directly in sputum samples remains the highest priority for direct clinical evaluation. Although reporter phages have been derived from TM4 (Jacobs *et al.*, 1993), D29 (Pearson *et al.*, 1996), and L5 (Sarkis *et al.*, 1995), none of these are specific to *M. tuberculosis*, and as with the PhaB assay, use of *M. tuberculosis*-specific phages would be advantageous.

2. Phage therapy?

Mycobacteriophages would seem to have some advantages for direct therapeutic treatment of pulmonary tuberculosis, especially in circumstances in which MDR-TB and XDR-TB infections respond poorly to antibiotic therapy. Delivery directly to the lung would seem feasible, and there are reports of evaluation in animal model systems (Koz'min-Sokolov and Vabilin, 1975; Sula *et al.*, 1981). The potential disadvantage is that the phage particles may not gain access to bacteria that are intracellular, or contained with granulomas, and therefore a therapeutic cure would seem improbable. Bronxmeyer and colleagues (2002) have explored successfully the possibility of using *M. smegmatis* as a surrogate to deliver TM4 to *M. tuberculosis*-infected macrophages, suggesting a novel route to killing intracellular bacteria (Broxmeyer *et al.*, 2002) and circumventing this problem. Phage resistance poses another potential concern, which could potentially be overcome by using either serial applications of phages to which different mechanisms of phage resistance occur or phage cocktails with broad combinations of phages. The actual number of phages currently available for such an application is rather small, with D29 being the most attractive candidate. If phage therapy is to be evaluated, it will be important to identify additional mycobacteriophages that infect both *M. tuberculosis* and *M. smegmatis* (for propagation purposes), kill a very high proportion of bacterial cells upon infection, and represent different patterns of host resistance responses. A related application is the possibility of using mycobacteriophages to interfere with active dissemination of tuberculosis from an actively infected person to household contacts, family members, and/or co-workers. Because dissemination likely involves forms of bacteria susceptible to phage infection, application of a suitable phage preparation by inhalation, aspiration, or nebulization could reduce the number of *M. tuberculosis* cells passing through the upper respiratory tract greatly and reduce the chances of transfer to an uninfected individual. An especially attractive configuration would be to use phages in a prophylactic form to protect those in close contact from acquisition from a patient, while enabling the infected person to undergo a normal course of antibiotic therapy. This would also minimize opportunities for the selection of phage-resistant mutants because the number of bacteria in contact with the phage is relatively small. However, success of this approach is anticipated to depend on the ability to deliver an effective quantity of phage particles, stability of the particles, and the likelihood that multiple doses over a period of time will be required for maximum effectiveness.

VIII. FUTURE DIRECTIONS

As the collection of sequenced mycobacteriophage genomes has grown, it has become abundantly clear how much we really do not understand about this fascinating group of viruses. For the most part, future directions are reasonably clear, and five major paths can be envisaged.

First, it is clear that much more needs to be learned about the genetic diversity of mycobacteriophages. As more mycobacteriophage genomes are sequenced, the numbers of more closely related phages have grown, but entirely new genomes continue to emerge, as well as phages related to those classified previously as singletons. The combination of an immensely powerful and high-impact integrated research and education platform for phage isolation, and the dramatic decline in genomic sequencing costs, will help fuel this ongoing effort in mycobacteriophage genome and discovery. It is not unreasonable to suppose that the collection of sequenced mycobacteriophage genomes could rise to more than 1000 within the next 5years. Presumably, at some point we will reach the point of genomic saturation where further genomic sequencing will provide diminishing returns, but it is unclear when that will be reached. We note that although isolating entirely new genomes is thrilling, the collection of groups of related phages provides powerful resources for understanding the detailed mechanisms of genome evolution. Current phages all share a common host in *M. smegmatis* mc^2155, but preliminary observations (C. Bowman and Graham F. Hatfull) show that some of these can discriminate between different substrains of *M. smegmatis*. It is thus likely that use of other *M. smegmatis* strains for phage isolation or different mycobacterial species will yet further expand the amazing diversity of the mycobacteriophage population. Moreover, it is critical that additional phages that infect *M. tuberculosis* be isolated using *M. tuberculosis* itself either as a host or as a surrogate that is much more closely related to it than *M. smegmatis*. Second, it will be important to establish the detailed host specificities of the sequenced mycobacteriophages. A plausible reason for their great diversity in nucleotide sequence, genome length, and GC% is that they share different but overlapping host ranges, with *M. smegmatis* mc^2155 being the common host. One approach would be to test the susceptibilities of known strains within Actinomycetales for infection by the mycobacteriophages, although it is important to recognize that these may poorly reflect the full diversity of bacteria in the environments from which the phages are isolated. Another approach would be to characterize the bacterial population of the samples from where phages are isolated more extensively, although this is complicated by the massive complexity of the soil biome and the likelihood that many of them, including potential mycobacteriophage hosts, are not cultivatable.

A third major area of focus should be on determination of what the many unknown mycobacteriophage genes do, using both functional genomic and structural genomic approaches. The BRED engineering technology provides a powerful means of constructing defined phage mutants, including gene knockouts and point mutations, and large numbers of mutants can be generated and characterized readily. Thus it is now possible to apply functional genomic approaches to whole genomes and to dissect them genetically. Structural genomic approaches will also be useful, especially as many of the phage-encoded genes are small, and the encoded proteins should be amenable to structural analysis by crystallography and nuclear magnetic resonance. Structural information should provide clues as to potential functions, but phage proteins may also be a rich source of novel protein folds, especially given their vast sequence diversity. These functional genomic and structural approaches are immensely powerful, although the sheer number of genes to analyze makes this an important but daunting prospect.

The fourth major direction is to characterize the patterns of mycobacteriophage genome expression, identify the signals for transcription initiation and termination, and elucidate the mechanisms of gene regulation. Little is known about the global patterns of mycobacteriophage gene expression, although the genomes should be amenable to transcriptome analysis using either microarrays or high throughput RNA-seq. In only a small number of examples have putative promoters been identified, and it is clear that many promoters cannot be readily identified bioinformatically. Investigation of gene expression and its regulation is expected to be especially rewarding, as previous studies reveal an abundance of novelty, such as with the remarkable stoperator system in Cluster A phages, and new systems for lytic–lysogenic decision systems in Cluster G phages. Moreover, it seems likely that some mycobacteriophage-encoded proteins are expressed from prophages with the capacity to influence host physiology, and a combination of expression and functional studies may provide important clues, especially in examples where phage-encoded proteins may influence pathogenicity. Finally, there are numerous potential routes to exploit the mycobacteriophages to develop additional tools for mycobacterial genetics. These range from additional integration-proficient vectors with novel *attB* target specificities, a suite of repressor-mediated selectable markers, and regulated expression systems to mycobacteriophage-specific packaging systems, mycobacteriophage-based antigen display systems, new tools for mutagenesis, and applications for diagnosis and therapy. The world is your oyster.

ACKNOWLEDGMENTS

I thank all of my colleagues in Pittsburgh for their long-standing and ongoing collaborations, including Roger Hendrix, Jeffrey Lawrence, Craig Peebles, Deborah Jacobs-Sera, Welkin Pope, Dan Russell, Bekah Dedrick, Greg Broussard, Anil Ojha, and Pallavi Ghosh, and the many graduate students and research assistants who have contributed to this work. I also thank Dr. Bill Jacobs and his colleagues at Albert Einstein College of Medicine and my colleagues at the HHMI Science Education Alliance, including Tuajuanda Jordan, Lucia Barker, Kevin Bradley, and Razi Khaja. I am especially grateful to the large number of individual high school and undergraduate phage hunters both at Pittsburgh and in the SEA-PHAGES programs that have contributed broadly to the advancement of our understanding of the mycobacteriophages. I extend special thanks to Dan Russell for help with Figure 2 and the GC% analysis and to Deborah Jacobs-Sera, Welkin Pope, and Roger Hendrix for helpful comments on the manuscript.

REFERENCES

Abedon, S. T. (2009). Phage evolution and ecology. *Adv. Appl. Microbiol.* **67**:1–45.

Albert, H., Heydenrych, A., Brookes, R., Mole, R. J., Harley, B., Subotsky, E., Henry, R., and Azevedo, V. (2002a). Performance of a rapid phage-based test, FASTPlaqueTB, to diagnose pulmonary tuberculosis from sputum specimens in South Africa. *Int. J. Tuberc. Lung Dis.* **6**(6):529–537.

Albert, H., Heydenrych, A., Mole, R., Trollip, A., and Blumberg, L. (2001). Evaluation of FASTPlaqueTB-RIF, a rapid, manual test for the determination of rifampicin resistance from *Mycobacterium tuberculosis* cultures. *Int. J. Tuberc. Lung Dis.* **5**(10):906–911.

Albert, H., Trollip, A., Seaman, T., and Mole, R. J. (2004). Simple, phage-based (FASTPplaque) technology to determine rifampicin resistance of *Mycobacterium tuberculosis* directly from sputum. *Int. J. Tuberc. Lung Dis.* **8**(9):1114–1119.

Albert, H., Trollip, A. P., Mole, R. J., Hatch, S. J., and Blumberg, L. (2002b). Rapid indication of multidrug-resistant tuberculosis from liquid cultures using FASTPlaqueTB-RIF, a manual phage-based test. *Int. J. Tuberc. Lung Dis.* **6**(6):523–528.

Aniukwu, J., Glickman, M. S., and Shuman, S. (2008). The pathways and outcomes of mycobacterial NHEJ depend on the structure of the broken DNA ends. *Genes Dev.* **22**(4):512–527.

Bandhu, A., Ganguly, T., Chanda, P. K., Das, M., Jana, B., Chakrabarti, G., and Sau, S. (2009). Antagonistic effects Na+ and Mg2+ on the structure, function, and stability of mycobacteriophage L1 repressor. *BMB Rep.* **42**(5):293–298.

Bandhu, A., Ganguly, T., Jana, B., Mondal, R., and Sau, S. (2010). Regions and residues of an asymmetric operator DNA interacting with the monomeric repressor of temperate mycobacteriophage L1. *Biochemistry* **49**(19):4235–4243.

Bardarov, S., Bardarov, S., Jr., Pavelka, M. S., Jr., Sambandamurthy, V., Larsen, M., Tufariello, J., Chan, J., Hatfull, G., and Jacobs, W. R., Jr. (2002). Specialized transduction: An efficient method for generating marked and unmarked targeted gene disruptions in *Mycobacterium tuberculosis. M. bovis BCG and M. smegmatis. Microbiology* **148** (Pt 10):3007–3017.

Bardarov, S., Kriakov, J., Carriere, C., Yu, S., Vaamonde, C., McAdam, R. A., Bloom, B. R., Hatfull, G. F., and Jacobs, W. R., Jr. (1997). Conditionally replicating mycobacteriophages: A system for transposon delivery to *Mycobacterium tuberculosis. Proc. Natl. Acad. Sci. USA* **94**(20):10961–10966.

Barsom, E. K., and Hatfull, G. F. (1996). Characterization of *Mycobacterium smegmatis* gene that confers resistance to phages L5 and D29 when overexpressed. *Mol. Microbiol.* **21**(1):159–170.

Berry, J., Summer, E. J., Struck, D. K., and Young, R. (2008). The final step in the phage infection cycle: The Rz and Rz1 lysis proteins link the inner and outer membranes. *Mol. Microbiol.* **70**(2):341–351.

Bhattacharya, B., Giri, N., Mitra, M., and Gupta, S. K. (2008). Cloning, characterization and expression analysis of nucleotide metabolism-related genes of mycobacteriophage L5. *FEMS Microbiol. Lett.* **280**(1):64–72.

Bibb, L. A., Hancox, M. I., and Hatfull, G. F. (2005). Integration and excision by the large serine recombinase phiRv1 integrase. *Mol. Microbiol.* **55**(6):1896–1910.

Bibb, L. A., and Hatfull, G. F. (2002). Integration and excision of the *Mycobacterium tuberculosis* prophage-like element, phiRv1. *Mol. Microbiol.* **45**(6):1515–1526.

Bisso, G., Castelnuovo, G., Nardelli, M. G., Orefici, G., Arancia, G., Laneelle, G., Asselineau, C., and Asselineau, J. (1976). A study on the receptor for a mycobacteriophage : phage phlei. *Biochimie* **58**(1–2):87–97.

Botstein, D., and Matz, M. J. (1970). A recombination function essential to the growth of bacteriophage P22. *J. Mol. Biol.* **54**(3):417–440.

Bowman, B., Jr. (1958). Quantitative studies on some mycobacterial phage-host systems. *J. Bacteriol.* **76**(1):52–62.

Bowman, B. U. (1969). Properties of mycobacteriophage DS6A. *I. Immunogenicity in rabbits. Proc. Soc. Exp. Biol. Med.* **131**(1):196–200.

Brown, K. L., Sarkis, G. J., Wadsworth, C., and Hatfull, G. F. (1997). Transcriptional silencing by the mycobacteriophage L5 repressor. *EMBO. J.* **16**(19):5914–5921.

Broxmeyer, L., Sosnowska, D., Miltner, E., Chacon, O., Wagner, D., McGarvey, J., Barletta, R. G., and Bermudez, L. E. (2002). Killing of *Mycobacterium avium* and *Mycobacterium tuberculosis* by a mycobacteriophage delivered by a nonvirulent mycobacterium: A model for phage therapy of intracellular bacterial pathogens. *J. Infect. Dis.* **186** (8):1155–1160.

Caruso, S. M., Sandoz, J., and Kelsey, J. (2009). Non-STEM undergraduates become enthusiastic phage-hunters. *CBE Life Sci. Educ.* **8**(4):278–282.

Casas, V., and Rohwer, F. (2007). *Phage metagenomics. Methods Enzymol.* **421:**259–268.

Catalao, M. J., Gil, F., Moniz-Pereira, J., and Pimentel, M. (2010). The mycobacteriophage Ms6 encodes a chaperone-like protein involved in the endolysin delivery to the peptidoglycan. *Mol. Microbiol.* **77**(3)**:**672–686.

Chattoraj, P., Ganguly, T., Nandy, R. K., and Sau, S. (2008). Overexpression of a delayed early gene hlg1 of temperate mycobacteriophage L1 is lethal to both *M. smegmatis* and *E. coli. BMB Rep.* **41**(5)**:**363–368.

Chen, J., Kriakov, J., Singh, A., Jacobs, W. R., Jr., Besra, G. S., and Bhatt, A. (2009). Defects in glycopeptidolipid biosynthesis confer phage I3 resistance in *Mycobacterium smegmatis. Microbiology* **155**(Pt 12)**:**4050–4057.

Cirillo, J. D., Barletta, R. G., Bloom, B. R., and Jacobs, W. R., Jr. (1991). A novel transposon trap for mycobacteria: isolation and characterization of IS1096. *J. Bacteriol.* **173**(24)**:**7772–7780.

Clark, A. J., Inwood, W., Cloutier, T., and Dhillon, T. S. (2001). Nucleotide sequence of coliphage HK620 and the evolution of lambdoid phages. *J. Mol. Biol.* **311**(4)**:**657–679.

Colangeli, R., Haq, A., Arcus, V. L., Summers, E., Magliozzo, R. S., McBride, A., Mitra, A. K., Radjainia, M., Khajo, A., Jacobs, W. R., Jr., Salgame, P., and Alland, D. (2009). The multifunctional histone-like protein Lsr2 protects mycobacteria against reactive oxygen intermediates. *Proc. Natl. Acad. Sci. USA* **106**(11)**:**4414–4418.

Colangeli, R., Helb, D., Vilcheze, C., Hazbon, M. H., Lee, C. G., Safi, H., Sayers, B., Sardone, I., Jones, M. B., Fleischmann, R. D., Peterson, S. N., Jacobs, W. R., Jr., *et al.* (2007). Transcriptional regulation of multi-drug tolerance and antibiotic-induced responses by the histone-like protein Lsr2 in *M. tuberculosis. PLoS Pathog* **3**(6)**:**e87.

Comeau, A. M., Hatfull, G. F., Krisch, H. M., Lindell, D., Mann, N. H., and Prangishvili, D. (2008). *Exploring the prokaryotic virosphere. Res. Microbiol.* **159**(5)**:**306–313.

Court, D. L., Sawitzke, J. A., and Thomason, L. C. (2002). Genetic engineering using homologous recombination. *Annu. Rev. Genet.* **36:**361–388.

Datta, H. J., Mandal, P., Bhattacharya, R., Das, N., Sau, S., and Mandal, N. C. (2007). The G23 and G25 genes of temperate mycobacteriophage L1 are essential for the transcription of its late genes. *J. Biochem. Mol. Biol.* **40**(2)**:**156–162.

Doke, S. (1960). Studies on mycobacteriophages and lysogenic mycobacteria. *J. Kumamoto Med. Soc.* **34:**1360–1373.

Donnelly-Wu, M. K., Jacobs, W. R., Jr., and Hatfull, G. F. (1993). Superinfection immunity of mycobacteriophage L5: Applications for genetic transformation of mycobacteria. *Mol. Microbiol.* **7**(3)**:**407–417.

Duda, R. L., Hempel, J., Michel, H., Shabanowitz, J., Hunt, D., and Hendrix, R. W. (1995). Structural transitions during bacteriophage HK97 head assembly. *J. Mol. Biol.* **247**(4)**:** 618–635.

Eltringham, I. J., Wilson, S. M., and Drobniewski, F. A. (1999). Evaluation of a bacteriophage-based assay (phage amplified biologically assay) as a rapid screen for resistance to isoniazid, ethambutol, streptomycin, pyrazinamide, and ciprofloxacin among clinical isolates of *Mycobacterium tuberculosis. J. Clin. Microbiol.* **37**(11)**:**3528–3532.

Engel, H. W. (1975). Phage typing of strains of "*M. tuberculosis*" in the Netherlands *Ann. Sclavo* **17**(4)**:**578–583.

Fineran, P. C., Blower, T. R., Foulds, I. J., Humphreys, D. P., Lilley, K. S., and Salmond, G. P. (2009). The phage abortive infection system, ToxIN, functions as a protein-RNA toxin-antitoxin pair. *Proc. Natl. Acad. Sci. USA* **106**(3)**:**894–899.

Fokine, A., Leiman, P. G., Shneider, M. M., Ahvazi, B., Boeshans, K. M., Steven, A. C., Black, L. W., Mesyanzhinov, V. V., and Rossmann, M. G. (2005). Structural and functional similarities between the capsid proteins of bacteriophages T4 and HK97 point to a common ancestry. *Proc. Natl. Acad. Sci. USA* **102**(20)**:**7163–7168.

Fomukong, N. G., and Dale, J. W. (1993). Transpositional activity of IS986 in *Mycobacterium smegmatis*. *Gene* **130**(1):99–105.

Ford, M. E., Sarkis, G. J., Belanger, A. E., Hendrix, R. W., and Hatfull, G. F. (1998a). Genome structure of mycobacteriophage D29: Implications for phage evolution. *J. Mol. Biol.* **279** (1):143–164.

Ford, M. E., Stenstrom, C., Hendrix, R. W., and Hatfull, G. F. (1998b). Mycobacteriophage TM4: Genome structure and gene expression. *Tuber. Lung Dis.* **79**(2):63–73.

Fraser, J. S., Yu, Z., Maxwell, K. L., and Davidson, A. R. (2006). Ig-like domains on bacteriophages: A tale of promiscuity and deceit. *J. Mol. Biol.* **359**(2):496–507.

Freitas-Vieira, A., Anes, E., and Moniz-Pereira, J. (1998). The site-specific recombination locus of mycobacteriophage Ms6 determines DNA integration at the tRNA(Ala) gene of *Mycobacterium* spp. *Microbiology* **144**(Pt 12):3397–3406.

Froman, S., Will, D. W., and Bogen, E. (1954). Bacteriophage active against *Mycobacterium tuberculosis. I. Isolation and activity. Am. J. Pub. Health* **44:**1326–1333.

Fullner, K. J., and Hatfull, G. F. (1997). Mycobacteriophage L5 infection of *Mycobacterium bovis* BCG: Implications for phage genetics in the slow-growing mycobacteria. *Mol. Microbiol.* **26**(4):755–766.

Furuchi, A., and Tokunaga, T. (1972). Nature of the receptor substance of *Mycobacterium smegmatis* for D4 bacteriophage adsorption. *J. Bacteriol.* **111**(2):404–411.

Ganguly, T., Bandhu, A., Chattoraj, P., Chanda, P. K., Das, M., Mandal, N. C., and Sau, S. (2007). Repressor of temperate mycobacteriophage L1 harbors a stable C-terminal domain and binds to different asymmetric operator DNAs with variable affinity. *Virol. J.* **4:**64.

Ganguly, T., Chanda, P. K., Bandhu, A., Chattoraj, P., Das, M., and Sau, S. (2006). Effects of physical, ionic, and structural factors on the binding of repressor of mycobacteriophage L1 to its cognate operator DNA. *Protein Pept. Lett.* **13**(8):793–798.

Ganguly, T., Chattoraj, P., Das, M., Chanda, P. K., Mandal, N. C., Lee, C. Y., and Sau, S. (2004). A point mutation at the C-terminal half of the repressor of temperate mycobacteriophage L1 affects its binding to the operator DNA. *J. Biochem. Mol. Biol.* **37**(6):709–714.

Garcia, M., Pimentel, M., and Moniz-Pereira, J. (2002). Expression of Mycobacteriophage Ms6 lysis genes is driven by two sigma(70)-like promoters and is dependent on a transcription termination signal present in the leader RNA. *J. Bacteriol.* **184**(11):3034–3043.

Ghosh, P., Bibb, L. A., and Hatfull, G. F. (2008). Two-step site selection for serine-integrase-mediated excision: DNA-directed integrase conformation and central dinucleotide proofreading. *Proc. Natl. Acad. Sci. USA* **105**(9):3238–3243.

Ghosh, P., Kim, A. I., and Hatfull, G. F. (2003). The orientation of mycobacteriophage Bxb1 integration is solely dependent on the central dinucleotide of attP and attB. *Mol. Cell.* **12** (5):1101–1111.

Ghosh, P., Pannunzio, N. R., and Hatfull, G. F. (2005). Synapsis in phage Bxb1 integration: Selection mechanism for the correct pair of recombination sites. *J. Mol. Biol.* **349**(2):331–348.

Ghosh, P., Wasil, L. R., and Hatfull, G. F. (2006). Control of phage Bxb1 Excision by a novel recombination directionality factor. *PLoS Biol.* **4**(6):e186.

Gil, F., Catalao, M. J., Moniz-Pereira, J., Leandro, P., McNeil, M., and Pimentel, M. (2008). The lytic cassette of mycobacteriophage Ms6 encodes an enzyme with lipolytic activity. *Microbiology* **154**(Pt 5):1364–1371.

Gil, F., Grzegorzewicz, A. E., Catalao, M. J., Vital, J., McNeil, M. R., and Pimentel, M. (2010). Mycobacteriophage Ms6 LysB specifically targets the outer membrane of *Mycobacterium smegmatis*. *Microbiology* **156**(Pt 5):1497–1504.

Giri, N., Bhowmik, P., Bhattacharya, B., Mitra, M., and Das Gupta, S. K. (2009). The mycobacteriophage D29 gene 65 encodes an early-expressed protein that functions as a structure-specific nuclease. *J. Bacteriol.* **191**(3):959–967.

Gomathi, N. S., Sameer, H., Kumar, V., Balaji, S., Dustackeer, V. N., and Narayanan, P. R. (2007). In silico analysis of mycobacteriophage Che12 genome: Characterization of genes required to lysogenise *Mycobacterium tuberculosis*. *Comput. Biol. Chem.* **31**(2)**:**82–91.

Guilhot, C., Gicquel, B., Davies, J., and Martin, C. (1992). Isolation and analysis of IS6120, a new insertion sequence from *Mycobacterium smegmatis*. *Mol. Microbiol.* **6**(1)**:**107–113.

Han, S., Craig, J. A., Putnam, C. D., Carozzi, N. B., and Tainer, J. A. (1999). Evolution and mechanism from structures of an ADP-ribosylating toxin and NAD complex. *Nat. Struct. Biol.* **6**(10)**:**932–936.

Hanauer, D. I., Jacobs-Sera, D., Pedulla, M. L., Cresawn, S. G., Hendrix, R. W., and Hatfull, G. F. (2006). Inquiry learning: Teaching scientific inquiry. *Science* **314**(5807)**:** 1880–1881.

Hassan, S., Mahalingam, V., and Kumar, V. (2009). Synonymous codon usage analysis of thirty two mycobacteriophage genomes. *Adv Bioinformatics*, 1–11.

Hatfull, G. F. (1994). Mycobacteriophage L5: A toolbox for tuberculosis. *ASM News* **60:**255–260.

Hatfull, G. F. (1999). Mycobacteriophages. *In* "Mycobacteria: Molecular Biology and Virulence" (C. Ratledge and J. Dale, eds.), pp. 38–58. Chapman and Hall, London.

Hatfull, G. F. (2000). Molecular genetics of mycobacteriophages. *In* "Molecular Genetics of the Mycobacteria" (G. F. Hatfull and W. R. Jacobs, Jr., eds.), pp. 37–54. ASM Press, Washington, DC.

Hatfull, G. F. (2004). Mycobacteriophages and tuberculosis. *In* "tuberculosis" (K. Eisenach, S. T. Cole, W. R. Jacobs, Jr., and D. McMurray, eds.), pp. 203–218. ASM Press, Washington, DC.

Hatfull, G. F. (2006). Mycobacteriophages. *In* "The Bacteriophages" (R. Calendar, ed.), pp. 602–620. Oxford University Press, New York.

Hatfull, G. F. (2008). *Bacteriophage genomics. Curr. Opin. Microbiol.* **11**(5)**:**447–453.

Hatfull, G. F. (2010). Mycobacteriophages: Genes and genomes. *Annu. Rev. Microbiol.* **64:**331–356.

Hatfull, G. F., Barsom, L., Chang, L., Donnelly-Wu, M., Lee, M. H., Levin, M., Nesbit, C., and Sarkis, G. J. (1994). Bacteriophages as tools for vaccine development. *Dev. Biol. Stand.* **82:**43–47.

Hatfull, G. F., Cresawn, S. G., and Hendrix, R. W. (2008). Comparative genomics of the mycobacteriophages: Insights into bacteriophage evolution. *Res. Microbiol.* **159**(5)**:** 332–339.

Hatfull, G. F., and Jacobs, W. R., Jr. (2000). Molecular Genetics of the Mycobacteria. ASM Press, Washington, DC.

Hatfull, G. F., and Jacobs, W. R., Jr. (1994). Mycobacteriophages: Cornerstones of mycobacterial research. *In* "Tuberculosis: Pathogenesis, Protection and Control" (B. R. Bloom, ed.), pp. 165–183. ASM, Washington, DC.

Hatfull, G. F., Jacobs-Sera, D., Lawrence, J. G., Pope, W. H., Russell, D. A., Ko, C. C., Weber, R. J., Patel, M. C., Germane, K. L., Edgar, R. H., Hoyte, N. N., Bowman, C. A., *et al.* (2010). Comparative genomic analysis of 60 mycobacteriophage genomes: Genome clustering, gene acquisition, and gene size. *J. Mol. Biol.* **397**(1)**:**119–143.

Hatfull, G. F., Pedulla, M. L., Jacobs-Sera, D., Cichon, P. M., Foley, A., Ford, M. E., Gonda, R. M., Houtz, J. M., Hryckowian, A. J., Kelchner, V. A., Namburi, S., Pajcini, K. V., *et al.* (2006). Exploring the mycobacteriophage metaproteome: Phage genomics as an educational platform. *PLoS Genet.* **2**(6)**:**e92.

Hatfull, G. F., and Sarkis, G. J. (2006). DNA sequence, structure and gene expression of mycobacteriophage L5: A phage system for mycobacterial genetics. *Mol. Microbiol.* **7**(3)**:** 395–405.

Hayes, C. S., and Keiler, K. C. (2010). Beyond ribosome rescue: tmRNA and co-translational processes. *FEBS Lett.* **584**(2)**:**413–419.

Hendrix, R. W. (2002). Bacteriophages: Evolution of the majority. *Theor. Popul. Biol.* **61**(4): 471–480.

Hendrix, R. W. (2003). *Bacteriophage genomics. Curr. Opin. Microbiol.* **6**(5):506–511.

Hendrix, R. W. (2005). Bacteriophage HK97: Assembly of the capsid and evolutionary connections. *Adv. Virus Res.* **64**:1–14.

Hendrix, R. W., Lawrence, J. G., Hatfull, G. F., and Casjens, S. (2000). The origins and ongoing evolution of viruses. *Trends Microbiol.* **8**(11):504–508.

Hendrix, R. W., Smith, M. C., Burns, R. N., Ford, M. E., and Hatfull, G. F. (1999). Evolutionary relationships among diverse bacteriophages and prophages: All the world's a phage. *Proc. Natl. Acad. Sci. USA* **96**(5):2192–2197.

Henry, M., Begley, M., Neve, H., Maher, F., Ross, R. P., McAuliffe, O., Coffey, A., and O'Mahony, J. M. (2010a). Cloning and expression of a mureinolytic enzyme from the mycobacteriophage TM4. *FEMS Microbiol. Lett.* **311**(2):126–132.

Henry, M., O'Sullivan, O., Sleator, R. D., Coffey, A., Ross, R. P., McAuliffe, O., and O'Mahony, J. M. (2010b). In silico analysis of Ardmore, a novel mycobacteriophage isolated from soil. *Gene* **453**(1–2):9–23.

Huff, J., Czyz, A., Landick, R., and Niederweis, M. (2010). Taking phage integration to the next level as a genetic tool for mycobacteria. *Gene* **468**(1–2):8–19.

Jacobs, W. R., Jr. (1992). Advances in mycobacterial genetics: New promises for old diseases. *Immunobiology* **184**(2–3):147–156.

Jacobs, W. R., Jr. (2000). *Mycobacterium tuberculosis*: A once genetically intractable organism. *In* "Molecular Genetics of the Mycobacteria" (G. F. Hatfull and W. R. Jacobs, Jr., eds.), pp. 1–16. ASM Press, Washington, DC.

Jacobs, W. R., Jr., Barletta, R. G., Udani, R., Chan, J., Kalkut, G., Sosne, G., Kieser, T., Sarkis, G. J., Hatfull, G. F., and Bloom, B. R. (1993). Rapid assessment of drug susceptibilities of *Mycobacterium tuberculosis* by means of luciferase reporter phages. *Science* **260**(5109):819–822.

Jacobs, W. R., Jr., Kalpana, G. V., Cirillo, J. D., Pascopella, L., Snapper, S. B., Udani, R. A., Jones, W., Barletta, R. G., and Bloom, B. R. (1991). *Genetic systems for mycobacteria. Methods Enzymol.* **204**:537–555.

Jacobs, W. R., Jr., Snapper, S. B., Tuckman, M., and Bloom, B. R. (1989). *Mycobacteriophage vector systems. Rev. Infect. Dis.* **11**(Suppl. 2):S404–S410.

Jacobs, W. R., Jr., Tuckman, M., and Bloom, B. R. (1987). Introduction of foreign DNA into mycobacteria using a shuttle phasmid. *Nature* **327**(6122):532–535.

Jain, S., and Hatfull, G. F. (2000). Transcriptional regulation and immunity in mycobacteriophage Bxb1. *Mol. Microbiol.* **38**(5):971–985.

Johnson, J. E. (2010). Virus particle maturation: Insights into elegantly programmed nanomachines. *Curr. Opin. Struct. Biol.* **20**(2):210–216.

Jones, W. D., Jr. (1975). Phage typing report of 125 strains of "Mycobacterium tuberculosis. *Ann. Sclavo* **17**(4):599–604.

Kalpana, G. V., Bloom, B. R., and Jacobs, W. R., Jr. (1991). Insertional mutagenesis and illegitimate recombination in mycobacteria. *Proc. Natl. Acad. Sci. USA* **88**(12):5433–5437.

Khoo, K. H., Suzuki, R., Dell, A., Morris, H. R., McNeil, M. R., Brennan, P. J., and Besra, G. S. (1996). Chemistry of the lyxose-containing mycobacteriophage receptors of *Mycobacterium phlei/Mycobacterium smegmatis. Biochemistry* **35**(36):11812–11819.

Kim, A. I., Ghosh, P., Aaron, M. A., Bibb, L. A., Jain, S., and Hatfull, G. F. (2003). Mycobacteriophage Bxb1 integrates into the *Mycobacterium smegmatis* groEL1 gene. *Mol. Microbiol.* **50**(2):463–473.

Kleckner, N., Bender, J., and Gottesman, S. (1991). Uses of transposons with emphasis on Tn10. *Methods Enzymol.* **204**:139–180.

Koz'min-Sokolov, B. N., and Vabilin, (1975). Effect of mycobacteriophages on the course of experimental tuberculosis in albino mice. *Probl. Tuberk* **4**:75–79.

Krisch, H. M., and Comeau, A. M. (2008). The immense journey of bacteriophage T4: From d'Herelle to Delbruck and then to Darwin and beyond. *Res. Microbiol.* **159**(5):314–324.
Krumsiek, J., Arnold, R., and Rattei, T. (2007). Gepard: A rapid and sensitive tool for creating dotplots on genome scale. *Bioinformatics* **23**(8):1026–1028.
Kumar, V., Loganathan, P., Sivaramakrishnan, G., Kriakov, J., Dusthakeer, A., Subramanyam, B., Chan, J., Jacobs, W. R., Jr., and Paranji Rama, N. (2008). Characterization of temperate phage Che12 and construction of a new tool for diagnosis of tuberculosis. *Tuberculosis (Edinb.)* **88**(6):616–623.
Kunisawa, T. (2000). Functional role of mycobacteriophage transfer RNAs. *J. Theor. Biol.* **205** (1):167–170.
Lai, X., Weng, J., Zhang, X., Shi, W., Zhao, J., and Wang, H. (2006). MSTF: A domain involved in bacterial metallopeptidases and surface proteins, mycobacteriophage tape-measure proteins and fungal proteins. *FEMS Microbiol. Lett.* **258**(1):78–82.
Lamichhane, G., Zignol, M., Blades, N. J., Geiman, D. E., Dougherty, A., Grosset, J., Broman, K. W., and Bishai, W. R. (2003). A postgenomic method for predicting essential genes at subsaturation levels of mutagenesis: Application to Mycobacterium tuberculosis. *Proc. Natl. Acad. Sci. USA* **100**(12):7213–7218.
Lawrence, J. G., Hatfull, G. F., and Hendrix, R. W. (2002). Imbroglios of viral taxonomy: Genetic exchange and failings of phenetic approaches. *J. Bacteriol.* **184**(17):4891–4905.
Lee, M. H., and Hatfull, G. F. (1993). Mycobacteriophage L5 integrase-mediated site-specific integration in vitro. *J. Bacteriol.* **175**(21):6836–6841.
Lee, M. H., Pascopella, L., Jacobs, W. R., Jr., and Hatfull, G. F. (1991). Site-specific integration of mycobacteriophage L5: Integration-proficient vectors for *Mycobacterium smegmatis, Mycobacterium tuberculosis,* and bacille Calmette-Guerin. *Proc. Natl. Acad. Sci. USA* **88** (8):3111–3115.
Lee, S., Kriakov, J., Vilcheze, C., Dai, Z., Hatfull, G. F., and Jacobs, W. R., Jr. (2004). Bxz1, a new generalized transducing phage for mycobacteria. *FEMS Microbiol. Lett.* **241**(2): 271–276.
Lehnherr, H., Maguin, E., Jafri, S., and Yarmolinsky, M. B. (1993). Plasmid addiction genes of bacteriophage P1: doc, which causes cell death on curing of prophage, and phd, which prevents host death when prophage is retained. *J. Mol. Biol.* **233**(3):414–428.
Levin, M. E., Hendrix, R. W., and Casjens, S. R. (1993). A programmed translational frameshift is required for the synthesis of a bacteriophage lambda tail assembly protein. *J. Mol. Biol.* **234**(1):124–139.
Lewis, J. A., and Hatfull, G. F. (2000). Identification and characterization of mycobacteriophage L5 excisionase. *Mol. Microbiol.* **35**(2):350–360.
Lewis, J. A., and Hatfull, G. F. (2001). Control of directionality in integrase-mediated recombination: Examination of recombination directionality factors (RDFs) including Xis and Cox proteins. *Nucleic Acids Res.* **29**(11):2205–2216.
Lewis, J. A., and Hatfull, G. F. (2003). Control of directionality in L5 integrase-mediated site-specific recombination. *J. Mol. Biol.* **326**(3):805–821.
Lima-Mendez, G., Toussaint, A., and Leplae, R. (2007). Analysis of the phage sequence space: The benefit of structured information. *Virology* **365**(2):241–249.
Lima-Mendez, G., Van Helden, J., Toussaint, A., and Leplae, R. (2008). Reticulate representation of evolutionary and functional relationships between phage genomes. *Mol. Biol. Evol.* **25**(4):762–777.
Marinelli, L. J., Piuri, M., Swigonova, Z., Balachandran, A., Oldfield, L. M., van Kessel, J. C., and Hatfull, G. F. (2008). BRED: A simple and powerful tool for constructing mutant and recombinant bacteriophage genomes. *PLoS One* **3**(12):e3957.
Martin, C., Mazodier, P., Mediola, M. V., Gicquel, B., Smokvina, T., Thompson, C. J., and Davies, J. (1991). Site-specific integration of the Streptomyces plasmid pSAM2 in *Mycobacterium smegmatis. Mol. Microbiol.* **5**(10):2499–2502.

Martinsohn, J. T., Radman, M., and Petit, M. A. (2008). The lambda red proteins promote efficient recombination between diverged sequences: Implications for bacteriophage genome mosaicism. *PLoS Genet.* **4**(5)**:**e1000065.

McNerney, R. (1999). TB: The return of the phage: A review of fifty years of mycobacteriophage research. *Int. J. Tuberc. Lung Dis.* **3**(3)**:**179–184.

McNerney, R., Kiepiela, P., Bishop, K. S., Nye, P. M., and Stoker, N. G. (2000). Rapid screening of Mycobacterium tuberculosis for susceptibility to rifampicin and streptomycin. *Int. J. Tuberc. Lung Dis.* **4**(1)**:**69–75.

McNerney, R., and Traore, H. (2005). Mycobacteriophage and their application to disease control. *J. Appl. Microbiol* **99**(2)**:**223–233.

Mediavilla, J., Jain, S., Kriakov, J., Ford, M. E., Duda, R. L., Jacobs, W. R., Jr., Hendrix, R. W., and Hatfull, G. F. (2000). Genome organization and characterization of mycobacteriophage Bxb1. *Mol. Microbiol.* **38**(5)**:**955–970.

Mizuguchi, Y. (1984). Mycobacteriophages. *In* "The Mycobacteria: A Sourcebook" (G. P. Kubica and L. G. Wayne, eds.), Vol. Part A, pp. 641–662. Marcel Dekker, New York.

Morris, P., Marinelli, L. J., Jacobs-Sera, D., Hendrix, R. W., and Hatfull, G. F. (2008). Genomic characterization of mycobacteriophage Giles: Evidence for phage acquisition of host DNA by illegitimate recombination. *J. Bacteriol.* **190**(6)**:**2172–2182.

Murphy, K. C. (1998). Use of bacteriophage lambda recombination functions to promote gene replacement in *Escherichia coli*. *J. Bacteriol.* **180**(8)**:**2063–2071.

Murry, J., Sassetti, C. M., Moreira, J., Lane, J., and Rubin, E. J. (2005). A new site-specific integration system for mycobacteria. *Tuberculosis (Edinb.)* **85**(5–6)**:**317–323.

Nesbit, C. E., Levin, M. E., Donnelly-Wu, M. K., and Hatfull, G. F. (1995). Transcriptional regulation of repressor synthesis in mycobacteriophage L5. *Mol. Microbiol.* **17**(6)**:** 1045–1056.

Ojha, A., Anand, M., Bhatt, A., Kremer, L., Jacobs, W. R., Jr., and Hatfull, G. F. (2005). GroEL1: A dedicated chaperone involved in mycolic acid biosynthesis during biofilm formation in mycobacteria. *Cell* **123**(5)**:**861–873.

Pai, M., Kalantri, S., Pascopella, L., Riley, L. W., and Reingold, A. L. (2005). Bacteriophage-based assays for the rapid detection of rifampicin resistance in Mycobacterium tuberculosis: A meta-analysis. *J. Infect.* **51**(3)**:**175–187.

Parish, T., Lewis, J., and Stoker, N. G. (2001). Use of the mycobacteriophage L5 excisionase in *Mycobacterium tuberculosis* to demonstrate gene essentiality. *Tuberculosis (Edinb.)* **81** (5–6)**:**359–364.

Pashley, C. A., and Parish, T. (2003). Efficient switching of mycobacteriophage L5-based integrating plasmids in *Mycobacterium tuberculosis*. *FEMS Microbiol. Lett.* **229**(2)**:**211–215.

Payne, K., Sun, Q., Sacchettini, J., and Hatfull, G. F. (2009). Mycobacteriophage Lysin B is a novel mycolylarabinogalactan esterase. *Mol. Microbiol.* **73**(3)**:**367–381.

Pearson, R. E., Jurgensen, S., Sarkis, G. J., Hatfull, G. F., and Jacobs, W. R., Jr. (1996). Construction of D29 shuttle phasmids and luciferase reporter phages for detection of mycobacteria. *Gene* **183**(1–2)**:**129–136.

Pedulla, M. L., Ford, M. E., Houtz, J. M., Karthikeyan, T., Wadsworth, C., Lewis, J. A., Jacobs-Sera, D., Falbo, J., Gross, J., Pannunzio, N. R., Brucker, W., Kumar, V., *et al.* (2003). Origins of highly mosaic mycobacteriophage genomes. *Cell* **113**(2)**:**171–182.

Pedulla, M. L., and Hatfull, G. F. (1998). Characterization of the mIHF gene of *Mycobacterium smegmatis*. *J. Bacteriol.* **180**(20)**:**5473–5477.

Pedulla, M. L., Lee, M. H., Lever, D. C., and Hatfull, G. F. (1996). A novel host factor for integration of mycobacteriophage L5. *Proc. Natl. Acad. Sci. USA* **93**(26)**:**15411–15416.

Pell, L. G., Gasmi-Seabrook, G. M., Morais, M., Neudecker, P., Kanelis, V., Bona, D., Donaldson, L. W., Edwards, A. M., Howell, P. L., Davidson, A. R., and Maxwell, K. L. (2010). The solution structure of the C-terminal Ig-like domain of the bacteriophage lambda tail tube protein. *J. Mol. Biol.* **403**(3)**:**468–479.

Pell, L. G., Kanelis, V., Donaldson, L. W., Howell, P. L., and Davidson, A. R. (2009). The phage lambda major tail protein structure reveals a common evolution for long-tailed phages and the type VI bacterial secretion system. *Proc. Natl. Acad. Sci. USA* **106** (11):4160–4165.

Peña, C. E., Kahlenberg, J. M., and Hatfull, G. F. (1998). The role of supercoiling in mycobacteriophage L5 integrative recombination. *Nucleic Acids Res.* **26**(17):4012–4018.

Peña, C. E., Kahlenberg, J. M., and Hatfull, G. F. (2000). Assembly and activation of site-specific recombination complexes. *Proc. Natl. Acad. Sci. USA* **97**(14):7760–7765.

Peña, C. E., Lee, M. H., Pedulla, M. L., and Hatfull, G. F. (1997). Characterization of the mycobacteriophage L5 attachment site, attP. *J. Mol. Biol.* **266**(1):76–92.

Peña, C. E., Stoner, J. E., and Hatfull, G. F. (1996). Positions of strand exchange in mycobacteriophage L5 integration and characterization of the attB site. *J. Bacteriol.* **178** (18):5533–5536.

Pham, T. T., Jacobs-Sera, D., Pedulla, M. L., Hendrix, R. W., and Hatfull, G. F. (2007). Comparative genomic analysis of mycobacteriophage Tweety: Evolutionary insights and construction of compatible site-specific integration vectors for mycobacteria. *Microbiology* **153**(Pt 8):2711–2723.

Pitcher, R. S., Brissett, N. C., and Doherty, A. J. (2007). Nonhomologous end-joining in bacteria: A microbial perspective. *Annu. Rev. Microbiol.* **61**:259–282.

Pitcher, R. S., Tonkin, L. M., Daley, J. M., Palmbos, P. L., Green, A. J., Velting, T. L., Brzostek, A., Korycka-Machala, M., Cresawn, S., Dziadek, J., Hatfull, G. F., Wilson, T. E., *et al.* (2006). Mycobacteriophage exploit NHEJ to facilitate genome circularization. *Mol. Cell* **23**(5):743–748.

Pitcher, R. S., Wilson, T. E., and Doherty, A. J. (2005). New insights into NHEJ repair processes in prokaryotes. *Cell Cycle* **4**(5):675–678.

Piuri, M., and Hatfull, G. F. (2006). A peptidoglycan hydrolase motif within the mycobacteriophage TM4 tape measure protein promotes efficient infection of stationary phase cells. *Mol. Microbiol.* **62**(6):1569–1585.

Piuri, M., Jacobs, W. R., Jr., and Hatfull, G. F. (2009). Fluoromycobacteriophages for rapid, specific, and sensitive antibiotic susceptibility testing of *Mycobacterium tuberculosis*. *PLoS ONE* **4**(3):e4870.

Popa, M. P., McKelvey, T. A., Hempel, J., and Hendrix, R. W. (1991). Bacteriophage HK97 structure: Wholesale covalent cross-linking between the major head shell subunits. *J. Virol.* **65**(6):3227–3237.

Pope, W. H., Jacobs-Sera, D., Russell, D. A., Peebles, C. L., Al-Atrache, Z., Alcoser, T. A., Alexander, L. M., Alfano, M. B., Alford, S. T., Amy, N. E., Anderson, M. D., Anderson, A. G., *et al.* (2011). Expanding the diversity of mycobacteriophages: Insights into genome architecture and evolution. *PLoS One* **6**(1):e16329.

Raj, C. V., and Ramakrishnan, T. (1970). Transduction in *Mycobacterium smegmatis*. *Nature* **228** (268):280–281.

Ramesh, G. R., and Gopinathan, K. P. (1994). Structural proteins of mycobacteriophage I3: Cloning, expression and sequence analysis of a gene encoding a 70-kDa structural protein. *Gene* **143**(1):95–100.

Redmond, W. B., and Ward, D. M. (1966). Media and methods for phage-typing mycobacteria. *Bull. World Health Organ.* **35**(4):563–568.

Rubin, E. J., Akerley, B. J., Novik, V. N., Lampe, D. J., Husson, R. N., and Mekalanos, J. J. (1999). In vivo transposition of mariner-based elements in enteric bacteria and mycobacteria. *Proc. Natl. Acad. Sci. USA* **96**(4):1645–1650.

Rybniker, J., Kramme, S., and Small, P. L. (2006). Host range of 14 mycobacteriophages in *Mycobacterium ulcerans* and seven other mycobacteria including *Mycobacterium tuberculosis*: Application for identification and susceptibility testing. *J. Med. Microbiol.* **55**(Pt 1):37–42.

Rybniker, J., Nowag, A., van Gumpel, E., Nissen, N., Robinson, N., Plum, G., and Hartmann, P. (2010). Insights into the function of the WhiB-like protein of mycobacteriophage TM4: A transcriptional inhibitor of WhiB2. *Mol. Microbiol.* **77**(3):642–657.

Rybniker, J., Plum, G., Robinson, N., Small, P. L., and Hartmann, P. (2008). Identification of three cytotoxic early proteins of mycobacteriophage L5 leading to growth inhibition in *Mycobacterium smegmatis*. *Microbiology* **154**(Pt 8):2304–2314.

Sahu, K., Gupta, S. K., and Ghosh, T. C. (2004). Synonymous codon usage analysis of the mycobacteriophage Bxz1 and its plating bacteria *M. smegmatis*: Identification of highly and lowly expressed genes of Bxz1 and the possible function of its tRNA species. (S. Sau, ed.), *J. Biochem. Mol. Biol.* **37**(4):487–492.

Sampson, T., Broussard, G. W., Marinelli, L. J., Jacobs-Sera, D., Ray, M., Ko, C. C., Russell, D., Hendrix, R. W., and Hatfull, G. F. (2009). Mycobacteriophages BPs, Angel and Halo: Comparative genomics reveals a novel class of ultra-small mobile genetic elements. *Microbiology* **155**(Pt 9):2962–2977.

Sarkis, G. J., Jacobs, W. R., Jr., and Hatfull, G. F. (1995). L5 luciferase reporter mycobacteriophages: A sensitive tool for the detection and assay of live mycobacteria. *Mol. Microbiol.* **15**(6):1055–1067.

Sassetti, C. M., Boyd, D. H., and Rubin, E. J. (2001). Comprehensive identification of conditionally essential genes in mycobacteria. *Proc. Natl. Acad. Sci. USA* **98**(22):12712–12717.

Sassetti, C. M., Boyd, D. H., and Rubin, E. J. (2003). Genes required for mycobacterial growth defined by high density mutagenesis. *Mol. Microbiol.* **48**(1):77–84.

Sau, S., Chattoraj, P., Ganguly, T., Lee, C. Y., and Mandal, N. C. (2004). Cloning and sequencing analysis of the repressor gene of temperate mycobacteriophage L1. *J. Biochem. Mol. Biol.* **37**(2):254–259.

Saviola, B., and Bishai, W. R. (2004). Method to integrate multiple plasmids into the mycobacterial chromosome. *Nucleic Acids Res.* **32**(1):e11.

Scollard, D. M., Adams, L. B., Gillis, T. P., Krahenbuhl, J. L., Truman, R. W., and Williams, D. L. (2006). The continuing challenges of leprosy. *Clin. Microbiol. Rev.* **19** (2):338–381.

Seoane, A., Navas, J., and Garcia Lobo, J. M. (1997). Targets for pSAM2 integrase-mediated site-specific integration in the *Mycobacterium smegmatis* chromosome. *Microbiology* **143** (Pt 10):3375–3380.

Sivanathan, V., Allen, M. D., de Bekker, C., Baker, R., Arciszewska, L. K., Freund, S. M., Bycroft, M., Lowe, J., and Sherratt, D. J. (2006). The FtsK gamma domain directs oriented DNA translocation by interacting with KOPS. *Nat. Struct. Mol. Biol.* **13**(11):965–972.

Smith, M. C., and Thorpe, H. M. (2002). Diversity in the serine recombinases. *Mol. Microbiol.* **44**(2):299–307.

Soding, J. (2005). Protein homology detection by HMM-HMM comparison. *Bioinformatics* **21** (7):951–960.

Springer, B., Sander, P., Sedlacek, L., Ellrott, K., and Bottger, E. C. (2001). Instability and site-specific excision of integration-proficient mycobacteriophage L5 plasmids: Development of stably maintained integrative vectors. *Int. J. Med. Microbiol.* **290**(8):669–675.

Stella, E. J., de la Iglesia, A. I., and Morbidoni, H. R. (2009). Mycobacteriophages as versatile tools for genetic manipulation of mycobacteria and development of simple methods for diagnosis of mycobacterial diseases. *Rev. Argent Microbiol.* **41**(1):45–55.

Stewart, C. R., Casjens, S. R., Cresawn, S. G., Houtz, J. M., Smith, A. L., Ford, M. E., Peebles, C. L., Hatfull, G. F., Hendrix, R. W., Huang, W. M., and Pedulla, M. L. (2009). The genome of *Bacillus subtilis* bacteriophage SPO1. *J. Mol. Biol.* **388**(1):48–70.

Sula, L., Sulova, J., and Stolcpartova, M. (1981). Therapy of experimental tuberculosis in guinea pigs with mycobacterial phages DS-6A, GR-21 T, My-327. *Czech. Med.* **4** (4):209–214.

Susskind, M. M., and Botstein, D. (1978). Molecular genetics of bacteriophage P22. *Microbiol. Rev.* **42**(2):385–413.

Timme, T. L., and Brennan, P. J. (1984). Induction of bacteriophage from members of the *Mycobacterium avium, Mycobacterium intracellulare. Mycobacterium scrofulaceum serocomplex. J. Gen. Microbiol.* **130**(Pt 8):2059–2066.

Tori, K., Dassa, B., Johnson, M. A., Southworth, M. W., Brace, L. E., Ishino, Y., Pietrokovski, S., and Perler, F. B. (2009). Splicing of the mycobacteriophage Bethlehem DnaB intein: Identification of a new mechanistic class of inteins that contain an obligate block F nucleophile. *J. Biol. Chem.* **285**(4):2515–2526.

Tyler, J. S., Mills, M. J., and Friedman, D. I. (2004). The operator and early promoter region of the Shiga toxin type 2-encoding bacteriophage 933W and control of toxin expression. *J. Bacteriol.* **186**(22):7670–7679.

van Kessel, J. C., and Hatfull, G. F. (2007). *Recombineering in Mycobacterium tuberculosis. Nat. Methods* **4**(2):147–152.

van Kessel, J. C., and Hatfull, G. F. (2008a). Efficient point mutagenesis in mycobacteria using single-stranded DNA recombineering: Characterization of antimycobacterial drug targets. *Mol. Microbiol.* **67**(5):1094–1107.

van Kessel, J. C., and Hatfull, G. F. (2008b). *Mycobacterial recombineering. Methods Mol. Biol.* **435**:203–215.

van Kessel, J. C., Marinelli, L. J., and Hatfull, G. F. (2008). Recombineering mycobacteria and their phages. *Nat. Rev. Microbiol.* **6**(11):851–857.

Vultos, T. D., Mederle, I., Abadie, V., Pimentel, M., Moniz-Pereira, J., Gicquel, B., Reyrat, J. M., and Winter, N. (2006). Modification of the mycobacteriophage Ms6 attP core allows the integration of multiple vectors into different tRNAala T-loops in slow- and fast-growing mycobacteria. *BMC Mol. Biol.* **7**:47.

Watterson, S. A., Wilson, S. M., Yates, M. D., and Drobniewski, F. A. (1998). Comparison of three molecular assays for rapid detection of rifampin resistance in *Mycobacterium tuberculosis. J. Clin. Microbiol.* **36**(7):1969–1973.

Wikoff, W. R., Liljas, L., Duda, R. L., Tsuruta, H., Hendrix, R. W., and Johnson, J. E. (2000). Topologically linked protein rings in the bacteriophage HK97 capsid. *Science* **289** (5487):2129–2133.

Wilson, S. M., al-Suwaidi, Z., McNerney, R., Porter, J., and Drobniewski, F. (1997). Evaluation of a new rapid bacteriophage-based method for the drug susceptibility testing of *Mycobacterium tuberculosis. Nat. Med.* **3**(4):465–468.

Xu, J., Hendrix, R. W., and Duda, R. L. (2004). Conserved translational frameshift in dsDNA bacteriophage tail assembly genes. *Mol. Cell.* **16**(1):11–21.

Zhang, Y., Buchholz, F., Muyrers, J. P., and Stewart, A. F. (1998). A new logic for DNA engineering using recombination in *Escherichia coli. Nat. Genet.* **20**(2):123–128.

Zhu, H., Yin, S., and Shuman, S. (2004). Characterization of polynucleotide kinase/phosphatase enzymes from Mycobacteriophages omega and Cjw1 and vibriophage KVP40. *J. Biol. Chem.* **279**(25):26358–26369.

Section 3
Interaction of Phages with Their Hosts

CHAPTER 8

Role of CRISPR/*cas* System in the Development of Bacteriophage Resistance

Agnieszka Szczepankowska

Contents

Department of Microbial Biochemistry, Institute of Biochemistry and Biophysics, Polish Academy of Sciences, Warsaw, Poland

Advances in Virus Research, Volume 82
ISSN 0065-3527, DOI: 10.1016/B978-0-12-394621-8.00011-X

Abstract

Acquisition of foreign DNA can be of advantage or disadvantage to the host cell. New DNAs can increase the fitness of an organism to certain environmental conditions; however, replication and maintenance of incorporated nucleotide sequences can be a burden for the host cell. These circumstances have resulted in the development of certain cellular mechanisms limiting horizontal gene transfer, including the immune system of vertebrates or RNA interference mechanisms in eukaryotes. Also, in prokaryotes, specific systems have been characterized, which are aimed especially at limiting the invasion of bacteriophage DNA, for example, adsorption inhibition, injection blocking, restriction/modification, or abortive infection. Quite recently, another distinct mechanism limiting horizontal transfer of genetic elements has been identified in procaryotes and shown to protect microbial cells against exogenous nucleic acids of phage or plasmid origin. This system has been termed CRISPR/*cas* and consists of two main components: (i) the CRISPR (**c**lustered, **r**egularly **i**nterspaced **s**hort **p**alindromic **r**egions) locus and (ii) *cas* genes, encoding **C**RISPR-**as**sociated (Cas) proteins. In simplest words, the mechanism of CRISPR/*cas* activity is based on the active integration of small fragments (proto-spacers) of the invading DNAs (phage or plasmids) into microbial genomes, which are subsequently transcribed into short RNAs that direct the degradation of foreign invading DNA elements. In this way, the host organism acquires immunity toward mobile elements carrying matching sequences. The CRISPR/*cas* system is regarded as one of the earliest defense system that has evolved in prokaryotic organisms. It is inheritable, but at the same time is unstable when regarding the evolutionary scale. Comparative sequence analyses indicate that CRISPR/*cas* systems play an important role in the evolution of microbial genomes and their predators, bacteriophages.

I. GENERAL BACKGROUND

Extensive genomic analyses suggested that noncoding sequences may be as important as sequences encoding specific proteins. In eukaryotes, transcripts of noncoding nucleotide regions were shown to be involved in the function and regulation of cellular processes through a mechanism based on RNA interference (RNAi). Also, for prokaryotes, similar systems of gene regulation via small RNA molecules have been described (Gottesman, 2004, 2005; Majdalani *et al.*, 2005; Storz *et al.*, 2004; Tang *et al.*, 2002, 2005). Based on the comparative genomics study performed by Makarova *et al.* (2006), CRISPR/*cas* systems were proposed to be a functional analogue of the eukaryotic RNAi systems. First reports on the exact function of CRISPR/*cas* systems came from Mojica *et al.* (2005), who proposed that its noncoding elements (CRISPR arrays), together with Cas-encoding sequences, constitute a prokaryotic defense system against foreign invasive DNA. This finding was later supported by other research groups (Bolotin *et al.*, 2005; Pourcel *et al.*, 2005). The mechanism of CRISPR/*cas*-conferred immunity against mobile DNAs is quite distinct from previously described mechanisms of resistance and is described in detail in this chapter.

II. ORGANIZATION OF CRISPR LOCI IN PROKARYOTIC ORGANISMS

The abundance of CRISPR/*cas* sequences in prokaryotic genomes is impressive and variable. However, it is suspected that not all CRISPR/*cas* systems present in the genome are active. In order to distinguish between functional and nonfunctional CRISPR cassettes, some general features of active CRISPR have been established. They include the noncoding CRISPR locus, containing direct repeats of identical sequence and divergent spacers, accompanied by two determinants: *cas* genes and leader sequence (Horvath *et al.*, 2008; Makarova *et al.*, 2006; Sorek *et al.*, 2008) (Fig. 1).

CRISPR sequences were first identified in the *Escherichia coli* genome (in 1987) by Ishino and colleagues as arrays of **c**lustered, **r**egularly **i**nterspaced **s**hort **p**alindromic **r**egions; one 24bp upstream of the *iap* gene, the second downstream of the *ygcF* gene, 24kb apart from the first one (Ishino *et al.*,1987; Nakata *et al.*, 1989). These two arrays contain identical 29-bp repeats (*iap* repeat), which, as proposed by Kunin *et al.* (2007), were classified to type 2 (CRISPR2) repeats. Moreover, one or two arrays comprising a different motif of 28bp (Ypest repeat) have also been reported and ascribed to type 4 (CRISPR4) repeats (Haft *et al.*, 2005; Kunin *et al.*, 2007). Four distinct CRISPR loci (CRISPR1–4) were later identified in *E. coli*, where

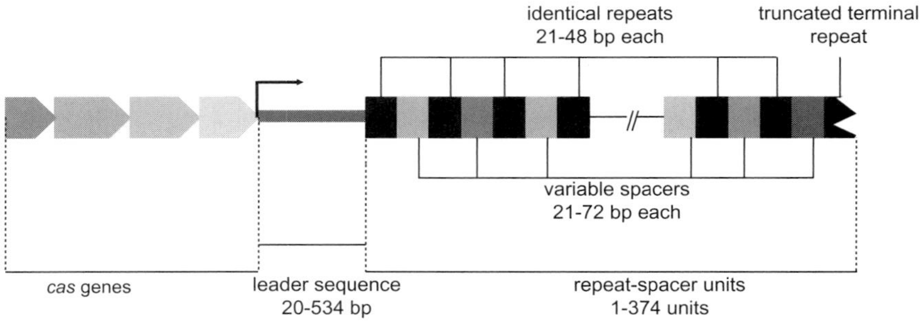

FIGURE 1 Characteristic organization of the clustered, regularly interspaced short palindromic repeat (CRISPR) locus. Black boxes indicate repeat sequences, which are conserved in size and sequence within a locus. Colored boxes indicate spacers of different sequences and sizes (within the given range). The repeat-spacer unit region is preceded by the leader sequence (blue bar), proposed to be the promoter region for CRISPR transcription. CRISPR-associated (*cas*) genes (gray arrows) are localized upstream (as on figure) or downstream of the repeat-spacer units. (See Page 31 in Color Section at the back of the book.)

CRISPR1/CRISPR2 and CRISPR3/CRISPR4 were found to share the same repeat sequences of 29bp (previously type 2) and 28bp (previously type 4), respectively (Touchon and Rocha, 2010).

Regions of similar organization were also found in other microorganisms, including *Haloferax mediterranei*, *Streptococcus pyogenes*, *Mycobacterium tuberculosis*, *Thermotoga maritima* (Hermans *et al.*, 1991; Haft *et al.*, 2005; Mojica *et al.*, 1995; Nelson *et al.*, 1999; Sorek *et al.*, 2008). Currently, nearly 40–70% of all bacteria and almost all archaea were shown to carry CRISPRs in their genomes (Godde and Bickerton, 2006; Grissa *et al.*, 2007a; Mojica *et al.*, 2005; Rousseau *et al.*, 2009). Based on these findings, CRISPRs are regarded as the most abundant family of noncoding sequences in prokaryotes. CRISPR loci were also detected within chromosome-residing prophage sequences (e.g., in *Clostridium difficile* prophage sequences and skin element) or on plasmids (e.g., in *Sulfolobus* sp. and *Thermus thermophilus*) (Agari *et al.*, 2010; Karginov and Hannon, 2010; Sebaihia *et al.*, 2006). The role of CRISPRs in such localizations is not fully understood. It could be that plasmid- or prophage-encoded CRISPRs prevent (or limit) the invasion of genetic elements that also contain functional CRISPR/*cas* systems. This way CRISPRs can prevent the spread of genes via horizontal transfer event, for example, multidrug resistance; thus, acting as an anti-CRISPR system against other mobile genetic elements (discussed also in Section IV.B).

A new more ordered classification of CRISPR/*cas* systems has been proposed by Makarova *et al.* (2011a,b), which, based on sequence and structural data of protein components, distinguishes three main types of CRISPR/*cas* systems further divided into subtypes (see Section II.C.6).

A. CRISPR array

A single CRISPR locus is built by two types of noncoding elements: (i) direct repeats (21–48bp) separated by (ii) distinct nonrepetitive linker sequences—spacers (21–72bp). Within each CRISPR region the length of both spacers and direct repeats is conserved, except the last repeat from the 3′ end, which appears to be truncated in 30% of cases (Horvath *et al.*, 2008; Jansen *et al.*, 2002a,b). Recognition of the last terminal repeat is a key factor in orienting the CRISPR locus and allows its proper annotation.

Based on a study of CRISPR arrays in enterobacteria, localization of a particular CRISPR locus within different genomes is always conserved (Touchon and Rocha, 2010). Around half of the prokaryotic organisms with available genome sequences in public databases were found to carry more than one CRISPR locus (Godde and Bickerton, 2006). The most CRISPR loci (20) were discovered in the genome of *Methanocaldococcus jannaschii* (Bult *et al.*, 1996; Deveau *et al.*, 2010). In case a genome contains multiple CRISPR loci, each of them differs from another in repeat as well as spacer sequences. Most genomes carry one CRISPR locus containing repeats of identical (or almost identical) sequence.

1. Repeats

Repeat sequences are strongly conserved within a locus and appear to be species specific (Deveau *et al.*, 2010; Goode and Bickerton, 2006). This regularity serves as a defining feature of individual CRISPR arrays. Sequence similarities between particular repeats allow dividing them into 12 main groups (Kunin *et al.*, 2007). Some groups of repeats were found to contain palindromic motifs, usually GTTTg/c and GAAAC, at their terminal ends, which were suggested to be implicated in the processing of CRISPR transcripts (discussed in later parts of this chapter)(Godde and Bickerton, 2006; Kunin *et al.*, 2007).

2. Spacers

In contrast to conserved repeat sequences, spacers of the same CRISPR region are usually very different and highly variable. Moreover, it appears that all spacers within a genome are unique, with very few exceptions (Grissa *et al.*, 2007a; Horvath and Barrangou, 2010). Also, the number of repeat-spacer elements present in a single CRISPR locus is different between species or even between strains of the same species. Based on current findings, a CRISPR array can hold from 2 to 375 (in thermophilic *Chloroflexus* sp.) repeat-spacer units (Deveau *et al.*, 2010; Grissa *et al.*, 2007a). For instance, CRISPR stretches identified in lactic acid bacteria contain on average 20 direct repeats, while in *E. coli* no more than 34 (Horvath and Barrangou, 2010; Touchon and Rocha, 2010).

Spacers identified in both bacterial and archeal CRISPR regions exhibit homology to various mobile elements, mainly phages and conjugative plasmids, indicating their origin (Bolotin *et al.*, 2005; Lillestøl *et al.*, 2006; Mojica *et al.*, 2005; Pourcel *et al.*, 2005). Based on a large study in *Streptococcus thermophilus*, 40% of the identified spacer sequences matched sequences of phage (75%) or plasmid (20%) DNA (Bolotin *et al.*, 2005). For the remaining cases, no significant homology to known sequences was found. Another extensive study analyzing a significant number of spacers (4500) from over 60 strains of various microbial species showed that only 2% of them have counterparts in phage genomes (Mojica *et al.*, 2005). This contrary outcome was discussed to be due to the underrepresentation of phage and plasmid sequences in available databases. However, the large representation of *S. thermophilus* spacers, which match phage sequences in databases, indicates the high activity of CRISPR loci, which could be related with the difficulty to gain phage resistance by mutations in phage receptor genes in these species. It also suggests that CRISPR-mediated immunity is the main defense system in these bacteria (Deveau *et al.*, 2010).

3. Dissemination of CRISPR arrays by horizontal transfer

Usually, CRISPR repeats from different genomes exhibit low similarity and are regarded as species specific. However, often distantly related species can carry similar, strongly conserved motifs, for example, *E. coli* and *Mycobacterium avium* contain similar CRISPR repeats, yet belong to different bacterial phyla (Jansen *et al.*, 2002b). This implies that CRISPR sequences might have evolved autonomously from the rest of the chromosome. The common presence of CRISPR sequences in archaea (90%) and less frequent in bacteria (40–70%) indicates that CRISPRs developed in ancestral archeal organisms and only later disseminated to bacteria, most likely by horizontal transmission. This implication is further reinforced by the fact that the GC content and codon bias of CRISPR loci differ from the rest of the chromosomal DNA. Transfer of these clusters to distantly related species is proposed to occur either via plasmids or prophages. Identification of CRISPR arrays within megaplasmids (e.g., from *Sulfolobus* sp.), as well as in prophages (e.g., from *C. difficile*), supports this assumption (Greve *et al.*, 2004; Sebaihia *et al.*, 2006).

B. The leader region

One of the determinants strictly connected with active CRISPR loci is the leader sequence. This noncoding, A–T-rich region is localized at the 5′ extremity of CRISPR arrays and, depending on the species, counts from

20 to 534 bp (Jansen *et al.*, 2002b; Lillestøl *et al.*, 2006). Similarly, as repeats, nucleotide sequences of leader regions preceding CRISPR loci within the same genome are identical up to 80%, but vary among species (Bult *et al.*, 1996; Klenk *et al.*, 1997; Smith *et al.*, 1997).

1. Integration of spacers

The role of the leader sequence has been proposed to involve acquisition of new spacers. The region was suggested to contain a binding site for specific proteins (likely encoded by *cas* genes), which play a part in duplication of repeats and/or spacer integration (Barrangou *et al.*, 2007; Pourcel *et al.*, 2005). A metagenomics study of two different natural CRISPR-containing *Leptospirillum* sp. populations allowed to establish a specific pattern by which spacers are introduced within the CRISPR array (Tyson and Banfield, 2008). Novel strain-specific spacers were incorporated into loci from the leader end, precisely between the leader and the first spacer unit, whereas older spacers, common for both populations, were localized closer to the distal end, some truncated. Similar regularity was noted for the integration of spacers within CRISPR loci of other species, for example, for laboratory *S. thermophilus* cultures (Barrangou *et al.*, 2007; Deveau *et al.*, 2008; Horvath *et al.*, 2008). The idea that insertion of new spacers occurs from the leader terminus is reinforced by the fact that repeat sequences in this region are highly homologous. Based on a study of CRISPR regions in 100 different *E. coli* strains, it was determined that degenerative repeat sequences are positioned at the distal end of the CRISPR array and are accompanied by conserved spacers that are usually inactive, while less degenerate repeats are identified closer to the leader region and are generally associated with spacers specific for a particular strain (Díez-Villaseñor *et al.*, 2010).

2. Role of the leader sequence

The leader region was suggested to function as a promoter for transcription of the CRISPR region (Brouns *et al.*, 2008; Hale *et al.*, 2008; Lillestøl *et al.*, 2006, 2009; Marraffini and Sontheimer, 2010a). This hypothesis was supported by the finding that CRISPR loci lacking a leader sequence do not incorporate new spacers and are regarded as inactive remnants (Lillestøl *et al.*, 2006). However, the activity of a functional CRISPR locus is manifested by mRNA transcription starting from the leader sequence. The resulting transcript is subsequently processed by specific Cas proteins (described later), generating short CRISPR (cr)RNAs. These molecules are regarded as the basis of CRISPR/*cas* systems, which function by interference mode (discussed later)(Brouns *et al.*, 2008).

C. *cas* genes

The majority of identified CRISPR loci are associated with a set of conserved protein-encoding genes termed *cas* (**C**RISPR-**as**sociated genes) (Haft *et al.*, 2005; Jansen *et al.*, 2002b; Makarova *et al.*, 2006). In addition to the leader region, they are a main determinant of CRISPR/*cas* systems. Current data show that *cas* genes are present only in genomes containing CRISPRs, which implies their tight correlation. The number of *cas* genes within a particular CRISPR region can vary from 4 to 20 (Barrangou *et al.*, 2007; Haft *et al.*, 2005; Sorek *et al.*, 2008). They can be positioned either upstream or downstream of repeat-spacer units, but always from the same side for a given CRISPR locus.

1. CRISPR repeat/*cas* gene coupling

Similar to CRISPR repeats, *cas* genes are locus specific and both determinants seem to be functionally coupled and exhibit similar clustering patterns (Haft *et al.*, 2005; Horvath *et al.*, 2008; Kunin *et al.*, 2007). Moreover, the orientation of *cas* genes is often in accordance with the orientation of CRISPR repeats. Also, a relation between the number of CRISPR repeats in a locus and the conserved organization of associated *cas* genes was noted. A study of CRISPR loci of *Escherichia* and *Salmonella* strains showed that when the *cas* region is intact, the number of repeats in the CRISPR array is high (Touchon and Rocha, 2010). However, when two or more CRISPR regions, containing the same repeat sequence, are present within one genome, only one of them is associated with the *cas* genes (Grissa *et al.*, 2007a). According to the CRISPR classification proposed by Makarova and colleagues (2011b), among 12 CRISPR repeats groups identified, 4 are clearly associated with specific CRISPR/*cas* subtypes.

2. Diversity of *cas* genes

Generally, *cas* genes are commonly distributed among archeal and bacterial genomes. Some are conserved and limited to certain species, whereas in other cases, similar sets of *cas* genes can be found in phylogenetically distant microorganisms, indicating their distribution via horizontal transfer. A wide-scale, bioinformatic analysis using hidden Markov models and multiple sequence alignments of over 200 available prokaryotic genomes has led to the identification of 45 Cas protein families (Haft *et al.*, 2005). Another group (Makarova *et al.*, 2006) reclassified Cas proteins into 23 groups comprising phylum-specific subfamilies. More than 65 distinct orthologous *cas* genes have now been identified based on nucleotide sequence analysis combined with extensive *in silico* and experimental studies (Makarova *et al.*, 2011b).

3. "Core" *cas* genes

Despite the great variability, initially four "core" *cas* genes (*cas1*–*cas4*) were distinguished and later two more genes, *cas5* and *cas6*, were added to the group (Bolotin *et al.*, 2005; Haft *et al.*, 2005; Jansen *et al.*, 2002b). Currently, 10 *cas* gene families have been identified and certainly more await identification in the future due to novel CRISPR sequences found in newly sequenced genomes (Makarova *et al.*, 2011a).

The "core" *cas* genes are widely distributed among genomes of often distant microbial species and generally lay in close proximity to CRISPR loci. Usually, CRISPR-containing genomes do not carry all "core" *cas* genes, but their various combinations, of which *cas1* and *cas2* appear most frequently (Haft *et al.*, 2005; Makarova *et al.*, 2006). The respective proteins, Cas1 and Cas2, are also the most conserved among all Cas proteins and appear universally in all active CRISPR/*cas* systems, while the rest of Cas proteins highly vary (Makarova *et al.*, 2011b). Due to the common occurrence of Cas1 in various microorganisms (except for *Pyrococcus abyssii*), it is regarded as a marker for detecting CRISPR/*cas* systems (Makarova *et al.*, 2006; Sorek *et al.*, 2008). Some of the "core" Cas proteins have been characterized at the biochemical as well as structural levels, whereas others await functional assignment. In general, the identified Cas proteins carry domains typical for nucleases, helicases, polymerases, or polynucleotide-binding proteins (Haft *et al.*, 2005; Jansen *et al.*, 2002b; Makarova *et al.*, 2006, 2011a,b).

i. Cas1 The Cas1 protein is universally identified in all CRISPR/*cas* systems. Biochemical characterization of Cas1 of *Pseudomonas aeruginosa* established it to be a metal-dependent, sequence-nonspecific DNA endonuclease/integrase (Wiedenheft *et al.*, 2009; Zegans *et al.*, 2009). A similar study of *Sulfolobus solfataricus* Cas1 led to characterization of its single-stranded (ss)/double-stranded (ds)RNA and ss/dsDNA-binding and -annealing activities; however, no nuclease activity has been detected (Han and Krauss, 2009). These findings allowed proposing Cas1 function to be implicated in recognition and/or cleavage of foreign DNA and spacer integration into the CRISPR locus (Deveau *et al.*, 2010). Interestingly, the role of the *E. coli* Cas1 protein has not been deciphered until recently and points to a novel function of Cas proteins. Babu and associates (2011) showed that *E. coli* Cas1 protein (YgbT) is a novel type of nuclease that *in vitro* acts on branched DNA substrates, such as Holliday junctions and replication forks. Comparison of YgbT to Cas1 from *S. solfataricus* (SSO1450) or *P. aeruginosa* (*Pa*Cas1) showed low sequence similarity and somewhat different biochemical activity. However, the function of *E. coli* Cas1 was proposed to also be involved in insertion/deletion of CRISPR spacers by recombination events (Babu *et al.*, 2011; Mojica *et al.*, 2009).

ii. Cas2 Studies on the second universally occurring Cas "core" protein, Cas2, in *S. solfataricus* showed it to have a ferredoxin-like domain and exhibit metal-dependent endonuclease activity (Beloglazova *et al.*, 2008). The protein was also suggested to cleave ssRNA within U-rich regions. Although its exact role in CRISPR/*cas* systems has yet to be established, most probably it is involved in the incorporation of new spacers into CRISPR loci.

iii. Cas3 The Cas3-encoding gene is a signature gene of type I CRISPR/*cas* systems (see later). In the *E. coli* K12 strain, the CRISPR/*cas* system is constituted by eight *cas* genes, encoding "core" Cas proteins, Cas1, Cas2, Cas3, and CasA-E, which constitute the Cascade complex described later in this chapter (Brouns *et al.*, 2008). *In vivo* studies in *E. coli* determined that the "core" Cas3 protein is essential in antiviral immunity of CRISPR/*cas* systems (Brouns *et al.*, 2008). However, no details on its exact function have been reported at that time. *In silico* analysis predicted Cas3 to have helicase activity and possess a nuclease domain (Haft *et al.*, 2005; Makarova *et al.*, 2002; van der Oost *et al.*, 2009). Only recently have the first biochemical experiments been performed revealing the exact properties of Cas3 proteins from *E. coli* and the *E. coli* subtype CRISPR4 system (currently CRISPR I-E subtype) of *S. thermophilus* (Howard *et al.*, 2011; Sinkunas *et al.*, 2011). In *S. thermophilus*, Cas3 was determined to execute multiple activities, including those of an ATP-dependent helicase, metal-dependent single-stranded DNA nuclease and finally ATPase stimulated by ssDNA. The purified *E. coli* Cas3 protein was also shown to display several functions. From one side, it was shown to catalyze ATP-independent, metal-dependent RNA–DNA annealing, while in the presence of ATP to act as a RNA-unwinding helicase (Howard *et al.*, 2011). Based on these observations, a mechanism of Cas3 action was proposed to involve recognition and subsequent processing of foreign DNA targets (Marraffini and Sontheimer, 2010; Sinkunas *et al.*, 2011; van der Oost *et al.*, 2009).

iv. Cas4 The Cas4 "core" protein was predicted to be a RecB-like nuclease most probably engaged in the digestion of invading DNAs (Makarova *et al.*, 2006).

v. Cas5, Cas6, and Cas7 *cas5*, *cas6*, and *cas7* genes encode proteins from the RAMP superfamily (see Section II.C.5)(Makarova *et al.*, 2006, 2011a). Studies characterizing four Cas6 proteins, namely Cas6 of *P. furiosus* and *S. sulfataricus*, *E. coli* CasE (current name Cas6e), and Csy4 (current name Cas6f) of *P. aeruginosa*, revealed that they are all metal-independent endoribonucleases. Their mechanism of action was proposed to be implicated in the cleavage of CRISPR RNA (crRNA) transcripts (Carte *et al.*, 2008; Makarova *et al.*, 2011a).

vi. Cas8 Cas8 proteins based on *in silico* analyses have been predicted to possess an inactive Cas10-like PALM polymerase domain (described later), engaged in nucleic acid binding. Based on these assumptions, the role of Cas8 could involve DNA binding and interaction with proteins from the RAMP superfamily, possibly during the incorporation of spacer sequences and during CRISPR-mediated interference (Makarova *et al.*, 2011b).

vii. Cas9 The Cas9 protein contains two nuclease domains—RuvC-like (RNaseH fold) and HNH (or McrA-like), which exact functions are yet to be determined. However, it seems that Cas9 alone can be responsible for cleaving DNA (HNH domain) as well as crRNA transcripts (RuvC-like domain), which found confirmation in *in vivo* studies in *S. thermophilus* (Barrangou *et al.*, 2007; Garneau *et al.*, 2010; Makarova *et al.*, 2011b).

viii. Cas10 The Cas10 protein contains a polymerase-PALM domain fused with a HD (metal-dependent phosphohydrolase) domain (Makarova *et al.*, 2011a). Cas10 is recognized as a component of the Cascade (Cmr) complex, and its activity was predicted to involve ssDNA cleavage and separation of DNA strands (Hale *et al.*, 2009; Makarova *et al.*, 2011a).

4. "Noncore" *cas* genes

Apart from "core" *cas* genes, other "noncore" *cas* genes have been identified. They are distributed more narrowly among CRISPR/*cas* systems. Their names derive from the organism in which they were found originally (e.g., *cse* stands for genes of the CRISPR system of *E. coli*; also known as *casA-casB-casE-casC-casD* genes). "Noncore" *cas* genes form clusters consisting of two to six separate genes, which are usually confined to a particular organism (Haft *et al.*, 2005). Generally, one or more of these "noncore" *cas* gene clusters accompany the "core" genes.

5. RAMP proteins

Another superfamily of Cas proteins identified by Makarova *et al.* (2002) has been updated and reanalyzed (Makarova *et al.*, 2011a). They were initially assigned as repair-associated mysterious proteins, as they were originally thought to be engaged in the DNA repair process. Only later was this group of Cas proteins redefined as **r**epeat-**a**ssociated **m**ysterious **p**roteins (RAMPs). The position of RAMP-encoding genes in reference to CRISPR repeat-spacer units is rather loose. They can be either closely or more distantly positioned to array (Haft *et al.*, 2005). From all Cas proteins, this family is most diverse and characterized by low sequence conservation (Makarova *et al.*, 2006). Based on recent sequence and structure analysis using refined computational methods, the RAMP superfamily was divided into three groups: Cas5, Cas6, and Cas7 (Makarova *et al.*, 2011a). Crystallization data of several RAMP proteins revealed that at the structural level they possess either one or two RNA-binding domains, otherwise known as the RNA recognition motif (RRM) or ferredoxin-fold domain, and a glycine-rich loop (G-loop) at the C′ end (Makarova *et al.*, 2006; 2011a). Cas5 proteins were divided into two distinct subgroups, depending on the number of RRM domains present (one or two). Cas6 proteins in general possess two RRM domains domain, whereas Cas7 proteins contain one RRM domain. A relationship between the CRISPR/*cas* subtype and the presence of Cas proteins from specific RAMP groups was also observed (see Section II.C.6).

RAMP proteins are nonautonomous and always appear in genomes carrying one of the eight CRISPR/*cas* subtypes. They were suggested to be involved in the recognition of specific targets during CRISPR-mediated interference (Hale *et al.*, 2009). Interestingly, the number of RAMPs present within a genome was found to be strongly correlated with the number of CRISPR spacer units, implying an essential biological link between these two determinants (Makarova *et al.*, 2006). Based on the aforementioned relationship and the fact that RAMPs are a highly diverse protein class, it was hypothesized that RAMPs distinguish inserts according to their length rather than by recognition of particular repeats. However, not all prokaryotic genomes encoding CRISPR/*cas* systems carry RAMP genes. In these organisms the role of RAMP proteins is presumably substituted by other Cas proteins.

Studies of CRISPR/*cas* systems in various species provided essential information on the two general roles performed by Cas proteins. Part of them are involved in the maintenance of CRISPR loci within microbial genomes and the incorporation of new invader-derived spacers in response to new infections, based on molecular interaction with CRISPR repeats. Other Cas proteins are responsible for conferring resistance against foreign genetic elements (Barrangou *et al.*, 2007). Another study also showed that certain Cas proteins apart from antiviral immunity can perform other functions. Further details on the specific role of Cas proteins are discussed in this chapter.

6. Classification of CRISPR/*cas* systems

A CRISPR repeat-spacer array, together with *cas* genes, constitutes an active CRISPR/*cas* system. First classification of CRISPR/*cas* systems based on Cas1 phylogenetic analysis and clustering of *cas* genes distinguished eight distinct subtypes—Ecoli, Ypest, Apern, Nmeni, Mtube, Tneap, Hmari, and Dvulg—identified respectively in specific strains of *E. coli*, *Yersinia pestis*, *Aeropyrum pernix*, *Neisseria meningitis*, *Mycobacterium tuberculosis*, *Thermotoga neapolitana*, *Haloarcula marismortui*, and *Desulfovibrio vulgaris* (Haft *et al.*, 2005; Makarova *et al.*, 2006). However, this classification now seems confusing in the light of increasing data on CRISPR/*cas* systems and their components, particularly that (i) CRISPRs often recombine, giving rise to hybrid systems; (ii) a single strain can have more than one CRISPR system; or (iii) CRISPR systems identified in various strains of the same species can vary. Currently, taking into account growing sequencing data on *cas* genes in various organisms and phylogenetic studies, a novel, more integrated classification of CRISPRs has been proposed (Makarova *et al.*, 2011a,b). In effect, three main types (types I–III) of CRISPR/*cas* systems have been proposed, in all of which the central core is constituted by *cas1* and *cas2* genes. Moreover, each type of CRISPR/*cas* system is characterized by its specific signature genes, respectively, type I by *cas3*, type II by *cas9*, and type III by *cas10* genes (Makarova e*t al.*, 2011a).

In type I CRISPR/*cas* systems, apart from the *cas3* gene, distinctive features are the *cas4* gene and genes encoding for RAMPs—one protein from each of the three RAMP families (Cas5, Cas6, and Cas7). Type I systems are further divided into subtypes, which include I-A (Apern or CASS5), I-B (Tneap–Hmari or CASS7), I-C (Dvulg or CASS1), I-D, I-E (Ecoli or CASS2), and I-F (Ypest or CASS3). Specific subtypes are distinguished by another signature *cas* gene (*cas8*)—*cas8a*, *cas8b*, and *cas8c*, respectively, for subtypes I-A, I-B, and I-C.

So far, type II CRISPR/cas systems have been identified solely in bacterial genomes and comprise two subtypes: II-A (Nmenni or CASS4) and II-B (Nmenni or CASS4a). Apart from the universally occurring *cas1* and *cas2* genes, the signature genes of type II systems are *cas4* and *cas9* (the latter previously termed *csn1* or *csx12*).

Type III CRISPR/*cas* systems are found most commonly in archaea. The signature genes encode for CRISPR polymerase (Cas10 with PALM domain) and RAMP proteins, including more than one Cas7-type protein and Cas6. Another signature gene specific for type III systems is *cas10*. Two subtypes of the type III CRISPR/cas systems have been recognized: subtype III-A (known otherwise as Mtube or CASS6) and subtype III-B (polymerase-RAMP module or Cmr system). Some type III systems lack *cas1*–*cas2* genes, but in such cases they always co-occur with other CRISPR loci (type I or type II), which have been suggested to supply these genes *in trans*.

Other CRISPR/*cas* systems that could not be classified to any of the three types are grouped as type U systems. In general, each subtype has been assigned a distinct signature gene, which defines and allows classification of individual CRISPR/*cas* systems (Makarova *et al.*, 2011a). Subtypes for which signature genes have not been identified are defined as I-U, II-U, or III-U.

III. BIOLOGICAL ROLE OF CRISPR/*CAS* SYSTEMS

Since their identification in 1987 by Ishino and colleagues, researchers have made extensive attempts to establish the biological function of CRISPR/*cas* systems. Many hypotheses were put forward on the role of

CRISPRs in microbial genomes, including DNA repair function or involvement in genome stability or replicon partitioning (Makarova *et al.*, 2002; Mojica *et al.*, 1995). First attempts to elucidate CRISPR/*cas* activity came in 2005, when, in consequence of *in silico* studies, spacers found in microbial genomes were reported to exhibit homology to plasmid or phage sequences (Bolotin *et al.*, 2005; Mojica *et al.*, 2005). However, details on the exact function of CRISPR/*cas* systems were revealed during an experimental study of culture growth of dairy lactic acid bacteria performed by Barrangou *et al.* (2007). The group examined the incident of lytic phage infection of *S. thermophilus*, one of the most commonly used strains in dairy fermentation processes. The experiment revealed that a small fraction of *S. thermophilus* cells of the infected culture survived phage attack and was resistant to subsequent infections. These phage-resistant cells appeared to occur spontaneously and were termed bacteriophage-insensitive mutants (BIMs). Analysis of genome sequences of the isolated BIMs revealed some differences compared with the wild-type genome. *S. thermophilus* cells that survived phage infection gained from one to four new sequences (spacers) in specific regions, later termed CRISPR loci. Further computational analysis of these spacer sequences exposed several interesting aspects.

First of all, a relationship between the presence of spacer sequences and phage resistance was determined. Moreover, the newly acquired sequences in BIM genomes were identical to sequences of the infecting phage. Such sequences that derive from phage genomes and match the sequence of CRISPR spacers were called *proto-spacers*. Interestingly, some of the resistant hosts, which contained spacers of different sequences, but originating from the same phage source, provided efficient protection to infection by this phage. Moreover, the resistance level of the surviving cells was shown to be correlated not only with the presence of new spacer elements, but also with their number. Furthermore, Barrangou *et al.* (2007) showed that introduction of a BIM-derived spacer conferring resistance to a particular phage into the CRISPR locus of a phage-sensitive *S. thermophilus* strain resulted in its subsequent resistance to this phage. Conversely, deletion of a spacer abolished phage resistance. Observations on acquisition of phage resistance by incorporation of spacers matching phage sequences were confirmed by experimental analysis of synthetic CRISPR loci in *E. coli*. Brouns *et al.* (2008) showed that introduction of an artificial CRISPR system, comprising *cas* genes and the CRISPR locus containing spacers homologous to the phage λ sequence, generated phage resistance *de novo*. The assay established the crucial components of CRISPR/*cas* activity (discussed elsewhere).

CRISPRs were also shown to prevent conjugative transfer of plasmids in *Staphylococcus epidermidis* (Marraffini and Sontheimer, 2008). Moreover, in the same study it was determined that CRISPR/*cas* confers protection not only against conjugative plasmids, but also against plasmids entering the cell by other routes, for example, transformation or electroporation. In all cases, the mechanism of protection against plasmid DNA invasion was shown to occur by an analogous way as for phage DNA and to be based on incorporation of spacers homologous to plasmid sequences.

Taking into account all data, a hypothesis was built that CRISPR/*cas* regions serve as an adaptive immunology system against foreign genetic elements, in which spacer sequences play a role of determinants responsible for specific recognition of invading DNAs (Barrangou *et al.*, 2007; Brouns *et al.*, 2008; Marraffini and Sontheimer, 2010b). It has been suggested that CRISPR/*cas* systems protect microbial cells from extracellular mobile elements by a gene-silencing mechanism. The process was proposed to rely on recognition and hybridization (base pairing) between CRISPR spacers and complementary sequence targets (proto-spacers) within sequences of the invading DNAs. The fact that CRISPR spacers seem to be species or even strain specific implies their rapid evolution and suggests that even closely related microorganisms are invaded by different phages (or plasmids). Expression studies in *T. thermophilus* have shown that phage infection upregulates several CRISPR/*cas* components: *cas* genes, RAMP module proteins, and CRISPR loci (Agari *et al.*, 2010). Based on this observation, CRISPR/*cas* systems can be regarded as a mechanism of sensing phage infections.

IV. MECHANISM OF CRISPR/*CAS*-CONFERRED PHAGE RESISTANCE

The mechanism of CRISPR/*cas* activity was suggested to be analogous to the eukaryotic mechanism of interfering RNA (RNAi). Moreover, both systems evolve quickly and show low primary level sequence similarities. However, although there are some mechanistic analogies, prokaryotic CRISPR/*cas* and eukaryotic RNA interference systems are not connected phylogenetically and differ in proteins and noncoding elements. Also the CRISPR mechanism of silencing foreign DNA occurs by its active recognition, whereas eukaryotic RNAi is a passive process. The CRISPR/*cas* system seems to have evolved as a protective function against extracellular DNA, whereas the eukaryotic system is directed toward endogenous DNA. Incorporation of new extracellular sequences (spacers) into the bacterial genome can be compared to a genetic type of memory, which protects the cell from later infections. Based on this aspect, the adaptive CRISPR/*cas* system imitates more the function of the immunological system of vertebrates rather than the RNAi activity model (Horvath and Barrangou, 2010).

A. Mode of action of CRISPR/*cas*-mediated resistance

The precise mechanism of acquiring resistance conditioned by the CRISPR system, as well as the mechanism of phage resistance itself, has not been fully elucidated. Current data allow determining, with all certainty, that the protection of bacterial cells against foreign invading genetic elements is conveyed by CRISPR spacer regions and proteins encoded by *cas* genes (Barrangou *et al.*, 2007; Jore *et al.*, 2011). These features are also responsible for the specificity of the protective response. The CRISPR/*cas* system is heritable, and CRISPR spacers are present in the progeny generations (van der Oost *et al.*, 2009).

Examination of CRISPR/*cas* systems identified in various bacterial and archeal species allowed determining the main stages of their mode of action, which are presented here (van der Oost *et al.*, 2009).

Stage 1. Immunization (or adaptation) phase, during which a new spacer(s) deriving from foreign DNA is incorporated into the CRISPR locus.

Stage 2. CRISPR expression, which involves transcription of CRISPR sequences and subsequent processing to small guide RNAs.

Stage 3. CRISPR interference effectuated by small guide RNAs (generated in stage 2) complexed with Cas proteins, which by binding and/or degradation activity eliminate invasive targets.

1. Adaptation phase (stage 1)

The first stage of CRISPR/*cas* activity is adaptation (or immunization). It is a step in which, in effect of foreign DNA invasion, the microbial genome acquires a new spacer(s) and gains resistance toward the infecting genetic element (Fig. 2). Spacers were previously considered to derive from mRNA. The fact that spacers originate either from sense (coding) or antisense (noncoding) strands was thought to be connected with the presence of a reverse transcriptase gene nearby the *cas* gene region (Makarova *et al.*, 2006). However, based on current knowledge, dsDNA is regarded as the source of spacers from both strands. This finding was confirmed for spacers of various species (e.g., from *Sulfolobus* sp., *Y. pestis*, *S. thermophilus*), with some exceptions (Barrangou *et al.*, 2007; Haft *et al.*, 2005; Lillestøl *et al.*, 2006; Pourcel *et al.*, 2005).

Studies performed on *Streptococcus mutans* and *S. thermophilus* revealed that phage infection of these bacterial cultures resulted in incorporation of new spacers within existing CRISPR loci (Deveau *et al.*, 2008; Horvath *et al.*, 2008; van der Ploeg, 2009). Detailed genome analyses of these bacterial species and, later, other microorganisms determined that the spot of integration of new spacers is almost always at the leader (proximal) end of the CRISPR locus, specifically between the leader and the first spacer (Barrangou *et al.*, 2007; Deveau *et al.*, 2008; Pourcel *et al.*, 2005).

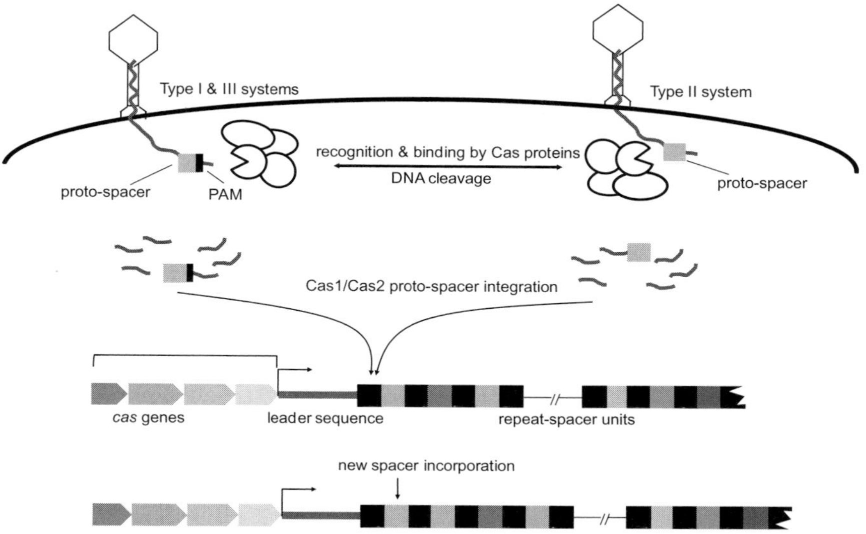

FIGURE 2 Adaptation stage of CRISPR/*cas* activity. CRISPR/*cas* machinery recognizes, binds, and subsequently cleaves invading DNA. In effect, phage-derived sequences (proto-spacers) are introduced into the CRISPR array in an event most probably promoted by Cas1 and Cas2 proteins. New spacers are added from the proximal (leader) end of the array, just after the leader sequence. In type I and II systems, selection of proto-spacers to be integrated into the repeat-spacer region possibly requires proto-spacer adjacent motifs (PAMs). (See Page 31 in Color Section at the back of the book.)

At the same time, repeat sequences that separate each spacer must be duplicated upon spacer integration. Duplication of CRISPR repeats is considered to proceed via a recombination event and resemble the integration of transposons (Steiniger-White *et al.*, 2004). The process is thought to initiate by the introduction of single-strand breaks at both ends of the repeat, subsequent insertion of a processed DNA fragment, and finally gap fill-in. Usually, one (up to four) spacer sequence is acquired at a time, but some exceptions to this rule have been reported (Barrangou *et al.*, 2007).

Acquisition of spacer sequences and duplication of repeat sequences were proposed to be catalyzed by certain Cas, as well as other host proteins. Among them, Cas1, Cas2, and Cas4 proteins were suggested to bind within the leader region and play main roles in the adaptation stage where Cas1 cleaves the invading DNA into fragments that serve as spacer precursors (van der Oost *et al.*, 2009; Wiedenheft *et al.*, 2009). Analyses performed in *S. thermophilus* suggested that the key element, which acts in this step of CRISPR/*cas* activity, is Csn2 (or Cas7) from the Cas family, subtype N. Its inactivation was determined to prevent the acquisition of new spacers (Deveau *et al.*, 2010). As Csn2 lacks homology to known proteins in databases, the exact mechanism of spacer integration is still unclear.

The CRISPR/*cas*-specific enzymatic machinery involved in the selection of phage (or plasmid) DNA fragments (proto-spacers) and their subsequent introduction into microbial remain to be established. Also, factors that recognize specific motifs and proteins engaged in ensuring proper spacer length need further studies. In a situation when no spacer is acquired, the phage lytic cycle or plasmid replication takes place.

i. Proto-spacer adjacent motifs Selection of foreign DNA fragments, which are to be integrated as spacers into the CRISPR arrays, seems not to be haphazard. An important determinant suggested to play a crucial role in attaining new spacers in type I and II CRISPR/*cas* systems are sequences termed PAMs (**p**roto-spacer **a**djacent **m**otif**s**). These several nucleotide-long sequences can be located either downstream or upstream in respect to the proto-spacer (Deveau *et al.*, 2008; Mojica *et al.*, 2009; Semenova *et al.*, 2009). Notably, after incorporation into the CRISPR locus, respective spacers lack the mentioned motifs. Thus, these specific motifs are a simple way for the CRISPR/*cas* system to distinguish between the target proto-spacer and the host spacer (lacking the motif).

The existence of PAMs was first reported in a study of randomly selected *S. thermophilus* BIMs (Deveau *et al.*, 2008). Analysis of the newly acquired spacer sequences revealed short motifs adjacent to proto-spacers in the genome of the infecting phage. Further examination of CRISPR loci in phage-resistant *S. thermophilus* showed that proto-spacers incorporated as spacers in the CRISPR1 locus are followed by a common motif NNAGAAW, while for CRISPR3 spacers a different downstream motif—NGGNG—was identified adjacent to their respective proto-spacers (Bolotin *et al.*, 2005; Deveau *et al.*, 2008; Horvath *et al.*, 2008). This observation allowed concluding that spacers of individual CRISPR loci are linked with particular PAM sequences. Different CRISPR/*cas* subtypes are accompanied by different PAMs, varying in sequence conservation and length (Lillestøl *et al.*, 2009; Mojica *et al.*, 2009). Similar motifs were also detected for proto-spacers of CRISPR loci found in other microbial genomes. For instance, in archaea (sulfolobales) a preference for 5′ PAMs was noted, in *S. mutans* either 3′ or 5′ PAMs were identified depending on CRISPR loci, while proto-spacers favored by the *Pelobacter carbinolicus* CRISPR locus were shown to be associated with a CTT motif at the 3′ end, typical also for *Geobacter sulfurreducens* CRISPR2 loci (Aklujkar and Lovely, 2010; Lillestøl *et al.*, 2009; van der Ploeg, 2009).

The fact that some spacers are more commonly represented than others suggests that PAMs might act in specific selection of proto-spacers or that these motifs occur more frequently on a particular strand (Barrangou *et al.*, 2007; Deveau *et al.*, 2008; Tyson and Banfield, 2008). Evidence also shows that PAMs influence the orientation by which a proto-spacer is introduced into the CRISPR array. As a general rule, proto-spacers accompanied by the same motif are always incorporated in the same direction. Although the mechanism of PAM activity in spacer acquisition remains to be described, the Cas3 protein has been proposed to be implicated in motif recognition (Sinkunas *et al.*, 2011; van der Oost *et al.*, 2009).

In the following steps of CRISPR/*cas* mode of action, the information stored in spacers is transcribed, processed, and subsequently used to prevent invasion of foreign DNA elements (described later as stages 2 and 3).

2. Transcription of CRISPR loci (stage 2)

Spacer elements are crucial determinants involved in the resistance of microbial cells to foreign DNAs. A study in *E. coli*, examining the mechanism of CRISPR/*cas* resistance, implied that in fact the key role in this stage is performed by transcripts of the entire CRISPR locus. Based on this finding, CRISPR expression delineates a distinct step of CRISPR/*cas* -mediated immunity (Fig. 3). During this phase, the CRISPR locus is transcribed into one long mRNA (pre-CRISPR RNA or pre-crRNA), extending over the leader region, CRISPR direct repeats, and multiple spacer sequences of the locus (Lillestøl *et al.*, 2006, 2009). The process was found to be constitutive and begin near (or in) the leader sequence. Based on the study in *Sulfolobus acidocaldarius*, differences detected in the length of CRISPR transcripts in the exponential stage vs stationary phase indicated that this specific mRNA is trimmed during cell growth (Lillestøl *et al.*, 2006). As a result of pre-crRNA processing, short mature products and a ladder of intermediate crRNAs are obtained (Brouns *et al.*, 2008; Carte *et al.*, 2008).

Also, observations of CRISPR transcription in *Archaeoglobus fulgidus* and *S. solfataricus* revealed formation of a long transcript (pre-crRNA) covering the entire CRISPR loci. The transcripts were determined to be subsequently processed into smaller and finally short mature RNA molecules (crRNAs). For *A. fulgidus*, depending on the CRISPR locus, the transcript is cleaved every 68 or 75bp, as the size of the detected crRNA fragments differed by these two lengths, whereas *S. sulfolobus* CRISPR mRNAs showed a regular 68-bp difference in length (Tang *et al.*, 2002, 2005). Similarly, apart from long pre-CRISPR transcripts, shorter RNAs were identified in several archeal and bacterial organisms.

Analysis of CRISPR expression in *E. coli*, *S. epidermidis*, and *Xanthomonas oryzae* showed that only the coding strand of CRISPR loci is transcribed and undergoes processing (Brouns *et al.*, 2008; Marraffini and Sontheimer, 2008; Semenova *et al.*, 2009), which is contrary to *P. carbinolicus*, for which RNAs representing both spacer strands were detected at

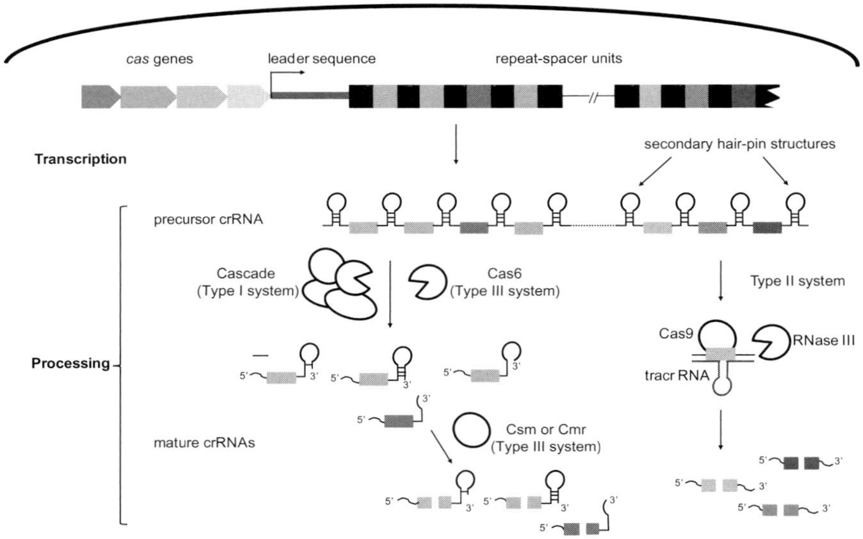

FIGURE 3 CRISPR expression and transcript processing. The CRISPR array is transcribed into one long RNA (precursor CRISPR RNA; pre-crRNA), starting from the leader sequence, suggested to constitute the CRISPR promoter region. Depending on the type of CRISPR/*cas* system, pre-crRNAs are processed differently. In type I and III systems, CRISPR transcripts are cleaved, respectively, by the Cascade or Cas6 protein into short CRISPR RNAs (crRNAs), composed of a spacer sequence, flanked from the 5′ side by a partial (eight to nine nucleotides) repeat and a heterogeneous 3′ end. In type III systems, crRNAs are processed further by Csm (subtype III-A) or Cmr (subtype III-B) Cas protein complexes. Alternatively, in type II CRISPR/*cas* systems, processing of pre-crRNA involves pairing with *trans*-encoded small RNA (tracrRNA). Such complexes are recognized by the housekeeping RNase III, which, in the presence of Cas9, cleaves the transcript within repeat sequences. Cas9 is also suspected to participate in the later maturation of crRNA. (See Page 32 in Color Section at the back of the book.)

similar levels in actively growing cultures, suggesting that both sense and antisense strands of the CRISPR region are transcribed (Lillestøl *et al.*, 2006, 2009). Also, observations in *S. acidocaldarius* showed that RNA molecules corresponding to both spacer strands are present, yet only during the stationary phase (Lillestøl *et al.*, 2006). For *P. furiosus* and *T. thermophilus*, RNA originating from transcription of the two strands was detected, although in unequal amounts (Agari *et al.*, 2010; Hale *et al.*, 2008). Whether this intriguing observation has any importance for CRISPR/*cas* function did not find any biological explanation, yet.

For type I CRISPR/*cas* systems, it was shown that crRNA interacts with Cas proteins, which implies participation of this class of proteins in transcript processing (Brouns *et al.*, 2008; Carte *et al.*, 2008). Indeed, a major role in this process is performed by a group of Cas proteins, termed Cascade (**Cas**-complex for **a**nti-virus **de**fense). Cascade recognizes specific secondary structures within the repeat sequences of pre-crRNA and digests them within specific sites. As a result, short crRNA products are generated. Formation of these secondary structures occurs within repeat sequences, containing palindromic regions. Palindromic sequences have been identified in repeats belonging to the most represented repeat groups (out of the 12 identified). By base pairing head to foot between palindromic sequences, stable hairpin structures are formed in the crRNA. The stem structure of the hairpin is strongly conserved; mutations within this region are compensated by other mutations in order to maintain the special structure (Kunin *et al.*, 2007; Makarova *et al.*, 2006). The cleavage site is localized downstream from the last nucleotide forming the hairpin.

First molecular studies on the expression of the CRISPR region and the role of Cas proteins in crRNA maturation were performed in *E. coli* strain K12 (type I system)(Brouns *et al.*, 2008). The Cascade complex was determined to be formed by "noncore" Cas proteins, CasABCDE (termed in different sources also as Cse1-Cse2-Cse4-Cas5e-Cse3 or Cas proteins of Cse type)(Deveau *et al.*, 2010; Jore *et al.*, 2011). Their general function was determined to involve cleavage of the precursor transcript within the repeat sequence. Mutations within *casE* (*cse3* or *cas6e* according to the most recently proposed nomenclature) and *casD* (*cas5e*) gene regions impaired pre-crRNA cleavage, which pointed to a particular engagement of these two respective proteins in the process. Furthermore, CasE, which can act independently of the Cascade complex, has been shown to be essential in proper recognition of the 5′ end of the pre-crRNA and its digestion into shorter crRNAs (Brouns *et al.*, 2008). In the *E. coli* type I CRISPR/*cas* system, mature crRNA remains associated to the Cascade complex after cleavage (Haurwitz *et al.*, 2010).

Cas proteins, including specific Cas endoRNases, involved in transcript processing have also been characterized in other organisms. For type III CRISPR/*cas* systems identified in *P. furiosus* and *T. thermophilus*, processing of crRNA is catalyzed not by a Cascade complex, but by a single protein, Cas6 and CasE (or Cas6e), respectively. Despite lack of sequence homology, both proteins exhibit similar structures, implying endonucleolytic function for both (Carte *et al.*, 2008; Ebihara *et al.*, 2006; van der Oost *et al.*, 2009). Specifically, the *P. furiosus* Cas6 protein was shown to interact with the 5′ end of crRNA and introduce a cut within the repeat sequence, eight nucleotides upstream of the spacer (Carte *et al.*, 2008). In type III systems, after initial processing, crRNA is then passed on to a Cascade complex (Cmr or Csm type), which catalyzes further cleavage at the 3′ end (Hale *et al.*, 2009; Wang *et al.*, 2011). For *T. thermophilus*, the newly obtained crystal structure of the Cascade Cse2 (CasB) protein revealed its α-helical structure with a positively charged surface. Its role was proposed to involve RNA binding (Agari *et al.*, 2008). *P. carbinolicus* homologues of the *E. coli* Cascade complex, responsible for processing of crRNA, have also been identified (Brouns *et al.*, 2008). Yet another Cas protein, Csy4 (or Cas6f according to the most recent nomenclature), of *P. aeruginosa* was crystallized (Haurwitz *et al.*, 2010). Similarly, as for *E. coli* CasE and Cas6 of *P. furiosus*, Csy4 was determined to be a metal ion-independent endonuclease responsible for crRNA cleavage. Despite lack of sequence homology, Csy4 exhibits a similar protein fold to CasE (*T. thermophilus*) and Cas6 (*P. furiosus*) proteins.

For the type II CRISPR/*cas* system, an alternative model of crRNA processing was described. Based on studies in *S. pyogenes*, cleavage of pre-crRNA within repeats was shown to be performed by the housekeeping RNAse III in the presence of Csn1 (Cas9) guided by *trans*-encoded small RNA (tracrRNA)(Deltcheva *et al.*, 2011). Further processing of crRNAs most probably involves Cas9 digestion within spacer regions.

Overall, processing of long precursor crRNAs into target-active RNA seems to be an unambiguous event in all CRISPR-containing organisms and a prerequisite for CRISPR-mediated resistance. The size of mono-spacer RNAs are, on average, 35–46 nucleotides long, with some minor differences (Hale *et al.*, 2008). In *E. coli*, in effect of Cascade activity, short crRNAs (57 nucleotides) are generated by cleavage within the repeat sequence, specifically 8 nucleotides upstream of the spacer. Mature crRNAs comprise a spacer sequence flanked by two partial repeats—a 5′ 8 nucleotide fragment and a 3′ fragment, which forms a hairpin structure. These final products of crRNA processing were proposed to act as guide sequences during the CRISPR interference step (presented as stage 3 in the following part of this chapter). In *S. acidocaldarius*, endonucleolytic digestion of CRISPR mRNA was shown to generate a group of 35–52 nucleotide RNA products (Lillestøl *et al.*, 2006). Also, for *P. furiosus*, two types of mature crRNA products are observed, 38–45 and 43–46 nucleotides long, which suggest that crRNA processing involves both endo- and exonucleolytic processing. The final *P. furiosus* mono-spacer crRNAs consist of a spacer and partial repeat sequence at the 5′ end only, while the 3′ ends are trimmed by a yet uncharacterized protein (Carte *et al.*, 2008; Hale *et al.*, 2008). In *S. epidermidis*, pre-crRNA was shown to be processed to mature crRNA, containing one spacer sequence flanked by an 8–9 nucleotide repeat at the 5′ termini and by a longer, heterogeneous repeat at the 3′ end (Marraffini and Sontheimer, 2010b). The Csy4 protein of *P. aeruginosa* was shown to cleave pre-crRNA into 60 nucleotide-long fragments, comprising a 32 nucleotide spacer region flanked by 20 and 8 nucleotide repeats from the 3′ and 5′ end, respectively (Haurwitz *et al.*, 2010).

A study of crRNA processing in *P. furiosus* revealed yet another interesting aspect. It was observed that mature crRNA molecules, corresponding to newly acquired spacers (near the leader end), appear more abundantly than crRNAs of older spacers. It would be interesting to examine whether these differences are associated with transcription, cleavage, or stability of the crRNA transcripts. It could be that crRNAs, comprising older spacers, which, as mentioned earlier, are often surrounded by more degenerate repeat sequences, are not processed properly and do not constitute the final pool of mature crRNAs. Generally, all mature crRNAs contain (i) a spacer, (ii) a short, eight to nine nucleotide-long repeat sequence at the 5′ end, and, optionally, depending on the CRISPR/*cas* subtype, (iii) a 3′ termini. Repeat sequences at the 5′ ends of mature crRNA molecules are regarded as tags, which are conserved among a specific group of organisms (Goode and Bickerton, 2006; Kunin *et al.*, 2007). The majority of mono-spacer crRNAs from CRISPR locus 8 of *P. furiosus* were identified to contain a 5′ seven nucleotide repeat motif AUUGAAG. In *E. coli*, mature crRNAs contain an eight nucleotide AUAAACCG repeat at their 5′ end, whereas *S. epidermidis* crRNAs carry an eight nucleotide ACGAGAAC 5′ motif (Brouns *et al.*, 2008; Kunin *et al.*, 2007; Marraffini and Sontheimer, 2010b). Moreover, a correlation between the 5′ tag sequence of crRNAs and specific Cas proteins has been determined (Kunin *et al.*, 2007). This finding strongly implies that the 5′ repeat tags are sites of recognition by specific Cascade(-like) proteins, which bind to them, forming CRISPR ribonucleoprotein (crRNP) complexes. However, it still needs to be established whether the partial repeat at the 3′ end of mature crRNAs has any significant role and whether variations in this region have any impact on their functionality or is just an artifact.

i. Regulation of CRISPR transcription In contrast to increasing data on Cas proteins and their function, the regulation of CRISPR loci transcription has not been studied in great detail. It has been implied that the transcription of *cas* genes seems to be regulated differently, depending on the microbial species or even strain (Deveau *et al.*, 2010). Among scarce data dealing with this issue is the microarray study performed for *T. thermophilus*. Results obtained in this analysis showed that depletion of glucose activates transcription of *cas* genes via the cAMP receptor protein (Shinkai *et al.*, 2007). Interestingly, an analogous experiment performed in *E. coli* did not reveal such an effect (Gosset *et al.*, 2004). In fact, it was shown in *E. coli* that expression of the CRISPR/*cas* operon is regulated by the global regulator—H-NS (heat-stable nucleoid-structuring) protein, which acts as a transcriptional repressor by binding to the *casA* transcriptional region, and LeuO, a LysR-type transcription factor, an activator of *cas* transcription and antagonist of H-NS (Westra *et al.*, 2010). Similarly, in *Salmonella enterica* serovar Typhi, H-NS, together with another regulator, LRP, plays a role of global repressor of CRISPR/*cas* transcription, while LeuO acts as a positive activator (Medina-Aparicio *et al.*, 2011). Data also suggest other regulatory factors that influence CRISPR/*cas* transcription in this bacterium. In *Myxococcus xanthus*, *cas1-4* and *cas6* genes are cotranscribed with *dev* genes, encoding key functions for *Myxococcus* development, and suggested to be negatively autoregulated by DevS (Viswanathan *et al.*, 2007). Also, upregulation of *cas* genes encoded by the CRISPR locus of the *S. mutans clp* mutant strain was demonstrated (Chattoraj *et al.*, 2010). In addition to these reports, no other communications elucidating CRISPR transcription have been made.

The final products of CRISPR expression are short mono-spacer crRNA fragments with 5′ partial repeats. These 5′ ends act as tags, which attract the associated Cas proteins (particularly nucleases) and direct them to foreign genetic elements, carrying sequences homologous to the spacers. Most likely, the crRNA molecule recognizes a particular sequence in the phage genome and tags it for degradation by the associated Cas protein complex. This activity is the quintessence of CRISPR-mediated interference, constituting the following step of CRISPR/*cas* mode of action (stage 3).

3. Resistance phase—CRISPR interference (stage 3)

The final phase of CRISPR-mediated activity is based on interference, which involves binding and/or degradation of the target foreign DNA (Fig. 4). The main role in this phase is performed by mature crRNAs, which serve as guides in recognizing and destroying invading elements with matching sequences. In type I CRISPR/*cas* systems, mature crRNA does not act alone, but forms a ribonucleoprotein (crRNP) complex with Cascade proteins and guides them to target DNA, cleaved subsequently by the Cas3 subunit or possibly Cas4 (in cases when the Cas4 RecB nuclease domain is fused to Cas1)(Sontheimer and Marraffini, 2010). Detection of matching invasive targets by crRNPs relies on a base-pairing process, which demands faithful complementation between a spacer and invading proto-spacer. Even a single point mutation within one of the sequences could abolish CRISPR interference (Deveau *et al.*, 2008; Marraffini and Sontheimer, 2010b). Most probably, proper recognition of target sequences in type I and II systems (but not type III systems) also involves PAMs. Indications of the interference between cRNA/target DNA were mentioned previously in the study performed by Barrangou *et al.* (2007) on *S. thermophilus*-infected cultures, which established that CRISPR/*cas* systems confer resistance to phages. Further studies on expression of synthetic CRISPRs in *E. coli* strain K12 revealed the detailed mechanism of the process (Brouns *et al.*, 2008). Introduction of specific CRISPR/*cas* components, including different *cas* genes and phage λ DNA-derived spacers into the bacterial genome, allowed determining crucial elements of CRISPR-mediated phage resistance. Cas proteins, specifically Cas1 and Cas2, together with spacer sequences alone, were

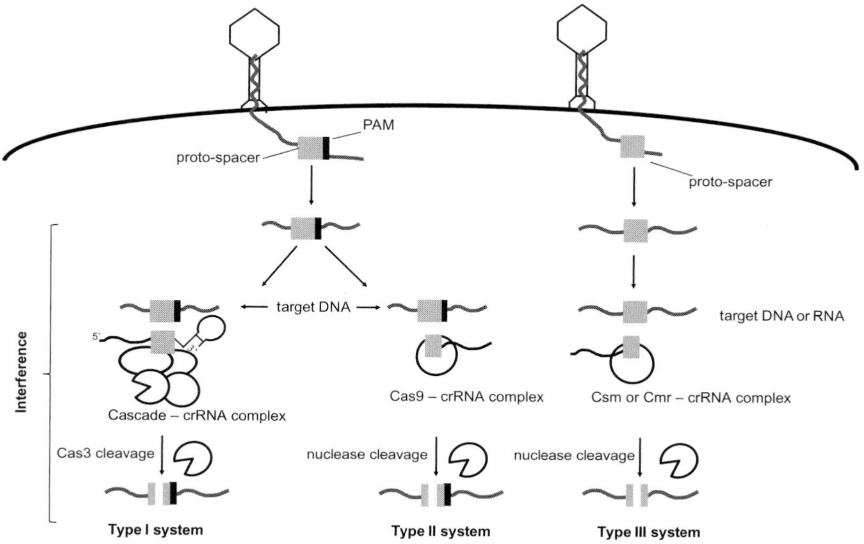

FIGURE 4 Interference stage of CRISPR/*cas* activity. Interference with foreign invading nucleic acids (DNA or RNA) occurs by complementary binding of the crRNA spacer with matching sequence target (proto-spacer) and subsequent cleavage of the latter by components of the CRISPR/*cas* system. In type I systems, crRNA binds with Cascade, forming a ribonucleoprotein (crRNP) complex that targets DNA with complementary sequences. Digestion of foreign DNA is catalyzed by Cas3. In type II systems, the crRNA–Cas9 complex also targets complementary DNA, which is then cleaved by Cas9 or another yet unidentified nuclease. Both systems (type I and type II) most probably require PAMs for proper target recognition. crRNA–Cas complexes in type III systems target DNA (subtype III-A) or RNA (subtype III-B), respectively. The specific nuclease(s) involved in this process remains to be identified.

insufficient in providing resistance to phage λ. However, the presence of Cascade (CasA-E), as well as the Cas3 protein, resulted in resistance of the CRISPR-containing strain to λ infection. Although cleavage of crRNA or assembly of the Cascade complex was shown to be independent of Cas3, its role in interference was proposed to involve base pairing of crRNA with a matching proto-spacer and its subsequent degradation (Brouns *et al.*, 2008; Karginov and Hannon, 2010). In biocomputational assays, Cas3 was identified as a helicase and suggested to specifically unwind target DNA, allowing hybridization of the proto-spacer with crRNA. Moreover, the majority of Cas3 proteins are fused with a nuclease domain (Cas2), most probably implicated in target degradation. A similar role of target destruction was proposed for Cas4, which was described as a RecB-like exonuclease (Makarova *et al.*, 2002). A Cas3 orthologue was identified in *S. solfataricus* and characterized as a nuclease specific on dsDNA, as well as dsRNA favoring GC pair digestions (Han and Krauss, 2009). The basis of Cas3 activity, similarly as in stage 1, was suggested to involve recognition of specific motifs adjacent to the proto-spacers (PAMs). However, not all proto-spacers have distinct motifs and not all CRISPR/*cas* systems contain Cas3 as a general component, which implies that there must be other proteins of analogous function.

Type II and III CRISPR/*cas* systems lack Cas3 orthologues, and nucleases of analogous function remain to be identified. For type II CRISPR/*cas* system of *S. thermophilus*, Cas9 (predicted nuclease) was proposed to be the functional analogue of both Cas3 and Cas4 (Haft *et al.*, 2005). Its deletion abolished the CRISPR immunity effect (van der Oost *et al.*, 2009). Most probably the crRNA–Cas9 complex interacts directly with the invading DNA, requiring PAMs for proper target recognition (Haurwitz *et al.*, 2010). For type III CRISPR/*cas* systems, the interference mechanism of the two subtypes is based on targeting different substrates. Subtype III-A systems target DNA, while III-B subtypes are directed against RNA targets (Hale *et al.*, 2009; Makarova *et al.*, 2011b; Marraffini and Sontheimer *et al.*, 2008).

The mechanism by which spacers induce resistance to a phage, which carries matching proto-spacers, is still under examination. Based on comparative genomic analysis, especially the diversity of Cas protein functions, it has been suggested that the interference mechanism of CRISPR/*cas* systems is analogous to eukaryotic RNAi systems (Makarova *et al.*, 2006). Despite lack of orthologous components, some functional analogies between Cas proteins and the enzymatic apparatus of the eukaryotic RNAi system were recognized, such as the presence of helicases, a broad spectrum of nucleases, specific polymerase, and a group of RNA-binding proteins.

i. Model of CRISPR-mediated interference—DNA vs RNA silencing Initially the interference of CRISPR transcripts (mature crRNA) was suggested to occur though pairing with invader-derived mRNAs carrying target sequences. Interaction of crRNA with target proto-spacers would initiate their degradation (or cause translation shutdown), thus acting like a RNA-silencing system, analogous to eukaryotic RNAi. However, studies performed for *E. coli* and *Staphylococcus* sp. suggested a somewhat different mechanism of interference. Based on obtained results, CRISPR-mediated interference appears to be based on DNA silencing (Brouns *et al.*, 2008; Marraffini and Sontheimer, 2010a). At present, the interference model of type I, type II, and type III-A CRISPR/*cas* systems considers DNA as target sequences, while the type III-B subtype recognizes and degrades RNA (Makarova *et al.*, 2011b).

The DNA interference model of the majority of CRISPR/*cas* systems is supported by several observations concerning spacer sequences. First of all, results obtained from the analysis of spacer regions of various species, including *S. solfataricus*, *S. mutans*, *S. thermophilus*, *Y. pestis*, and *X. oryzae*, showed that they can originate from coding (sense) as well as noncoding (anti-sense) strands of the invading DNA (plasmid or phage)(Cui *et al.*, 2008; Deveau *et al.*, 2008; Horvath *et al.*, 2008; Lillestøl *et al.*, 2006; Semenova *et al.*, 2009; van der Ploeg *et al.*, 2009). The earlier mentioned study in *E. coli* with the use of artificial λ-derived spacers confirmed this observation; spacer sequences of mature crRNAs matched sequences from both strands of the phage λ genome, while for *S. thermophilus* BIMs, more spacers were detected for the coding strand (Barrangou *et al.*, 2007; Brouns *et al.*, 2008). All of the aforementioned findings indicate that crRNP complexes can recognize antisense sequences; thus, it is more probable that they target dsDNA rather than mRNA. Also, no spacers matched sequences from RNA phages, which could be interpreted as an additional argument for the DNA interference model (Mojica *et al.*, 2009). However, this could also be due to the limited amount of RNA phage genome sequences present in databases.

Another indication on DNA interference was provided by studies performed on a clinically isolated *S. epidermidis* strain, RP62a. The CRISPR locus of this strain was determined to carry a spacer against the nickase (*nes*) gene present on a conjugative plasmid. Nickase activity is essential in donor cells during the conjugation process. If the hypothesis of mRNA targeting would apply, then targeting of nickase mRNA would abolish the donor function of the strain, but the cell should maintain its recipient abilities. The nickase spacer, however, inhibited entry of the plasmid, thus providing proof that both donor and recipient functions of the strain were impaired. However, interruption of the nickase proto-spacer by a self-splicing intron did not affect plasmid conjugation into RP62a. The fact that the intron disrupts the nickase gene sequence in the DNA, which is reconstituted in the RNA, strongly implies that DNA is the aim of CRISPR interference. Also, introduction of a nickase proto-spacer into a nonconjugative plasmid (pC194) prevented plasmid transformation into the RP62a strain. This proved that the observed effects are independent of the mode of plasmid entry (Marraffini and Sontheimer, 2008). Similarly, as for phage-derived spacers, the orientation of the target region in the plasmid was shown to be irrelevant for CRISPR interference. All of the observations just given reinforced the DNA-targeting model. The DNA interference model also explains the lack of *cas* genes for some CRISPR regions, which, in the light of this theory, was suggested to be due to self-interference of *cas* genes with *cas*-derived spacers, acquired by an ancestral organism, which led to deletion of these *cas* gene targets.

Nevertheless, experimental data cannot completely exclude RNA targeting as the mechanism of action of other CRISPR/*cas* systems. This scenario of CRISPR interference seems to apply for systems that encode the RAMP module (described earlier) and was considered in light of the biochemical characterization of crRNP complexes from *P. furiosus* (subtype III-B CRISPR/*cas* system)(Hale *et al.*, 2009). The study showed that RAMP proteins, Cmr1–Cmr6, interact with mature crRNAs possessing a common 8 nucleotide 5′ repeat tag. This ribonucleoprotein complex was determined to have endonucleolytic activity toward RNA targets with sequence matching endogenous crRNAs. At the same time, no DNA cleavage was detected. Within the complex, the Cmr2 protein is a predicted nuclease; Cmr1, Cmr3, Cmr4, and Cmr6 are ribonucleases; and Cmr5 is a putative RNA-binding protein (Beloglazova *et al.*, 2008; Brouns *et al.*, 2008; Carte *et al.*, 2008; Makarova *et al.*, 2006; Sakamoto *et al.*, 2009). Moreover, Cmr proteins were shown to form complexes with both species of mature crRNA identified in *P. furiosus* (38–45 and 43–46 nucleotides). Detailed studies determined that endonucleolytic cleavage of RNA targets occurs precisely at a distance of 14 nucleotides opposite the 3′ end of the crRNA. Truncation of mature crRNA from the 5′ end does not influence cleavage efficiency significantly, in contrast to 3′ truncations. The RNA-targeting model was confirmed using synthetic crRNAs and recombinant Cmr proteins. Obtained results established that crRNAs, together with the RAMP module (Cmr proteins), direct the cleavage of target RNA in *P. furiosus* (Hale *et al.*, 2009). The proposed model of CRISPR interference awaits its confirmation *in vivo*, including determination of whether the proto-spacers are sense oriented or insensitive to disruption by an intron.

The activity of CRISPR/*cas* systems based on RNA interference could also apply to other species expressing RAMP module proteins. Cmr proteins have been identified in both archaea, including *Archaeoglobus* sp. and *Sulfolobus* sp., and such bacterial species as *Bacillus* and *Myxococcus*. However, functional studies *in vivo* have not been performed so far to confirm this model. Moreover, identification in RAMP-containing *B. halodurans* and *S. solfataricus* CRISPR spacers that are sense and antisense oriented in reference to proto-spacers of foreign elements suggests that DNA interference in these species may also be possible. However, it cannot be excluded that in other species, which do not encode Cmr proteins, RNA target sequences can also be destroyed by an RNAi mechanism. In such cases, it is speculated that the role of Cmr proteins is fulfilled by other (Cas) proteins. Given all of the aforementioned, the large diversity of CRISPR loci components and encoded Cas proteins implies various CRISPR-mediated mechanisms of silencing invading genetic elements.

ii. Discrimination between self and nonself sequences An interesting aspect of CRISPR interference is the mechanism of discriminating between own and foreign sequences. It has been established that CRISPR-conferred immunity is based on recognition of the invading DNA by crRNA–Cas ribonucleoprotein complexes and involves direct interaction of a crRNA spacer with it counterpart proto-spacer target

DNA. However, crRNAs also match with spacer sequences present in the CRISPR locus. This situation evokes a question as to how the CRISPR/*cas* system targets proto-spacers without degrading its own host spacers. It was suggested that identification of self/nonself sequences was linked to repeat regions flanking the spacers. The detailed answer to this problem was provided by a study performed in *Staphylococcus epidermidis* (Marraffini and Sontheimer, 2008). Introduction of a CRISPR spacer homologous to the *nes* gene flanked from either side by 200-bp sequences into the pC194 plasmid, which normally cannot be transformed into CRISPR-containing cells, resulted in its effective transformation. Contrarily, the presence of the matching proto-spacer sequence on the same plasmid was insufficient to introduce the molecule into staphylococcal cells. This assay provided evidence that spacers, in contrast to proto-spacers, are somehow excluded from CRISPR-conferred immunity. Shortening the spacer-flanking sequences established that protection from the CRISPR autoimmunity effect was connected with sequences laying outside the spacer sequences, precisely the 5′ repeat region localized upstream of the spacer sequence. This region is identical for both crRNA and CRISPR DNA, which permits perfect base pairing between the two molecules. In contrast, invasive targets lack this complementarity and are subjected to CRISPR immunity. It is here where the crucial point of self/nonself sequence discrimination lies.

Further mutational studies in *S. epidermidis* established that especially crucial for homologous base pairing are 8bp closest to the spacer from the 5′ end, particularly three positions (−4, −3 and −2). Point mutations in these sites generated mismatches between the spacer and the 5′ terminal partial repeat of crRNA. This, in effect, triggered cleavage of the spacer by the crRNA–Cas machinery (Marraffini and Sontheimer, 2010b). Subsequent examinations determined that the nucleotide sequence of the terminal motif itself, including positions −4, −3 and −2, has an inferior importance and that, in fact, only a perfect complementarity between crRNA and the 5′ repeat sequence of CRISPR DNA is crucial for preventing the autoimmune response.

The mechanism of self/nonself discrimination of DNA targets has been described in only several microorganisms. However, the fact that the pairing of crRNA and target DNA is a common element of interference of all CRISPR/*cas* systems allows inferring that similar mechanisms of preventing autoimmunity could apply in other microbial species as well.

B. Evasion of CRISPR/*cas*-mediated resistance

The constant interplay between bacteria and their infecting phages leads eventually to selection of phages that are apt at evading CRISPR/*cas*-mediated resistance. The study of *S. thermophilus* BIMs containing CRISPR loci revealed that a fraction of the phage population was still able to infect such cells. Further analysis of CRISPR/*cas* sequences of BIMs showed that this was the effect of single point mutations or deletions within the phage proto-spacer regions (Barrangou *et al.*, 2007). Other phages able to infect BIM cells carried mutations within the CRISPR motif sequence (PAM) adjacent to proto-spacers. Further studies determined that even a single point mutation in these regions can impair the CRISPR defense mechanism sufficiently (Deveau *et al.*, 2008; Lillestøl *et al.*, 2009; Mojica *et al.*, 2009). It should be mentioned that this means of escaping CRISPR interference has been observed for proto-spacers associated with PAMs, as in the case of proto-spacers identified for *S. thermophilus* (Deveau *et al.*, 2008). However, not all proto-spacers possess distinguishable PAMs. This also includes proto-spacers of the aforementioned *S. epidermidis*. Therefore, based on current knowledge, it seems that evasion of CRISPR interference may involve different mechanisms, depending on the system.

The ability of phages to escape CRISPR/*cas* immunity led researchers to set forward a hypothesis on the existence of phage-encoded **anti-CRISPR systems**. Studies in *S. solfataricus* P2 suggested that its residing prophage encodes a protein of putative anti-CRISPR activity. The purified protein was determined to bind preferentially to CRISPR repeats and induce a conformational change in the DNA, forming an open structure near the center of the repeat (Peng *et al.*, 2003; Sorek *et al.*, 2008). The exact role of this protein in anti-CRISPR activity remains to be established; however, it can be speculated that binding of the protein to the DNA within the repeat sequence disrupts accurate base pairing between crRNA and proto-spacers necessary for target degradation. Genes encoding homologues of *Sulfolobus* protein were also discovered in other bacterial genomes, always within prophage sequences, indicating that similar putative anti-CRISPR systems may function in other microbial species as well (Sorek *et al.*, 2008).

CRISPR arrays were also identified on plasmids and in prophage sequences. The biological sense of such localization is still under discussion; however, it seems plausible that these loci act as anti-CRISPR mechanisms directed against CRISPRs present on invasive genetic elements, which, in effect, eliminates their entry into the cell. It has also been suggested that anti-CRISPR mechanisms are also encoded in microbial chromosomes as a protection against CRISPR/*cas*-containing invading elements. The presence of plasmid-encoded CRISPR/*cas* systems, carrying spacers matching chromosomal CRISPR sequences, could lead to serious cellular interference effects, including alteration of gene expression. An example of such a chromosomal anti-CRISPR system is the CRISPR3 *cas*-free region, identified in a majority of *E. coli* strains (Touchon and Rocha, 2010). Apart from the lack of *cas* genes, CRISPR3 was found to contain spacers matching sequences of several types of *cas* genes, connected with a specific CRISPR/*cas* subtype (Ypest). Interestingly, the genomes of CRISPR3-containing *E. coli* strains possess no *cas* genes from the Ypest subtype. This indicates that the anti-CRISPR interference activity is directed against invasion of foreign DNA elements carrying active CRISPR/Cas systems.

Another phage CRISPR-evading strategy reported by Andersson and Banfield (2008) is based on genome rearrangements. Examination of phages from two biofilm *Leptospirillum* sp. populations revealed significant variations between their genome sequences. The observed differences were implied to be due to extensive homologous recombination events in response to CRISPR immunity. Individual phages shared short (<25 nucleotides) sequence homologies suggested to occur as a result of reshuffling of polymorphic loci. These rearrangements were claimed to be sufficient to evade interference by the 28–54 nucleotide CRISPR spacers. Recombination was proposed to be a more efficient anti-CRISPR strategy compared to sequence mutations as it is less likely to lead to changes affecting protein function.

V. ADDITIONAL ROLES OF CRISPR/*CAS* SYSTEMS

Apart from the main function of CRISPR/*cas* systems, which is providing resistance against foreign genetic elements, other functions have also been implied. Based on genome analysis of archaea species, CRISPRs were suggested to be involved in **regulatory** rather than inhibitory **functions**. Many archeal CRISPR spacers are homologous to sequences of their own resident prophages or plasmids. This potential regulatory activity of archeal CRISPR could explain the low plasmid copy numbers and the fact that various archeal phages are nonlytic (Deveau *et al.*, 2010; Shah *et al.*, 2009). Also, in bacteria, the CRISPR/*cas* system was suggested to regulate the expression of genes engaged in various cellular processes. This idea seems to be supported by genomic data, which revealed that 7 to 35% of CRISPR spacers are homologous to bacterial DNA (Bolotin *et al.*, 2005; Horvath *et al.*, 2008; Mojica *et al.*, 2005).

In *P. aeruginosa*, the presence of the CRISPR locus was determined to be implicated in **inhibition of biofilm formation and swarming motility**, which are the main characteristics of this bacterial species (Zegans *et al.*, 2009). The aforementioned functions in this bacterium are inhibited by a lysogenic phage DMS3. Mutations or deletions within the CRISPR array or in certain *cas* genes were shown to reestablish these activities. It was suggested that gene regulation is based on the presence of spacers complementary to housekeeping genes. Overall, these observations imply a CRISPR-mediated modification of the *P. aeruginosa* lysogeny state and suggest a distinct role of CRISPRs in resistance to lytic phages vs lysogenic phages. CRISPR-mediated impairment of biofilm formation and swarming upon lysogeny can be a way of reducing phage dissemination in the bacterial population. This can be seen as an altruistic behavior of microorganisms that alters their physiological features to limit the infection of a larger bacterial community.

A study using *P. carbinolicus* revealed that the CRISPR/*cas* system can also **influence expression of a housekeeping gene**, *hisS* (Aklujkar and Lovely, 2010). The *P. carbinolicus hisS* gene, encoding for histydyl-tRNA synthase, is important for cellular growth and physiology. An experiment involving introduction of a *P. carbinolicus*-derived CRISPR spacer and its potential target, *hisS*, into the genome of a related species, *G. sulfurreducens*, was shown to impair the growth of this microorganism. This strongly suggested that the CRISPR spacer interferes with *hisS* gene expression. It is also appeared that no other determinant (e.g., Cas proteins), other than the associated CRISPR repeat, was necessary for the observed *P. carbinolicus* spacer–*hisS* gene interaction. Such activity reduced amounts of histydyl-tRNA in the cell greatly. *P. carbinolicus* adapted to the CRISPR-mediated interference with its endogenous *hisS* gene by selecting genes that demand fewer histidines and eliminating genes in which histidine codons appear frequently. This example illustrates how important the role of the CRISPR/*cas* system can be in the evolutionary selection of housekeeping genes, which could, in effect, lead to significant changes in the genomes and physiology of microorganisms.

Based on experimental studies performed in *M. xanthus*, the CRISPR/*cas* system was also shown to be involved in **spore formation** (Viswanathan *et al.*, 2007). The *M. xanthus devTRS* locus, responsible for fruiting body development upon starvation, was shown to be cotranscribed with the CRISPR region. Specifically, DevR and DevS were determined to be part of the CRISPR locus and identified as Cas proteins, respectively, from the Cst2 and Cas5 families (Haft *et al.*, 2005). Moreover, detailed examinations revealed that DevR interacts with FruA. This example shows that CRISPR/*cas* regions can play a role in alternative pathways and interact with other proteins involved in cellular regulation.

Another suggested alternative activity of CRISPR/*cas* systems is a **role in replicon maintenance and segregation**. Studies in *Haloferax volcanii* showed that addition of new repeats to already existing CRISPR loci led to changes in segregation and lowered cell viability (Mojica *et al.*, 1995). This role of the CRISPR/*cas* system was further supported by a study of two conjugative plasmids of *Sulfolobus* sp. carrying repeat sequences. Deletion of these repeats from the plasmid sequence was shown to affect their stability (Greve *et al.*, 2004).

From a general point of view, CRISPR repeat sequences are also regarded as **hot spots for recombination**. Studies performed in *Thermotoga* species have shown that CRISPR-mediated rearrangements are as frequent as those occurring within tRNA genes (DeBoy *et al.*, 2006). This delineates the important role of CRISPRs in the evolution of microbial genome sequences.

A study performed by Babu *et al.* (2011) in *E. coli* suggests that certain components of the CRISPR/*cas* system are involved not only in antiviral immunity, but also in **DNA repair–recombination**, as implied in earlier works (Makarova *et al.*, 2002). Particularly, *E. coli* Cas1 (YgbT) was found to interact with DNA repair and recombination proteins, whereas cells lacking Cas1 were impaired in chromosome segregation and DNA repair. These results indicate that the universally occurring Cas protein may play various roles, which is further confirmed by the fact that many prokaryotic organisms encode for more than one Cas1 paralog.

VI. CRISPR/*CAS* SYSTEMS IN VARIOUS MICROBIAL SPECIES

A. *Streptococcus thermophilus*

Over the last decade, CRISPR/*cas* systems have been identified and studied in various microorganisms. One of the first bacterial species in which CRISPR arrays were observed was *S. thermophilus*, which genome was found to carry two distinct CRISPR loci, CRISPR1 and CRISPR2 (Bolotin *et al.*, 2005). A study of *S. thermophilus* bacteriophage-insensitive mutants by Barrangou *et al.* (2007) allowed linking CRISPR spacers with resistance to phages. Further comparative analysis allowed examining the diversity of CRISPR arrays in 124 various *S. thermophilus* strains and identified a third CRISPR locus, CRISPR3 (Horvath *et al.*, 2008). The study also showed that among the identified spacer sequences (over 3600 spacers) in the genomes of those strains, 77% matched phage and 16% plasmid-derived sequences. Additionally, a small fraction of spacers was found homologous to chromosomal sequences, specifically *dtpT* and *rexA* genes, suggesting a potential role of CRISPR loci in regulating the expression of genes encoded on the chromosome. CRISPR3 was detected in the majority of analyzed strains, whereas CRISPR1 was confined to only a few streptococci. Moreover, based on spacer polymorphism, conservation of their size, and high number within a given locus, CRISPR1 was determined to be the most diverse and, thus, suggested to be the most active locus, whereas CRISPR2 was thought to be truncated. Specific motifs, PAMs, were detected for proto-spacers linked respectively to CRISPR1 (NNAGAAW motif) and CRISPR3 (NGGNG motif). This finding confirmed earlier speculations that each *S. thermophilus* CRISPR locus has a unique CRISPR motif, recognizable by specific Cas enzymatic machinery.

B. *Escherichia coli* and *Salmonella*

An analysis of over 50 genome sequences of *E. coli* and *Salmonella* revealed 125 CRISPR arrays, localized in four separate loci: CRISPR1–4 (Touchon and Rocha, 2010). The majority of *E. coli* genomes carried three (CRISPR1–3) and *Salmonella* genomes two different CRISPR regions (CRISPR1–2). CRISPR4 was implied to have evolved quite recently, as it was found only in genomes of a distinct *E. coli* group (B2), which at the same time lacked the CRISPR1 locus. Further results indicated frequent replacement or loss of *cas* genes. CRISPR regions, containing spacers matching plasmid-encoded genes responsible for replication, conjugation, and antirestriction, were identified. In total, from the 49 spacers exhibiting homology, 14% matched phage, 42% plasmids, and 53% chromosomal sequences.

Detailed analysis of CRISPR arrays in *E. coli* allowed assigning them into two functional groups: CRISPR1/CRISPR2 and CRISPR3/CRISPR4. The main features distinguishing the two groups are (i) colocalization in the genome, (ii) homologous repeats (29 and 28bp, respectively) with palindromic sequences able to form stable secondary structures in the RNA, and (iii) similar adjacent *cas* gene subtypes (Touchon and Rocha, 2010). The latter observation led to a conclusion that *cas* genes are specialized toward activity on a particular type of repeat sequence. Therefore, they could act also *in trans* on CRISPRs that carry identical repeats. It was also shown that CRISPR1 and CRISPR2 expression is coregulated by the global H-NS regulator (Pul *et al.*, 2010). Moreover, CRISPR1 and CRISPR2 were determined to contain spacers that match sequences of phage origin, whereas CRISPR3 and CRISPR4 spacers correspond to sequences of plasmids. Phylogenetic analyses performed by Touchon and Rocha (2010) further revealed some interesting aspects. Results of the study implied that CRISPRs identified in the 51 examined enterobacterial genomes are quite unchangeable. The number of CRISPR loci found in the examined enterobacteria did not exceed 3, their localization was always conserved, and the number of spacers was fairly low, no more than 34 spacers per locus. Older spacers, which, based on previous communications, are known to be the most rapidly changing sequences in the microbial genomes and undergo frequent loss from CRISPR loci, were shown to be conserved in most of the *E. coli* CRISPRs. Moreover, less than 10% of the numerous coliphage genome sequences present in databases matched CRISPR spacers of the examined enterobacterial strains (Touchon and Rocha, 2010). Additionally, it was reported that cells that survived phage attack did not acquire new spacers in the CRISPR arrays (Díez-Villaseñor *et al.*, 2010). This indicates that the examined CRISPR arrays do not necessarily provide protection against the existing phages. Furthermore, the strong conservation of spacer sequences among CRISPRs of various *E. coli* strains suggests an alternative function of the CRISPR/*cas* system.

C. Multidrug-resistant enterococci

A study of enterococcal strains reported a statistically significant relationship between CRISPR/*cas* systems and antibiotic resistance (Palmer and Gilmore, 2010). Mobile genetic elements constitute up to 25% of the genomes of multidrug *Enterococcus faecalis* strains, which are the etiological factor of hospital-acquired bacterial infections. Resistance toward antibiotics is usually connected with the transfer of resistance genes via mobile DNA elements, such as plasmids or transposons. A frequent and often hasty use of antibiotics is considered to influence the development of enterococcal strains that readily gain new mobile elements, including antibiotic resistance determinants. Thus, the activity of CRISPR/*cas* systems limiting the entry of mobile elements would be in opposition to the development of multidrug resistance (MDR).

Overall, over 50 different enterococcal strains have been examined for the presence of CRISPR loci and their correlation with drug resistance. As a result of this analysis, two CRISPR regions, CRISPR1/*cas* and CRISPR3/*cas*, were identified among antibiotic-sensitive strains and shown to be absent in MDR strains, including all vancomycin-resistant strains. The two CRISPR loci were determined to differ in repeat sequences and *cas* genes, and their presence varied among the examined enterococcal strains. Additionally, another, but truncated locus, CRISPR2, lacking the associated *cas* gene region, was recognized in the genomes of all of the studied strains. The spacer sequences of CRISPR1/*cas* and CRISPR2 were determined to be identical; however, the link between these two loci has not been determined in detail (Bourgogne *et al.*, 2008; Palmer and Gilmore, 2010). The persistence of orphan CRISPR2 loci in the genome of both sensitive and multidrug-resistant enterococcal strains suggested that it was not responsible for the transmission of immunity phenotype. Analysis of CRISPR spacers indicated that they often derive from pheromone-responsive plasmids or phages. The pheromone-responsive plasmids encode functions, which are engaged in the induction of conjugation events and direct production of cellular signals. Such plasmids not only stimulate their own transfer, but also the transfer of other mobile elements and antibiotic resistance determinants encoded on the chromosome (Manson *et al.*, 2010). The abundance of sequences identified as spacers deriving from these plasmids in the genome of enterococcal cells indicated their frequent contact and facile integration into CRISPR arrays. In contrast, there were no identified CRISPR spacers with homology to conjugative transposons, for example, Tn*916*, which are often the carriers of drug resistance in enterococci. This implied that CRISPRs do not protect the cell from these conjugative elements. This supposition was strengthened by the co-occurrence of CRISPRs and *tetM* genes, often harbored by the Tn*916*, in the genomes of antibiotic-sensitive *E. faecalis* strains. CRISPR-containing enterococci also lacked spacers homologous to sequences found in the Inc18 plasmids, which play a key role in the spread of vancomycin resistance from enterococci to methicillin-resistant *Staphylococcus aureus*. Overall, the absence of spacers matching the broad host range Tn*916* and Inc18 elements was speculated to be the consequence of infrequent or ineffective transfer of such sequences between species, for example, due to the lack of pheromone-responsive plasmids. Moreover, it has been argued that the observed lack of CRISPR/*cas* sequences, which by limiting horizontal transfer act to stabilize bacterial genomes, in favor of mobile genetic elements (e.g., pheromone-responsive plasmids), seems to be due to the common use of antibiotics. Drug therapy evoked a change in the dynamic stability of bacterial populations, allowing the entry of foreign genetic elements, including antibiotic-resistance determinants, and, in effect, increasing genome diversity. This phenomenon undoubtedly requires further study, as some cases were in contrast to these general observations. Some antibiotic-sensitive enterococci possessed the *tetM* gene and, contrarily, two MDR strains were found to have CRISPR/*cas* loci.

D. Lactic acid bacteria

The presence of CRISPRs was also explored in lactic acid bacteria (LAB), including *Lactobacillus*, *Bifidobacterium*, and *Streptococcus*. Overall, in over 100 examined genomes of 26 LAB species, more than 60 CRISPR loci were identified and grouped into eight distinct families in respect to the CRISPR repeat sequences and *cas* genes (Horvath *et al.*, 2009). The majority of CRISPR loci were chromosomally located, and only in one case was a plasmid-encoded CRISPR locus noted (*E. faecium* pHT beta plasmid) (Tomita and Ike, 2005). Such species as *Lactobacillus*, *Streptococcus*, or *Bifidobacterium* were determined to carry more than one CRISPR locus in their genomes, which specific number varied depending on the strain. In contrast and quite unexpectedly, no CRISPR sequences have been found for one of the best characterized LAB bacteria, *Lactococcus*, neither for *Leuconostoc*, *Carnobacterium*, *Pediococcus*, nor *Oenococcus*. The absence of CRISPR loci was suggested to be strain dependent and due to the insufficient amount of genome sequence data for these species in public databases. Other strains of these species should be examined in order to fully resolve the issue on the existence of CRISPR/*cas* systems in these LAB. From over 100 spacers identified within LAB CRISPRs, 26% of them matched phage, 47% prophage, 0.05% plasmid, and 22% chromosomal sequences. The remaining 5% of spacer sequences did not exhibit homology to known sequences and remained uncharacterized. Interestingly, LAB CRISPRs do not cluster according to the classical phylogenetic correlations observed between LAB phyla. Instead, it has been proposed that CRISPR loci were disseminated to separate lineages by horizontal gene transfer events and further evolved under the selective pressure imposed by phage infections. The wide polymorphism of LAB CRISPRs makes them a good candidate for strain typing and comparative analyses.

VII. APPLICATION POTENTIAL OF CRISPR/*CAS* SYSTEMS

The wide variety of CRISPR sequences present in the microbial world offers an opportunity to exploit them in various applications. CRISPR loci were determined to be among the most rapidly evolving structures. Even strains of the same species were observed to vary in the number and sequence of repeat-spacer units (Fabre *et al.*, 2004; Groenen *et al.*, 1993; Hoe *et al.*, 1999; Jansen *et al.*, 2002a; Kamerbeek *et al.*, 1997).

A. Strain typing

Strain typing, based on comparative analyses of CRISPR spacers, was first applied by Groenen and associates (1993) to differentiate among a number of *M. tuberculosis* strains. Later, this technique was adapted for other bacterial species, including *Corynebacterium diphtheria* and *Y. pestis*, and has acquired a general term of spacer- or spoligotyping (Kamerbeek *et al.*, 1997; Mokrousov *et al.*, 2007; Vergnaud *et al.*, 2007). Currently, spoligotyping of *M. tuberculosis* strains is one of the main techniques used in the diagnostics and monitoring of tuberculosis epidemics (Brudey *et al.*, 2006; Driscoll, 2009). Other (nonspoligo)typing methods based on sequence comparison of whole CRISPR arrays are applied to distinguish among strains of such bacteria as *Campylobacter jejuni*, *T. neapolitana*, and *Lactobacillus* sp. (DeBoy *et al.*, 2006; Schouls *et al.*, 2003; Sorek *et al.*, 2008).

However, given that some loci are relatively active and quite diverse, whereas others are quite conserved, the potential use of CRISPRs for typing and epidemiological studies must be assessed separately. For instance, in *Mycobacterium* and *Campylobacter*, the high variability of CRISPR spacers seems useful for typing strains, while the low diversity of spacers detected in the genomes of enterobacteria speaks against this method of differentiating between bacteria belonging to this phylum (Touchon and Rocha, 2010).

Overall, the high variability of CRISPR loci proved to be valuable for strain typing as well as comparative analyses and phylogenetic studies of microbial populations.

B. Phylogenetic studies of microbial populations

Examination and comparative analyses of CRISPR sequences also lay the ground for studies of bacterial evolution (DeBoy *et al.*, 2006). The link between CRISPRs and the dynamics of various microbial populations (*E. coli* and *Salmonella, S. thermophilus, Sulfolobus* sp., *Synechcoccus thermophilus,* etc.) has been examined by several research groups (Heidelberg *et al.*, 2009; Held *et al.*, 2010; Horvath *et al.*, 2008; Touchon and Rocha, 2010). For example, Kunin *et al.* (2008) analyzed two sludge bioreactors, with *Candidatus* Accumulibacter phosphatis as a predominant species. The study revealed that no common CRISPR spacer was shared between the two samples, whereas other loci did not differ significantly. This distinct divergence in spacer sequences was proposed to be due to locally specific phages, which induced changes within CRISPR loci. In general, CRISPR arrays were determined to evolve quite rapidly compared to the rest of the genome, except rare cases of strong conservation of older spacers. Additionally, the development of metagenomics allowed studying CRISPR-mediated coevolution of various microbial populations, including noncultivable bacteria. Results of metagenomic studies of thermophilic *S. thermophilus* and acidophilic *Leptospirillum* sp. bacteria showed some interesting differences between the microbial biofilm populations of these two species (Heidelberg *et al.*, 2009; Tyson and Banfield, 2008). Analysis of two almost identical *Leptospirillum* sp. populations revealed a similar pattern in organization of their CRISPR arrays. The oldest spacers present at the distal end of the CRISPR array were conserved among both populations, whereas spacers at the leader end were strain specific (Tyson and Banfield, 2008). Meanwhile, only a few common spacers were detected within more complex microbial populations, containing *S. thermophilus* (Heidelberg *et al.*, 2009). A similar study of a hot spring *Sulfolobus islandicus* population also showed a large diversity of CRISPR spacers among the 39 strains examined (Held *et al.*, 2010). The observed CRISPR diversity in *S. islandicus* and *S. thermophilus* populations was suggested to result from an independent acquisition of different spacers from the same phage, which allowed the strains to compete with each other in their environmental niche. In turn, this prevented a shift in the population diversity toward dominance of one or several strains. The same conclusion was made by Barrangou *et al.* (2007) when examining *S. thermophilus* BIMs. The study reported that phage infection results in the generation of different resistant hosts, containing different spacers that render them resistant to the same phage. The environmental impact of this is that a whole range of BIMs is resistant to a given phage and there is no selection of one resistant genotype. In this aspect, spacers are regarded as the element that promotes diversity among strains of the same population. Differences in CRISPR spacer diversity observed between *Leptospirillum* and *Sulfolobus* sp., as well as *S. islandicus*, could be explained by the fact that perhaps *Leptospirillum* sp. populations encounter more virulent phages, which inflict a stronger selection of spacer sequences, or that the intensity of interactions among the strains of these three species varies in their natural environmental niches.

Identification and detailed characterization of CRISPR sequences have been facilitated by the development of bioinformatics tools, which allow extensive analysis of sequenced genomes. Currently available are several bioinformatics programs that serve to distinguish CRISPR repeats as well as whole CRISPR/*cas* systems in the microbial genomes (e.g., CRISPRFinder, CRISPRI)(Grissa *et al.*, 2007b; Rousseau *et al.*, 2009). Moreover, all CRISPR arrays identified to data can be found in the CRISPR database CRISPRdb (Grissa *et al.*, 2007a).

C. Engineered defense against viruses

As CRISPR/*cas* systems provide immunity to host cells against phage infection, they are quite of interest for industries, such as dairy plants, where microbial production plays a significant role. Phage attack in such production environments causes fermentation failure, decreases the quality of the final dairy product, and, as a consequence, leads to great economic losses. Thus, there is the constant pursuit of natural strains, as well as attempts to construct artificially, by molecular manipulations, new strains able to sustain phage infection. Among the different approaches used to generate phage-resistant strains is the exploitation of CRISPR/*cas* systems (Barrangou *et al.*, 2007; Horvath *et al.*, 2008; Mills *et al.*, 2010). Natural selection of CRISPR-containing BIM cells of industrially applied bacteria could be an interesting alternative solution for obtaining resistant strains, without deliberate genetic modifications. Protocols of selecting such strains have already been developed for dairy *S. thermophilus*, applied in the manufacturing of cheese and yogurts (Mills *et al.*, 2007). The presence of naturally acquired spacer sequences from one side renders the strain resistant to phage infections, while, on the other side, maintains the features of the initial starter culture. Another possibility of constructing phage-resistant strains could be by integration of synthetic spacers matching conserved sequences of industrially occurring phages into the CRISPR array of starter bacteria. However, this approach would involve certain molecular manipulations at the DNA level.

Further on, it could also be imagined that CRISPR-mediated protection against conjugative plasmids could be exploited to limit the spread of antibiotic-resistant strains in hospitals and health clinics.

D. Selective silencing of endogenous genes

Observation of the interference of the *P. carbinolicus* spacer sequence with a housekeeping gene (*hisS*) laid the groundwork for a novel approach in gene silencing. As it has been established, CRISPR spacers hybridize with homologous proto-spacers, promoting their degradation (Makarova *et al.*, 2006). Application of specifically designed synthetic spacers would allow silencing selected regions (genes) without manipulation on the level of the original microbial genome. The spacer could be introduced into the chosen microorganism on a plasmid harboring a functional CRISPR locus. In fact, such a CRISPR-based system could enable introduction of an array of spacer sequences. As a result of their activity, simultaneous silencing of several endogenous genes could be obtained. For eukaryotes, comparable interference-based techniques have already found effective application.

VIII. ROLE OF CRISPR/*CAS* SYSTEMS IN HOST:PHAGE EVOLUTION

The constant interplay between host bacteria and bacteriophages is a key factor in determining the diversity of the microbial world in its natural surroundings. The influence of CRISPR interference in microbial populations and the environmental equilibrium between phages and their hosts cannot be underestimated. Moreover, studies of environmental microbial strains and their phages provide an interesting set of information on their coevolution.

A. CRISPR/*cas* limit horizontal gene transfer and strain lysogenization

Bacteriophages influence microbial populations by both (i) infection incidences, which control the size and structure of bacterial communities, and (ii) transfer of genetic material between bacterial strains. Vice versa, microorganisms by various mechanisms, including CRISPR/*cas* systems, prevent entry into the host of foreign DNAs (Barrangou *et al.*, 2007; Marraffini and Sontheimer, 2008). Based on studies of multidrug-resistant *S. aureus* and *S. epidermidis* bacteria, it was shown that the CRISPR/*cas* system can limit the spread of mobile genetic elements, such as conjugative plasmids carrying resistance and other virulence factors (Marraffini and Sontheimer, 2008). Therefore, the CRISPR/*cas* system can be regarded as a mechanism of limiting horizontal gene transfer in archaea and bacteria, which is one of the main driving forces of microbial evolution, shaping the genomes of both bacteria and phages. In this aspect, the activity of CRISPR arrays prevents the acquisition of new functions (genes) that might provide selective advantage for the cells, such as the earlier-mentioned multidrug resistance, virulence factors, or other determinants. However, incorporation of phage proto-spacers within CRISPR arrays is a way for microorganisms to escape death (Kunin *et al.*, 2008). These observations show the multiple roles of CRISPRs in microbial evolution, which, at the same time, limit infections by lytic phages and lysogenization by prophages.

Despite extensive data confirming the CRISPR role in preventing horizontal transfer by lytic phages or other mobile genetic elements, only a limited number of studies have demonstrated the effect of CRISPR on strain lysogenization and temperate phage development. Implication of CRISPR/*cas* systems in these events is well illustrated by a study performed on group A streptococci, which revealed that strains carrying CRISPR sequences against prophages had no or only a few residing prophages, in opposition to strains without CRISPRs (Banks *et al.*, 2002; Grissa *et al.*, 2007a). Also, Edgar and Qimron (2010) showed that CRISPR affects (i) lysogenization, (ii) maintenance of the lysogenic state, and (iii) induction of prophages existing in the bacterial genome. Introduction of a plasmid carrying spacers homologous to phage λ genes into the *E. coli* strain resulted in 100 times lower lysogenization than for cells harboring a control plasmid. Data suggested that the CRISPR/*cas* system impairs the lysogenic phage life cycle, disallowing its integration into the host genome. As in other described cases, here also CRISPR interference seems to be spacer specific, as acquisition of DNA from a temperate phage (phage P1) lacking homology to phage λ spacers was not affected. CRISPR/*cas* activity was also observed against prophages already existing in the bacterial genome. Introduction of two plasmids, one encoding *cas* genes under the control of an inducible promoter and the other carrying spacers homologous to phage λ, into a previously lysogenized CRISPR-free *E. coli* strain, rendered a toxic effect on the strain, resulting in the death of cells containing the spacer-encoding plasmid. It has been proposed that the massive cell death (98%) is the effect of CRISPR-mediated digestion of target prophage DNA, which leads further to significant genomic DNA damage. This implies that lysogenic DNA is also somehow recognized by CRISPR/*cas* as nonself DNA, similarly as in the case of invading foreign DNA (lytic phages, conjugative plasmids). In microbial communities the described mechanism is regarded as an altruistic strategy of preventing lysogen spread. Similar behavior was observed for *P. aeruginosa* lysogen, for which the CRISPR/*cas* system was shown to limit swarming and biofilm formation (Pul *et al.*, 2010). In specific conditions, CRISPRs were also shown to rescue bacteria from lysis due to prophage induction. Laboratory attempts to simultaneously induce prophages and plasmid-encoded *cas* genes resulted in increased (if only slightly) cell survival, suggesting a mechanism of prophage elimination, most probably involving the degradation of excised DNA. Indeed, cells that survived prophage induction were shown to lack the integrated prophage. Activity of CRISPR/*cas* demands precise regulation, which, in addition to the global regulator H-NS and LeuO protein, may be executed by stress response factors, such as σ^{32} and others (Pul *et al.*, 2010; Wade *et al.*, 2006; Westra *et al.*, 2010).

B. Evolution of CRISPR arrays in the face of phage infections

Analysis of CRISPR diversity among microorganisms provides valuable information on the relationship between prokaryotes and their environments. Attainment of novel spacer sequences permits the cell to escape lysis evoked by continuous phage infections. Moreover, infections provoked by newly appearing phage types trigger swift and dynamic changes in CRISPR regions (Andersson and Banfield, 2008; Barrangou *et al.*, 2007; Tyson and Banfield, 2008). In general, CRISPR loci are under the constant selective pressure associated with genetic variations within the highly dynamic phage populations. Also, host defense systems, including CRISPR/*cas*, inflict selective pressure on phages, which is the force driving the evolution of their genome sequences and adaptation to specific environmental conditions. Although the evolution of CRISPR loci seems to fit the Darwinian law—the fittest survive—changes within CRISPR/*cas* systems are based on random mutations and not on more specific mechanisms.

The mechanistic basis of modifications within CRISPR arrays relies mainly on the constant acquisition (e.g., by horizontal transfer or duplication) and loss (e.g., by mutation or deletion) of CRISPR spacers (Barrangou *et al.*, 2007; Tyson and Banfield, 2008). These integration/deletion events occur at different ends of the CRISPR array. Most integrations of new spacers are observed at the leader end, while the distal end of CRISPR loci carrying old spacers is the region of frequent deletions. Mutational changes within CRISPR regions affect the activity of the system. A single nucleotide mutation in the CRISPR spacer region can impair resistance, while incorporation of another phage proto-spacer into the CRISPR locus can restore it (Deveau *et al.*, 2010).

Even in closely related species, spacer sequences are largely diverse. This phenomenon is well illustrated in a study of *S. thermophilus*, which revealed a generally low genetic polymorphism among the different strains, except for CRISPR loci, which exhibit significant variability (Horvath *et al.*, 2008). Although there are a few exceptions (e.g., *E. coli* CRISPRs appear to be rather conserved), the great diversity of spacers is a sign of their rapid replacement on the evolutionary scale. This observation indicates that closely related microbial populations encounter different dominant phages and plasmids. Analysis of CRISPR-containing genomes from two biofilm-forming *Leptospirillum* sp. populations allowed determining, as a general rule, that the most recently acquired spacers match genome sequences of phages that are currently present in the ecosystem, whereas old spacers are a sign of past infections (Andersson and Banfield, 2008).

C. CRISPRs provide short-term immunity

Rapid changes within CRISPR sequences indicate that this defense system provides a short-term immunity to bacterial populations. Constant adaptive changes are needed to defend the cell from phage infections and to generate new resistant bacteria (Levin, 2010). Therefore, the overall cost of harboring and replicating an excess amount of DNA (novel spacers) is overgrown by the advantage of sustaining phage infection. This, in effect, leads to a stable maintenance of CRISPR elements. However, researchers indicate that there must be some limiting cost that CRISPR-containing cells have to bear in order to keep and replicate the newly acquired spacer DNA. The longer the CRISPR array, the higher the cost and time of its replication, which correlates with slower cell growth. Also, the fidelity of replicating longer CRISPR regions is lower and can, as a result, lead to accumulation of mutational changes. Their removal, most probably through spontaneous recombination events between repeats, is a natural process that limits the growth of the CRISPR locus (Aranaz *et al.*, 2004; Pourcel *et al.*, 2005). The constantly evolving environment, including changes within phage genomes, makes older spacers (signs of former infections) less efficient in providing protection, thus their maintenance is of no advantage for the cell.

D. Significance of CRISPR/*cas* defense systems for microbial populations

A study of *E. coli* CRISPR arrays showed that the majority of spacer sequences derive from phage genomes. However, in environments where phages are limited (e.g., colon or urogenital tract), CRISPR sequences seem to be less vital for protection against phages, and some of the spacers may degenerate and become inactive (Díez-Villaseñor *et al.*, 2010). Moreover, in the same study, no correlation between the number of CRISPR spacers and resistance to phage infection was observed. Infection of *E. coli* strains by different coli-type phages did not result in the acquisition of new spacer sequences in surviving cells. This can indicate either that other phage resistance mechanisms (e.g., adsorption inhibition, restriction-modification) act before CRISPR/*cas* or that CRISPR/*cas* systems play other alternative roles than conferring resistance, which still need to be established.

E. Questions to be answered

Despite multiple results on host:phage coevolution, several issues still need to be examined. For instance, intriguing is the aspect of how quickly microbial populations can acquire a phage-derived spacer into a CRISPR locus and if this happens fast enough to prevent lysis of a large part of the population. Another problem to be resolved is the question if and how CRISPR/*cas* systems cooperate with other phage defense mechanisms. Some of these aspects have been addressed in a study performed by Levin (2010), who used mathematical modeling to establish under what conditions CRISPR sequences provide selective advantage to bacteria, under the pressure of lytic phages or competing bacteria harboring conjugative plasmids, and what determines their stable maintenance in the cell. However, the validity of these models would be hard to assess in natural conditions.

Metagenomics data obtained from natural microbial populations show that the evolution of CRISPR spacers, due to phage infections, and respectively of phage genomes, in order to overcome CRISPR immunity, is strongly correlated (Vale and Little, 2010). It needs to be assessed how long this race between phages and CRISPR-carrying microbial cells can persist before finally a resistant bacteria dominates or the mutant phage overcomes the whole bacterial population.

REFERENCES

Agari, Y., Sakamoto, K., Tamakoshi, M., Oshima, T., Kuramitsu, S., and Shinkai, A. (2010). Transcription profile of *Thermus thermophilus* CRISPR systems after phage infection. *J. Mol. Biol.* **395:**270–281.

Agari, Y., Yokoyama, S., Kuramitsu, S., and Shinkai, A. (2008). X-ray crystal structure of a CRISPR-associated protein, Cse2, from *Thermus thermophilus* HB8. *Proteins* **73:**1063–1067.

Aklujkar, M., and Lovley, D. R. (2010). Interference with histidyl-tRNA synthetase by a CRISPR spacer sequence as a factor in the evolution of *Pelobacter carbinolicus*. *BMC Evol. Biol.* **10:**230.

Andersson, A. F., and Banfield, J. F. (2008). Virus population dynamics and acquired virus resistance in natural microbial communities. *Science* **320:**1047–1050.

Aranaz, A., Romero, B., Montero, N., Alvarez, J., Bezos, J., de Juan, L., Mateos, A., and Domínguez, L. (2004). Spoligotyping profile change caused by deletion of a direct variable repeat in a *Mycobacterium tuberculosis* isogenic laboratory strain. *J. Clin. Microbiol.* **42:**5388–5391.

Babu, M., Beloglazova, N., Flick, R., Graham, C., Skarina, T., Nocek, B., Gagarinova, A., Pogoutse, O., Brown, G., Binkowski, A., Phanse, S., Joachimiak, A., *et al.* (2011). A dual function of the CRISPR-Cas system in bacterial antivirus immunity and DNA repair. *Mol. Microbiol.* **79:**484–502.

Banks, D. J., Beres, S. B., and Musser, J. M. (2002). The fundamental contribution of phages to GAS evolution, genome diversification and strain emergence. *Trends Microbiol.* **10:**515–521.

Barrangou, R., Fremaux, C., Deveau, H., Richards, M., Boyaval, P., Moineau, S., Romero, D. A., and Horvath, P. (2007). CRISPR provides acquired resistance against viruses in prokaryotes. *Science* **315:**1709–1712.

Beloglazova, N., Brown, G., Zimmerman, M. D., Proudfoot, M., Makarova, K. S., Kudritska, M., Kochinyan, S., Wang, S., Chruszcz, M., Minor, W., Koonin, E. V., Edwards, A. M., *et al.* (2008). A novel family of sequence-specific endoribonucleases associated with the clustered regularly interspaced short palindromic repeats. *J. Biol. Chem.* **283:**20361–20371.

Bolotin, A., Quinquis, B., Sorokin, A., and Ehrlich, S. D. (2005). Clustered regularly interspaced short palindrome repeats (CRISPRs) have spacers of extrachromosomal origin. *Microbiology* **151:**2551–2561.

Bourgogne, A., Garsin, D. A., Qin, X., Singh, K. V., Sillanpaa, J., Yerrapragada, S., Ding, Y., Dugan-Rocha, S., Buhay, C., Shen, H., Chen, G., Williams, G., *et al.* (2008). Large scale variation in *Enterococcus faecalis* illustrated by the genome analysis of strain OG1RF. *Genome Biol.* **9:**R110.

Brouns, S. J., Jore, M. M., Lundgren, M., Westra, E. R., Slijkhuis, R. J., Snijders, A. P., Dickman, M. J., Makarova, K. S., Koonin, E. V., and van der Oost, J. (2008). Small CRISPR RNAs guide antiviral defense in prokaryotes. *Science* **321:**960–964.

Brudey, K., Driscoll, J. R., Rigouts, L., Prodinger, W. M., Gori, A., Al-Hajoj, S. A., Allix, C., Aristimuño, L., Arora, J., Baumanis, V., Binder, L., Cafrune, P., *et al.* (2006). *Mycobacterium tuberculosis* complex genetic diversity: Mining the fourth international spoligotyping database (SpolDB4) for classification, population genetics and epidemiology. *BMC Microbiol.* **6:**23.

Bult, C. J., White, O., Olsen, G. J., Zhou, L., Fleischmann, R. D., Sutton, G. G., Blake, J. A., FitzGerald, L. M., Clayton, R. A., Gocayne, J. D., Kerlavage, A. R., Dougherty, B. A., *et al.* (1996). Complete genome sequence of the methanogenic archaeon, *Methanococcus jannaschii*. *Science* **273:**1058–1073.

Carte, J., Wang, R., Li, H., Terns, R. M., and Terns, M. P. (2008). Cas6 is an endoribonuclease that generates guide RNAs for invader defense in prokaryotes. *Genes Dev.* **22:**3489–3496.

Chattoraj, P., Banerjee, A., Biswas, S., and Biswas, I. (2010). ClpP of *Streptococcus mutans* differentially regulates expression of genomic islands, mutacin production and antibiotics tolerance. *J. Bacteriol.* **192:**1312–1323.

Cui, Y., Li, Y., Gorgé, O., Platonov, M. E., Yan, Y., Guo, Z., Pourcel, C., Dentovskaya, S. V., Balakhonov, S. V., Wang, X., Song, Y., Anisimov, A. P., *et al.* (2008). Insight into microevolution of *Yersinia pestis* by clustered regularly interspaced short palindromic repeats. *PLoS ONE* **3:**e2652.

DeBoy, R. T., Mongodin, E. F., Emerson, J. B., and Nelson, K. E. (2006). Chromosome evolution in the *Thermotogales*: Large scale inversions and strain diversification of CRISPR sequences. *J. Bacteriol.* **188:**2364–2374.

Deltcheva, E., Chylinski, K., Sharma, C. M., Gonzales, K., Chao, Y., Pirzada, Z. A., Eckert, M. R., Vogel, J., and Charpentier, E. (2011). CRISPR RNA maturation by *trans*-encoded small RNA and host factor RNase III. *Nature* **471:**602–607.

Deveau, H., Barrangou, R., Garneau, J. E., Labonté, J., Fremaux, C., Boyaval, P., Romero, D. A., Horvath, P., and Moineau, S. (2008). Phage response to CRISPR-encoded resistance in *Streptococcus thermophilus*. *J. Bacteriol.* **190:**1390–1400.

Deveau, H., Garneau, J. E., and Moineau, S. (2010). CRISPR/Cas system and its role in phage-bacteria interactions. *Annu. Rev. Microbiol.* **64:**475–493.

Díez-Villaseñor, C., Almendros, C., García-Martínez, J., and Mojica, F. J. M. (2010). Diversity of CRISPR loci in *Escherichia coli*. *Microbiology* **156:**1351–1361.

Driscoll, J. R. (2009). Spoligotyping for molecular epidemiology of the *Mycobacterium tuberculosis* complex. *Methods Mol. Biol.* **551:**117–128.

Ebihara, A., Yao, M., Masui, R., Tanaka, I., Yokoyama, S., and Kuramitsu, S. (2006). Crystal structure of hypothetical protein TTHB192 from *Thermus thermophilus* HB8 reveals a new protein family with an RNA recognition motif-like domain. *Protein Sci.* **15:**1494–1499.

Edgar, R., and Qimron, U. (2010). The *Escherichia coli* CRISPR system protects from λ lysogenization, lysogens, and prophage induction. *J. Bacteriol.* **192:**6291–6294.

Fabre, M., Koeck, J. L., Le Fleche, P., Simon, F., Herve, V., Vergnaud, G., and Pourcel, C. (2004). High genetic diversity revealed by variable-number tandem repeat genotyping and analysis of hsp65 gene polymorphism in a large collection of "*Mycobacterium canettii*" strains indicates that the *M. tuberculosis* complex is a recently emerged clone of "*M. canettii*. *J. Clin. Microbiol.* **42:**3248–3255.

Garneau, J. E., Dupuis, M.È., Villion, M., Romero, D. A., Barrangou, R., Boyaval, P., Fremaux, C., Horvath, P., Magadán, A. H., and Moineau, S. (2010). The CRISPR/Cas bacterial immune system cleaves bacteriophage and plasmid DNA. *Nature* **468:**67–71.

Godde, J. S., and Bickerton, A. (2006). The repetitive DNA elements called CRISPRs and their associated genes: Evidence of horizontal transfer among prokaryotes. *J. Mol. Evol.* **62:**718–729.

Gosset, G., Zhang, Z., Nayyar, S., Cuevas, W. A., and Saier, M. H., Jr. (2004). Transcriptome analysis of CRP-dependent catabolite control of gene expression in *Escherichia coli. J. Bacteriol.* **186:**3516–3524.

Gottesman, S. (2004). The small RNA regulators of *Escherichia coli*: Roles and mechanisms. *Annu. Rev. Microbiol.* **58:**303–328.

Gottesman, S. (2005). Micros for microbes: Non-coding regulatory RNAs in bacteria. *Trends Genet.* **21:**399–404.

Greve, B., Jensen, S., Brugger, K., Zillig, W., and Garrett, R. A. (2004). Genomic comparison of archaeal conjugative plasmids from *Sulfolobus*. *Archaea* **1:**231–239.

Grissa, I., Vergnaud, G., and Pourcel, C. (2007a). The CRISPRdb database and tools to display CRISPRs and to generate dictionaries of spacers and repeats. *BMC Bioinformatics* **8:**172.

Grissa, I., Vergnaud, G., and Pourcel, C. (2007b). CRISPRFinder: A web tool to identify clustered regularly interspaced short palindromic repeats. *Nucleic Acids Res.* **35:**W52–57.

Groenen, P. M., Bunschoten, A. E., van Soolingen, D., and van Embden, J. D. (1993). Nature of DNA polymorphism in the direct repeat cluster of *Mycobacterium tuberculosis*; application for strain differentiation by a novel typing method. *Mol. Microbiol.* **10:**1057–1065.

Haft, D. H., Selengut, J., Mongodin, E. F., and Nelson, K. E. (2005). A guild of 45 CRISPR-associated (Cas) protein families and multiple CRISPR/Cas subtypes exist in prokaryotic genomes. *PloS Comput. Biol.* **1:**e60.

Hale, C., Kleppe, K., Terns, R. M., and Terns, M. P. (2008). Prokaryotic silencing (psi) RNAs in *Pyrococcus furiosus*. *RNA* **14:**2572–2579.

Hale, C. R., Zhao, P., Olson, S., Duff, M. O., Graveley, B. R., Wells, L., Terns, R. M., and Terns, M. P. (2009). RNA-guided RNA cleavage by a CRISPR RNA-Cas protein complex. *Cell* **139:**945–956.

Han, D., and Krauss, G. (2009). Characterization of the endonuclease SSO2001 from *Sulfolobus solfataricus* P2. *FEBS Lett.* **583:**771–776.

Haurwitz, R. E., Jinek, M., Wiedenheft, B., Zhou, K., and Doudna, J. A. (2010). Sequence- and structure-specific RNA processing by a CRISPR endonuclease. *Science* **329:**1355–1358.

Heidelberg, J. F., Nelson, W. C., Schoenfeld, T., and Bhaya, D. (2009). Germ warfare in a microbial mat community: CRISPRs provide insights into the co-evolution of host and viral genomes. *PLoS One* **4:**e4169.

Held, N. L., Herrera, A., Cadillo-Quiroz, H., and Whitaker, R. J. (2010). CRISPR associated diversity within a population of *Sulfolobus islandicus*. *PLoS One* **5:**e12988.

Hermans, P. W., van Soolingen, D., Bik, E. M., de Haas, P. E., Dale, J. W., and van Embden, J. D. (1991). Insertion element IS987 from *Mycobacterium bovis* BCG is located in a hot-spot integration region for insertion elements in *Mycobacterium tuberculosis* complex strains. *Infect. Immun.* **59:**2695–2705.

Hoe, N., Nakashima, K., Grigsby, D., Pan, X., Dou, S. J., Naidich, S., Garcia, M., Kahn, E., Bergmire-Sweat, D., and Musser, J. M. (1999). Rapid molecular genetic subtyping of serotype M1 group A *Streptococcus* strains. *Emerg. Infect. Dis.* **5:**254–263.

Horvath, P., Romero, D. A., Coûté-Monvoisin, A. C., Richards, M., Deveau, H., Moineau, S., Boyaval, P., Fremaux, C., and Barrangou, R. (2008). Diversity, activity, and evolution of CRISPR loci in *Streptococcus thermophilus*. *J. Bacteriol.* **190:**1401–1412.

Horvath, P., and Barrangou, R. (2010). CRISPR/Cas, the immune system of Bacteria and Archaea. *Science* **327:**167–170.

Howard, J. A., Delmas, S., Ivančić-Baće, I., and Bolt, E. L. (2011). Helicase dissociation and annealing of RNA-DNA hybrids by *Escherichia coli* Cas3 protein. *Biochem. J.* **439:**85–95.

Ishino, Y., Shinagawa, H., Makino, K., Amemura, M., and Nakata, A. (1987). Nucleotide sequence of the iap gene, responsible for alkaline phosphatase isozyme conversion in *Escherichia coli*, and identification of the gene product. *J. Bacteriol.* **169:**5429–5433.

Jansen, R., Embden, J. D., Gaastra, W., and Schouls, L. M. (2002a). Identification of a novel family of sequence repeats among prokaryotes. *OMICS* **6:**23–33.

Jansen, R., Embden, J. D., Gaastra, W., and Schouls, L. M. (2002b). Identification of genes that are associated with DNA repeats in prokaryotes. *Mol. Microbiol.* **43:**1565–1575.

Jore, M. M., Lundgren, M., van Duijn, E., Bultema, J. B., Westra, E. R., Waghmare, S. P., Wiedenheft, B., Pul, U., Wurm, R., Wagner, R., Beijer, M. R., Barendregt, A., *et al.* (2011). Structural basis for CRISPR RNA-guided DNA recognition by Cascade. *Nat. Struct. Mol. Biol.* **18:**529–536.

Kamerbeek, J., Schouls, L., Kolk, A., van Agterveld, M., van Soolingen, D., Kuijper, S., Bunschoten, A., Molhuizen, H., Shaw, R., Goyal, M., and van Embden, J. (1997). Simultaneous detection and strain differentiation of *Mycobacterium tuberculosis* for diagnosis and epidemiology. *J. Clin. Microbiol.* **35:**907–914.

Karginov, F. V., and Hannon, G. J. (2010). The CRISPR system: Small RNA-guided defense in bacteria and archaea. *Mol. Cell.* **37:**7–19.

Klenk, H. P., Clayton, R. A., Tomb, J. F., White, O., Nelson, K. E., Ketchum, K. A., Dodson, R. J., Gwinn, M., Hickey, E. K., Peterson, J. D., Richardson, D. L., Kerlavage, A. R., *et al.* (1997). The complete genome sequence of the hyperthermophilic, sulphate-reducing archaeon *Archaeoglobus fulgidus*. *Nature* **390:**364–370.

Kunin, V., Sorek, R., and Hugenholtz, P. (2007). Evolutionary conservation of sequence and secondary structures in CRISPR repeats. *Genome Biol.* **8:**R61.

Kunin, V., He, S., Warnecke, F., Peterson, S. B., Garcia Martin, H., Haynes, M., Ivanova, N., Blackall, L. L., Breitbart, M., Rohwer, F., McMahon, K. D., and Hugenholtz, P. (2008). A bacterial metapopulation adapts locally to phage predation despite global dispersal. *Genome Res.* **18:**293–297.

Levin, B. R. (2010). Nasty viruses, costly plasmids, population dynamics, and the conditions for establishing and maintaining CRISPR-mediated adaptive immunity in bacteria. *PLoS Genet.* **6:**e1001171.

Lillestøl, R. K., Redder, P., Garrett, R. A., and Brugger, K. (2006). A putative viral defense mechanism in archaeal cells. *Archaea* **2:**59–72.

Lillestøl, R. K., Shah, S. A., Brügger, K., Redder, P., Phan, H., Christiansen, J., and Garrett, R. A. (2009). CRISPR families of the crenarchaeal genus *Sulfolobus*: Bidirectional transcription and dynamic properties. *Mol. Microbiol.* **72:**259–272.

Majdalani, N., Vanderpool, C. K., and Gottesman, S. (2005). Bacterial small RNA regulators. *Crit. Rev. Biochem. Mol. Biol.* **40:**93–113.

Makarova, K. S., Aravind, L., Grishin, N. V., Rogozin, I. B., and Koonin, E. V. (2002). A DNA repair system specific for thermophilic archaea and bacteria predicted by genomic context analysis. *Nucleic Acids Res.* **30:**482–496.

Makarova, K. S., Aravind, L., Wolf, Y. I., and Koonin, E. V. (2011a). Unification of Cas protein families and a simple scenario for the origin and evolution of CRISPR-Cas systems. *Biol Direct.* **6:**38.

Makarova, K. S., Grishin, N. V., Shabalina, S. A., Wolf, Y. I., and Koonin, E. V. (2006). A putative RNA-interference-based immune system in prokaryotes: Computational analysis of the predicted enzymatic machinery, functional analogies with eukaryotic RNAi, and hypothetical mechanisms of action. *Biol. Direct.* **1:**7.

Makarova, K. S., Haft, D. H., Barrangou, R., Brouns, S. J., Charpentier, E., Horvath, P., Moineau, S., Mojica, F. J., Wolf, Y. I., Yakunin, A. F., van der Oost, J., and Koonin, E. V. (2011b). Evolution and classification of the CRISPR-Cas systems. *Nat. Rev. Microbiol.* **9:**467–477.

Manson, J. M., Hancock, L. E., and Gilmore, M. S. (2010). Mechanism of chromosomal transfer of *Enterococcus faecalis* pathogenicity island, capsule, antimicrobial resistance, and other traits. *Proc. Natl. Acad. Sci. USA* **107:**12269–12274.

Marraffini, L. A., and Sontheimer, E. J. (2008). CRISPR interference limits horizontal gene transfer in staphylococci by targeting DNA. *Science* **322:**1843–1845.

Marraffini, L. A., and Sontheimer, E. J. (2010a). CRISPR interference: RNA-directed adaptive immunity in bacteria and archaea. *Nat. Rev. Genet.* **11:**181–190.

Marraffini, L. A., and Sontheimer, E. J. (2010b). Self vs. non-self discrimination during CRISPR RNA-directed immunity. *Nature* **463:**568–574.

Medina-Aparicio, L., Rebollar-Flores, J. E., Gallego-Hernandez, A. L., Vazquiez, A., Olvera, L., Gutiérrez-Ríos, R. M., Calva, E., and Hernández-Lucas, I. (2011). The CRISPR/Cas immune system is an operon regulated by LeuO, H-NS, and leucine-responsive regulatory protein in *Salmonella enterica* serovar Typhi. *J. Bacteriol.* **193:**2396–2407.

Mills, S., Coffey, A., McAuliffe, O. E., Meijer, W. C., Hafkamp, B., and Ross, R. P. (2007). Efficient method for generation of bacteriophage insensitive mutants of *Streptococcus thermophilus* yoghurt and mozzarella strains. *J. Microbiol. Methods* **70:**159–164.

Mills, S., Griffin, C., Coffey, A., Meijer, W. C., Hafkamp, B., and Ross, R. P. (2010). CRISPR analysis of bacteriophageinsensitive mutants (BIMs) of industrial *Streptococcus thermophilus*: Implications for starter design. *J. Appl. Microbiol.* **108:**945–955.

Mojica, F. J., Díez-Villaseñor, C., García-Martínez, J., and Almendros, C. (2009). Short motif sequences determine the targets of the prokaryotic CRISPR defence system. *Microbiology* **155:**733–740.

Mojica, F. J., Díez-Villaseñor, C., García-Martínez, J., and Soria, E. (2005). Intervening sequences of regularly spaced prokaryotic repeats derive from foreign genetic elements. *J. Mol. Evol.* **60:**174–182.

Mojica, F. J., Ferrer, C., Juez, G., and Rodriguez-Valera, F. (1995). Long stretches of short tandem repeats are present in the largest replicons of the Archaea *Haloferax mediterranei* and *Haloferax volcanii* and could be involved in replicon partitioning. *Mol. Microbiol.* **17:**85–93.

Mokrousov, I., Limeschenko, E., Vyazovaya, A., and Narvskaya, O. (2007). *Corynebacterium diphtheriae* spoligotyping based on combined use of two CRISPR loci. *Biotechnol. J.* **2:**901–906.

Nakata, A., Amemura, M., and Makino, K. (1989). Unusual nucleotide arrangement with repeated sequences in the *Escherichia coli* K-12 chromosome. *J. Bacteriol.* **171:**3553–3556.

Nelson, K. E., Clayton, R. A., Gill, S. R., Gwinn, M. L., Dodson, R. J., Haft, D. H., Hickey, E. K., Peterson, J. D., Nelson, W. C., Ketchum, K. A., McDonald, L., Utterback, T. R., *et al.* (1999). Evidence for lateral gene transfer between archaea and bacteria from genome sequence of *Thermotoga maritima*. *Nature* **399:**323–329.

Palmer, K. L., and Gilmore, M. S. (2010). Multidrug-resistant enterococci lack CRISPR-cas. *MBio.* **1:**e00227–10.

Peng, X., Brugger, K., Shen, B., Chen, L., She, Q., and Garrett, R. A. (2003). Genus-specific protein binding to the large clusters of DNA repeats (short regularly spaced repeats) present in *Sulfolobus genomes*. *J. Bacteriol.* **185:**2410–2417.

Pourcel, C., Salvignol, G., and Vergnaud, G. (2005). CRISPR elements in *Yersinia pestis* acquire new repeats by preferential uptake of bacteriophage DNA, and provide additional tools for evolutionary studies. *Microbiology* **151:**653–663.

Pul, U., Wurm, R., Arslan, Z., Geißen, R., Hofmann, N., *et al.* (2010). Identification and characterization of *E. coli* CRISPR-cas promoters and their silencing by HNS. *Mol. Microbiol.* **75:**1495–1512.

Rousseau, C., Nicolas, J., and Gonnet, M. (2009). CRISPI: A CRISPR Interactive database. *Bioinformatics* **25:**3317–3318.

Sakamoto, K., Agari, Y., Agari, K., Yokoyama, S., Kuramitsu, S., and Shinkai, A. (2009). X-ray crystal structure of a CRISPR-associated RAMP superfamily protein, Cmr5, from *Thermus thermophilus* HB8. *Proteins* **75:**528–532.

Schouls, L. M., Reulen, S., Duim, B., Wagenaar, J. A., Willems, R. J., Dingle, K. E., Colles, F. M., and Van Embden, J. D. (2003). Comparative genotyping of *Campylobacter jejuni* by amplified fragment length polymorphism, multilocus sequence typing, and short repeat sequencing: Strain diversity, host range, and recombination. *J. Clin. Microbiol.* **41:**15–26.

Sebaihia, M., Wren, B. W., Mullany, P., Fairweather, N. F., Minton, N., Stabler, R., Thomson, N. R., Roberts, A. P., Cerdeño-Tárraga, A. M., Wang, H., Holden, M. T., Wright, A., *et al.* (2006). The multidrug-resistant human pathogen *Clostridium difficile* has a highly mobile, mosaic genome. *Nat. Genet.* **38:**779–786.

Semenova, E., Nagornykh, M., Pyatnitskiy, M., Artamonova, I. I., and Severinov, K. (2009). Analysis of CRISPR system function in plant pathogen *Xanthomonas oryzae*. *FEMS Microbiol. Lett.* **296:**110–116.

Shah, S. A., Hansen, N. R., and Garrett, R. A. (2009). Distribution of CRISPR spacer matches in viruses and plasmids of crenarchaeal acidothermophiles and implications for their inhibitory mechanism. *Biochem. Soc. Trans.* **37:**23–28.

Shinkai, A., Kira, S., Nakagawa, N., Kashihara, A., Kuramitsu, S., and Yokoyama, S. (2007). Transcription activation mediated by a cyclic AMP receptor protein from *Thermus thermophilus* HB8. *J. Bacteriol.* **189:**3891–3901.

Sinkunas, T., Gasiunas, G., Fremaux, C., Barrangou, R., Horvath, P., and Siksnys, V. (2011). Cas3 is a single-stranded DNA nuclease and ATP-dependent helicase in the CRISPR/Cas immune system. *EMBO J* **30:**1335–1342.

Smith, D. R., Doucette-Stamm, L. A., Deloughery, C., Lee, H., Dubois, J., Aldredge, T., Bashirzadeh, R., Blakely, D., Cook, R., Gilbert, K., Harrison, D., Hoang, L., *et al.* (1997). Complete genome sequence of *Methanobacterium thermoautotrophicum* deltaH: Functional analysis and comparative genomics. *J. Bacteriol.* **179:**7135–7155.

Sontheimer, E. J., and Marraffini, L. A. (2010). Slicer for DNA. *Nature* **468:**45–46.

Sorek, R., Kunin, V., and Hugenholtz, P. (2008). CRISPR: A widespread system that provides acquired resistance against phages in bacteria and archaea. *Nat. Rev. Microbiol.* **6:**181–186.

Steiniger-White, M., Rayment, I., and Reznikoff, W. S. (2004). Structure/function insights into Tn5 transposition. *Curr. Opin. Struct. Biol.* **14:**50–57.

Storz, G., Opdyke, J. A., and Zhang, A. (2004). Controlling mRNA stability and translation with small, noncoding RNAs. *Curr. Opin. Microbiol.* **7:**140–144.

Tang, T. H., Bachellerie, J. P., Rozhdestvensky, T., Bortolin, M. L., Huber, H., Drungowski, M., Elge, T., Brosius, J., and Huttenhofer, A. (2002). Identification of 86 candidates for small non-messenger RNAs from the archaeon *Archaeoglobus fulgidus*. *Proc. Natl. Acad. Sci. USA* **99:**7536–7541.

Tang, T. H., Polacek, N., Zywicki, M., Huber, H., Brugger, K., Garrett, R., Bachellerie, J. P., and Huttenhofer, A. (2005). Identification of novel non-coding RNAs as potential antisense regulators in the archaeon *Sulfolobus solfataricus*. *Mol. Microbiol.* **55:**469–481.

Tomita, H., and Ike, Y. (2005). Genetic analysis of transfer-related regions of the *vancomycin* resistance *Enterococcus* conjugative plasmid pHTbeta: Identification of oriT and a putative relaxase gene. *J. Bacteriol.* **187:**7727–7737.

Touchon, M., and Rocha, E. P. (2010). The small, slow and specialized CRISPR and anti-CRISPR of *Escherichia* and *Salmonella*. *PLoS One* **5:**e11126.

Tyson, G. W., and Banfield, J. F. (2008). Rapidly evolving CRISPRs implicated in acquired resistance of microorganisms to viruses. *Environ. Microbiol.* **10:**200–207.

Vale, P. F., and Little, T. J. (2010). CRISPR-mediated phage resistance and the ghost of coevolution past. *Proc. R. Soc. B* **277:**2097–2103.

van der Oost, J., Jore, M. M., Westra, E. R., Lundgren, M., and Brouns, S. J. (2009). CRISPR-based adaptive and heritable immunity in prokaryotes. *Trends Biochem. Sci.* **34:**401–407.

van der Ploeg, J. R. (2009). Analysis of CRISPR in *Streptococcus mutans* suggests frequent occurrence of acquired immunity against infection by M102-like bacteriophages. *Microbiology* **155:**1966–1976.

Vergnaud, G., Li, Y., Gorgé, O., Cui, Y., Song, Y., Zhou, D., Grissa, I., Dentovskaya, S. V., Platonov, M. E., Rakin, A., Balakhonov, S. V., Neubauer, H., *et al.* (2007). Analysis of the three *Yersinia pestis* CRISPR loci provides new tools for phylogenetic studies and possibly for the investigation of ancient DNA. *Adv. Exp. Med. Biol.* **603:**327–338.

Viswanathan, P., Murphy, K., Julien, B., Garza, A. G., and Kroos, L. (2007). Regulation of dev, an operon that includes genes essential for *Myxococcus xanthus* development and CRISPR-associated genes and repeats. *J. Bacteriol.* **189:**3738–3750.

Wade, J. T., Roa, D. C., Grainger, D. C., Hurd, D., Busby, S. J., Struhl, K., and Nudler, E. (2006). Extensive functional overlap between sigma factors in *Escherichia coli*. *Nat. Struct. Mol. Biol.* **13:**806–814.

Wang, R., Preamplume, G., Terns, M. P., Terns, R. M., and Li, H. (2011). Interaction of the Cas6 riboendonuclease with CRISPR RNAs: Recognition and cleavage. *Structure* **19:**257–264.

Westra, E. R., Pul, U., Heidrich, N., Jore, M. M., Lundgren, M., Stratmann, T., Wurm, R., Raine, A., Mescher, M., Van Heereveld, L., Mastop, M., Wagner, E. G., *et al.* (2010). H-NS-mediated repression of CRISPR-based immunity in *Escherichia coli* K12 can be relieved by the transcription activator LeuO. *Mol. Microbiol.* **77:**1380–1393.

Wiedenheft, B., Zhou, K., Jinek, M., Coyle, S. M., Ma, W., and Doudna, J. A. (2009). Structural basis for DNase activity of a conserved protein implicated in CRISPR-mediated genome defense. *Structure* **17:**904–912.

Zegans, M. E., Wagner, J. C., Cady, K. C., Murphy, D. M., Hammond, J. H., and O'Toole, G. A. (2009). Interaction between bacteriophage DMS3 and host CRISPR region inhibits group behaviors of *Pseudomonas aeruginosa*. *J. Bacteriol.* **191:**210–219.

CHAPTER 9

Pseudolysogeny

Marcin Łoś[*,†,‡] and **Grzegorz Węgrzyn**[*]

Contents

Abstract

Pseudolysogeny can be defined as the stage of stalled development of a bacteriophage in a host cell without either multiplication of the phage genome (as in lytic development) or its replication synchronized with the cell cycle and stable maintenance in the cell line (as in lysogenization), which proceeds with no viral genome degradation, thus allowing the subsequent restart of virus development. This phenomenon is usually caused by unfavorable growth conditions for the host cell (such as starvation) and is terminated with initiation of either true lysogenization or lytic growth when growth conditions improve. Pseudolysogeny has been known for tens of years; however, its role has often been underestimated. Currently, it is being considered more often as an important aspect of phage–host interactions. The reason for this is mostly an increased interest in phage–host interactions in the natural environment. Pseudolysogeny seems to play an important role in phage survival, as bacteria in a natural environment are starved or their growth is very slow. This phenomenon can be an important aspect of phage-dependent bacterial mortality and may influence the virulence of some bacterial strains.

[*] Department of Molecular Biology, University of Gdańsk, Gdańsk, Poland
[†] Institute of Physical Chemistry, Polish Academy of Sciences, Warsaw, Poland
[‡] Phage Consultants, Gdańsk, Poland

Advances in Virus Research, Volume 82
ISSN 0065-3527, DOI: 10.1016/B978-0-12-394621-8.00019-4

I. INTRODUCTION

According to Whitman *et al.* (1998), the number of prokaryotic cells on earth is estimated to be about 5×10^{30}. It was suggested that the number of bacteriophages may be an order of magnitude higher. These numbers do not seem very striking at first glance, as phages under laboratory conditions may yield as many as 100–1000 progeny particles per infected cell. However, in a natural environment, the situation looks totally different. In natural habitats, bacteriophages have to cope with extremely variable conditions, which impact the ability of the host to produce phage progeny. Among them, there are factors determining host availability and host quality. The former ones influence the period of time necessary for phage progeny to find suitable hosts to perform the next infection. Such factors are important, as viral particles, in natural environment, may be exposed to damaging conditions (such as ultraviolet radiation and oxidative agents), which cause a rapid decay of phage virions (Miller, 2006). The latter ones influence the ability of the host to produce phage progeny particles. Some phages (e.g., T4) cannot produce progeny in starved host cells (Kutter *et al.*, 1994), whereas others can develop, but usually the burst size is severely limited (Schrader *et al.*, 1997; Yin, 1993). In oligotrophic environments, phage hosts may be of "low quality" and may also be dispersed, which would prevent phage progeny from starting a successful infection cycle; thus, the presence of phages in the environment may be endangered. Even in relatively nutritionally rich habitats, bacterial growth rates differ considerably from those obtained under laboratory conditions. For example, while the generation time for *Escherichia coli* "wild-type" laboratory strains cultured at 37°C in rich medium is about 20min, this parameter estimated for *E. coli* in mammalian gut spans from 40–80 min (Poulsen *et al.*, 1995), through 4h (Freter *et al.*, 1983) and up to 140h (Abedon, 1989).

Bacteriophages had to develop effective strategies to be able to sustain their presence in the environment. Two of such major strategies, virulence (lytic development) and lysogeny, have been studied and described extensively in the literature. However, there is a third way, called pseudolysogeny. One can suspect that even Twort (1915) might have actually observed this phenomenon in his studies, although it is clear that he could not understand the processes he found. After indication that bacteriophages indeed exist, the processes of their development appeared unclear. One of the problems was the contamination of partially susceptible bacterial strains by various bacteriophages. Under such conditions, it is difficult to make solid conclusions from experiments on bacteriophage development. Perhaps such problems led d'Herelle (1930) to describe a putative process called "symbiosis between bacteria and bacteriophages" (*symbiose bacterie-bacteriophage*, in French). This discoverer of phages suspected that a bacterial population consists of cells, among which there is a continuous spectrum of their susceptibility to phage infection. Moreover, he supposed that a population of bacteriophages consists of virions with a wide spectrum of virulence. He speculated that the most virulent phages may multiply on the most susceptible bacteria, whereas the probability of multiplication of less virulent ones on less susceptible hosts would be significantly lower. In such a way, bacteriophages and bacteria might coexist in the environment, which was called bacterium–bacteriophage symbiosis (d'Herelle, 1930). These early speculations by d'Herelle have indirectly influenced the understanding of pseudolysogeny as currently understood. Namely, a hypothesis related to that of d'Herelle (1930) was proposed by Delbrück (1946), who called the phenomenon "pseudolysogenesis." This term described the process of reproduction of a bacteriophage contaminating the phage-resistant bacterial population due to the appearance of phage-susceptible mutants, as it was understood at that time. Lwoff (1953) named such "pseudolysogenic" bacterial strains "the carrier strains." This terminology led to some confusion after describing the pseudolysogeny in the current meaning by Romig and Brodetsky (1961)(in fact, Romig and Brodetsky described unstable lysogeny, which was simply manifested by prophage loss from a population that was derived from a lysogenic cell; however, they did not use the term "pseudolysogeny"). Later, the pseudolysogeny phenomenon was defined by Baess (1971) as an unstable interaction between phage and host cell, which is not productive and eventually resolves into true lysogeny or into virulent growth. This definition is very close to our current understanding of pseudolysogeny and should not be confused with the process described by d'Herelle (1930), Delbrück (1946), and Lwoff (1953).

Despite its discovery in the early 1960s, pseudolysogeny remains a largely unexplored phenomenon. Generally, pseudolysogeny itself is a nonproductive stage. In this stage, the viral genome may be maintained for a potentially long period of time and is sometimes called "preprophage" (Miller and Ripp 2002). The importance of pseudolysogeny has been recognized fairly recently, but only relatively few researchers decided to explore this subject. Some of the reasons for this are (i) a relative instability of pseudolysogeny, (ii) a need for the use of cultivation conditions that are far from so-called "standard laboratory conditions," and (iii) problems in the recognition of pseudolysogeny itself. Moreover, studies on pseudolysogeny often require the use of specialized equipment, such as a chemostat, turbidostat, or fermenter, which allows for cultivation of bacteria under nutrient limitation conditions. Perhaps these factors caused pseudolysogeny to be mostly an unexplored subject.

II. CURRENT DEFINITIONS OF PSEUDOLYSOGENY

Currently, there is no single, commonly accepted definition of pseudolysogeny, as some researchers may prefer a quite minimalistic description of this phenomenon, whereas others accept a broad scope of the phage–host interactions that could be covered by such a definition. Thus, there is still confusion about what pseudolysogeny is. The simplest explanation is that it is neither virulent phage development nor lysogenization. However, this definition appears to be incomplete and it describes what is not pseudolysogeny rather than what it is. Therefore, we propose the following definition of pseudolysogeny, which—in our opinion—summarizes and collects other current proposals: pseudolysogeny is the stage of stalled development of a bacteriophage in the host cell, without either multiplication of phage genome or its replication synchronized with the cell cycle and stable maintenance in the cell line, which proceeds with no viral genome degradation, thus allowing the subsequent restart and resumption of virus development.

One should remember that due to some technical problems in analyzing interactions between bacteria and bacteriophages, certain phage developmental pathways, including pseudolysogeny, may be misrecognized. For example, spontaneous phage release, which is often attributed to pseudolysogeny, may occur in both lysogens and pseudolysogens and, by definition, occurs in the case of virulent infection. Another feature that is often considered a characteristic of pseudolysogens, the lack of prophage induction by mitomycin C or a SOS-triggering factor, does not necessary mean that an actual pseudolysogenic relationship occurs between phage and host, as a large group of temperate phages is not inducible by the SOS response; examples are phages 18, W, 62, 299 (Garen *et al.*, 1965), P2, and P3 (Bertani and Bertani, 1971). An ability to cure a bacterium from a (pro)phage should also be considered a suggestion rather than proof for pseudolysogeny. There are many examples of strains cured from a prophage, such as the *E. coli* K12 lineage, which is the progeny of bacterium that was previously lysogenic for phage λ. However, one should note that in the case of pseudolysogeny, phage-free cells should constitute a major fraction of bacterial cells' population due to a lack of phage genome replication.

A possibility to misidentify or overlook pseudolysogeny is enhanced because both lysogeny and virulence are studied most extensively on phages that often represent extreme cases of their kind. Phages λ and P1, which are the best studied examples of temperate phages, are extremely stable in the form of a prophage. Phage λ, due to chromosome integration, is stably segregated to virtually all daughter cells. Moreover, the frequency of spontaneous induction of λ prophage is low relative to other phages, for example, ϕ80 (Rotman *et al.*, 2010). P1 as a representative of episomal prophages shows extreme stability due to two mechanisms: partitioning and postsegregational killing of daughter cells deprived of a prophage copy (Lehnherr *et al.*, 1993). Nevertheless, such prophages do not necessarily constitute the majority in natural ecosystems. An example of an unstable plasmid prophage has been described by Sakaguchi *et al.* (2005). It seems that there is a whole spectrum of temperate phage strategies in the field of stability of the interaction with host cells.

Not all prophage maintenance strategies must rely solely on the ability of the prophage to be inherited by all progeny cells. In a natural environment, some prophages may rely more on lysogenic conversion genes, which, under laboratory conditions, are usually useless for bacterial isolates. This may create a situation where ostensible prophage instability, observed in laboratory, does not actually occur in real life, and the instability in an artificial habitat takes place due to a lack of selective pressure (which is always present in a natural environment).

A high growth rate of a bacterial host is another factor that may affect prophage stability. In a natural environment, bacterial growth is usually significantly slower than that achieved under "idealized" conditions ensured by investigators who aim to obtain results as rapidly as possible. Therefore, the mechanisms of prophage maintenance may be effective enough in natural habitats but not under laboratory conditions. In the latter case, bacterial hosts are grown under conditions optimal for their rapid reproduction, but because such conditions are never encountered naturally by both phage and host, no proper adaptations were formed during their evolution. Thus, a special precaution should be taken when interpreting unstable lysogenic development of temperate phage, as unstable lysogeny is a totally different phenomenon than pseudolysogeny.

There is also another possibility to maintain lysogeny without stringent mechanisms of partitioning and/or postsegregational killing. A phage can be maintained in a bacterial population due to a constitutive spontaneous induction rate of the corresponding prophage. One of *Borrelia burgdorferi* phages is a good example of such a phenomenon (Eggers and Samuels, 1999). In this case, free phages, present in the population, may assure lytic propagation on part, and relysogenization on the remaining population of cells naturally cured from prophage. Such a strategy, being a mixture of virulent and temperate lifestyles, could provide benefits of both rapid evolution of a virulent phage (due to frequent replication and genetic recombination cycles) and a relatively safe lifestyle of a prophage, with its presence in the environment assured by trading lysogenic conversion features for stability in the bacterial strain. Thus, the level of prophage stability may vary depending on several factors.

In phages that integrate their genomes into host chromosomes, the stability of lysogeny depends mostly on specific repressor stability and the ability of the repressor to sense and react to different inducing agents

(Węgrzyn and Węgrzyn, 2005). In phages that form episomal prophages, the regulation is more complex (Yarmolinsky, 2004). However, due to an observed broad spectrum of prophage stabilities in different bacterial hosts, it is sometimes difficult to draw a line that clearly separates lysogens from pseudolysogens. Almost all features postulated so far for a practical determination of the occurrence of pseudolysogeny can also be found in lysogens. The frequent selection of prophage-cured cells was observed not only in pseudolysogens but also in prophages of genera *Clostriduim* (Sakaguchi *et al.*, 2005) and *Vibrio* (Khemayan *et al.*, 2006). The ability to yield a high titer of phage during the restarting of host culture growth, characteristic for pseudolysogeny, was found in *E. coli* phage ϕ80 (Rotman *et al.*, 2010). Note that detection of this ability may be impaired by the low efficiency of plating and/or plaque formation by some temperate phages (Łoś *et al.*, 2008). Some authors suggested that pseudolysogens do not release phages spontaneously under environmental conditions simulated in a laboratory (Ripp and Miller, 1997). Finally, an inability to induce prophages using SOS-triggering agents, often mentioned as the pseudolysogen feature, is well known for P2 and P2-like prophages (Garen *et al.*, 1965; Bertani *et al.*, 1971). In this light, a lack of immunity to superinfection and superinfection exclusion, combined with the presence of a phage genome in the host cell, may be considered a suitable indicator for pseudolysogeny (Miller and Ripp, 2002; Wommack and Cowell, 2000). This phenomenon arises from the lack of repressor and/or immunity genes' products due to a lack of expression of early genes in pseudolysogens.

For many years, some researchers often considered that it was easier to describe what is not a pseudolysogeny than providing unambiguous features of this. Because of such problems, Ripp and Miller (1997) proposed a simple definition of pseudolysogeny, which appeared convincing to many people working on bacteriophages. According to this definition, pseudolysogeny is a state of phage development before resolving into true lysogeny or virulent growth. Although minimalistic, this definition shows a clear cutoff: pseudolysogeny is terminated upon appearance of the repressor in the case of prophage or active phage replication in the case of virulent development. This feature allows excluding all types of lysogeny, even if it differs from canonical ones represented by λ or P1, as well as less effective lytic growth in relation to host capabilities, which may be observed when environmental phages are to be tested. These phages, even if belonging to families known for their very aggressive lytic behavior, such as T1-like phages of *E. coli*, often show reduced virulence and effectiveness of propagation, when isolated, but select very quickly for more virulent mutants during propagation under laboratory conditions (unpublished data). As we gained the majority of our knowledge on phage lytic development by working on phages propagated under laboratory conditions for decades, the conclusions should be formulated with care when working on fresh isolates. The problem is that knowing little about different variants of regulatory mechanisms and different lifestyles of as yet unknown phages, the definition of pseudolysogeny based on its distinction from lysogeny and virulent development may be potentially not enough for some newly discovered phages.

In summary, the definition of pseudolysogeny proposed by Ripp and Miller (1997) is simple, elegant, and sufficient to most, if not all, already known phages; however, we propose a wider and more precise definition (although not so simple and elegant), presented in the first paragraph of this section, that should be effective for any pseudolysogens found under any conditions.

III. EXAMPLES OF PSEUDOLYSOGENY

Pseudolysogeny is observed in various bacterial species and can be triggered by different factors. Usually, it is triggered by conditions that cause suboptimal growth of bacteria or starvation. The latter case is better documented in the literature, despite the fact that such conditions are relatively seldom used in laboratories working on bacteriophages. To study pseudolysogeny in starved bacterial hosts, a chemostat appears to be the most suitable apparatus, as it allows for achieving almost any type of bacterial growth limitation, with a precisely set bacterial growth rate and a very homogeneous population (Natrian and Sirenc, 2000). The most commonly used limitation is carbon source limitation, which restricts energy production; however, other types of limitations, such as nitrogen or phosphorus limitations, are also possible.

A classical example of bacteriophages able to produce pseudolysogens is phage T4. This typically virulent phage can enter pseudolysogeny when infecting starved, nongrowing host cells (Kutter *et al.*, 1994). It was also demonstrated that T4 can form pseudolysogens when infecting *E. coli* cells growing in a chemostat under conditions supporting slow growth and at a temperature of 25°C. Under such conditions, the ability to start pseudolysogeny was dependent on the activity of the *rI* gene (Łoś *et al.*, 2003).

The method of nutrient limitation was also used to study pseudolysogeny in *Pseudomonas aeruginosa* (Ripp and Miller, 1998). When infecting very slowly growing cells, temperate phage F116 and virulent phage UT1 went into pseudolysogeny in fractions of cells, which were strongly correlated with the concentration of nutrients in the media, namely the lower the nutrient concentration, the higher the proportion of pseudolysogenic cells. Those results were obtained in cells cultured with the same growth rate but with different availabilities of nutrients. Employing another approach, namely analyzing bacterial cells grown with various generation times but with the same nutrient concentration in the culture medium, it was possible to show a similar pattern, that is, the slower the growth, the larger the fraction of pseudolysogens (Ripp and Miller, 1998).

It has been postulated that pseudolysogeny may play an important role in phage–bacteria interactions in water environments. The reason for this seems to be relatively low concentrations of nutrients and their seasonal variability. Ripp and Miller (1997) demonstrated the importance of pseudolysogeny in maintaining the presence of phages for a prolonged time in natural ecosystems. The ability of some virulent phages to form pseudolysogens allows minimizing the pressure of the phage on the bacterial population, especially when conditions of bacterial growth are unfavorable. This, in turn, allows for their coexistence in an ecological niche without mutual destruction when resources are very limited (Bohannan and Lenski, 1997; Ripp and Miller, 1997). The ability to stop phage development in a starved host seems to also be widespread among temperate phages, as showed experimentally on virulent mutants of P1 and λ phages (Łoś *et al.*, 2007). The reason for this may be as postulated by Ripp and Miller (1998), who suggested that the extremely low energy supply in a severely starved host cannot support early stages of phage development, such as the establishment of lysogeny. In phage λ, this step requires initial replication of phage DNA (Węgrzyn and Węgrzyn, 2005), which may not be possible if bacteria are starved for a carbon or phosphorus source. Starvation for a nitrogen source may impair early protein production and thus may impair the establishment of lysogeny.

Contrary to conditions of starvation, under which pseudolysogeny seems to be widespread and relatively easy to invoke, the pseudolysogeny observed in rich media, where all compounds necessary for phage development are available, seems to be significantly less understood. In some cases, other conditions may influence the establishment of pseudolysogeny. A good example is salinity of the environment, as in the case of *Halobacterium salinarium* phage Hs1, which tends to form large fractions of pseudolysogens upon infection, when NaCl concentration is close to the upper tolerance level for this bacterial species (Torsvik and Dundas, 1980).

Some phages show pseudolysogenic or rather pseudolysogenic-like behavior when propagated in hosts cultured under optimal growth conditions. Among them there are some temperate phages, such as phage c-st of *Clostridium botulinum* (Sakaguchi *et al.*, 2005) and phage VHS1 of *Vibrio harveyi* (Khemayan *et al.*, 2006). These phages encode lysogenic conversion genes necessary for pathogenicity of the bacterial species. They form prophages as episomal elements, and thus the loss of prophages may occur even if proper lysogeny was established. Under laboratory conditions they prove to be very unstable in strains and are lost easily. Moreover, other characteristics of their infection may resemble pseudolysogeny, namely strains of the *V. harveyi* cured from VHS1 show phenotypic conversion, despite the fact that the prophage genome is absent from the cell (Khemayan *et al.*, 2006). In fact, *C. botulinum* toxin-encoding phages were recognized previously as unstable pseudolysogenic elements (Oguma, 1976; Oguma *et al.*, 1976). However, at least in the latter case, unstable lysogeny is likely caused by the inefficient partitioning of plasmid prophages (Sakaguchi *et al.*, 2005).

IV. FUTURE PROSPECTS

Pseudolysogeny seems to be one of the most interesting and relevant features in phage–host interactions occurring in a natural environment. The rather low number of studies on this subject arises perhaps from a very problematic methodology necessary to study pseudolysogeny in starved hosts. Moreover, the instability of this interaction in growing hosts makes this subject tricky to study. However, as general knowledge about the roles of phages in a natural environment, with special emphasis on water, increased enormously in recent years, it is plausible that the need to understand the role of pseudolysogeny will increase over time. Real progress in elucidating the mechanisms of this process can be brought by different "*omics*" approaches, as well as by methods allowing for the investigation of single cells in a population. It would also require a broader use of cultivation methods, which reflect conditions found in the natural environment. Furthermore, investigation of pseudolysogeny in environmental samples may be more intensive soon, as seasonal changes in lysogen content were found in many habitats (Jiang and Paul, 1998; Laybourn-Parry *et al.*, 2007; Paul, 2008).

ACKNOWLEDGMENTS

This work was partially supported by the European Union within European Regional Development Fund through grant Innovative Economy (POIG.01.01.02-00-008/08) and the University of Gdańsk.

REFERENCES

Abedon, S. T. (1989). Selection for bacteriophage latent period length by bacterial density: A theoretical examination. *Microb. Ecol.* **18:**79–88.

Baess, I. (1971). Report on a pseudolysogenic mycobacterium and a review of the literature concerning pseudolysogeny. *Acta Path. Microbiol. Scand.* **79:**428–434.

Bertani, L. E., and Bertani, G. (1971). Genetics of P2 and related phages. *Adv. Genet.* **16:**199–237.
Bohannan, B. J. M., and Lenski, R. E. (1997). Effect of resource enrichment on a chemostat community of bacteria and bacteriophage. *Ecology* **78:**2303–2315.
Delbrüeck, M. (1946). Bacterial viruses or bacteriophages. *Biol. Rev.* **21:**30–40.
d'Herelle, F. (1930). Elimination du bacteriophage dans les symbioses bacterie-bacteriophage. *Compt. Rend. Soc. Biol.* **104:**1254.
Eggers, C. H., and Samuels, D. S. (1999). Molecular evidence for a new bacteriophage of *Borrelia burgdorferi. J. Bacteriol.* **181:**7308–7313.
Garen, A., Garen, S., and Wilhelm, R. C. (1965). Suppressor genes for nonsense mutations. *J. Mol. Biol.* **14:**167–178.
Jiang, S. C., and Paul, J. H. (1998). Significance of lysogeny in the marine environment: Studies with isolates and an ecosystem model. *Microb. Ecol.* **35:**235–243.
Khemayan, K., Pasharawipas, T., Puiprom, O., Sriurairatana, S., Suthienkul, O., and Flegel, T. W. (2006). Unstable lysogeny and pseudolysogeny in *Vibrio harveyi* siphovirus-like phage 1. *Appl. Environ. Microbiol.* **72:**1355–1363.
Kutter, E., Kellenberger, E., Carlson, K., Eddy, S., Neitzel, J., Messinger, L., North, J., and Guttman, B. (1994). Effects of bacterial growth condition and physiology on T4 infection. *In* "Molecular Biology of Bacteriophage T4" (J. D. Karam, ed.), pp. 406–418. American Society for Microbiology, Washington, DC.
Laybourn-Parry, J., Marshall, W. A., and Madan, N. J. (2007). Viral dynamics and patterns of lysogeny in saline Antarctic lakes. *Polar Biol.* **30:**351–358.
Lehnherr, H., Maguin, E., Jafri, S., and Yarmolinsky, M. B. (1993). Plasmid addiction genes of bacteriophage P1: doc, which causes cell death on curing of prophage, and phd, which prevents host death when prophage is retained. *J. Mol. Biol.* **233:**414–428.
Łoś, J. M., Golec, P., Węgrzyn, G., Węgrzyn, A., and Łoś, M. (2008). Simple method for plating *Escherichia coli* bacteriophages forming very small plaques or no plaques under standard conditions. *Appl. Environ. Microbiol.* **74:**5113–5120.
Łoś, M., Węgrzyn, G., and Neubauer, P. (2003). A role for bacteriophage T4 *rI* gene function in the control of phage development during pseudolysogeny and in slowly growing host cells. *Res. Microbiol.* **154:**547–552.
Lwoff, A. (1953). Lysogeny. *Bacteriol. Rev.* **17:**269–337.
Miller, R. V. (2006). Marine phages. *In* "The Bacteriophages" (R. Calendar and S. T. Abedon, eds.), pp. 534–544. Oxford University Press, Oxford.
Miller, R. V., and Ripp, S. A. (2002). Pseudolysogeny: A bacteriophage strategy for increasing longevity *in situ*. *In* "Horizontal Gene Transfer" (M. Syvanen and C. I. Kado, eds.), 2nd edn., pp. 81–91. Academic Press, San Diego, CA.
Oguma, K. (1976). The stability of toxigenicity in *Clostridium botulinum* types C and D. *J. Gen. Microbiol.* **92:**67–75.
Oguma, K., Iida, H., Shiozaki, M., and Inoue, K. (1976). Antigenicity of converting phages obtained from *Clostridium botulinum* types C and D. *Infect. Immun.* **13:**855–860.
Paul, J. H. (2008). Prophages in marine bacteria: Dangerous molecular time bombs or the key to survival in the seas? *ISME J.* **2:**579–589.
Poulsen, L. K., Licht, T. R., Rang, C., Krogfelt, K. A., and Molin, S. (1995). Physiological state of *Escherichia coli* BJ4 growing in the large intestines of streptomycin-treated mice. *J. Bacteriol.* **177:**5840–5845.
Ripp, S., and Miller, R. V. (1997). The role of pseudolysogeny in bacteriophage-host interactions in a natural freshwater environment. *Microbiology* **143:**2065–2070.
Ripp, S., and Miller, R. V. (1998). Dynamics of the pseudolysogenic response in slowly growing cells of *Pseudomonas aeruginosa. Microbiology* **144:**2225–2232.
Romig, W. R., and Brodetsky, A. M. (1961). Isolation and preliminary characterization of bacteriophages of *Bacillus* subtilis. *J. Bacteriol.* **82:**135–141.

Rotman, E., Amado, L., and Kuzminov, A. (2010). Unauthorized horizontal spread in the laboratory environment: The tactics of Lula, a temperate lambdoid bacteriophage of *Escherichia coli*. *PLoS One* **5:**e11106.

Sakaguchi, Y., Hayashi, T., Kurokawa, K., Nakayama, K., Oshima, K., Fujinaga, Y., Ohnishi, M., Ohtsubo, E., Hattori, M., and Oguma, K. (2005). The genome sequence of *Clostridium botulinum* type C neurotoxin-converting phage and the molecular mechanisms of unstable lysogeny. *Proc. Natl. Acad. Sci. USA* **102:**17472–17477.

Schrader, H. S., Schrader, J. O., Walker, J. J., Wolf, T. A., Nickerson, K. W., and Kokjohn, T. A. (1997). Bacteriophage infection and multiplication occur in *Pseudomonas aeruginosa* starved for 5 years. *Can. J. Microbiol.* **43:**1157–1163.

Torsvik, T., and Dundas, I. D. (1980). Persisting phage infection in *Halobacterium salinarium*. *J. Gen. Virol.* **47:**29–36.

Twort, F. W. (1915). An investigation on the nature of ultramicroscopic viruses. *Lancet* **2:**1241–1243.

Węgrzyn, G., and Węgrzyn, A. (2005). Genetic switches during bacteriophage lambda development. *Prog. Nucleic Acid Res. Mol. Biol.* **79:**1–48.

Whitman, W. B., Coleman, D. C., and Wiebe, W. J. (1998). Prokaryotes: The unseen majority. *Proc. Natl. Acad. Sci. USA* **95:**6578–6583.

Wommack, K. E., and Colwell, R. R. (2000). Virioplankton: Viruses in aquatic ecosystems. *Microbiol. Mol. Biol. Rev.* **64:**69–114.

Yarmolinsky, M. B. (2004). Bacteriophage P1 in retrospect and in prospect. *J. Bacteriol.* **186:**7025–7028.

Yin, J. (1993). Evolution of bacteriophage T7 in a growing plaque. *J. Bacteriol.* **175:**1272–1277.

CHAPTER 10

Role of Host Factors in Bacteriophage ϕ29 DNA Replication

Daniel Muñoz-Espín,* Gemma Serrano-Heras,† and Margarita Salas*

Contents

* Instituto de Biología Molecular "Eladio Viñuela" (CSIC), Centro de Biología Molecular "Severo Ochoa" (CSIC-UAM), Universidad Autónoma, Madrid, Spain
† Experimental Research Unit, General University Hospital of Albacete, Albacete, Spain

Advances in Virus Research, Volume 82
ISSN 0065-3527, DOI: 10.1016/B978-0-12-394621-8.00020-0

Abstract During the course of evolution, viruses have learned to take advantage of the natural resources of their hosts for their own benefit. Due to their small dimension and limited size of genomes, bacteriophages have optimized the exploitation of bacterial host factors to increase the efficiency of DNA replication and hence to produce vast progeny. The *Bacillus subtilis* phage ϕ29 genome consists of a linear double-stranded DNA molecule that is duplicated by means of a protein-primed mode of DNA replication. Its genome has been shown to be topologically constrained at the size of the bacterial nucleoid and, as to avoid generation of positive supercoiling ahead of the replication forks, the bacterial DNA gyrase is used by the phage. In addition, the *B. subtilis* actin-like MreB cytoskeleton plays a crucial role in the organization of ϕ29 DNA replication machinery in peripheral helix-like structures. Thus, in the absence of an intact MreB cytoskeleton, ϕ29 DNA replication is severely impaired. Importantly, MreB interacts directly with the phage membrane protein p16.7, responsible for attaching ϕ29 DNA at the cell membrane. Moreover, the ϕ29-encoded protein p56 inhibits host uracil–DNA glycosylase activity and has been proposed to be a defense mechanism developed by the phage to prevent the action of the base excision repair pathway if uracil residues arise in replicative intermediates. All of them constitute incoming examples on how viruses have profited from the cellular machinery of their hosts.

LIST OF ABBREVIATIONS

AP	apurinic/apyrimidinic
BER	base excision repair
DBP	double-stranded DNA binding protein
ds	double-stranded
DSP	dithiobis[succinimidylpropionate]
GFP	green fluorescent protein
PAGE	polyacrylamide gel electrophoresis
ss	single-stranded
SSB	single-stranded DNA binding protein
TP	terminal protein

UDG	Uracil DNA glycosylase
YFP	yellow fluorescent protein

I. ϕ29 PROTEIN-PRIMED MODE OF DNA REPLICATION

A. General features of bacteriophage ϕ29

Bacteriophage ϕ29 is a lytic phage that belongs to the Podoviridae family, which is composed of phages with bilateral symmetry containing a short tail and an icosaedric capsid (Ackermann, 1998). Its natural host is the Gram-positive bacterium *Bacillus subtilis* [Reilly and Spizizen (1965); for a general review, see Graumann (2007)], an endospore-forming organism that normally inhabits the soil or decaying plant material. ϕ29 is one of the smallest *Bacillus* phages isolated so far and is among the smallest known phages containing double-stranded DNA (dsDNA)(Anderson *et al.*, 1966). The phage ϕ29 particle consists of a prolate head, a neck formed by a conector (required for head assembly) and a lower collar from which the appendages are attached (required for phage adsorption to the cell wall), and a tail knob.

Phage ϕ29 has been the object of extensive research studies, many of which have contributed to the understanding of several molecular mechanisms of general biological processes, such as regulation of transcription, viral DNA packaging, viral morphogenesis, and, particularly, DNA replication. Initiation of phage ϕ29 DNA replication utilizes a protein-primed mode. Such a mode is a common mechanism to initiate DNA synthesis of linear dsDNAs that contain a specific protein linked covalently to the two 5′ DNA ends, the so-called terminal protein (TP). In this case, the primer is the hydroxyl group of a specific serine, threonine, or tyrosine residue of the TP. Thus, the genome of phage ϕ29 is composed of a linear dsDNA 19,285bp long (Garvey *et al.*, 1985; Vlcek and Paces, 1986; Yoshikawa and Ito, 1982) with a TP (parental TP) linked covalently at each 5′ end (Ito, 1978; Salas *et al.*, 1978; Yehle, 1978).

Development of an *in vitro* replication system using purified DNA and proteins from phage ϕ29 laid the foundation for studying the protein-primed mode of DNA replication and makes ϕ29 a paradigm to unravel the intrinsic mechanisms that govern this process [for review, see Salas (1991)].

B. *In vitro* ϕ29 DNA replication

It has been determined that four ϕ29 DNA replication proteins constitute the essential system to amplify ϕ29 TP-DNA *in vitro* (for a genetic and transcriptional map, see Fig. 1A): DNA polymerase (p2, encoded by

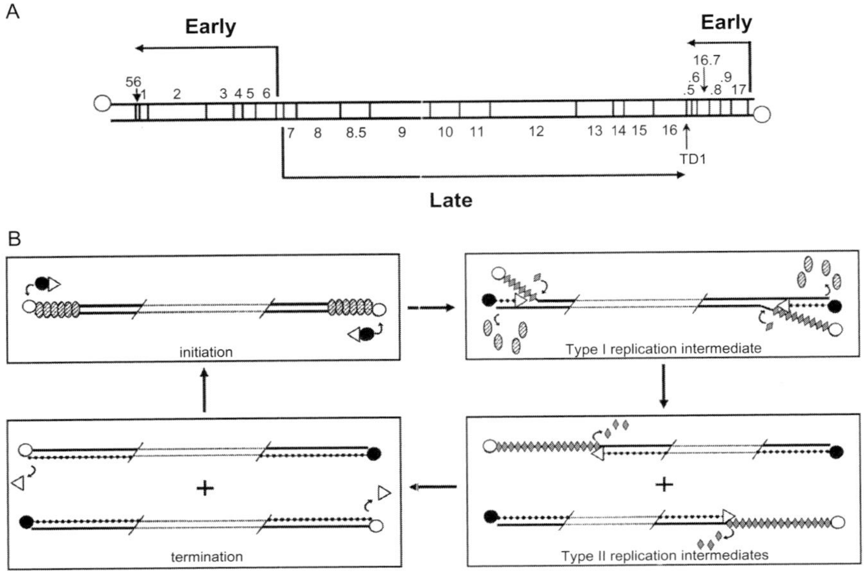

FIGURE 1 Genetic and transcriptional map of the ϕ29 genome and mechanism of *in vitro* ϕ29 DNA replication. (A) Map of the ϕ29 genome. The direction of transcription and length of the transcripts are indicated by arrows, and positions of genes are indicated with numbers. TD1 corresponds to the bidirectional transcriptional terminator located in between the convergently transcribed late and right-side early operons. White circles represent the terminal protein linked covalently to the 5′ DNA ends. (B) Overview of *in vitro* ϕ29 DNA replication mechanism. Replication starts by recognition of the p6-nucleoprotein complexed origins of replication by a TP/DNA polymerase heterodimer. The DNA polymerase then catalyses the addition of the first dAMP to the TP present in the heterodimer complex. Next, after a transition step, these two proteins dissociate and the DNA polymerase continues processive elongation until replication of the nascent DNA strand is completed. Replication is coupled to strand displacement. The ϕ29-encoded SSB protein p5 binds to the displaced ssDNA strands and is removed by the DNA polymerase during later stages in the replication process. Continuous polymerization results in the generation of two fully replicated ϕ29 genomes. White circles, parental TP; black circles, primer TP; triangles, DNA polymerase; ovals, replication initiator protein p6; diamonds, SSB protein p5; *de novo* synthesized DNA is shown as beads on a string.

gene 2), TP (p3, encoded by gene 3), single-stranded DNA (ssDNA)-binding protein or SSB (p5, encoded by gene 5), and double-stranded DNA-binding protein or DBP (p6, encoded by gene 6). By employing the appropriate amounts of DNA polymerase, TP, SSB, and DBP it has been possible to amplify *in vitro* small amounts of ϕ29 TP-DNA (0.5ng) over 1000-fold. The amplified DNA was shown to retain its biological activity and a high degree of fidelity, as demonstrated by the fact that its infectivity in transfection experiments was identical to that of natural ϕ29 DNA obtained from virions (Blanco *et al.*, 1994).

Initiation of ϕ29 DNA replication (for details, see Fig. 1B) starts with formation of a heterodimer between the ϕ29 DNA polymerase and a free TP molecule (priming TP), which recognizes and interacts with the replication origin (containing a parental TP linked covalently to the DNA) at both ends of the genome. Crystallographic data have revealed that both intermediate and C-terminal domains of the priming TP make extensive contacts with several domains and subdomains of the DNA polymerase, giving rise to a functional heterodimeric complex (Kamtekar *et al.*, 2006). The ϕ29 DBP protein p6 forms a nucleoprotein complex at the replication origins that would help open the DNA ends (Serrano *et al.*, 1994). Protein p6, described as a histone-like protein, binds DNA as a dimer every 24 nucleotides (Prieto *et al.*, 1988) in a cooperative way and interacts with the viral DNA through the minor groove (Freire *et al.*, 1994; Serrano *et al.*, 1990), restraining its positive supercoiling (Prieto *et al.*, 1988; Serrano *et al.*, 1993). This nucleoprotein complex facilitates formation of a covalent linkage between the first inserted nucleotide (dAMP) and the hydroxyl group of serine 232 of the priming TP, which is catalyzed by the ϕ29 DNA polymerase (Blanco and Salas, 1984; Hermoso *et al.*, 1985). The DNA ends of ϕ29 have an inverted terminal repeat of six nucleotides (3′-TTTCAT), and formation of the first TP–dAMP covalent complex is directed by the second nucleotide at the 3′ end of the template strand. Then, the TP–dAMP complex slides back one position to recover the terminal nucleotide, and the second 3′ terminal nucleotide acts again as a template to direct the incorporation of the second dAMP residue (Méndez *et al.*, 1992). Next, the ϕ29 DNA polymerase synthesizes a short elongation product of about six to nine nucleotides and dissociates from the priming TP (Méndez *et al.*, 1997). Replication, starting at both DNA ends, is coupled to strand displacement (Blanco *et al.*, 1989) and results in the generation of so-called type I replication intermediates, which consist of full-length ϕ29 dsDNA molecules with one or more ssDNA branches of different lengths. The large amounts of ssDNA stretches generated are bound by the SSB protein p5 in a cooperative way, which stimulates dNTP incorporation (Martín and Salas, 1988) and increases the elongation rate of the DNA polymerase (Soengas *et al.*, 1992). When two converging DNA polymerases merge, a type I replication intermediate becomes physically separated into two type II replication intermediates, consisting of full-length ϕ29 DNA molecules partially double-stranded and partially single-stranded (Gutiérrez *et al.*, 1991; Inciarte *et al.*, 1980). DNA polymerases continue processive elongation until replication of the parental DNA strands is completed, obtaining two fully replicated ϕ29 genomes.

C. *In vivo* φ29 DNA replication

Electron microscopic analysis of DNA replicative intermediates synthesized in *B. subtilis* cells infected with phage φ29 showed the presence of type I and type II replication intermediates (Harding and Ito, 1980; Inciarte *et al.*, 1980; Sogo *et al.*, 1982). Analysis of these replication intermediates indicated that replication starts at both DNA ends (although not at the same time) and proceeds by strand displacement toward the opposite end.

It is well known that genes *1, 2, 3, 5, 6, 16.7, 17*, and *56* (see Fig. 1A) play a role in the synthesis of viral DNA *in vivo* (Carrascosa *et al.*, 1976; Hagen *et al.*, 1976; Meijer *et al.*, 2001b; Prieto *et al.*, 1989; Serrano-Heras *et al.*, 2006; Talavera *et al.*, 1972). From these, φ29 genes *2, 3, 5*, and *6*, which are conserved in φ29-related phages such as B103 and GA-1, are essential for *in vivo* phage DNA replication. As mentioned earlier, DNA polymerase (product of gene *2*), TP (encoded by gene *3*), and DBP (encoded by gene *6*) are involved in the initiation of DNA replication, whereas SSB protein (encoded by gene *5*) is mainly concerned with the elongation step.

Protein p1 (product of gene *1*) assembles into long protofilaments forming bidimensional sheets (Bravo and Salas, 1998) in association with the bacterial membrane *in vivo* (Serrano-Heras *et al.*, 2003). Phage φ29 DNA replication was shown to be reduced very significantly when nonsuppressor *B. subtilis* cells were infected with mutant phage sus*1*(629) at 37°C (Bravo and Salas, 1997; Prieto *et al.*, 1989). In addition, protein p1 was also shown to interact with the viral TP *in vitro* (Bravo *et al.*, 2005). These and other results suggest that protein p1 is a component of a membrane-associated structure that could play a role in the organization of φ29 DNA replication by providing an anchoring site for the replication machinery.

Protein p16.7 (encoded by gene *16.7*) is an integral membrane protein that possesses nonspecific DNA-binding capacity (Meijer *et al.*, 2001b), forms multimers *in vivo*, and is able to interact with the viral TP (Serna-Rico *et al.*, 2003). A model has been proposed in which p16.7 would be involved in the attachment of replicating φ29 TP–DNA molecules at the membrane of infected *B. subtilis* cells [see Section III and Serna-Rico e*t al.* (2003)].

Protein p17 (encoded by gene *17*) has been shown to be necessary *in vivo* for efficient DNA replication (Carrascosa *et al.*, 1976; Crucitti *et al.*, 1998; González-Huici *et al.*, 2004b) and *in vitro* enhances the binding of protein p6 to DNA (Crucitti *et al.*, 2003). Additionally, p17 stimulates φ29 DNA amplification *in vitro* (Crucitti *et al.*, 1998) and interacts with the viral protein p6 (Crucitti *et al.*, 2003). Results suggest that p17 could participle in maintenance of the proper φ29 DNA topology required for efficient DNA replication (González-Huici *et al.*, 2004b).

The role of protein p56, encoded by gene *56*, is described in Section IV.

D. Organization of ϕ29 DNA replication machinery within the cell

Remarkably little was known about the *in vivo* organization of viral DNA replication proteins in bacteria. However, over the last decade, the development of fluorescence microscopy and immunomicroscopy techniques has led to a better comprehension of the intrinsic functional and structural mechanisms that govern viral DNA replication in live cells. In this respect, our laboratory has been pioneer in applying fluorescence microscopy techniques to gain insight into the organization of viral DNA replication machineries in bacteria. For this purpose, *B. subtilis*-infecting bacteriophage ϕ29 has been used as a model system.

Bacteriophages produce high numbers of progeny within a small period of time during their lytic cycle. In the case of phage ϕ29, it develops a lytic cycle of only 40–50 min and hence, after infection, the phage DNA replication machinery is expected to be organized rapidly to allow simultaneous amplification of multiple templates in a short period of time. It has been demonstrated that once injection of the ϕ29 genome takes place in *B. subtilis* cells, the parental TP (which is linked covalently to each 5′ end of the genome) associates with the bacterial nucleoid (Muñoz-Espín *et al.*, 2010)(for a model, see Fig. 2). It was determined previously that the ϕ29 TP possesses dsDNA-binding capacity *in vitro* (Zaballos *et al.*, 1989), and novel results show that the dsDNA-binding capacity of the TP is responsible for its nucleoid association (Muñoz-Espín *et al.*, 2010). This association occurs independently of the priming TP (Fig. 2A) and, therefore, the parental TP directs the viral genome to a

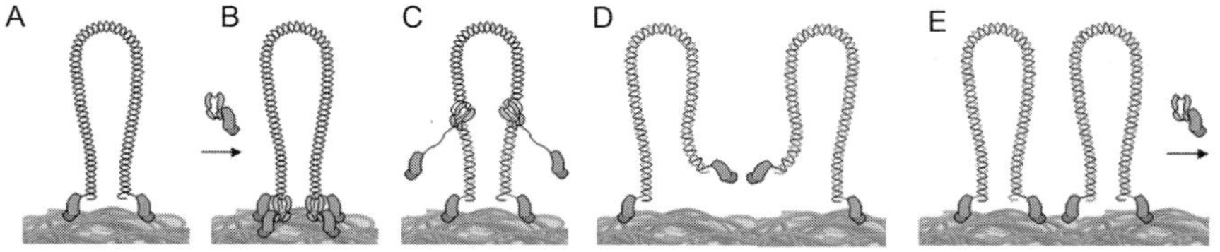

FIGURE 2 Model of nucleoid-associated early ϕ29 DNA replication organized by the TP. (A) A complete ϕ29 TP-DNA molecule (linear dsDNA shown as a double-helix loop) is shown to be attached to the bacterial nucleoid surface (bottom grey mass) by the N-terminal domain (colored red) of the two parental TPs (colored red and green). (B) The priming TP interacts with phage DNA polymerase (colored cyan) forming a heterodimer that associates to the nucleoid via the TP N-terminal domain and recognizes the origins of replication. (C) After a transition step, DNA polymerases dissociate and continue processive elongation of the nascent DNA strands (red lines) coupled to strand displacement. (D and E) Once DNA replication is completed, two ϕ29 TP-DNA molecules are ready for another round of replication. For simplicity, other viral proteins involved in DNA replication are not drawn. (See Page 32 in Color Section at the back of the book.)

specific site within the cell, where rapid transcription of viral genes and initiation of DNA replication may take place. In this respect, the *B. subtilis* RNA polymerase has been shown to localize principally within the bacterial nucleoid (Lewis *et al.*, 2000) and to colocalize with the ϕ29 TP (Muñoz-Espín *et al.*, 2010).

It has been determined that a green fluorescent protein fusion of the ϕ29 DNA polymerase localizes uniformly throughout the cell in the absence of other phage-encoded proteins (Muñoz-Espín *et al.*, 2009). In sharp contrast, a yellow fluorescent protein fusion of the ϕ29 TP (which mimics the priming TP) is able to distribute at the bacterial nucleoid by itself (Muñoz-Espín *et al.*, 2010). Importantly, when fluorescent fusion proteins of ϕ29 DNA polymerase and TP are expressed simultaneously in *B. subtilis* cells, the DNA polymerase relocalizes to the bacterial chromosome (Muñoz-Espín *et al.*, 2010). These results, together with the sequence-independent DNA-binding capacity of the TP, represent a strong body of evidence that once both ϕ29 TP and DNA polymerase are synthesized, they form a heterodimeric complex that associates with the bacterial nucleoid through the DNA-binding domain of the TP (see Fig. 2B). Thus, the ϕ29 priming TP would recruit the DNA polymerase to the specific sites for the initiation of DNA replication and recognize the replication origins at both DNA ends by means of specific interactions with the parental TP (González-Huici *et al.*, 2000; Serna-Rico *et al.*, 2000). After a transition step, the DNA polymerase dissociates and continues processive elongation coupled to strand displacement (Fig. 2C) until the nascent DNA is completed (Fig. 2D). Once DNA replication is completed, two TP-DNA molecules would be prepared to initiate a new replication cycle (Fig. 2E). Additionally to ϕ29, the TP of phage PRD1, infecting *Escherichia coli*, also localizes at the bacterial nucleoid independently of other viral-encoded proteins, suggesting a new functional property of viral TPs in bacteria (Muñoz-Espín *et al.*, 2010).

Resolution of the crystallographic structure of the ϕ29 priming TP/DNA polymerase heterodimer showed that the TP has an elongated three-domain structure composed of an N-terminal domain (TP-Nt), an intermediate domain (TP-I) that makes extensive contacts with the phage DNA polymerase, and a C-terminal priming domain (TP-Ct) that mimics duplex product DNA in its binding site of the polymerase (Kamtekar *et al.*, 2006). From these, the N-terminal domain retained the ability to bind dsDNA in a sequence-independent way and is responsible for nucleoid localization of the TP and, ultimately, for that of ϕ29 early DNA replication machinery (Muñoz-Espín *et al.*, 2010). Importantly, in the absence of the N-terminal domain of the TP, ϕ29 DNA replication is affected severely, indicating that this domain is important for the *in vivo* organization of phage ϕ29 DNA replication (see Fig. 2).

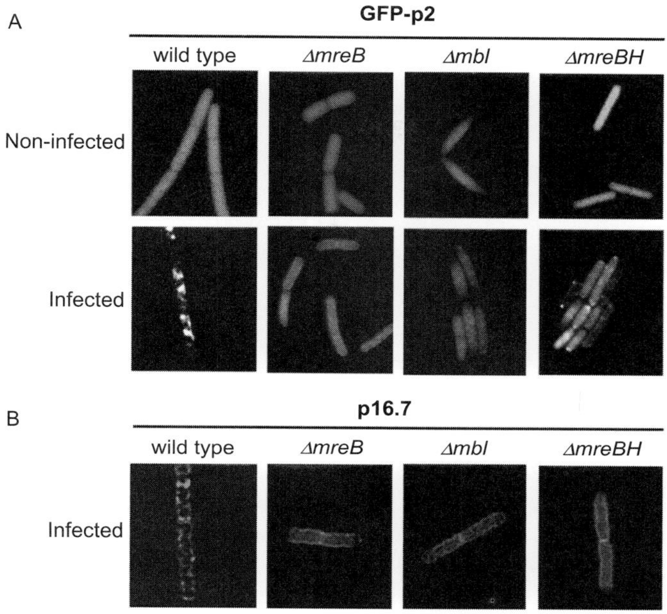

FIGURE 3 Helical pattern of DNA polymerase (GFP-p2) and protein p16.7 is lost in *mreB* cytoskeleton mutant strains. (A) Wild-type, Δ*mreB*, Δ*mbl*, and Δ*mreBH B. subtilis* strains containing a xylose-inducible copy of *gfp-p2* at the *amyE* locus were grown to mid-exponential phase in LB medium and infected with *sus2*(513) mutant phage (Moreno *et al.*, 1974). Samples were withdrawn and analyzed 50 min after infection. (B) Wild-type, Δ*mreB*, Δ*mbl*, and Δ*mreBH B. subtilis* strains were grown in LB medium and infected with phage *sus14*(1242)(Jiménez *et al.*, 1977). Samples were harvested 20 min postinfection and subjected to immunofluorescence analysis using polyclonal antibodies against p16.7. Cells are shown after deconvolution of an image stack, as a "max projection."

At middle infection times, φ29 DNA replication machinery associates with the bacterial chromosome, and while the cell enlarges prior to division, the DNA polymerase (Fig. 3A), TP, and viral dsDNA are reorganized, adopting a peripheral helix-like distribution toward the poles of the cell (Muñoz-Espín *et al.*, 2009, 2010). At this stage of the phage cycle, φ29 DNA polymerase and TP were shown to colocalize substantially. Because the newly synthesized DNA of the replicating *B. subtilis* chromosome is translocated via a helical structure from midcell positions toward the poles (Berlatzky *et al.*, 2008), it has been postulated that the φ29 DNA replication machinery may take advantage of the intrinsic dynamic behavior provided by the chromosome segregation process to be redistributed at multiple peripheral sites (Muñoz-Espín *et al.*, 2010).

ϕ29 membrane protein p16.7, which also localizes in a helix-like configuration (Muñoz-Espín *et al.*, 2009)(Fig. 3B), is able to bind ssDNA and dsDNA in a nonspecific way (Albert *et al.*, 2005; Serna-Rico *et al.*, 2002) and has been proposed to organize the viral-replicating intermediates at numerous peripheral locations. It has been demonstrated that protein p16.7 interacts physically with the actin-like cytoskeleton protein MreB, which ultimately directs the helix-like distribution of p16.7 and is required for efficient ϕ29 DNA replication [see Section III for details and Muñoz-Espín *et al.* (2009)].

II. PHAGE ϕ29 USES BACTERIAL DNA GYRASE

A. *Bacillus subtilis* topoisomerases: DNA gyrase

Chromosomes are compacted to fit into the bacterial cell, and DNA topology must be regulated to allow DNA replication to take place. Due to the helical nature of DNA, duplication of a DNA double helix will lead to superhelical torsions ahead of the replication fork, caused through action of the DNA helicase. Topoisomerases are enzymes that facilitate DNA replication by removing torsions in the DNA in a variety of ways. Like most other bacteria, *B. subtilis* contains four topoisomerases: Topo I, Topo II (DNA gyrase), Topo III, and Topo IV. Three of them are essential (Topo I, DNA gyrase, and Topo IV), as depletion leads to the formation of large cells with single nucleoids and the formation of anucleate cells that arise through division within the spaces that are free of DNA separate from the nucleoid (Kato *et al.*, 1990; Steck and Drlica, 1984; Tadesse *et al.*, 2005; Zhu *et al.*, 2001). These three enzymes are therefore indispensable for chromosome segregation and have been shown to localize at the nucleoid of growing cells.

Topo I and III are composed of a single polypeptide and change supercoiling through breakage, strand passage, and resealing of a single DNA strand. They are able to relax negative supercoils (Espeli and Marians, 2004), and Topo III can also decatenate circular DNA molecules (Dekker *et al.*, 2002). Differently, DNA gyrase and Topo IV, which are composed of two different polypeptides, break both strands of the DNA duplex and move an unbroken DNA strand through this gap before releasing. These polypeptides are encoded by *gyrA* and *gyrB* in the case of DNA gyrase and by *parC* and *parE* for Topo IV. DNA gyrase and Topo IV can relax positive supercoils ahead of the DNA polymerase and can also remove entanglements between sister chromosomes, leading to a complete chromosome segregation toward the daughter cells (Espeli and Marians, 2004). Although standard measurements that average the action of different topoisomerases on different substrate molecules demonstrated that DNA gyrase and Topo IV catalyzed comparable rates of positive supercoiling removal, the Topo IV reaction was distributive, whereas that of DNA gyrase was processive (Hiasa and Marians, 1996). Instead, Topo IV has much higher decatenation activity than DNA gyrase at the end of the replication/segregation cycle.

It is thought that DNA gyrase works mainly at the replication forks as it forms discrete accumulations and foci at the central replisome in growing cells (Stone *et al.*, 2003; Tadesse and Graumann, 2006). DNA gyrase has been studied extensively by the use of different drugs such as novobiocin or nalidixic acid. The gyrA polypeptide constitutes the target for nalidixic acid, which inhibits its nicking-closing activity. However, the gyrB polypeptide is the target for novobiocin, affecting the ATPase subunit.

B. The ϕ29 genome is topologically constrained *in vivo*

The histone-like phage ϕ29-encoded protein p6 is essential for DNA replication *in vivo*, as a ϕ29 sus *6* mutation prevents viral DNA synthesis (Bravo *et al.*, 1994; Camacho and Salas, 2000; Carrascosa *et al.*, 1976; Escarmís *et al.*, 1989; Schachtele *et al.*, 1973) and activates the initiation step *in vitro* (Blanco *et al.*, 1986; Pastrana *et al.*, 1985). Its small size and abundance in infected cells (about 700,000 copies/cell)(Abril *et al.*, 1997) are features expected for proteins with architectural roles. Initiation of ϕ29 DNA replication *in vitro* requires the formation of a p6–DNA nucleoprotein complex at the DNA ends (containing the replication origins, see earlier discussion), in which the DNA forms a right-handed toroidal superhelix around a multimeric protein core (Serrano *et al.*, 1990, 1993). However, it has been determined that protein p6 binds to most, if not all, the viral genome *in vivo*, although with higher affinity for both DNA ends (González-Huici *et al.*, 2004c). In fact, protein p6 binds all along ϕ29 DNA *in vivo* with a much higher affinity than for plasmid DNA (González-Huici *et al.*, 2004a), which, similarly to the bacterial chromosome, consists of a circular DNA molecule. Because protein p6 is known to restrain positive supercoiling of the DNA *in vitro* (Prieto *et al.*, 1988; Serrano *et al.*, 1993) and negative supercoiling of the DNA impairs p6 binding *in vivo* (González-Huici *et al.*, 2004c), it has been postulated that protein p6 would allow discrimination between bacterial DNA and ϕ29 DNA due to a lower negative superhelicity of the latter. In agreement with this assumption, binding of protein p6 to ϕ29 DNA increases when cultures of *B. subtilis*-infected cells are supplemented with novobiocin (González-Huici *et al.*, 2004a), a DNA gyrase inhibitor that produces a loss of negative supercoiling [see earlier discussion and Osburne *et al.* (1988)], as expected for a DNA with topological restrictions. In contrast, nalidixic acid, which also inhibits DNA gyrase but has no topological effects, did not increase protein p6–DNA binding. These results strongly suggest that ϕ29 DNA, although not covalently closed, is topologically constrained *in vivo*. This issue, as described in the following section, has been investigated further measuring the efficiency of ϕ29 DNA replication under different conditions of DNA gyrase inhibitors.

C. DNA gyrase relaxes supercoiling of ϕ29 DNA

Although the genome of phage ϕ29 is composed of a linear dsDNA with a parental TP linked covalently at the 5′ ends and therefore is not covalently closed, it associates with the bacterial nucleoid through the DNA-binding domain of the parental TPs (Muñoz-Espín *et al.*, 2010). In this scenario, progression of the two replication forks would generate positive supercoiling ahead due to the DNA unwinding, which in turn would impair replication if the viral DNA were not allowed to rotate freely. To avoid this situation, a bacterial topoisomerase actively introducing negative supercoils would be required for efficient replication of a topologically restricted ϕ29 DNA. To study the view of a replicating ϕ29 -DNA as a molecule with topological restrictions, the efficiency of ϕ29 DNA replication was analyzed by agarose gel electrophoresis and real-time polymerase chain reaction (PCR) in cultures supplemented with the DNA gyrase inhibitors novobiocin or nalidixic acid (González-Huici *et al.*, 2004a). Results showed that both inhibitors, especially novobiocin, affected ϕ29 DNA replication substantially, constituting additional evidence that the phage genome is topologically constrained *in vivo* and showing that DNA gyrase plays an important role in ϕ29 DNA replication.

III. THE MREB CYTOSKELETON ORGANIZES ϕ29 DNA REPLICATION

A. MreB family of proteins

MreB and MreB homologues are actin-like cytoskeletal proteins that play an important role in a broad variety of cellular functions in bacteria, including regulation of cell shape, chromosome segregation, cell polarity, macromolecular trafficking, and organization of membranous organelles (Carballido-López, 2006). The crystallographic structure of MreB showed that it has the same fold as eukaryotic actin; moreover, crystal packing revealed polymerization of actin-like protofilaments in which the subunits are translated in one dimension (Van den Ent *et al.*, 2001). Different than eukaryotic actin, whose assembly is favored in the presence of ATP over GTP, MreB assembly can be triggered by both GTP and ATP, indicating that MreB is an equally effective ATPase and GTPase (Esue *et al.*, 2006).

MreB has been demonstrated to be widespread in bacteria with nonspherical shapes, but is absent from most bacteria displaying coccoid morphologies (Jones *et al.*, 2001). Thus, these proteins are present in both Gram-negative and Gram-positive bacteria and also in some archaebacteria and mollicutes. MreB appears to be essential in all bacteria studied so far, including *B. subtilis, E. coli, Caulobacter crescentus, Salmonella enterica, Streptomyces coelicolor*, and *Rhodobacter sphaeroides* (Carballido-López, 2006). Whereas a single copy of *mreB* is present in Gram-negative genomes, several *mreB* homologues are usually present in Gram-positive genomes. In the case of *B. subtillis* there are three *mreB* genes: *mreB, mbl* (*mreB*-like), and *mreBH*.

In *B. subtilis*, MreB proteins have been shown to form helix-like filamentous structures *in vivo* that are essential for the control of rod-shaped morphologies (Jones *et al.*, 2001). It has been determined by time-lapse microscopy that these filaments are flexible and highly dynamic structures moving constantly through live *B. subtilis* cells (Carballido-López and Errington, 2003a,b; Defeu Soufo and Graumann, 2004). Colocalization studies in living cells have shown that all three MreB orthologues form a single helical structure (Carballido-López *et al.*, 2006), and direct interactions between cytoskeleton homologues are inferred from observations that they affect each other's mode of filament formation (Defeu Soufo and Graumann, 2005, 2006). At present, it is unknown whether the single helical structure results from noncovalent association of the monomeric MreB isoforms into homogeneous or mixed filaments or from the interaction of monomeric MreBs with a pre-existing bacterial helix-like structure.

The three *B. subtilis* MreB homologues align along analogous motifs and display a high degree of amino acid sequence identity and similarity. Mutants of *mreB, mbl*, and *mreBH* genes display overlapping although distinct cell shape defects. Depletion of *mreB* gives rise to enlarged cells affected in width, which become progressively swollen over several generations and, eventually, lyse. Mutants of *mbl* also display distorted morphologies, with twisted, bent, and irregularly shaped cells; different than MreB, Mbl seems to be more important in maintaining the linearity of the longitudinal axis of growth rather than cell width (Jones *et al.*, 2001). It has been pointed out that, at least in *B. subtilis*, synthesis of the cell wall takes place in a helical manner over the cylindrical part of the cell, especially at the septum site in dividing cells. Because helical incorporation of fluorescein-labeled vancomycin, a cell wall synthesis inhibitor that binds specifically to peptidoglycan inetermediates, is abolished in a strain lacking Mbl and not in a strain lacking MreB (Daniel and Errington, 2003), Mbl appears to be specifically involved in directing lateral wall synthesis in *B. subtilis*. Finally, deletion mutant strains in *mreBH* led to a mild cell shape defect, with altered cell width, and occasional formation of curved and vibrio-shaped cells (Soufo and Graumann, 2003). In a different way than MreB and Mbl, MreBH has an important role in cell morphogenesis by controlling autolytic activity over the lateral wall (Carballido-López *et al.*, 2006). In a two-hybrid screening, MreBH was found to interact specifically with the cell wall hydrolase domain of an extracellular autolysin, LytE, which was confirmed in a pull-down assay using a purified recombinant MreBH protein. Hence, targeting of LytE to the side wall of the cell was dependent on MreBH and not on the other two isologues, MreB and Mbl. A coordinated helical mode of cell wall hydrolysis and cell wall insertion depending on MreBH and Mbl, respectively, is currently believed to control elongation of *B. subtilis* cells during growth.

B. Efficient ϕ29 DNA replication requires an intact MreB cytoskeleton

Because components of ϕ29 DNA replication machinery (i.e., DNA polymerase, TP, and viral dsDNA) are reorganized at middle infection stages adopting a peripheral helix-like distribution remarkably similar to that of the MreB actin-like proteins (Muñoz-Espín *et al.*, 2009, 2010), a possible role of the MreB cytoskeleton in ϕ29 DNA replication has been studied. By using a *B. subtilis* strain containing *cfp-mreB* and *yfp-p2* fusions under a xylose inducible promoter, it has been established that MreB and ϕ29 DNA polymerase follow similar helical paths at the periphery of infected cells, displaying a substantial degree of colocalization. In addition, the subcelullar localization of ϕ29 DNA polymerase and that of viral dsDNA has been examined in *mreB*, *mbl*, and *mreBH* single deletion strains (Muñoz-Espín *et al.*, 2009). From middle infection stages, and contrary to wild-type infected cells, the phage DNA polymerase did not adopt a helical configuration in any of the three cytoskeleton mutant strains analyzed (see Fig. 3A). Instead, the pattern was much more uniform throughout the cell periphery in infected *mreB*, *mbl* or *mreBH* mutant cells. A similar trend was seen for the organization of viral dsDNA, which mislocalized throughout the cell periphery in infected *mreB*, *mbl*, or *mreBH* mutant cells. Similarly, ϕ29 TP did not redistribute in a helix-like manner in an *mreB* deletion mutant strain (Muñoz-Espín *et al.*, 2010). Rather, fluorescence was observed throughout the cell or in the central region of the cell.

Because proper subcellular localization of ϕ29 DNA replication machinery requires an intact MreB cytoskeleton, the efficiency of *in vivo* ϕ29 DNA replication in *mreB*, *mbl*, and *mreBH* mutant strains has been analyzed and compared with that in a wild-type background (Muñoz-Espín *et al.*, 2009). Analyses by real-time PCR and agarose gel electrophoresis at different times after infection revealed that the amount of ϕ29 DNA accumulated in the cytoskeleton *mreB*, *mbl*, and *mreBH* mutant strains was much lower than that in the wild-type strain, demonstrating that the efficiency of ϕ29 DNA replication is affected severely in the absence of any of the three cytoskeleton proteins. It has been suggested that the fact that the three *B. subtilis* MreB orthologues colocalize in a single helical structure [see earlier discussion and Carballido-López *et al.* (2006)], and may have coordinated overlapping roles, could explain that a single deletion mutant in any the three cytoskeleton genes affects the organization and efficiency of ϕ29 DNA replication machinery.

C. Role of MreB proteins in ϕ29 DNA replication

As indicated previously, ϕ29 membrane protein p16.7 displays a helix-like configuration at the membrane of infected cells (Muñoz-Espín *et al.*, 2009). Available data strongly indicate that protein p16.7 is involved in the organization of *in vivo* membrane-associated ϕ29 DNA replication by interacting with phage DNA, and crystallographic experiments revealed that one dsDNA-binding unit is formed by three p16.7 dimers that are arranged forming a deep cavity that interacts with viral dsDNA by means of an electropositive interface (Albert *et al.*, 2005; Serna-Rico *et al.*, 2002). Similar to other components of ϕ29 DNA replication machinery, the helical distribution of protein p16.7 is lost in the absence of an intact MreB cytoskeleton [see Fig. 3B and Muñoz-Espín *et al.* (2009)], and hence p16.7 localizes uniformly at the membrane of *ΔmreB*, *Δmbl*, or *ΔmreBH* mutant cells. Importantly, pull-down assays using a *B. subtilis* strain expressing His-tagged MreB revealed that protein p16.7 (and not other components of ϕ29 DNA replication machinery, such as the DNA polymerase or the TP) associates specifically with the purified MreB fraction. Further experiments using bacterial two-hybrid techniques indicated that protein p16.7 interacts physically with the actin-like cytoskeleton protein MreB, which is in agreement with the result that p16.7 and MreB substantially colocalize and with the fact that p16.7 forms helix-like structures independently of other phage-encoded proteins (Muñoz-Espín *et al.*, 2009). In light of these findings, it has been suggested that MreB would contribute to efficient ϕ29 DNA replication by recruiting protein p16.7 to the appropriate sites at the cell membrane, thus allowing simultaneous replication of multiple templates at numerous peripheral locations. The possibility cannot, however, be excluded that MreB and/or the other two orthologues (Mbl and MreBH) may play a role in ϕ29 DNA replication by recruiting another protein(s).

Alternatively to a role of the MreB cytoskeleton as a scaffold to compartmentalize phage DNA replication, it has been speculated that MreB might have a mitotic-like role in *B. subtilis* chromosome segregation, which in turn would be responsible in redistributing phage ϕ29 DNA replication machinery at multiple peripheral sites. As stated previously, newly synthesized DNA of a replicating *B. subtilis* chromosome is translocated via a helical structure toward the poles of the cell (Berlatzky *et al.*,

2008) and it seems possible that phage DNA replication machinery adopts this helical conformation by a direct association between the TP and the bacterial DNA. In *E. coli*, convincing evidence has been presented that MreB plays a leading role in chromosome segregation (Kruse *et al.*, 2003, 2006). In addition, it has been concluded that the origin-proximal regions of the *C. crescentus* chromosome associate to MreB and segregate through an MreB-dependent mechanism (Gitai *et al.*, 2005). Nevertheless, the view of a direct role of MreB in chromosome dynamics remains controversial in *B. subtilis* and has proven difficult to demonstrate (Defeu Soufo and Graumann, 2005; Formstone and Errington, 2005; Soufo and Graumann, 2003). At present, it cannot be assessed whether the effect of the MreB cytoskeleton on *B. subtilis* chromosome partitioning is direct or a consequence of an indirect sequence of events. Future efforts in research will be required to obtain conclusive evidence in this field, but this hypothesis would attractively explain the observations that the helix-like reorganization of essential ϕ29 DNA replication components (i.e., TP and DNA polymerase) and, in turn, the efficiency of DNA replication are compromised in MreB cytoskeleton mutant strains.

D. MreB and other phages

For decades, it has been known that the cytoskeleton is exploited by many eukaryotic viruses modulating its intrinsic dynamic behaviour [for review, see Greber and Way (2006)], which allows them to get to their site of replication or to establish a route for newly assembled progeny to leave the infected cell. In analogy to eukaryotic viruses, and in addition to bacteriophage ϕ29, the novel role of the actin-like MreB cytoskeleton organizing phage DNA replication has been investigated for the distantly related phages SPP1 and PRD1. SPP1 belongs to the Siphoviridae family of phages, it has a long tail and its genome is formed by a circular dsDNA. *B. subtilis* is the natural host of both phages, ϕ29 and SPP1. However, phage PRD1 belongs to the Tectiviridae family and is a nontailed phage containing a linear dsDNA. PRD1 infects different hosts such as *E. coli*, *S. enterica*, or *Pseudomonas aeruginosa*. Whereas PRD1 and ϕ29 share a similar protein-primed mechanism of DNA replication, phage SPP1 replicates its DNA initially via a theta mode and later via a rolling circle mechanism [for a general review, see Calendar (2006)]. Strikingly, and similarly to ϕ29, real-time PCR and agarose gel electrophoresis experiments revealed that the efficiency of DNA replication of phages SPP1 and PRD1 is severely affected in *ΔmreB* single deletion strains (Muñoz-Espín *et al.*, 2009). Replicated intracellular phage DNA was detected early after infection of both wild-type and mutant cells, indicating that the cytoskeleton mutation had no or little effect on phage DNA injection. Hence, these results suggest that the bacterial MreB cytoskeleton is required for efficient phage DNA replication in Gram-positive and Gram-negative bacteria, independently of different viral DNA replication mechanisms. Of note, the new role of the MreB actin-like cytoskeleton providing efficient viral DNA replication in prokaryotes constitutes an example on how viruses are able optimize the cell resources for their own benefit. It has been speculated that this acquired ability might have been conserved during evolution of a prokaryotic viral ancestor to eukaryotic variants (Muñoz-Espín *et al.*, 2009).

IV. ϕ29 PROTEIN P56 INHIBITS URACIL–DNA GLYCOSYLASE

A. Base excision repair of uracil residues in DNA

Damage to DNA arises continually throughout the cell cycle and must be recognized and repaired prior to the next round of replication to maintain the genomic integrity of the cell. The presence of uracil is one of the most common lesions in DNA (Lindahl, 1974). It is a natural component of RNA but can occur in DNA by spontaneous or enzymatic deamination of an inherently unstable base, cytosine. The chemical instability of cytosine in DNA (Shapiro, 1980) results in slow hydrolytic deamination that generates premutagenic G:U mispairs. If such changes were left unrepaired, they may cause G:C to A:T transitions because DNA polymerases efficiently incorporate A opposite U in the template (Waters and Swann, 2000). Indeed, these transitions are the most common mutations in human cells and are also found very frequently in human tumors (Krokan *et al.*, 2002).The number of cytosine deaminations has been calculated to be in the order of 60–500 per human genome per day.

However, cytosine in DNA is also a target for gene-specific enzymatic deamination (Kavli *et al.*, 2007). Several members of the APOBEC1/AID family of cytidine deaminases have been shown to directly target C:G base pairs in DNA (Harris *et al.*, 2002). AID was initially described as a B cell-specific protein proposed to be involved in RNA editing (Muramatsu *et al.*, 1999) and also induces changes in immunoglobulin genes (Muramatsu *et al.*, 2000).

The other main source of uracil in DNA is by incorporation of dUMP instead of dTMP during DNA replication (Krokan *et al.*, 2002) due to the fact that a great number of DNA polymerases use dTTP and dUTP with similar efficiency (Dube *et al.*, 1979; Serrano-Heras *et al.*, 2008; Shlomai and Kornberg, 1978). dUMP is a normal intermediate in the biosynthesis of dTMP and dTTP and is converted to dUTP by the same kinases that form dTTP from dTMP. The level of dUTP is kept very low by an efficient deoxyuridine triphosphatase (dUTPase). Despite the action of dUTPase, incorporation of dUMP by replicative DNA polymerases is thought to be the most abundant source of genomic uracil in eukaryotic cells (Andersen *et al.*, 2005). The content of such a residue appears to be at least 10^3-fold lower than that reported for prokaryotic systems, mainly because these cells lack dCTP deaminase, which converts dCTP directly to dUTP, and produce abundant dUTPase (Goulian *et al.*, 1980; Lari *et al.*, 2006). Incorporation of dUMP results in U:A pairs that themselves are not mutagenic. However, chromosomal abasic sites resulting from the removal of uracil are mutagenic and cytotoxic (Auerbach *et al.*, 2005). Interestingly, it has been reported that incorporated uracils in yeast are probably the major cause of formation of spontaneous abasic sites, whereas deaminated cytosines appear to be less important for the generation of abasic sites (Guillet and Boiteux, 2003). Further, uracil in an A:U pair in the DNA sequence can impede their recognition by the cognate regulatory proteins (Mosbaugh and Bennett, 1994).

Therefore, uracilation of genomes represents a constant threat to survival of many organisms, including viruses. As potentially mutagenic and deleterious, uracil is eliminated rapidly from DNA in most prokaryotic and eukaryotic cells by the base excision repair (BER) pathway, which is initiated by uracil–DNA glycosylase (UDG). UDGs hydrolyze the N-glycosidic bond between the uracil and the deoxyribose sugar, leaving an apurinic–apyrimidinic (AP) site, which is then repaired through the sequential action of AP endonuclease, DNA polymerase, and DNA ligase.

The UDG superfamily is divided into four protein families. They possess the same fold and have probably evolved from the same ancestral gene (Aravind and Koonin, 2000). Family-1 enzymes are the most ubiquitous, share similar biochemical properties, and are highly conserved at both the amino acid and structural levels (Xiao *et al.*, 1999). The first UDG activity (family-1) reported was purified from *E. coli* cells (Lindahl, 1974). Since then, UDGs have been identified in numerous organisms, including human cells, yeast, herpesvirus, and poxvirus (Pearl, 2000). Family-1 UDGs are relatively small monomeric proteins that usually do not require cofactors or ions for their activity. They are able to excise uracil efficiently from ssDNA and dsDNA regardless of the partner base on the second strand. Such enzymes have been reported to be exquisitely specific for uracil in DNA and show negligible activity toward the natural DNA bases cytosine or thymine, or uracil in RNA (Pearl, 2000).

B. Inhibition of host factors by bacteriophage proteins

Over the course of evolution, bacteriophages have developed unique proteins that bind to and inactivate critical cellular proteins, shutting off key processes. Phages are the most abundant microorganisms on the planet and are also possibly the most diversified. This diversity is mostly driven by their dynamic adaptation when facing selective pressure, such as phage resistance mechanisms. One of the most important antiviral mechanisms, which are widespread in bacterial hosts, is the DNA restriction system. It consists of endonucleolytic cleavage of the viral genome by DNA restriction enzymes when phage entry into the cells takes place (Labrie *et al.*, 2010). Despite that, bacteriophages have learned to live with the restriction systems of their hosts by developing a wide range of different mechanisms to avoid the worst effects of restriction. Antirestriction mechanisms have been found in practically every phage that has been examined (Krüger and Bickle, 1983). The most thoroughly investigated example of active antirestriction is that exerted by the closely related phages T3 and T7. Both phages encode an Ocr protein, which is able to inhibit the restriction enzyme of *E. coli* (Krüger *et al.*, 1978).

On the other hand, UDGs are emerging as an attractive therapeutic target due to their role in a wide range of biological processes. Hence, the discovery of small molecules able to inhibit the activity of particular UDGs has a great interest. UDGs are inhibited by free uracil and some of its derivatives (Krokan *et al.*, 1997). Specifically, uracil-based ligands have been designed to inactivate selectively human UDG-2 and herpes simplex virus type-1 UDG (Jiang *et al.*, 2005; Sekino *et al.*, 2000). Another category of UDG inhibitors is represented by the *B. subtilis* phage PBS-1/PBS-2-encoded protein Ugi (Cone *et al.*, 1980) and a phage T5-induced UDG inhibitor that has not yet been identified (Warner *et al.*, 1980). Ugi, a highly acidic protein (84 amino acids), inactivates family-1 UDGs from *B. subtilis, Micrococcus luteus, Saccharomyces cerevisae*, rat liver, herpes simplex virus, and human, but not other DNA glycosylases (Cone *et al.*, 1980; Karran *et al.*, 1981; Wang and Mosbaugh, 1989). Identification of a novel natural inhibitor of *B. subtilis* UDG has been reported. This inhibitor, named p56, is a small acidic protein (56 amino acids) encoded by phage ϕ29 (Serrano-Heras *et al.*, 2006, 2007, 2008).

C. Phage ϕ29 encodes an inhibitor of cellular UDG

Analysis of the nucleotide sequence downstream gene *1* (see Fig. 1A) revealed the existence of an open reading frame (ORF56) that would encode an acidic protein of 56 amino acids (protein p56). ORF56 would be mainly transcribed from two strong early promoters, named *A2c* and *A2b*, into a polycistronic RNA. Both promoters are partially repressed at late times of infection (Rojo *et al.*, 1998). In addition, ORF56 would be transcribed from the weak early promoter *A1IV*, which is located within gene *2* (Fig. 1A)(Barthelemy *et al.*, 1986; Bravo *et al.*, 2000; Sogo *et al.*, 1984). To analyze whether protein p56 is synthesized during ϕ29 infection, *B. subtilis* cells were infected with phage ϕ29 and total proteins were analyzed by immunoblotting using antibodies against p56. Results indicated that protein p56 accumulates throughout the ϕ29 infective cycle. In particular, it was present in 10^4 molecules per cell at early stages of infection (15 min) and increased up to 10^5 molecules per cell at late stages of infection (50 min)(Serrano-Heras *et al.*, 2006).

As an approach to determine the role of viral protein p56, protein–protein interaction experiments were carried out. Specifically, infected cells and cells constitutively synthesizing protein p56 were treated with a homobifunctional cross-linker [dithiobis(succinimidylpropionate)(DSP)], which reacts with the amine of lysine residues. Chemical cross-linking assays showed that viral protein p56 is able to interact with a host protein both during the infective process and in the absence of viral components. Taking into account the molecular mass of p56 (6.6 kDa) and DSP (0.4 kDa), the molecular mass of the host protein would be ~28 kDa (Serrano-Heras *et al.*, 2006). To identify the cell target for protein p56, gene *56* was engineered to encode a FLAG-tagged p56 protein (p56FLAG). Using affinity chromatography, we found that *B. subtilis* UDG coeluted with p56FLAG when extracts of *B. subtilis* cells producing p56FLAG and anti-FLAG affinity columns were used. In addition, protein p56 also formed a complex with *E. coli* UDG *in vitro*, as determined by native polyacrylamide gel electrophoresis (Serrano-Heras *et al.*, 2006). *B. subtilis* UDG has strong sequence homology to *E. coli* UDG (Glaser *et al.*, 1993) and both enzymes belong to family-1 UDG (Pearl, 2000).

Subsequently, the ability of protein p56 to inhibit UDG was examined. To this end, a uracil-containing substrate was incubated with purified *E. coli* UDG in the presence of different amounts of purified p56. A decrease in the amount of the cleavage product was detected when p56 was added to the reaction (Serrano-Heras *et al.*, 2006, 2007), indicating that protein p56 acts as an inhibitor of UDG. Additional experiments also showed that p56 is able to inactivate other family-1 UDGs, such as purified *B. subtilis* UDG (Pérez-Lago *et al.*, 2011) and the enzyme activity present in human cell extracts, but it failed to inhibit MUG enzyme from *E. coli* belonging to family-2 UDGs (unpublished results).

Inhibition of cellular UDG activity after phage infection was established previously in two systems. The first was inhibition of the host UDG after infection with phage PBS2 (Friedberg *et al.*, 1975). In *B. subtilis*, uracil-containing phage PBS2 was shown to inhibit UDG activity beginning at about 2 min after infection, and inhibition was complete by 5 min (Katz *et al.*, 1976). Another case of inactivation of host UDG has been described in T5 phage-infected cells. By 2 min after infection almost 40% of the UDG activity has been lost, and by 20 min after infection only 10% of the activity remains (Warner *et al.*, 1980). On the other hand, extracts from ϕ29-infected cells showed a drastic reduction in UDG activity, whereas nearly 90% of the activity remained after incubation with extracts from noninfected cells (Serrano-Heras *et al.*, 2006).

D. Protein p56: A novel naturally occurring DNA mimicry

Protein p56 has nine aspartic, three glutamic, three arginine, and two lysine residues, resulting in an acidic protein with a low theoretical isoelectric point (4.17), which is very similar to that (4.2) reported for another UDG inhibitor, Ugi (Acharya *et al.*, 2002; Serrano-Heras *et al.*, 2007). The oligomerization state of protein p56 in solution was determined by sedimentation equilibrium experiments to be a dimer. In addition, sedimentation velocity techniques revealed that the hydrodynamic behavior of the p56 dimer deviates from the one corresponding to a rigid spherical particle, strongly suggesting that p56 has an ellipsoidal shape (Serrano-Heras *et al.*, 2007).

Proteins can mimic DNA structures as a mechanism to block DNA-binding enzymes. At present, only a small number of DNA mimic proteins have been discovered. Although proteins in this class are structurally diverse, they tend to resemble some DNA structural features, such as the phosphate backbone of DNA or the hydrogen-bonding properties of the nucleotide bases (Dryden and Tock, 2006; Putnam and Tainer, 2005). The Ocr protein from the *E. coli* phage T7, which binds tightly to type I DNA restriction and modification enzymes (Mark and Studier, 1981), is an example of DNA mimicry. It binds tightly to type I DNA restriction and modification enzymes (Mark and Studier, 1981). Biophysical and crystallographic studies revealed that the Ocr dimer mimics 24 bp of B-form DNA containing a central bend (Atanasiu *et al.*, 2001). Moreover, Ugi acts as a UDG inhibitor by mimicking electronegative and structural features of duplex DNA (Mol *et al.*, 1995; Putnam *et al.*, 1999; Savva *et al.*, 1995). A more recent example of DNA mimicry is Mfpa, a *Mycobacterium tuberculosis* protein that binds to DNA gyrase. MfpA exhibits a highly unusual right-handed β helix fold (Hegde *et al.*, 2005).

Several features of protein p56 suggest that it could be novel naturally occurring DNA mimicry. First, protein p56 competes with DNA for binding to UDG. Second, as observed previously for Ugi (Bennett and Mosbaugh, 1992), the interaction between p56 and UDG blocks the interaction between UDG and DNA. Third, Ugi, which occupies the DNA-binding groove of UDG, is able to replace protein p56 bound previously to UDG, strongly suggesting that both inhibitors have totally or partially overlapping binding sites within UDG (Serrano-Heras *et al.*, 2007). Finally, like Ocr and Ugi, p56 is a highly acidic protein. In most of the known DNA mimic proteins, carboxylates from the side chains of aspartates and glutamates generate an overall charge distribution that resembles the DNA phosphate backbone (Putnam and Tainer, 2005). Further resolution of the three-dimensional structure of both protein p56 and the UDG–p56 complex will reveal whether protein p56 functions as a structural mimic of the DNA substrate recognized by UDG.

E. Role of protein p56 in φ29 DNA replication

In *B. subtilis*, the uracil-containing phage PBS2 was shown to inhibit UDG activity at the beginning of the infection (Friedberg *et al.*, 1975). The need for this inhibition was obvious because the DNA of PBS2 contains uracil instead of thymine residues and it would be difficult to synthesize uracil-containing progeny phage DNA in the presence of UDG activity.

However, viral protein p56 is the first example of an UDG inhibitor encoded by a nonuracil-containing viral DNA. We have suggested that this inhibition is related to the mechanism of φ29 DNA replication. As mentioned previously, the genome of phage φ29 is a linear dsDNA with a TP linked covalently at both 5′ends, which replicates by a protein-priming mechanism (Salas, 1991). Synthesis of the linear φ29 DNA starts at both ends, where the replication origins are located, and proceeds by a strand displacement mechanism, generating replicative intermediates with long stretches of ssDNA. Hence, we proposed that inhibition of the host UDG by protein p56 is likely a defense mechanism developed by φ29 to prevent damaging action of the BER pathway if uracils arise in their replicative intermediates (Serrano-Heras *et al.*, 2006). The presence of these residues in ssDNA intermediates, arising either by deamination of cytosine or by misincorporation of dUMP during the previous replication round, could recruit components of the cellular BER pathway, such as UDGs and AP endonucleases. As depicted in Figure 4, the subsequent action of these activities would introduce a nick into the phosphodiester backbone with accompanying loss of the terminal region.

In bacteria, a considerable amount of dUTP is produced because it is an obligatory intermediate in the *de novo* synthesis of dTMP (Lari *et al.*, 2006). This fact, together with the ability of numerous prokaryotic DNA polymerases to use dUTP in place of dTTP during DNA synthesis (Hitzeman and Price, 1978; Mosbaugh, 1988), makes incorporation of uracil into the bacterial chromosome unavoidable.

We have demonstrated that the φ29 DNA polymerase is able to insert uracil residues into φ29 DNA when both dUTP and dTTP are present in the replication reaction. We also found that φ29 DNA polymerase is capable of elongating efficiently the dA:dUMP pair and reading through uracil, giving full-length DNA. Moreover, *in vivo* experiments revealed the presence of uracil residues in φ29 DNA molecules synthesized during the infective cycle (Serrano-Heras *et al.*, 2008). It seems likely that most of the uracils arise from misinsertion of dUMP rather than by deamination of cytosines, as phage DNA polymerase efficiently incorporates dUMP into DNA. Misincorporation of dUMP into DNA has also been observed in a number of viruses, including polyomavirus (Brynolf *et al.*, 1978) and adenovirus (Ariga and Shimojo, 1979), as well as in bacteria (Shlomai and Kornberg, 1978; Tamanoi and Okazaki, 1978). Indeed, it has

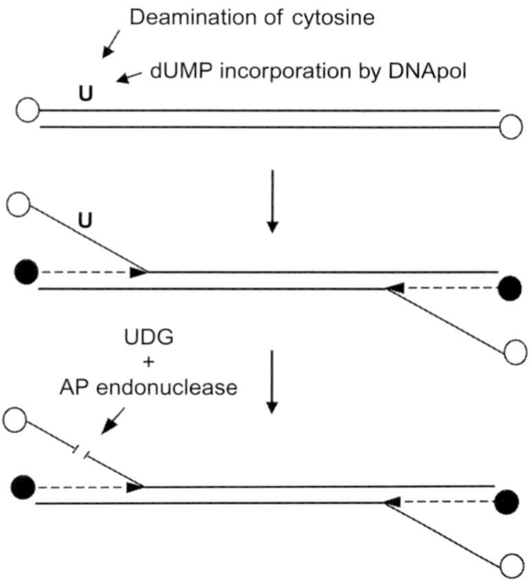

FIGURE 4 Model of the effect of UDG on ϕ29 DNA replication when uracil residues appear in ssDNA regions of replicative intermediates either by deamination of cytosine or by misinsertion of dUMP by DNA polymerase during the previous replication round. Parental TP (white circles) and primer TP (black circles) are indicated. Dashed lines indicate newly synthesized viral DNA.

been estimated that under normal physiological conditions, *E. coli* incorporates one uracil residue into DNA for every 300 thymine nucleotides polymerized (Shlomai and Kornberg, 1978).

On the other hand, the consequences of uracil misincorporation on *in vitro* ϕ29 DNA replication when UDG is present were examined. Removal of uracil residues incorporated into the phage genome by UDG caused a drastic reduction in the efficiency of ϕ29 DNA replication. Furthermore, it is interesting to note that the presence of ϕ29 DNA-binding proteins SSB and DBP during viral DNA amplification did not inhibit uracil excision by UDG. In fact, this activity could have been enhanced by viral SSB binding to ssDNA, as interactions between SSBs and UDGs from *E. coli* and *M. tuberculosis* have been reported to result in increased efficiency of uracil excision from structured substrates (Purnapatre *et al.*, 1999). In addition, protein p56 prevented the ϕ29 DNA replication impairment caused by the action of UDG on the uracil-containing ϕ29 DNA. In agreement, the transfection activity of uracil-containing ϕ29 DNA was significantly higher in cells that constitutively synthesized protein p56 than in cells lacking this protein.

Altogether, these findings provided an important validation of the model proposed for the *in vivo* role of viral protein p56 in DNA synthesis.

According to this model, bacteriophage ϕ29 encodes an UDG inhibitor (protein p56) to prevent the impairment of DNA synthesis produced by the excision of uracils from the viral genome. It has been reported consistently that the major measurable consequences of high levels of uracil in DNA are lesions created by the attempt of cells to remove this nucleotide (Richards *et al.*, 1984). For example, it has been found that a vicious circle of excision and repair of the DNA, caused by high intracellular levels of dUTP, led to strand breaks, strand exchanges, and eventually cell death as a result of chromosomal aberrations (Richards *et al.*, 1984). *In vivo* studies revealed that *E. coli* with dUTPase mutations transiently accumulates small DNA fragments during replication fork progression, probably because of the UDG-based repair process (Olivera, 1978; Tye *et al.*, 1978). In the case of viruses, such as polyoma virus and adenovirus, the removal of uracil causes strand breakage and may contribute to the formation of Okazaki fragments in the viral genome (Ariga and Shimojo, 1979; Brynolf *et al.*, 1978). Interestingly, the genomes of many human viruses are attacked by host cell cytidine deaminases (APOBEC family) that play important roles in protection against infection (Priet *et al.*, 2006). In particular, it has been shown that some cytidine deaminases are able to block HIV replication in macrophages and lymphocytes. This is likely due to the fact that the uracil residues are removed by UDG, and subsequent cleavage of the resulting AP sites by human AP endonuclease leads to viral DNA degradation (Yang *et al.*, 2007). To counteract excessive cytosine deamination in the viral DNA and, thus, to maintain genome integrity, HIV encodes a specific protein, Vif (virion infectivity factor), that mediates APOBEC degradation by polyubiquitinylation and proteasomal degradation (Sousa *et al.*, 2007). Moreover, a mutant of phage T5, containing significant amounts of uracil in its DNA, failed to produce plaques unless the plating host was deficient in UDG activity (Swart and Warner, 1985).

Considering the relevance of the protecting function of p56 and that the generation of replicative intermediates containing long stretches of ssDNA is a feature of the protein-primed mechanism of DNA replication, it is not surprising that gene *56* is conserved in the genome of other ϕ29-related phages. In the case of phage B103, which belongs to the second group of ϕ29-like phages, there is an open reading frame that would encode a 56 amino acid protein, whose deduced sequence has a high level of homology to protein p56 (64% identity and 75% similarity) (Meijer *et al.*, 2001a). Furthermore, in phage GA-1, which is the most distantly ϕ29-related phage, there is an open reading frame that would encode a 130 amino acid protein. The region of this putative protein spanning amino acids 27 and 82 shows 23% identity and 27% similarity with protein p56, suggesting that they may have similar functions. In fact, we have shown that the putative p56-like protein of phage GA-1 inhibits *B. subtilis* UDG activity similarly to the ϕ29 p56 protein (Pérez-Lago *et al.*, 2011).

V. CONCLUSIONS AND PERSPECTIVES

Important advances on the subject of virus–host relationships have been made in the last decade. This chapter provided a comprehensive overview of the organization of *B. subtilis* bacteriophage ϕ29 DNA replication and how it takes advantage of different cellular factors to increase the efficiency of this biological process. Available data show that once ϕ29 infection takes place, it directs its TP-DNA to the bacterial nucleoid, where the phage genome becomes constrained topologically. This topological arrangement may be, at least in part, caused by the interaction of parental TPs (linked covalently at each 5′ DNA end) with bacterial DNA. Although the ϕ29 genome consists of a linear dsDNA molecule, association with the bacterial nucleoid would give rise to a "closed" structure with topological restrictions. Under this scenario, the *B. subtilis* DNA gyrase is required to eliminate the positive supercoiling generated by the approach of the two replication forks. Thus, when the DNA gyrase is inhibited, ϕ29 DNA replication is importantly impaired.

After the ϕ29 DNA replication machinery associates with the bacterial nucleoid, the actin-like MreB cytoskeleton plays a crucial role in organizing phage DNA replication in peripheral helix-like structures, which are lost in the absence of any of the three *B. subtilis* MreB isoforms. Hence, efficient ϕ29 DNA replication requires an intact bacterial cytoskeleton. It has been determined that MreB interacts with protein p16.7, which is responsible for attaching the viral DNA at peripheral sites, thus providing insights into the mechanism by which MreB plays a role in ϕ29 DNA replication. Importantly, MreB also appears to play a similar role in bacteriophages replicating their genomes by a different mode of DNA replication than ϕ29 or infecting a different host than *B. subtilis*, such as phage SPP1 or PRD1, respectively.

Finally, phage ϕ29 protein p56 is essential in the viral DNA replication cycle, as it prevents the impairment caused by the host UDG. Inhibition of UDG has been proposed to be a defense mechanism developed by ϕ29 to prevent the formation of abortive replicative intermediates. This is the first case reported of an UDG inhibitor encoded by a nonuracil containing DNA.

To conclude, the mechanistic details of how *B. subtilis* DNA gyrase, MreB, or UDG are specifically employed by phage ϕ29 and how they are organized temporally and spatially may be main directions for future experiments. Because bacteriophages contain genomes with a limited size due to their small dimensions, it is expected that new bacterial proteins interacting with viral components may be discovered. A major challenge is to identify these novel targets that might be used by bacteriophages to optimize the production of high numbers of progeny.

ACKNOWLEDGMENTS

This investigation was supported by Grants BFU2008-00215 and Consolider-Ingenio 2010 24717 from the Spanish Ministry of Science and Innovation to MS and by an institutional grant from Fundación Ramón Areces to the Centro de Biología Molecular "Severo Ochoa." DM-E was a recipient of an I3P postdoctoral contract from the Spanish National Research Council.

REFERENCES

Abril, A. M., Salas, M., Andreu, J. M., Hermoso, J. M., and Rivas, G. (1997). Phage ϕ29 protein p6 is in a monomer-dimer equilibrium that shifts to higher association states at the millimolar concentrations found *in vivo*. *Biochemistry* **36:**11901–11908.

Ackermann, H. W. (1998). Tailed bacteriophages: The order *Caudovirales*. *Adv. Virus Res.* **51:**135–201.

Albert, A., Muñoz-Espín, D., Jiménez, M., Asensio, J. L., Hermoso, J. A., Salas, M., and Meijer, W. J. J. (2005). Structural basis for membrane anchorage of viral ϕ29 DNA during replication. *J. Biol. Chem.* **280:**42486–42488.

Andersen, S., Heine, T., Sneve, R., König, I., Krokan, H. E., Epe, B., and Nilsen, H. (2005). Incorporation of dUMP into DNA is a major source of spontaneous DNA damage, while excision of uracil is not required for cytotoxicity of fluoropyrimidines in mouse embryonic fibroblasts. *Carcinogenesis* **26:**547–555.

Anderson, D. L., Hickman, D. D., and Reilly, B. E. (1966). Structure of *Bacillus subtilis* bacteriophage ϕ29 and the length of ϕ 29 deoxyribonucleic acid. *J. Bacteriol.* **91:**2081–2089.

Aravind, L., and Koonin, E. V. (2000). The alpha/beta fold uracil DNA glycosylases: A common origin with diverse fates. *Genome Biol* **1:** RESEARCH0007.

Ariga, H., and Shimojo, H. (1979). Incorporation of uracil into the growing strand of adenovirus 12 DNA. *Biochem. Biophys. Res. Commun.* **87:**588–597.

Atanasiu, C., Byron, O., McMiken, H., Sturrock, S. S., and Dryden, D. T. (2001). Characterisation of the structure of ocr, the gene 0.3 protein of bacteriophage T7. *Nucleic Acids Res* **29:**3059–3068.

Auerbach, P., Bennett, R. A., Bailey, E. A., Krokan, H. E., and Demple, B. (2005). Mutagenic specificity of endogenously generated abasic sites in *Saccharomyces cerevisiae* chromosomal DNA. *Proc. Natl. Acad. Sci. USA* **102:**17711–17716.

Barthelemy, I., Salas, M., and Mellado, R. P. (1986). *In vivo* transcription of bacteriophage ϕ29 DNA: Transcription initiation sites. *J. Virol.* **60:**874–879.

Bennett, S. E., and Mosbaugh, D. W. (1992). Characterization of the *Escherichia coli* uracil-DNA glycosylase inhibitor protein complex. *J. Biol. Chem.* **267:**22512–22521.

Berlatzky, I. A., Rouvinski, A., and Ben-Yehuda, S. (2008). Spatial organization of a replicating bacterial chromosome. *Proc. Natl. Acad. Sci. USA* **105:**14136–14140.

Blanco, L., Bernad, A., Lázaro, J. M., Martín, G., Garmendia, C., and Salas, M. (1989). Highly efficient DNA synthesis by the phage ϕ29 DNA polymerase: Symmetrical mode of DNA replication. *J. Biol. Chem.* **264:**8935–8940.

Blanco, L., Gutiérrez, J., Lázaro, J. M., Bernad, A., and Salas, M. (1986). Replication of phage ϕ29 DNA *in vitro*: Role of the viral protein p6 in initiation and elongation. *Nucleic Acids Res.* **14:**4923–4937.

Blanco, L., Lázaro, J. M., De Vega, M., Bonnin, A., and Salas, M. (1994). Terminal protein-primed DNA amplification. *Proc. Natl. Acad. Sci. USA* **91:**12198–12202.

Blanco, L., and Salas, M. (1984). Characterization and purification of a phage ϕ29-encoded DNA polymerase required for the initiation of replication. *Proc. Natl. Acad. Sci. USA* **81:**5325–5329.

Bravo, A., Hermoso, J. M., and Salas, M. (1994). A genetic approach to the identification of functional amino acids in protein p6 of *Bacillus subtilis* phage ϕ29. *Mol. Gen. Genet.* **245:**529–536.

Bravo, A., Illana, B., and Salas, M. (2000). Compartmentalization of phage ϕ29 DNA replication: Interaction between the primer terminal protein and the membrane-associated protein p1. *EMBO J.* **19:**5575–5584.

Bravo, A., and Salas, M. (1997). Initiation of bacteriophage ϕ29 DNA replication *in vivo:* Assembly of a membrane-associated multiprotein complex. *J. Mol. Biol.* **269:**102–112.

Bravo, A., and Salas, M. (1998). Polymerization of bacteriophage ϕ29 replication protein p1 into protofilament sheets. *EMBO J.* **17:**6096–6105.

Bravo, A., Serrano-Heras, G., and Salas, M. (2005). Compartmentalization of prokaryotic DNA replication. *FEMS Microbiol. Rev.* **29:**25–47.

Brynolf, K., Eliasson, R., and Reichard, P. (1978). Formation of Okazaki fragments in polyoma DNA synthesis caused by misincorporation of uracil. *Cell* **13:**573–580.

Calendar, R. (2006). The Bacteriophages. Oxford University Press, New York.

Camacho, A., and Salas, M. (2000). Pleiotropic effect of protein p6 on the viral cycle of bacteriophage ϕ29. *J. Bacteriol.* **182:**6927–6932.

Carballido-López, R. (2006). The bacterial actin-like cytoskeleton. *Microbiol. Mol. Biol. Rev.* **70:**888–909.

Carballido-López, R., and Errington, J. (2003a). The bacterial cytoskeleton: *In vivo* dynamics of the actin-like protein Mbl of *Bacillus subtilis. Dev. Cell.* **4:**19–28.

Carballido-López, R., and Errington, J. (2003b). A dynamic bacterial cytoskeleton. *Trends Cell Biol.* **13:**577–583.

Carballido-López, R., Formstone, A., Li, Y., Noirot, P., and Errington, J. (2006). Actin homolog MreBH governs cell morphogenesis by localization of the cell wall hydrolase LytE. *Dev. Cell.* **11:**399–409.

Carrascosa, J. L., Camacho, A., Moreno, F., Jiménez, F., Mellado, R. P., Viñuela, E., and Salas, M. (1976). *Bacillus subtilis* phage ϕ29; characterization of gene products and functions. *Eur. J. Biochem.* **66:**229–241.

Cone, R., Bonura, T., and Friedberg, E. C. (1980). Inhibitor of uracil-DNA glycosylase induced by bacteriophage PBS2: Purification and preliminary characterization. *J. Biol. Chem.* **255:**10354–10358.

Crucitti, P., Abril, A. M., and Salas, M. (2003). Bacteriophage ϕ29 early protein p17. Self-association and hetero-association with the viral histone-like protein p6. *J. Biol. Chem.* **278:**4906–4911.

Crucitti, P., Lázaro, J. M., Benes, V., and Salas, M. (1998). Bacteriophage ϕ29 early protein p17 is conditionally required for the first rounds of viral DNA replication. *Gene J.* **223:**135–142.

Daniel, R. A., and Errington, J. (2003). Control of cell morphogenesis in bacteria: Two distinct ways to make a rod-shaped cell. *Cell* **113:**767–776.

Defeu Soufo, H. J., and Graumann, P. L. (2004). Dynamic movement of actin-like proteins within bacterial cells. *EMBO Rep.* **5:**789–794.

Defeu Soufo, H. J., and Graumann, P. L. (2005). *Bacillus subtilis* actin-like protein MreB influences the positioning of the replication machinery and requires membrane proteins MreC/D and other actin-like proteins for proper localization. *BMC Cell Biol.* **6:**10.

Defeu Soufo, H. J., and Graumann, P. L. (2006). Dynamic localization and interaction with other *Bacillus subtilis* actin-like proteins are important for the function of MreB. *Mol. Microbiol.* **62:**1340–1356.

Dekker, N. H., Rybenkov, V. V., Duguet, N. J., Crisona, N. J., Cozzarelli, N. R., Bensimon, D., and Croquette, V. (2002). The mechanism of type IA topoisomerases. *Proc. Natl. Acad. Sci. USA* **99:**12126–12131.

Dryden, D. T., and Tock, M. R. (2006). DNA mimicry by proteins. *Biochem. Soc. Trans.* **34:**317–319.

Dube, D. K., Kunkel, T. A., Seal, G., and Loeb, L. A. (1979). Distinctive properties of mammalian DNA polymerases. *Biochem. Biophys. Acta* **561:**369–382.

Escarmís, C., Guirao, D., and Salas, M. (1989). Replication of recombinant ϕ29 DNA molecules in *Bacillus subtilis* protoplasts. *Virology* **169:**152–160.

Espeli, O., and Marians, K. J. (2004). Untangling intracellular DNA topology. *Mol. Microbiol.* **52:**925–931.

Esue, O., Wirtz, D., and Tseng, Y. (2006). GTPase activity, structure, and mechanical properties of filaments assembled from bacterial cytoskeleton protein MreB. *J. Bacteriol.* **188:**968–976.

Formstone, A., and Errington, J. (2005). A magnesium-dependent *mreB* null mutant: Implications for the role of *mreB* in *Bacillus subtilis*. *Mol. Microbiol.* **55:**1646–1657.

Freire, R., Salas, M., and Hermoso, J. M. (1994). A new protein domain for binding to DNA through the minor groove. *EMBO J.* **13:**4353–4360.

Friedberg, E. C., Ganesan, A. K., and Minton, K. (1975). N-Glycosidase activity in extracts of *Bacillus subtilis* and its inhibition after infection with bacteriophage PBS2. *J. Virol.* **16:**315–321.

Garvey, K. J., Yoshikawa, H., and Ito, J. (1985). The complete sequence of the *Bacillus* phage ϕ29 right early region. *Gene* **40:**301–309.

Gitai, Z., Dye, N. A., Reisenauer, A., Wachi, M., and Shapiro, L. (2005). MreB actin-mediated segregation of a specific region of a bacterial chromosome. *Cell* **120:**329–341.

Glaser, P., Kunst, F., Arnaud, M., Coudart, M. P., Gonzales, W., Hullo, M. F., Ionescu, M., Lubochinsky, B., Marcelino, L., Moszer, I., *et al.* (1993). *Bacillus subtilis* genome project: Cloning and sequencing of the 97 kb region from 325 degrees to 333 degrees. *Mol. Microbiol.* **10:**371–384.

González-Huici, V., Alcorlo, M., Salas, M., and Hermoso, J. M. (2004a). Binding of phage ϕ29 architectural protein p6 to the viral genome: Evidence for topological restriction of the phage linear DNA. *Nucleic Acids Res.* **32:**3493–3502.

González-Huici, V., Alcorlo, M., Salas, M., and Hermoso, J. M. (2004b). Phage ϕ29 proteins p1 and p17 are required for efficient binding of architectural protein p6 to viral DNA *in vivo* J. *Bacteriol.* **186:**8401–8406.

González-Huici, V., Lázaro, J. M., Salas, M., and Hermoso, J. M. (2000). Specific recognition of parental terminal protein by DNA polymerase for initiation of protein-primed DNA replication. *J. Biol. Chem.* **275:**14678–14683.

González-Huici, V., Salas, M., and Hermoso, J. M. (2004c). Genome wide, supercoiling-dependent *in vivo* binding of a viral protein involved in DNA replication and transcriptional control. *Nucleic Acids Res.* **32:**2306–2314.

Goulian, M., Bleile, B., and Tseng, B. (1980). Methotrexate-induced misincorporation of uracil into DNA. *Proc. Natl. Acad. Sci. USA* **77:**1956–1960.

Graumann, P. L. (2007). *Bacillus*, Cellular and Molecular Biology. Caister Academic Press, Norfolk.

Greber, U. F., and Way, M. (2006). A superhighway to virus infection. *Cell* **124:**741–754.

Guillet, M., and Boiteux, S. (2003). Origin of endogenous DNA abasic sites in *Saccharomyces cerevisiae*. *Mol. Cell. Biol.* **23:**8386–8394.

Gutiérrez, C., Sogo, J. M., and Salas, M. (1991). Analysis of replicative intermediates produced during bacteriophage ϕ29 DNA replication *in vitro*. *J. Mol. Biol.* **222:**983–994.

Hagen, E. W., Reilly, B. E., Tosi, M. E., and Anderson, D. L. (1976). Analysis of gene function of bacteriophage ϕ29 of *Bacillus subtilis*: Identification of cistrons essential for viral assembly. *J. Virol.* **19:**501–517.

Harding, N. E., and Ito, J. (1980). DNA replication of bacteriophage ϕ29: Characterization of the intermediates and location of the termini of replication. *Virology* **104:**323–338.

Harris, R. S., Petersen-Mahrt, S. K., and Neuberger, M. S. (2002). RNA editing enzyme APOBEC1 and some of its homologs can act as DNA mutators. *Mol. Cell.* **10:**1247–1253.

Hegde, S. S., Vetting, M. W., Roderick, S. L., Mitchenall, L. A., Maxwell, A., Takiff, H. E., and Blanchard, J. S. (2005). A fluoroquinolone resistance protein from *Mycobacterium tuberculosis* that mimics DNA. *Science* **308:**1480–1483.

Hermoso, J. M., Méndez, E., Soriano, F., and Salas, M. (1985). Location of the serine residue involved in the linkage between the terminal protein and the DNA of ϕ29. *Nucleic Acids Res.* **13:**7715–7728.

Hiasa, H., and Marians, K. J. (1996). Two distinct modes of strand unlinking during theta-type DNA replication. *J. Biol. Chem.* **271:**21529–21535.

Hitzeman, R. A., and Price, A. R. (1978). Relationship of *Bacillus subtilis* DNA polymerase III to bacteriophage PBS2-induced DNA polymerase and to the replication of uracil-containing DNA. *J. Virol.* **28:**697–709.

Inciarte, M. R., Salas, M., and Sogo, J. M. (1980). The structure of replicating DNA molecules of *Bacillus subtilis* phage ϕ29. *J. Virol.* **34:**187–190.

Ito, J. (1978). Bacteriophage ϕ29 terminal protein: Its association with the 5′ termini of the ϕ29 genome. *J. Virol.* **28:**895–904.

Jiang, Y. L., Krosky, D. J., Seiple, L., and Stivers, J. T. (2005). Uracil-directed ligand tethering: an efficient strategy for uracil DNA glycosylase (UNG) inhibitor development. *J. Am. Chem. Soc.* **127:**17412–17420.

Jiménez, F., Camacho, A., de la Torre, J., Viñuela, E., and Salas, M. (1977). Assembly of *Bacillus subtilis* phage phi29. 2 mutants in the cistrons coding for the non-structural proteins. *Eur. J. Biochem.* **73:**57–72.

Jones, L. J., Carballido-López, R., and Errington, J. (2001). Control of cell shape in bacteria: helical, actin-like filaments in *Bacillus subtilis*. *Cells* **104:**913–922.

Kamtekar, S., Berman, A. J., Wang, J., Lázaro, J. M., De Vega, M., Blanco, L., Salas, M., and Steitz, T. A. (2006). The ϕ29 DNA polymerase:protein-primer structure suggests a model for the initiation to elongation transition. *EMBO J.* **25:**1335–1343.

Karran, P., Cone, R., and Friedberg, E. C. (1981). Specificity of the bacteriophage PBS2 induced inhibitor of uracil-DNA glycosylase. *Biochemistry* **20:**6092–6096.

Kato, J., Nishimura, Y., Imamura, R., Niki, H., Hiraga, S., and Suzuki, H. (1990). New topoisomerase essential for chromosome segregation in *E. coli*. *Cell* **63:**393–404.

Katz, G. E., Price, A. R., and Pomerantz, M. J. (1976). Bacteriophage PBS2-induced inhibition of uracil-containing DNA degradation. *J. Virol.* **20:**535–538.

Kavli, B., Otterlei, M., Slupphaug, G., and Krokan, H. E. (2007). Uracil in DNA: General mutagen, but normal intermediate in acquired immunity. *DNA Repair (Amst.)* **6:**505–516.

Krokan, H. E., Drablos, F., and Slupphaug, G. (2002). Uracil in DNA: Occurrence, consequences and repair. *Oncogene* **21:**8935–8948.

Krokan, H. E., Standal, R., and Slupphaug, G. (1997). DNA glycosylases in the base excision repair of DNA. *Biochem. J.* **325:**1–16.

Krüger, D. H., and Bickle, T. A. (1983). Bacteriophage survival: Multiple mechanisms for avoiding the deoxyribonucleic acid restriction systems of their hosts. *Microbiol. Rev.* **47:**345–360.

Krüger, D. H., Chernin, L. S., Hansen, S., Rosenthal, H. A., and Goldfarb, D. M. (1978). Protection of foreign DNA against host-controlled restriction in bacterial cells. I. Protection of F′ plasmid DNA by preinfection with bacteriophages T3 or T7. *Mol. Gen. Genet.* **159:**107–110.

Kruse, T., Blagoev, B., Lobner-Olesen, A., Wachi, M., Sasaki, K., Iwai, N., Mann, M., and Gerdes, K. (2006). Actin homolog MreB and RNA polymerase interact and are both required for chromosome segregation in *Escherichia coli*. *Genes Dev.* **20:**113–124.

Kruse, T., Moller-Jensen, J., Lobner-Olesen, A., and Gerdes, K. (2003). Dysfunctional MreB inhibits chromosome segregation in *Escherichia coli*. *EMBO J.* **22:**5283–5292.

Labrie, S. J., Samson, J. E., and Moineau, S. (2010). Bacteriophage resistance mechanisms. *Nat. Rev. Microbiol.* **8:**317–327.

Lari, S. U., Chen, C. Y., Vertéssy, B. G., Morré, J., and Bennett, S. E. (2006). Quantitative determination of uracil residues in *Escherichia coli* DNA: Contribution of *ung*, *dug*, and *dut* genes to uracil avoidance. *DNA repair (Amst.)* **5:**1407–1420.

Lewis, P. J., Thaker, S. D., and Errington, J. (2000). Compartmentalization of transcription and translation in *Bacillus subtilis*. *EMBO J.* **19:**710–718.

Lindahl, T. (1974). An N-glycosidase from *Escherichia coli* that releases free uracil from DNA containing deaminated cytosine residues. *Proc. Natl. Acad. Sci. USA* **71:**3649–3653.

Mark, K. K., and Studier, F. W. (1981). Purification of the gene 0.3 protein of bacteriophage T7, an inhibitor of the DNA restriction system of *Escherichia coli*. *J. Biol. Chem.* **256:**2573–2578.

Martín, G., and Salas, M. (1988). Characterization and cloning of gene *5* of *Bacillus subtilis* phage ϕ29. *Gene J.* **67:**193–201.

Meijer, W. J. J., Horcajadas, J. A., and Salas, M. (2001a). ϕ29-family of phages. *Microbiol. Mol. Biol. Rev.* **65:**261–287.

Meijer, W. J. J., Serna-Rico, A., and Salas, M. (2001b). Characterization of the bacteriophage ϕ29-encoded protein p16.7: A membrane protein involved in phage DNA replication. *Mol. Microbiol.* **39:**731–746.

Méndez, J., Blanco, L., Esteban, J. A., Bernad, A., and Salas, M. (1992). Initiation of ϕ29 DNA replication occurs at the second 3′ nucleotide of the linear template: A sliding-back mechanism for protein-primed DNA replication. *Proc. Natl. Acad. Sci. USA* **89:**9579–9583.

Méndez, J., Blanco, L., and Salas, M. (1997). Protein-primed DNA replication: A transition between two modes of priming by a unique DNA polymerase. *EMBO J.* **16:**2519–2527.

Mol, C. D., Arvai, A. S., Slupphaug, G., Kavli, B., Alseth, I., Krokan, H. E., and Tainer, J. A. (1995). Crystal structure and mutational analysis of human uracil-DNA glycosylase: Structural basis for specificity and catalysis. *Cell* **80:**869–878.

Moreno, F., Camacho, A., Viñuela, E., and Salas, M. (1974). Suppressor-sensitive mutants and genetic map of *Bacillus subtilis* bacteriophage ϕ29. *Virology* **62:**1–16.

Mosbaugh, D. W. (1988). Purification and characterization of porcine liver DNA polymerase gamma: Utilization of dUTP and dTTP during *in vitro* DNA synthesis. *Nucleic Acids Res.* **16:**5645–5659.

Mosbaugh, D. W., and Bennett, M. F. (1994). Uracil-excision DNA repair. *Prog. Nuclic. Acid. Res. Mol. Biol.* **48:**315–370.

Muñoz-Espín, D., Daniel, R., Kaway, Y., Carballido-López, R., Castilla-Llorente, V., Errington, J., Meijer, W. J. J., and Salas, M. (2009). The actin-like MreB cytoskeleton organizes viral DNA replication in bacteria. *Proc. Natl. Acad. Sci. USA* **106:**13347–13352.

Muñoz-Espín, D., Holguera, I., Ballesteros-Plaza, D., Carballido-López, R., and Salas, M. (2010). Viral terminal protein directs early organization of phage DNA replication at the bacterial nucleoid. *Proc. Natl. Acad. Sci. USA* **107:**16548–16553.

Muramatsu, M., Kinoshita, K., Fagarasan, S., Yamada, S., Shinkai, Y., and Honjo, T. (2000). Class switch recombination and hypermutation require activation-induced cytidine deaminase (AID), a potential RNA editing enzyme. *Cell* **102:**553–563.

Muramatsu, M., Sankaranand, V. S., Anant, S., Sugai, M., Kinoshita, K., Davidson, N. O., and Honjo, T. (1999). Specific expression of activation-induced cytidine deaminase (AID), a novel member of the RNA-editing deaminase family in germinal center B cells. *J. Biol. Chem.* **274:**18470–18476.

Olivera, B. M. (1978). DNA intermediates at the *Escherichia coli* replication fork: Effect of dUTP. *Proc. Natl. Acad. Sci. USA* **75:**238–242.

Osburne, M. S., Zavodny, S. M., and Peterson, G. A. (1988). Drug-induced relaxation of supercoiled plasmid DNA in *Bacillus subtilis* and induction of the SOS response. *J. Bacteriol.* **170:**442–445.

Pastrana, R., Lázaro, J. M., Blanco, L., García, J. A., Méndez, E., and Salas, M. (1985). Overproduction and purification of protein p6 of *Bacillus subtilis* phage ϕ29: Role in the initiation of DNA replication. *Nucleic Acids Res.* **13:**3083–3100.

Pearl, L. H. (2000). Structure and function in the uracil-DNA glycosylase superfamily. *Mutat. Res.* **460:**165–181.

Pérez-Lago, L., Serrano-Heras, G., Banos, B., Lazaro, J. M., Alcorlo, M., Villar, L., and Salas, M. (2011). Characterization of *Bacillus subtilis* uracil-DNA glycosylase and its inhibition by phage ϕ29 protein p56. *Mol. Microbiol.* **80:**1657–1666.

Priet, S., Sire, J., and Querat, G. (2006). Uracils as a cellular weapon against viruses and mechanisms of viral escape. *Curr. HIV Res.* **4:**31–42.

Prieto, I., Méndez, E., and Salas, M. (1989). Characterization, overproduction and purification of the product of *gene1* of *Bacillus subtilis* phage ϕ29. *Gene J.* **77:**204.

Prieto, I., Serrano, M., Lázaro, J. M., Salas, M., and Hermoso, J. M. (1988). Interaction of the bacteriophage ϕ29 protein p6 with double-stranded DNA. *Proc. Natl. Acad. Sci. USA* **85:**314–318.

Purnapatre, K., Handa, P., Venkatesh, J., and Varshney, U. (1999). Differential effects of single-stranded DNA binding proteins (SSBs) on uracil DNA glycosylases (UDGs) from *Escherichia coli* and mycobacteria. *Nucleic Acids Res.* **27:**3487–3492.

Putnam, C. D., Shroyer, M. J., Lundquist, A. J., Mol, C. D., Arvai, A. S., Mosbaugh, D. W., and Tainer, J. A. (1999). Protein mimicry of DNA from crystal structures of the uracil-DNA glycosylase inhibitor protein and its complex with *Escherichia coli* uracil-DNA glycosylase. *J. Mol. Biol.* **287:**331–346.

Putnam, C. D., and Tainer, J. A. (2005). Protein mimicry of DNA and pathway regulation. *DNA Repair (Amst.)* **4:**1410–1420.

Reilly, B. E., and Spizizen, J. (1965). Bacteriophage deoxyribonucleate infection of competent *Bacillus subtilis. J. Bacteriol.* **89:**782–790.

Richards, R. G., Sowers, L. C., Laszlo, J., and Sedwick, W. D. (1984). The occurrence and consequences of deoxyuridine in DNA. *Adv. Enzyme Regul.* **22:**157–185.

Rojo, F., Mencía, M., Monsalve, M., and Salas, M. (1998). Transcription activation and repression by interaction of a regulator with the α subunit of RNA polymerase: The model of phage ϕ29 protein p4. *Prog. Nucleic Acid Res. Mol. Biol.* **60:**29–46.

Salas, M. (1991). Protein-priming of DNA replication. *Annu. Rev. Biochem.* **60:**39–71.

Salas, M., Mellado, R. P., Viñuela, E., and Sogo, J. M. (1978). Characterization of a protein covalently linked to the 5′ termini of the DNA of *Bacillus subtilis* phage ϕ29. *J. Mol. Biol.* **119:**269–291.

Savva, R., McAuley-Hecht, K., Brown, T., and Pearl, L. (1995). The structural basis of specific base-excision repair by uracil-DNA glycosylase. *Nature* **373:**487–493.

Schachtele, C. F., Reilly, B. E., De Sain, C. V., Whittington, M. O., and Anderson, D. L. (1973). Selective replication of bacteriophage ϕ29 deoxyribonucleic acid in 6-(p-hydroxyphenylazo)-uracil-treated *Bacillus subtilis. J. Virol.* **11:**153–155.

Sekino, Y., Bruner, S. D., and Verdine, G. L. (2000). Selective inhibition of herpes simplex virus type-1 uracil-DNA glycosylase by designed substrate analogs. *J. Biol. Chem.* **275:**36506–36508.

Serna-Rico, A., Illana, B., Salas, M., and Meijer, W. J. J. (2000). The putative coiled coil domain of the ϕ29 terminal protein is a major determinant involved in recognition of the origin of replication. *J. Biol. Chem.* **275:**40529–40538.

Serna-Rico, A., Muñoz-Espín, D., Villar, L., Salas, M., and Meijer, W. J. J. (2003). The integral membrane protein p16.7 organizes *in vivo* ϕ29 DNA replication through interaction with both the terminal protein and ssDNA. *EMBO J* **22:**2297–2306.

Serna-Rico, A., Salas, M., and Meijer, W. J. J. (2002). The *Bacillus subtilis* phage ϕ29 protein p16.7, involved in ϕ29 DNA replication, is a membrane-localized single-stranded DNA-binding protein. *J. Biol. Chem.* **277:**6733–6742.

Serrano, M., Gutiérrez, C., Freire, R., Bravo, A., Salas, M., and Hermoso, J. M. (1994). Phage ϕ29 protein p6: A viral histone-like protein. *Biochimie* **76:**981–991.

Serrano, M., Gutiérrez, C., Salas, M., and Hermoso, J. M. (1993). Superhelical path of the DNA in the nucleoprotein complex that activates the initiation of phage ϕ29 DNA replication. *J. Mol. Biol.* **230:**248–259.

Serrano, M., Salas, M., and Hermoso, J. M. (1990). A novel nucleoprotein complex at a replication origin. *Science* **248:**1012–1016.

Serrano-Heras, G., Ruiz-Masó, J. A., Del Solar, G., Espinosa, M., Bravo, A., and Salas, M. (2007). Protein p56 from the *Bacillus subtilis* phage ϕ29 inhibits DNA-binding ability of uracil-DNA glycosylase. *Nucleic Acids Res.* **35:**5393–5401.

Serrano-Heras, G., Salas, M., and Bravo, A. (2003). *In vivo* assembly of phage ϕ29 replication protein p1 into membrane-associated multimeric structures. *J. Biol. Chem.* **278:**40771–40777.

Serrano-Heras, G., Salas, M., and Bravo, A. (2008). Phage ϕ29 protein p56 prevents viral DNA replication impairment caused by uracil excision activity of uracil-DNA glycosylase. *Proc. Natl. Acad. Sci. USA* **105:**19044–19049.

Serrano-Heras, S., Salas, M., and Bravo, A. (2006). A uracil-DNA glycosylase inhibitor encoded by a non-uracil containing viral DNA. *J. Biol. Chem.* **281:**7068–7074.

Shapiro, R. (1980). Chromosome Damage and Repair. *In* (E. Seeberg and K. Kleppe, eds.), pp. 3–18. Plenum Press, New York.

Shlomai, J., and Kornberg, A. (1978). Deoxyuridine triphosphatase of *Escherichia coli*: Purification, properties, and use as a reagent to reduce uracil incorporation into DNA. *J. Biol. Chem.* **253:**3305–3312.

Soengas, M. S., Esteban, J. A., Lázaro, J. M., Bernad, A., Blasco, M. A., Salas, M., and Blanco, L. (1992). Site-directed mutagenesis at the Exo III motif of ϕ29 DNA polymerase; overlapping structural domains for the 3′-5′ exonuclease and strand-displacement activities. *EMBO J.* **11:**4227–4237.

Sogo, J. M., Inciarte, M. R., Corral, J., Viñuela, E., and Salas, M. (1982). Structure of protein-containing replicative intermediates of *Bacillus subtilis* phage ϕ29 DNA. *Virology* **116:**1–18.

Sogo, J. M., Lozano, M., and Salas, M. (1984). *In vitro* transcription of the *Bacillus subtilis* phage ϕ29 DNA by *Bacillus subtilis* and *Escherichia coli* RNA polymerases. *Nucleic Acids Res.* **24:**1943–1960.

Soufo, H. J., and Graumann, P. L. (2003). Actin-like proteins MreB and Mbl from *Bacillus subtilis* are required for bipolar positioning of replication origins. *Curr. Biol.* **13:**1916–1920.

Sousa, M. M., Krokan, H. E., and Slupphaug, G. (2007). DNA-uracil and human pathology. *Mol. Aspects Med.* **28:**276–306.

Steck, T. R., and Drlica, K. (1984). Bacterial chromosome segregation: Evidence for DNA gyrase involvement in decatenation. *Cell* **36:**1081–1088.

Stone, M. D., Bryant, Z., Crisona, N. J., Smith, S. B., Vologodskii, A., Bustamante, C., and Cozzarelli, N. R. (2003). Chirality sensing by *Escherichia coli* topoisomerase IV and the mechanism of type II topoisomerases. *Proc. Natl. Acad. Sci. USA* **100:**8654–8659.

Swart, J. R. J., and Warner, H. R. (1985). Isolation and partial characterization of a bacteriophage T5 mutant unable to induce thymidylate synthetase and its use in studying the effect of uracil incorporation into DNA on early gene expression. *J. Virol.* **54:**86–91.

Tadesse, S., and Graumann, P. L. (2006). Differential and dynamic localization of toposiomerases in *Bacillus subtilis*. *J. Bacteriol.* **188:**3002–3011.

Tadesse, S., Mascarenhas, J., Kösters, B., Hasilik, A., and Graumann, P. L. (2005). Genetic interaction of the SMC complex with topoisomerase IV in *Bacillus subtilis*. *Microbiology* **151:**3729–3737.

Talavera, A., Salas, M., and Viñuela, E. (1972). Temperature-sensitive mutants of bacteriophage ϕ29 of *Bacillus subtilis*. *Eur. J. Biochem.* **31:**367–371.

Tamanoi, F., and Okazaki, T. (1978). Uracil incorporation into nascent DNA of thymine-requiring mutant of *Bacillus subtilis* 168. *Proc. Natl. Acad. Sci. USA* **75:**2195–2199.

Tye, B. K., Chien, J., Lehman, I. R., Duncan, B. K., and Warner, H. R. (1978). Uracil incorporation: A source of pulse-labeled DNA fragments in the replication of the *Escherichia coli* chromosome. *Proc. Natl. Acad. Sci. USA* **75:**233–237.

Van den Ent, F., Amos, L. A., and Löwe, J. (2001). Prokaryotic origin of the actin cytoskeleton. *Nature* **413:**39–44.

Vlcek, C., and Paces, V. (1986). Nucleotide sequence of the late region of *Bacillus* phage ϕ29 completes the 19285-bp sequence of ϕ29: Comparison with the homologous sequence of phage PZA. *Gene* **46:**215–225.

Wang, Z., and Mosbaugh, D. W. (1989). Uracil-DNA glycosylase inhibitor gene of bacteriophage PBS2 encodes a binding protein specific for uracil-DNA glycosylase. *J. Biol. Chem.* **264:**1163–1171.

Warner, H. R., Johnson, L. K., and Snustad, D. P. (1980). Early events after infection of *Escherichia coli* by bacteriophage T5. III. Inhibition of uracil-DNA glycosylase activity. *J. Virol.* **33:**535–538.

Waters, T. R., and Swann, P. F. (2000). Thymine-DNA glycosylase and G to A transition mutations at CpG sites. *Mutat. Res.* **462:**137–147.

Xiao, G., Tordova, M., Jagadeesh, J., Drohat, A. C., Stivers, J. T., and Gilliland, G. L. (1999). Crystal structure of *Escherichia coli* uracil DNA glycosylase and its complexes with uracil and glycerol: Structure and glycosylase mechanism revisited. *Proteins* **35:**13–24.

Yang, H., Chen, K., Zhang, C., Huang, S., and Zhang, H. (2007). Virion-associated uracil DNA glycosylase-2 and apurinic/apyrimidinic endonuclease are involved in the degradation of APOBEC3G-edited nascent HIV-1 DNA. *J. Biol. Chem.* **282:**11667–11675.

Yehle, C. O. (1978). Genome-linked protein associated with the 5′ termini of bacteriophage ϕ29 DNA. *J. Virol.* **27:**776–783.

Yoshikawa, H., and Ito, J. (1982). Nucleotide sequence of the major early region of bacteriophage ϕ29. *Gene* **17:**323–335.

Zaballos, A., Lázaro, J. M., Méndez, E., and Salas, M. (1989). Effects of internal deletions on the priming activity of the phage ϕ29 terminal protein. *Gene J.* **83:**187–195.

Zhu, Q., Pongpech, P., and DiGate, R. J. (2001). Type I topoisomerase activity is required for proper chromosomal segregation in *Escherichia coli*. *Proc. Natl. Acad. Sci. USA* **98:**9766–9771.

INDEX

Note: Page numbers followed by "*f*" indicate figures, and "*t*" indicate tables.

C

F

G

H

I

R

S

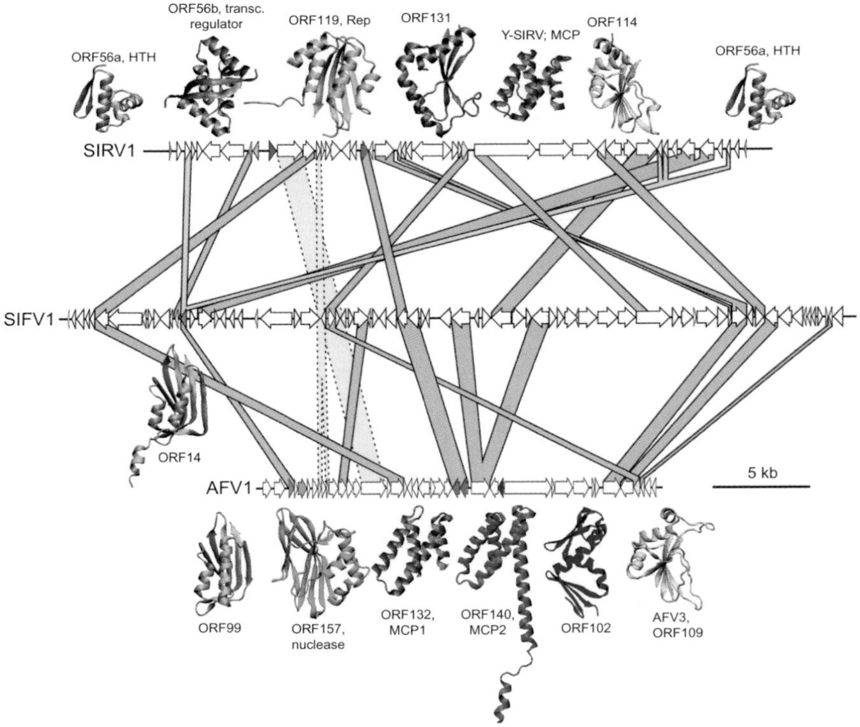

Figure 4, Mart Krupovic *et al.* (See Page 46 of this Volume)

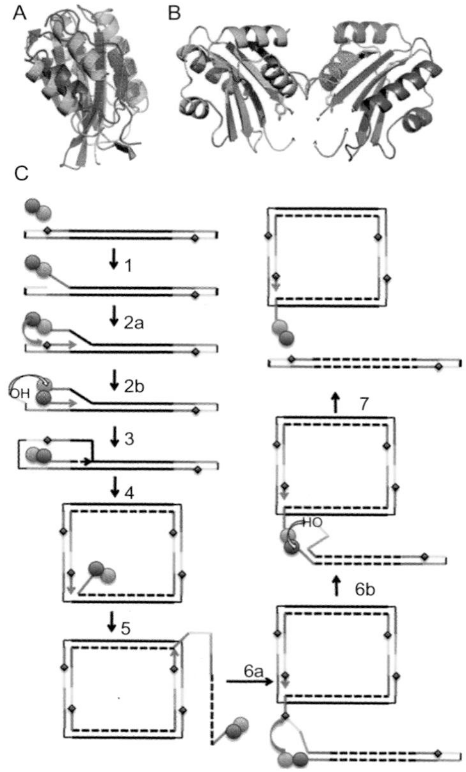

Figure 5, Mart Krupovic *et al.* (See Page 50 of this Volume)

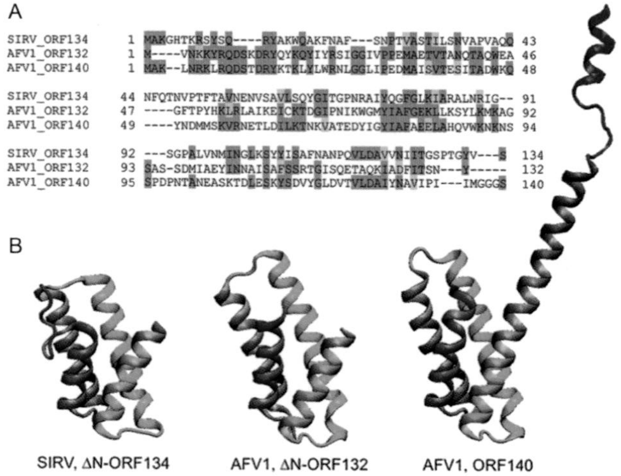

Figure 6, Mart Krupovic *et al.* (See Page 52 of this Volume)

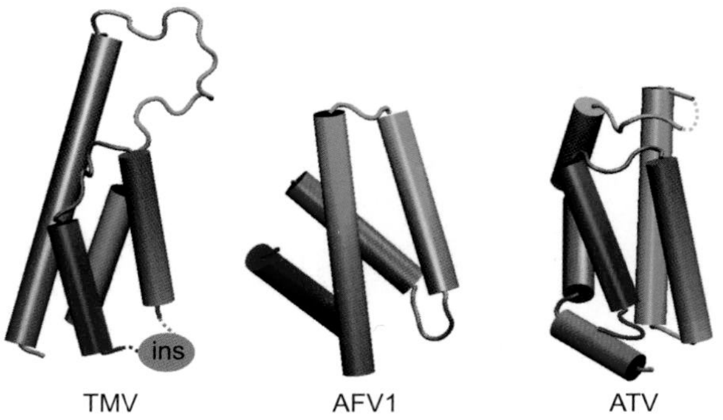

Figure 7, Mart Krupovic *et al.* (See Page 53 of this Volume)

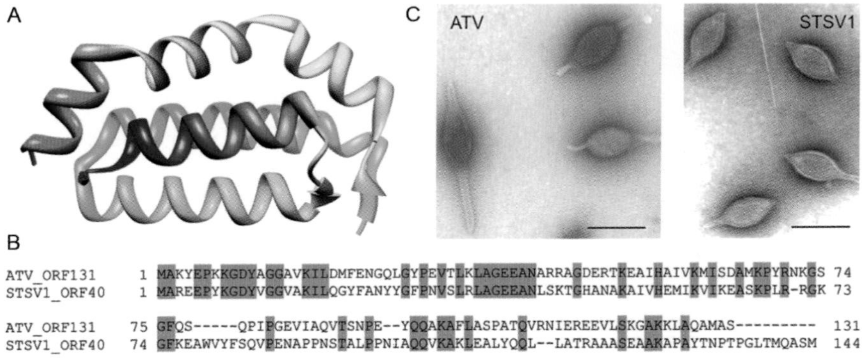

Figure 8, Mart Krupovic *et al.* (See Page 54 of this Volume)

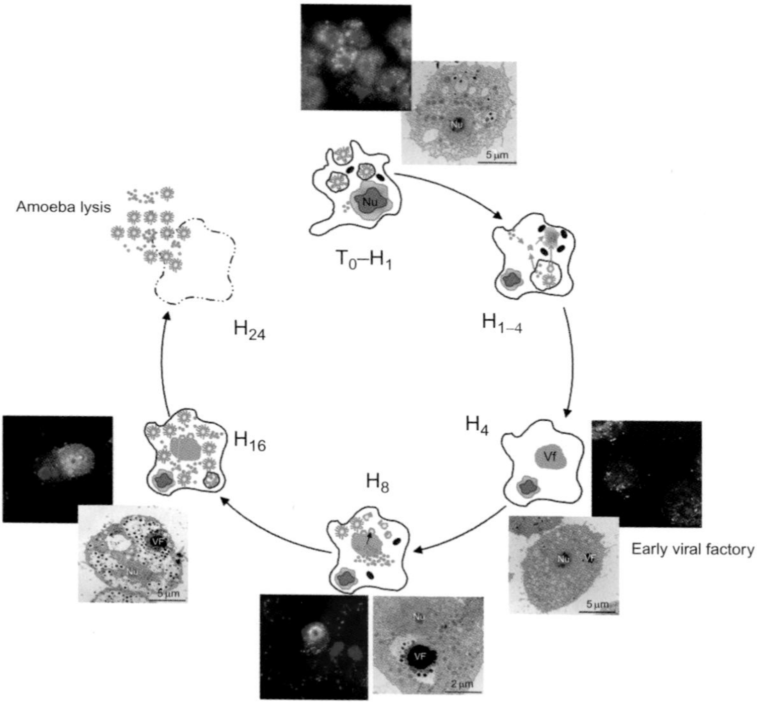

Figure 1, Christelle Desnues *et al.* (See Page 67 of this Volume)

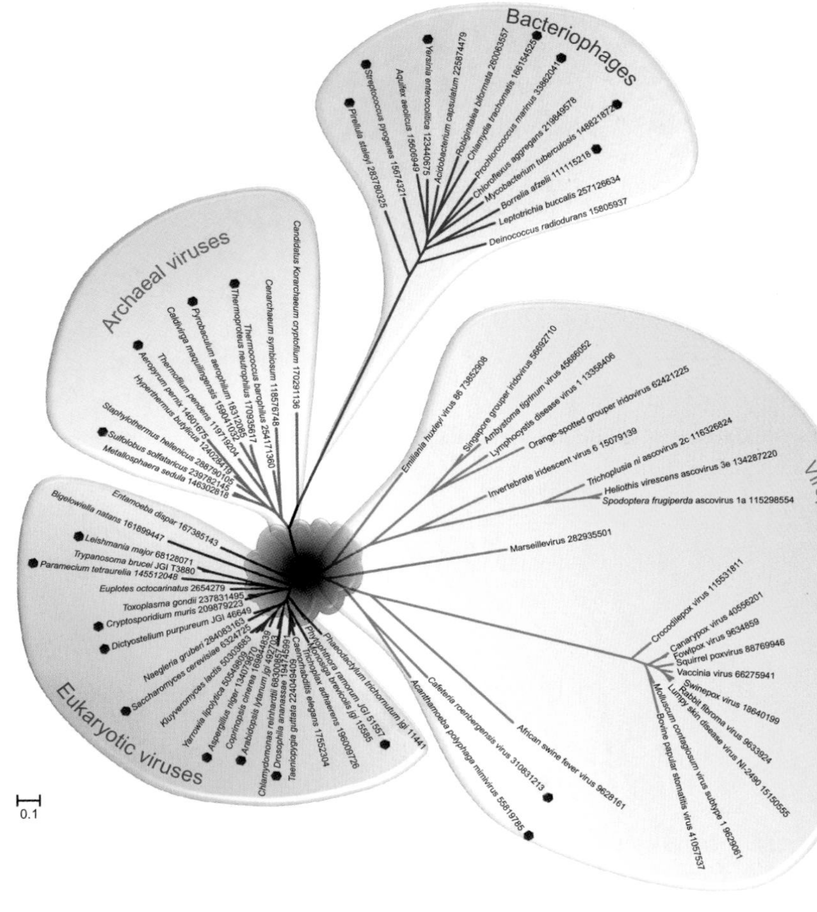

Figure 4, Christelle Desnues *et al.* (See Page 84 of this Volume)

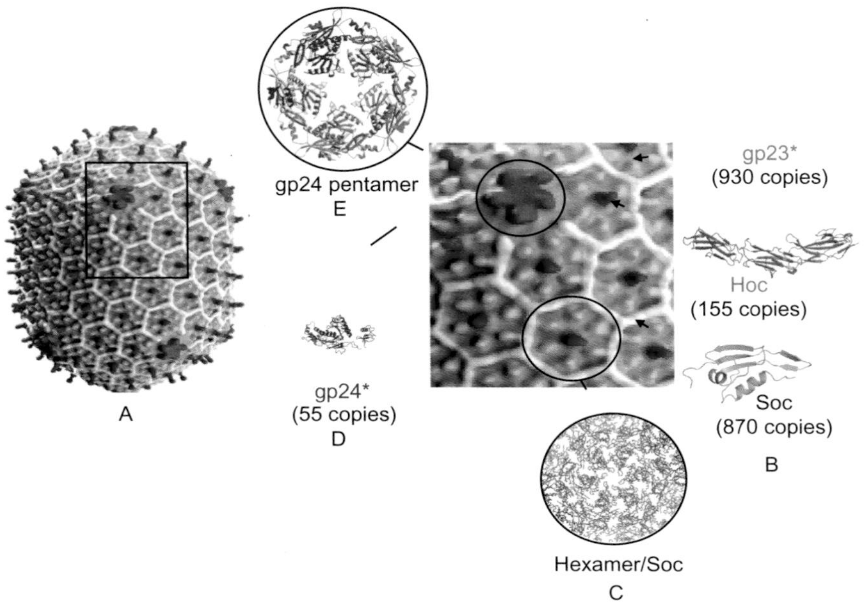

Figure 1, Lindsay W. Black and Venigalla B. Rao (See Page 122 of this Volume)

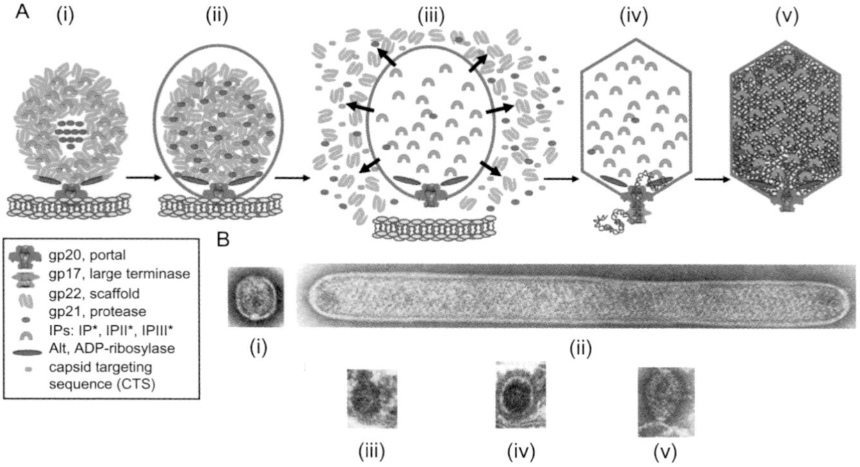

Figure 2, Lindsay W. Black and Venigalla B. Rao (See Page 123 of this Volume)

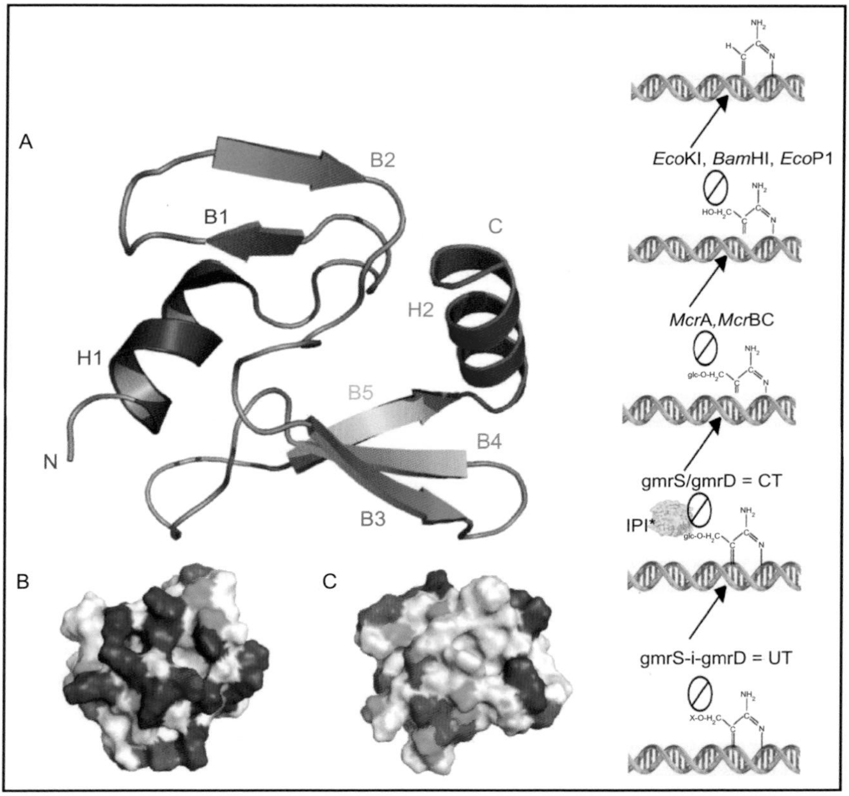

Figure 4, Lindsay W. Black and Venigalla B. Rao (See Page 128 of this Volume)

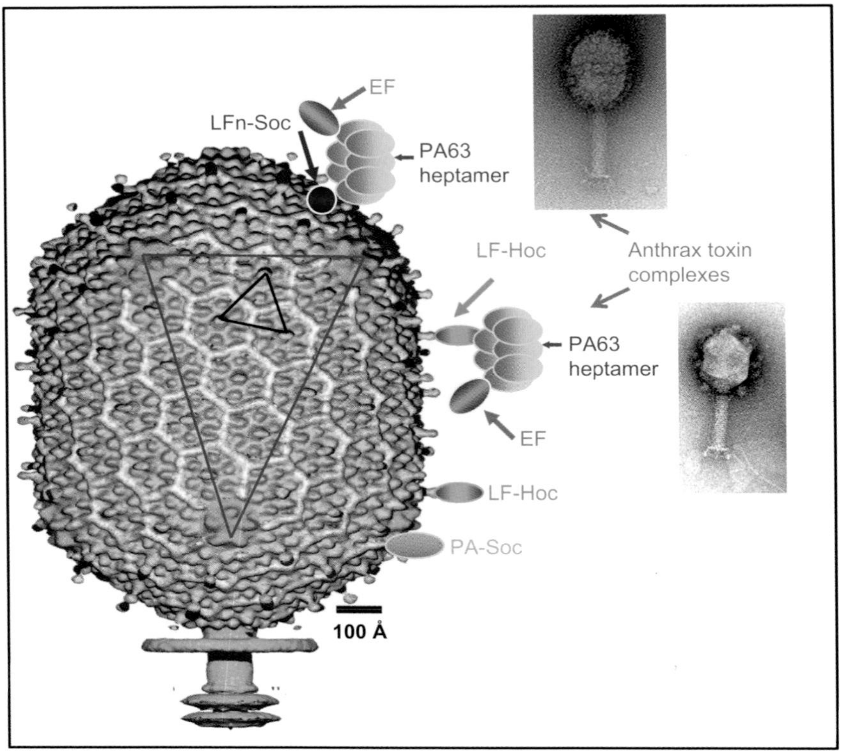

Figure 5, Lindsay W. Black and Venigalla B. Rao (See Page 131 of this Volume)

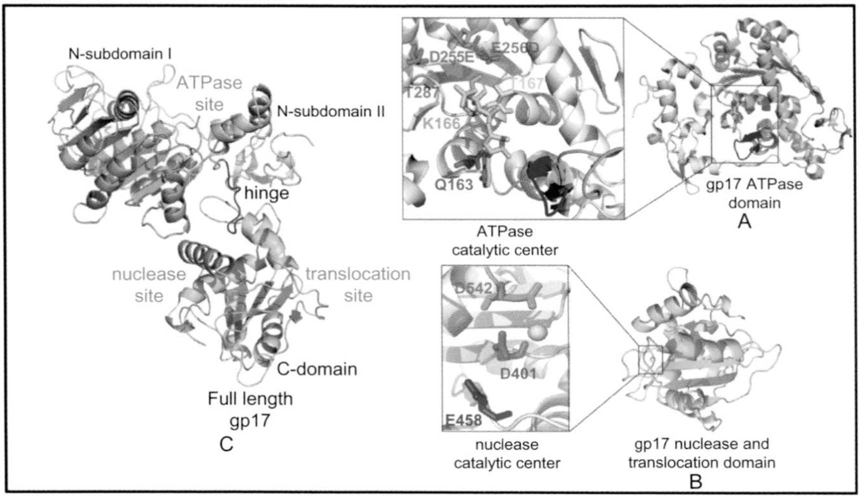

Figure 7, Lindsay W. Black and Venigalla B. Rao (See Page 137 of this Volume)

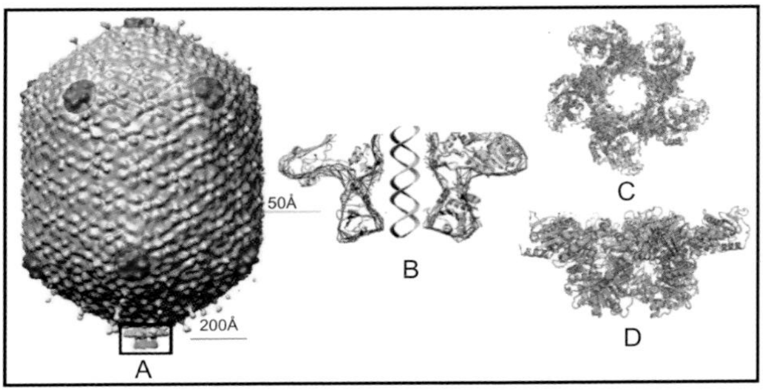

Figure 8, Lindsay W. Black and Venigalla B. Rao (See Page 141 of this Volume)

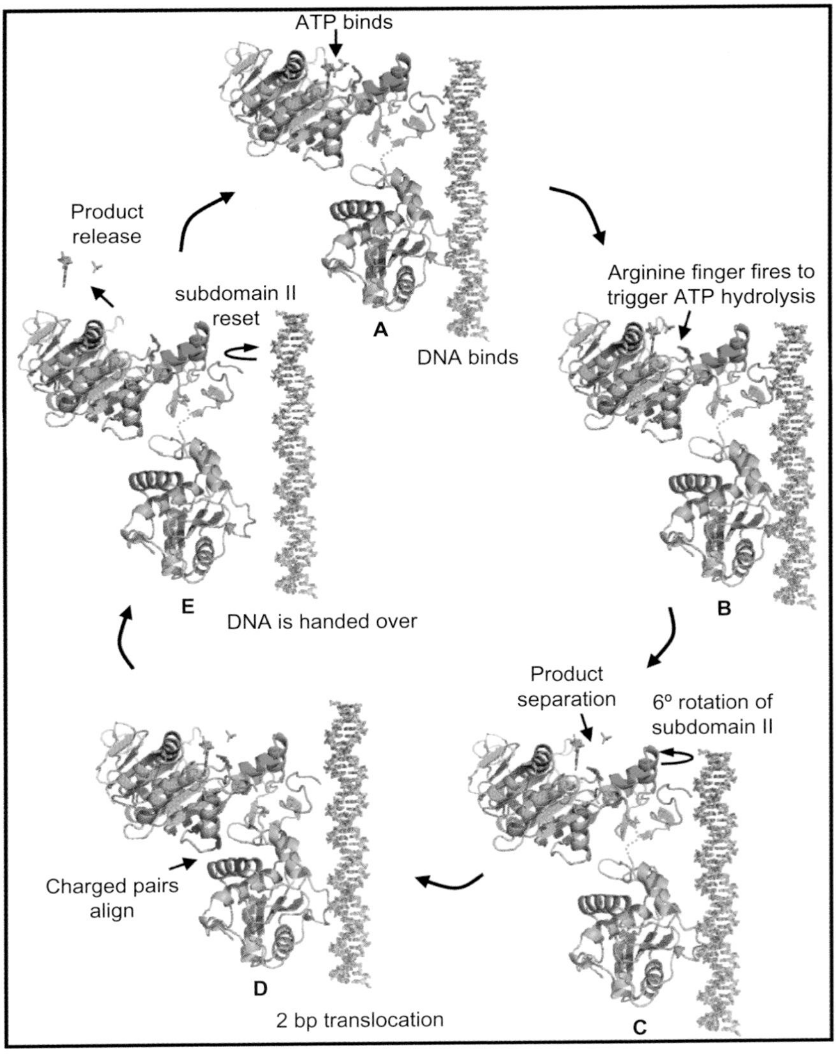

Figure 9, Lindsay W. Black and Venigalla B. Rao (See Page 143 of this Volume)

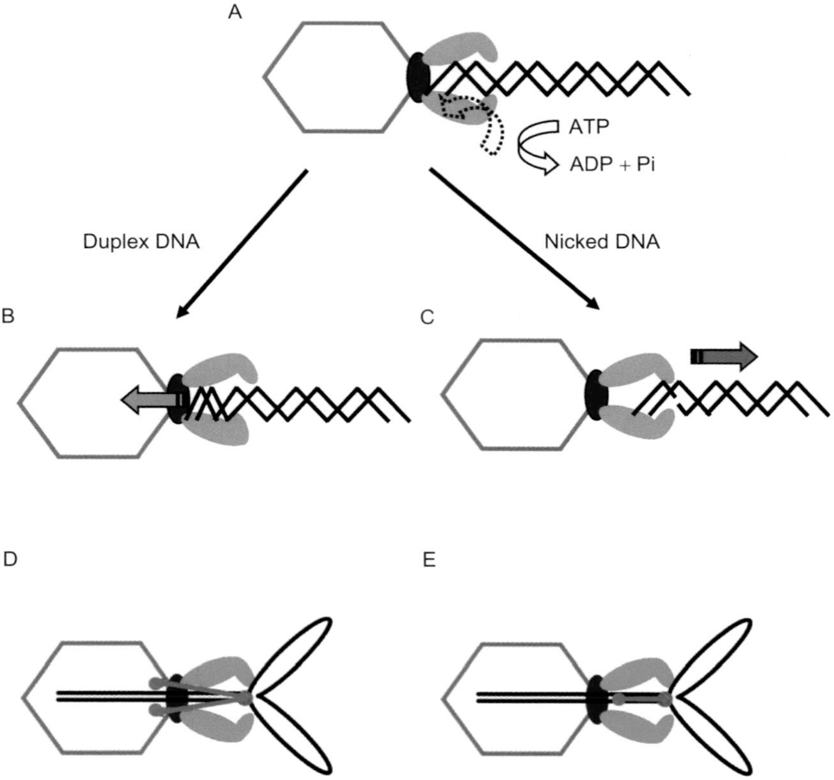

Figure 10, Lindsay W. Black and Venigalla B. Rao (See Page 145 of this Volume)

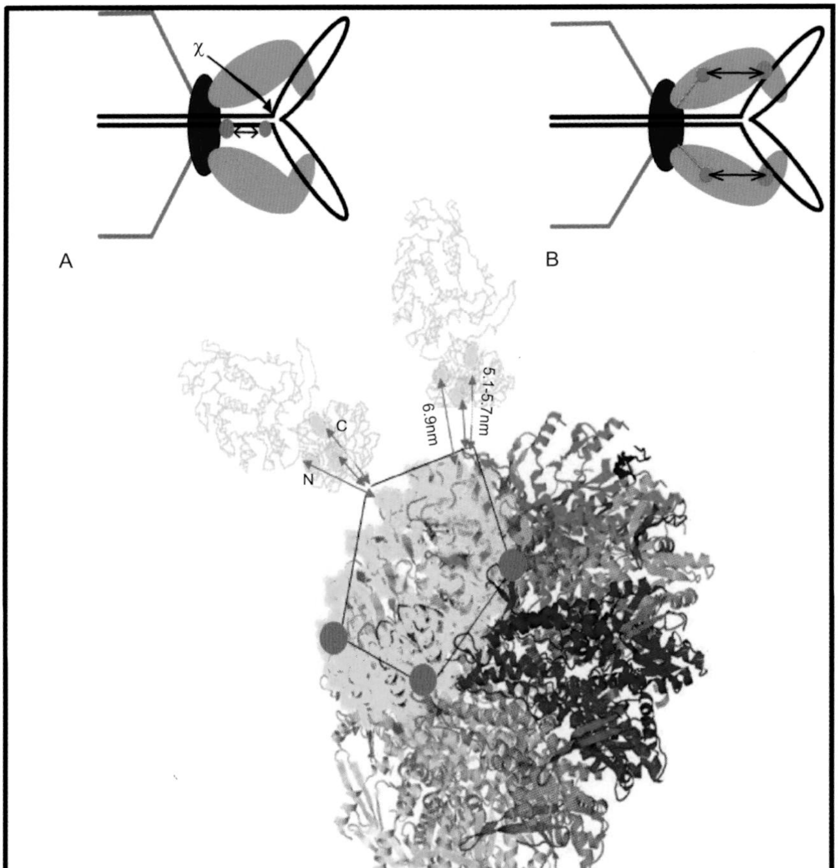

Figure 11, Lindsay W. Black and Venigalla B. Rao (See Page 146 of this Volume)

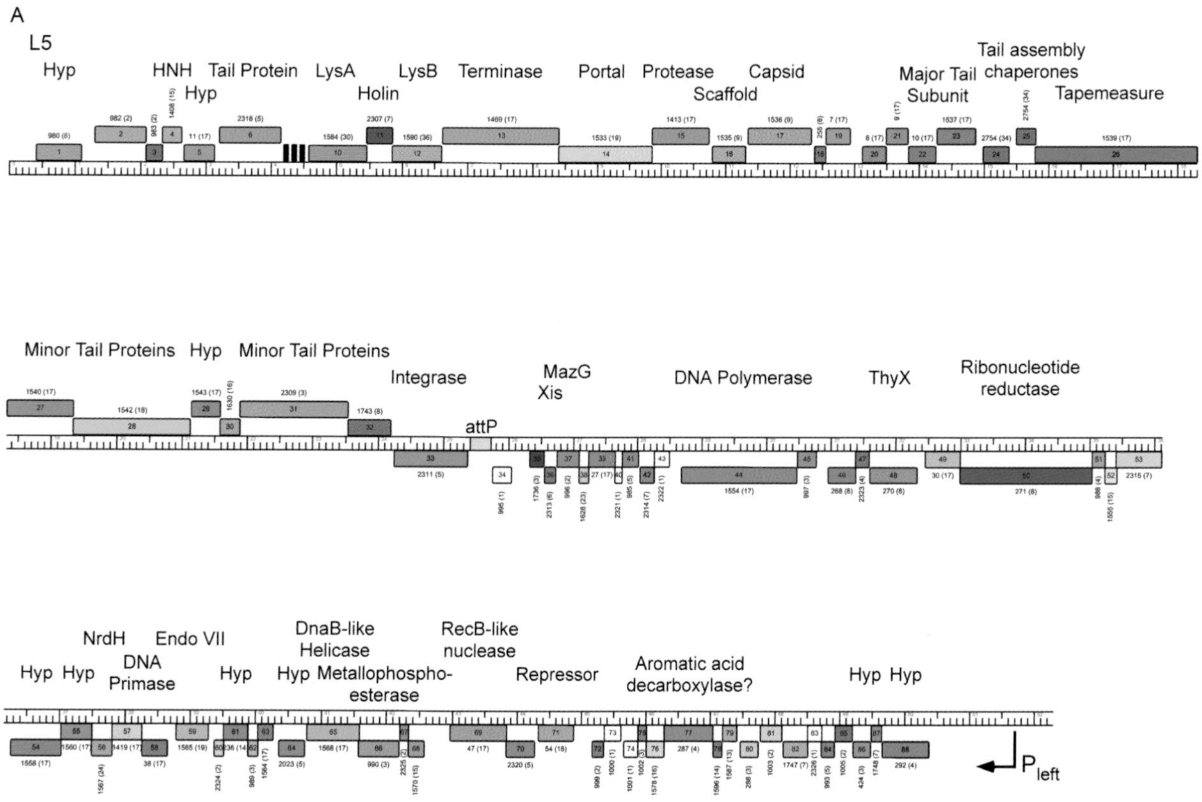

Figure 3A, Graham F. Hatfull (See Page 194 of this Volume)

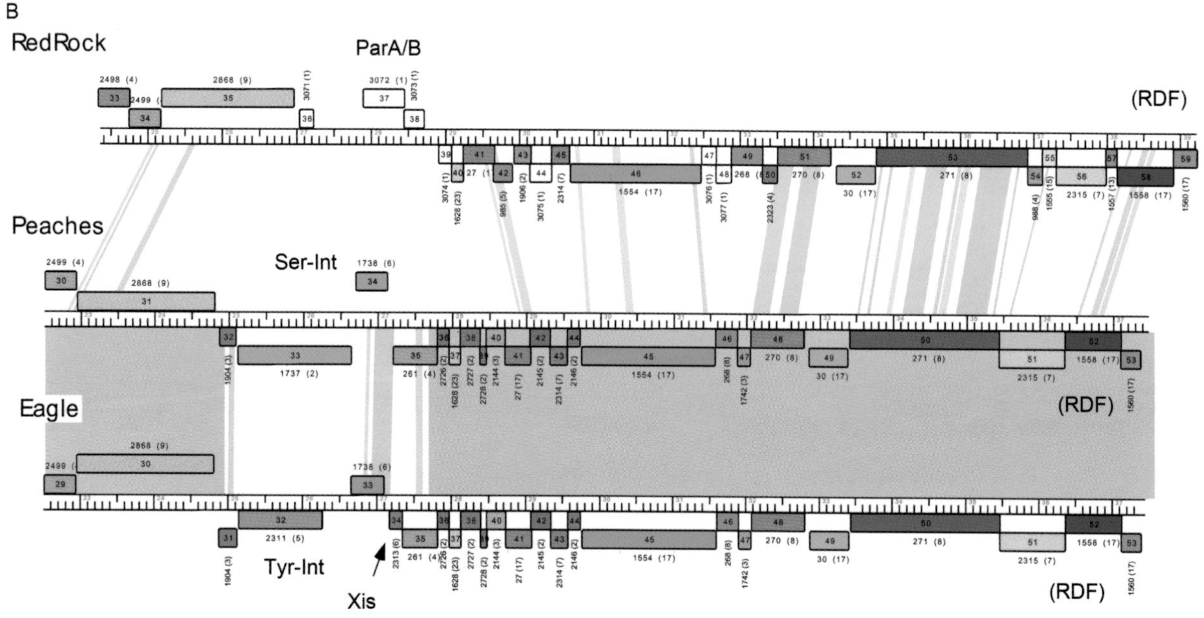

Figure 3B, Graham F. Hatfull (See Page 195 of this Volume)

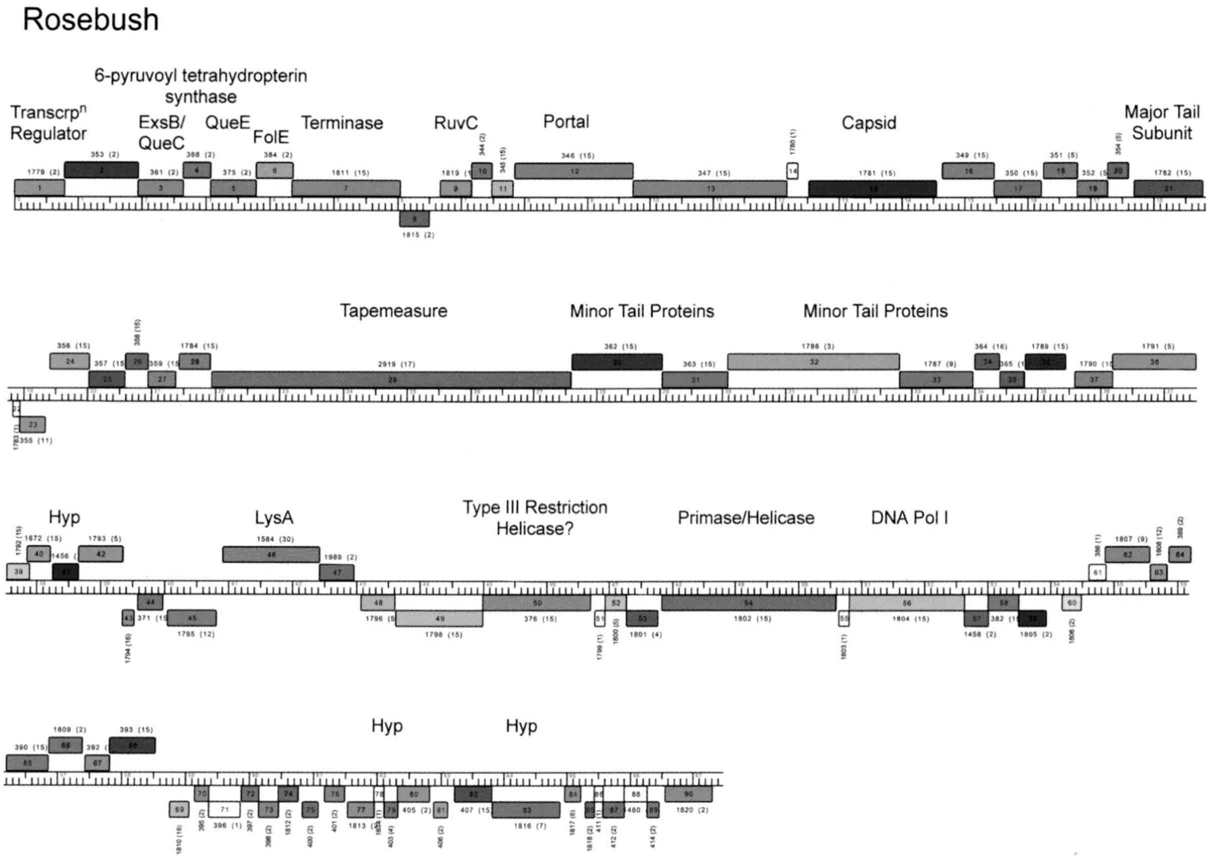

Figure 4, Graham F. Hatfull (See Page 201 of this Volume)

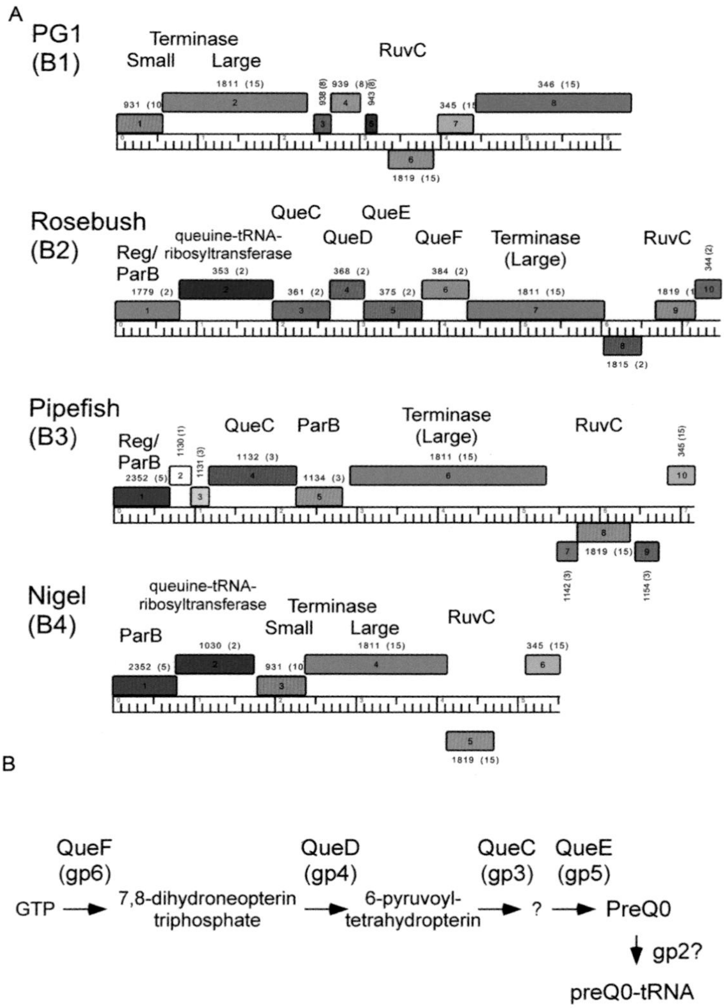

Figure 5, Graham F. Hatfull (See Page 203 of this Volume)

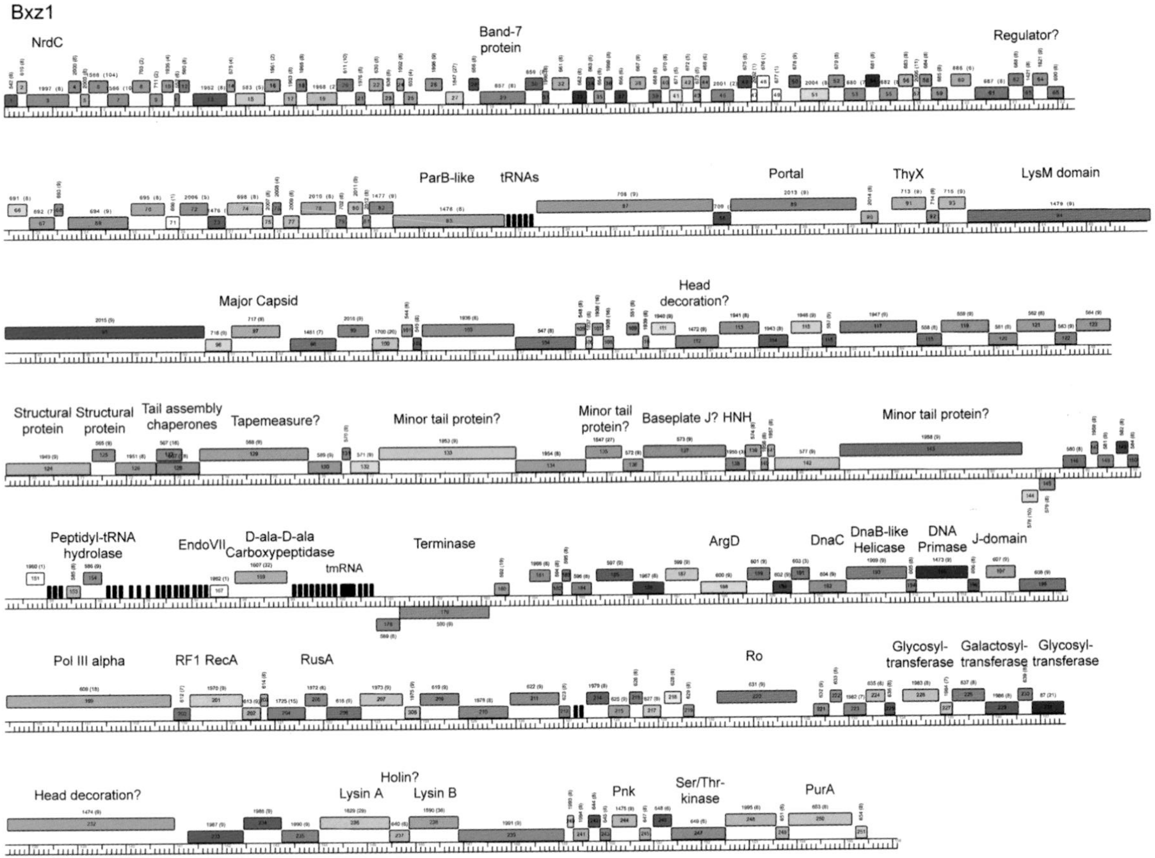

Figure 6, Graham F. Hatfull (See Page 206 of this Volume)

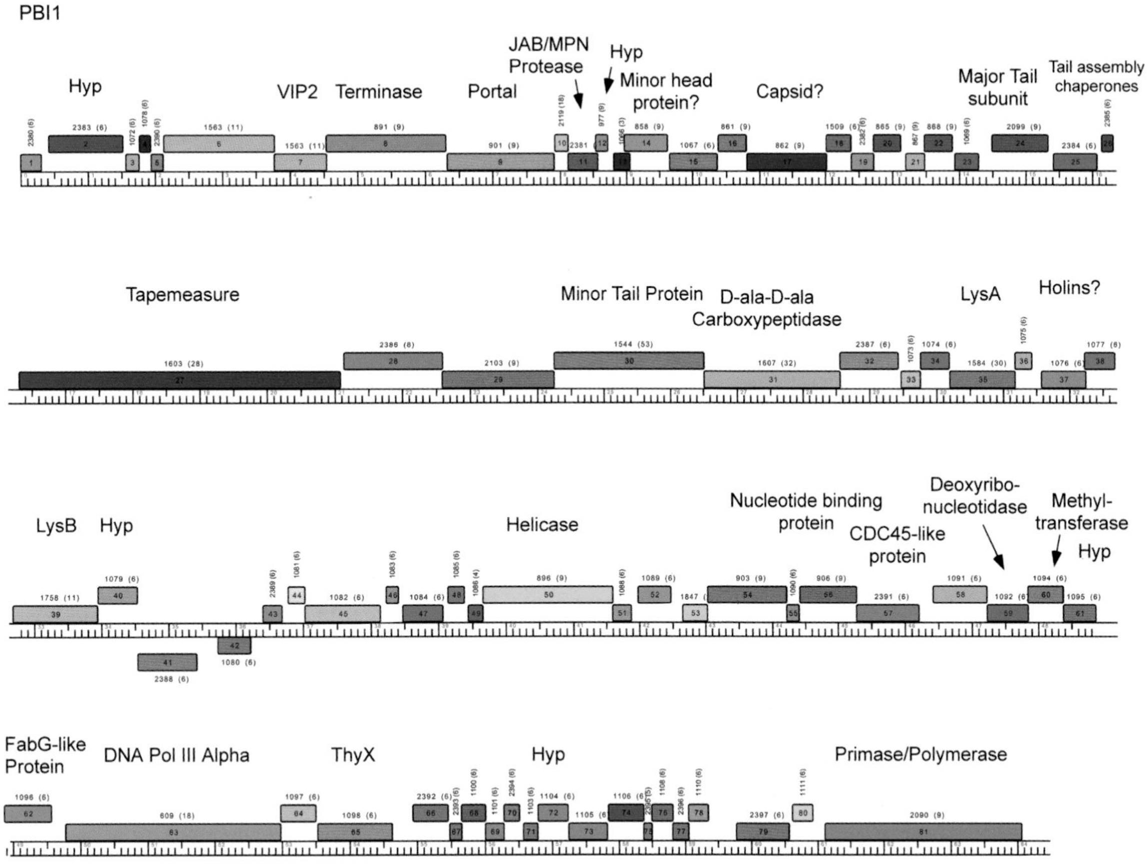

Figure 7, Graham F. Hatfull (See Page 209 of this Volume)

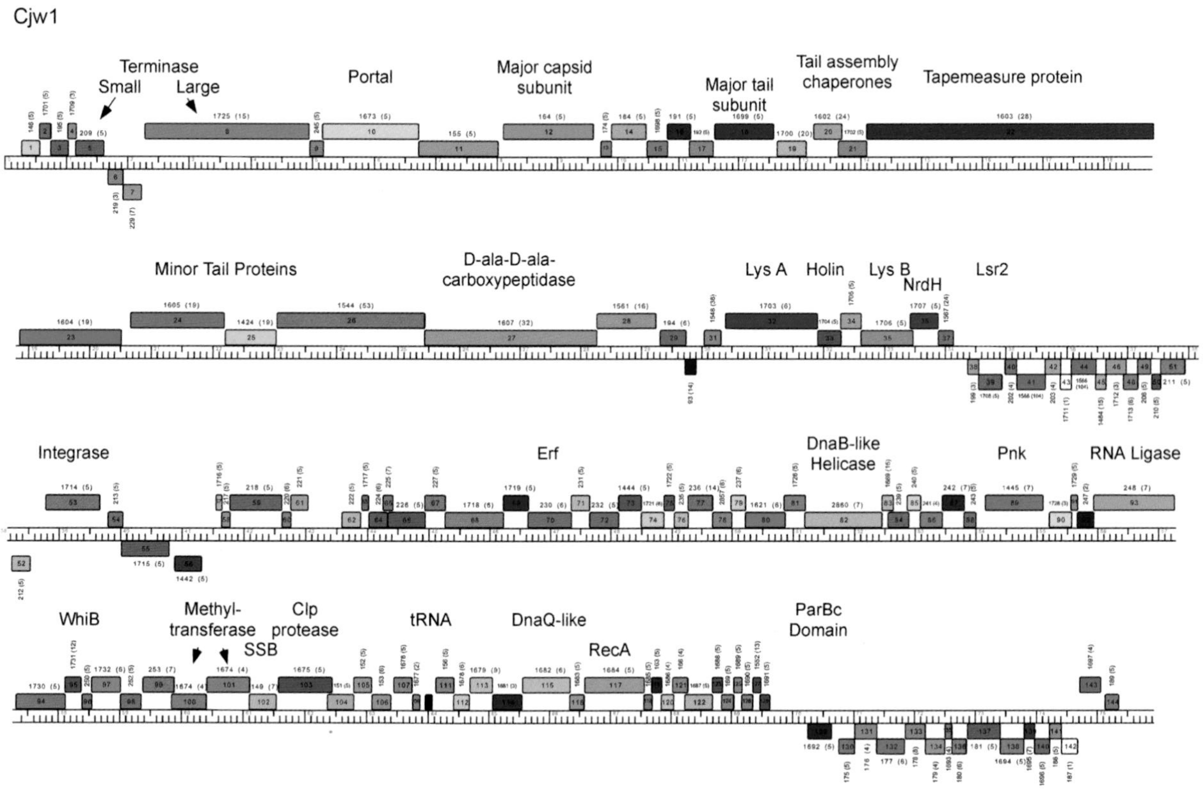

Figure 8, Graham F. Hatfull (See Page 211 of this Volume)

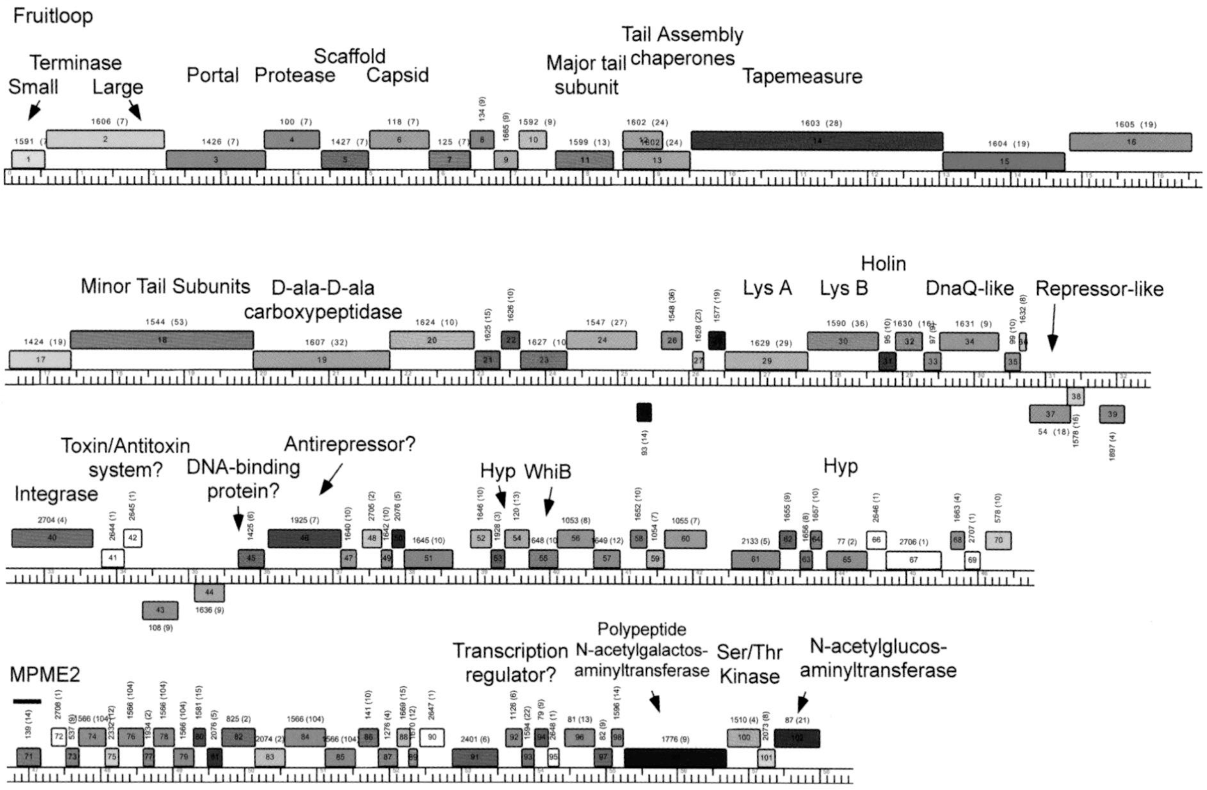

Figure 9, Graham F. Hatfull (See Page 212 of this Volume)

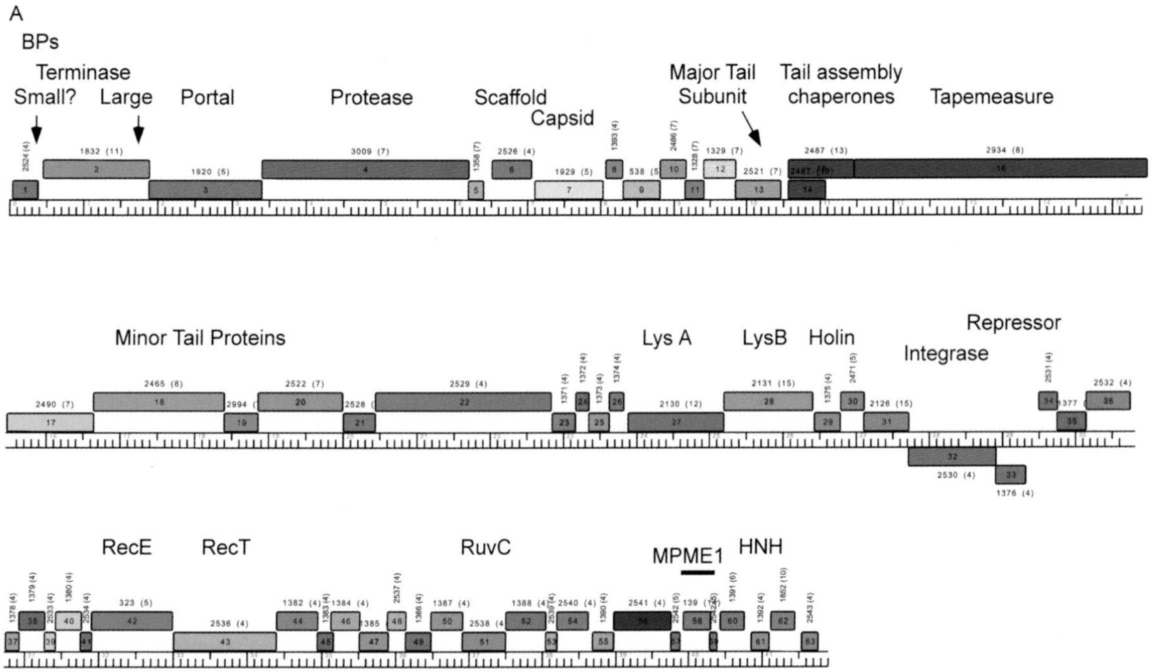

Figure 10A, Graham F. Hatfull (See Page 217 of this Volume)

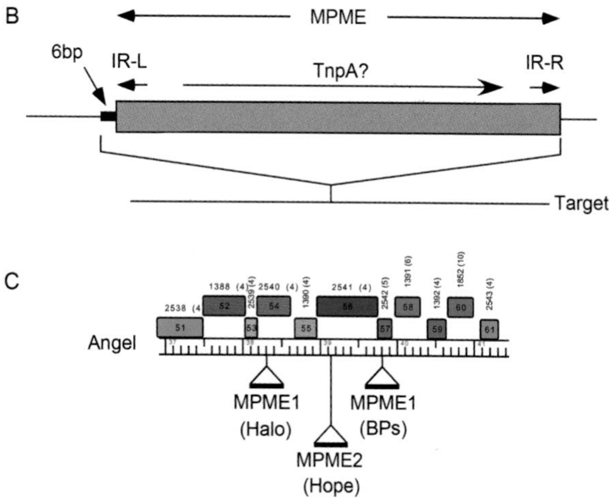

Figure 10B,C, Graham F. Hatfull (See Page 218 of this Volume)

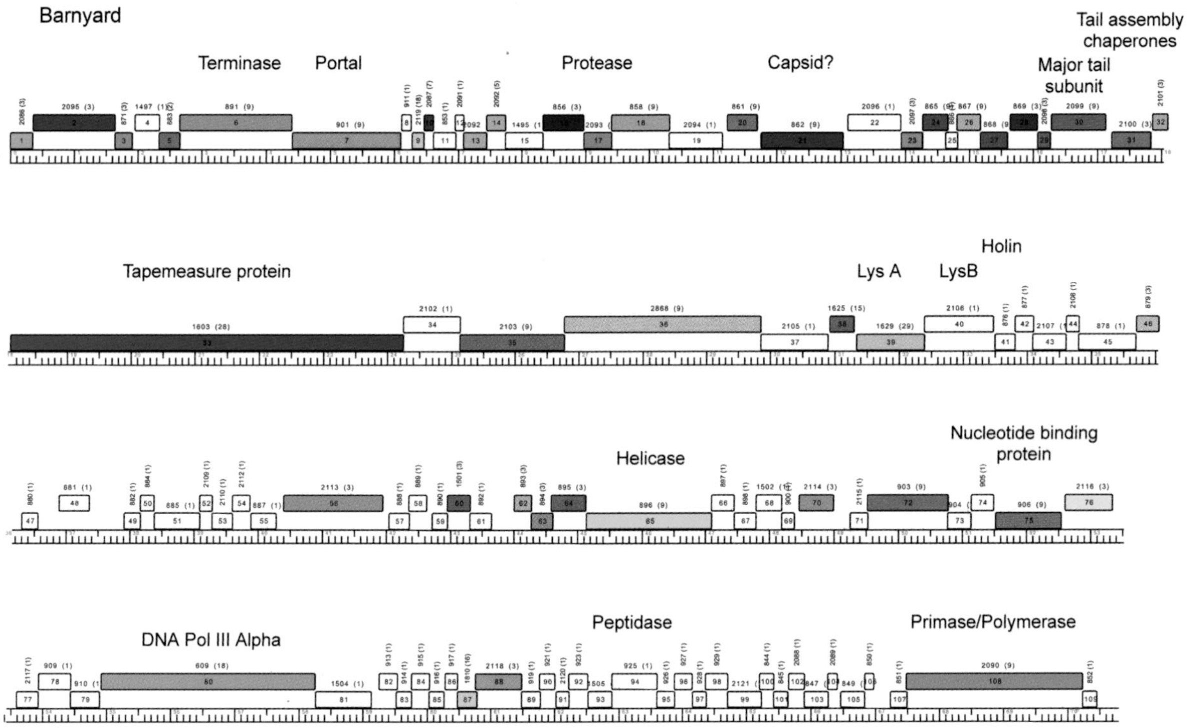

Figure 11, Graham F. Hatfull (See Page 221 of this Volume)

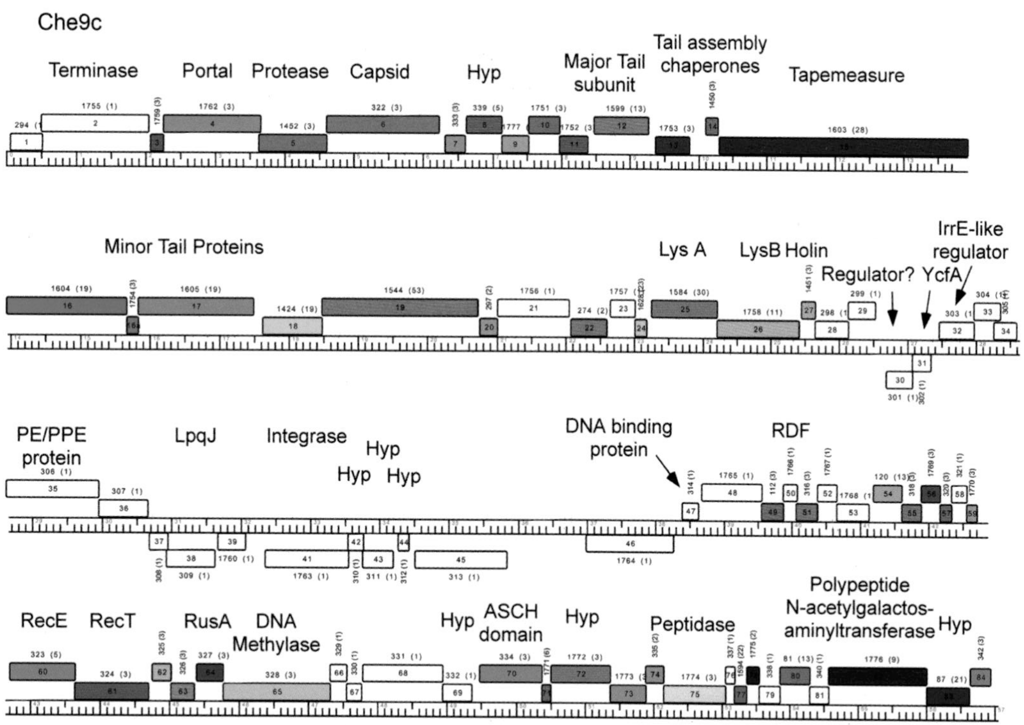

Figure 12, Graham F. Hatfull (See Page 223 of this Volume)

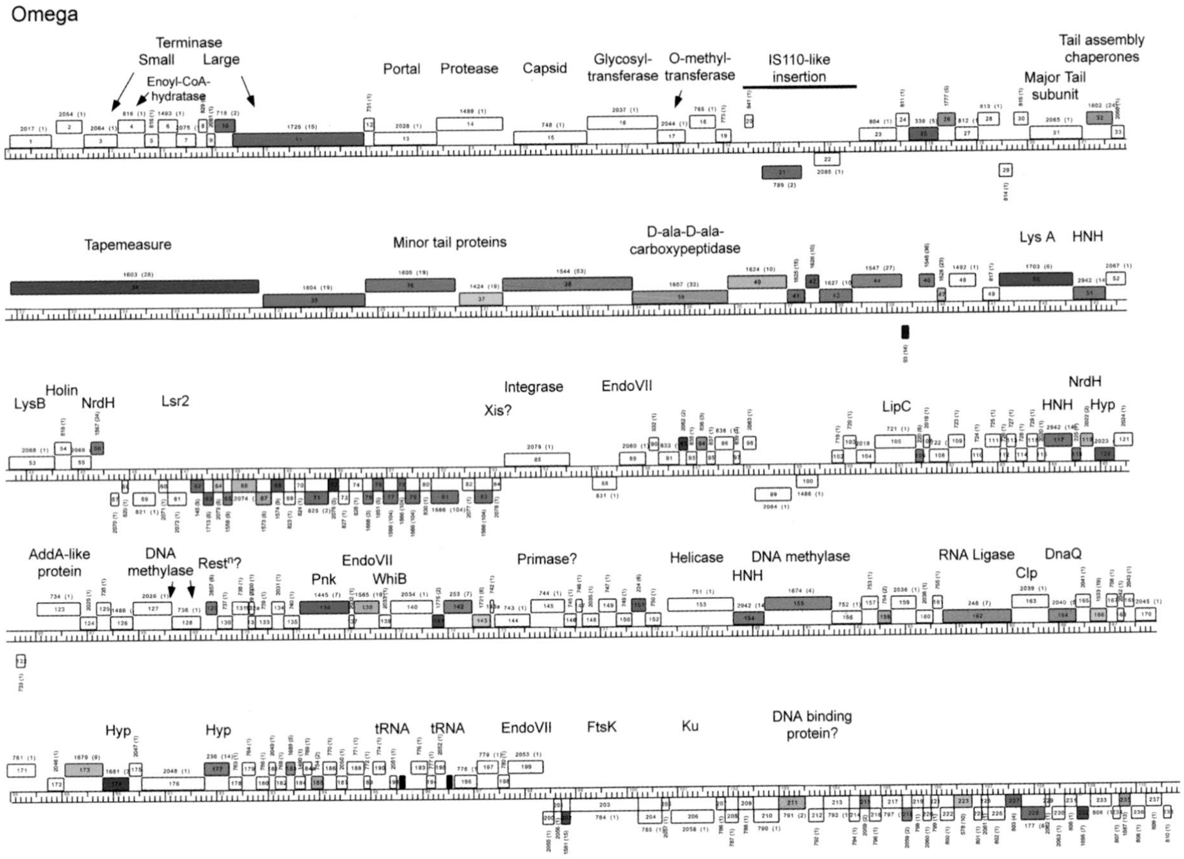

Figure 13, Graham F. Hatfull (See Page 226 of this Volume)

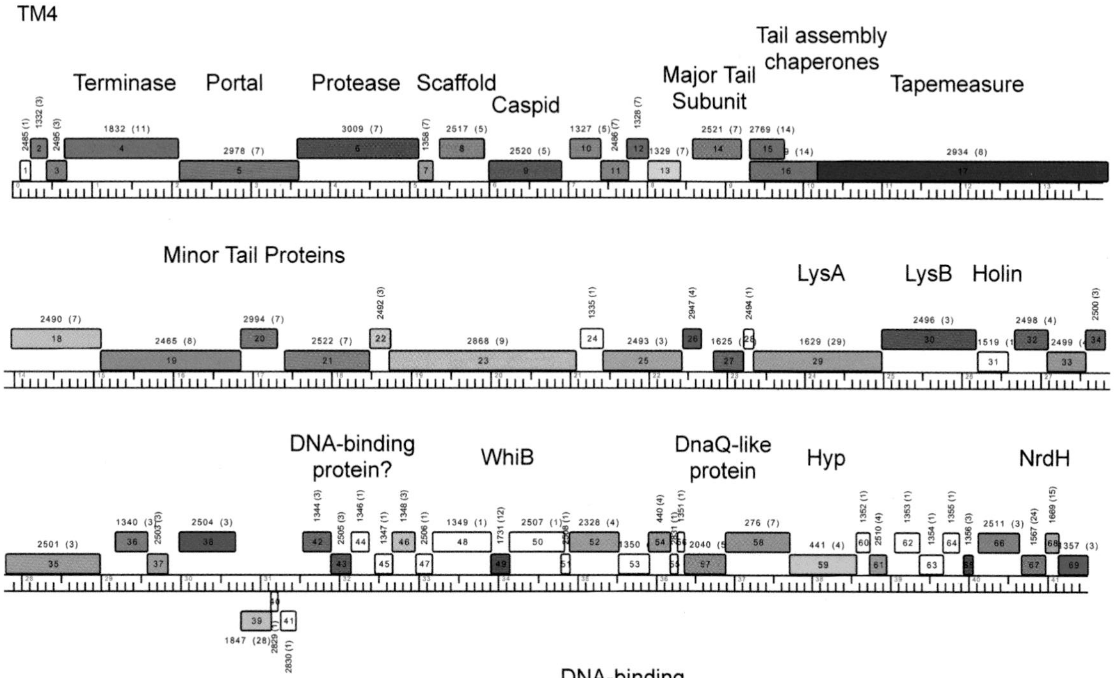

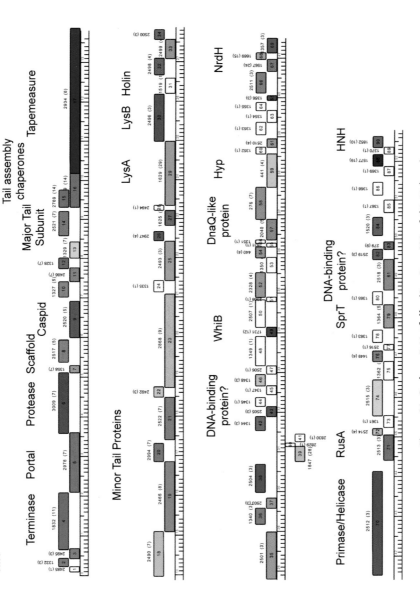

Figure 14, Graham F. Hatfull (See Page 230 of this Volume)

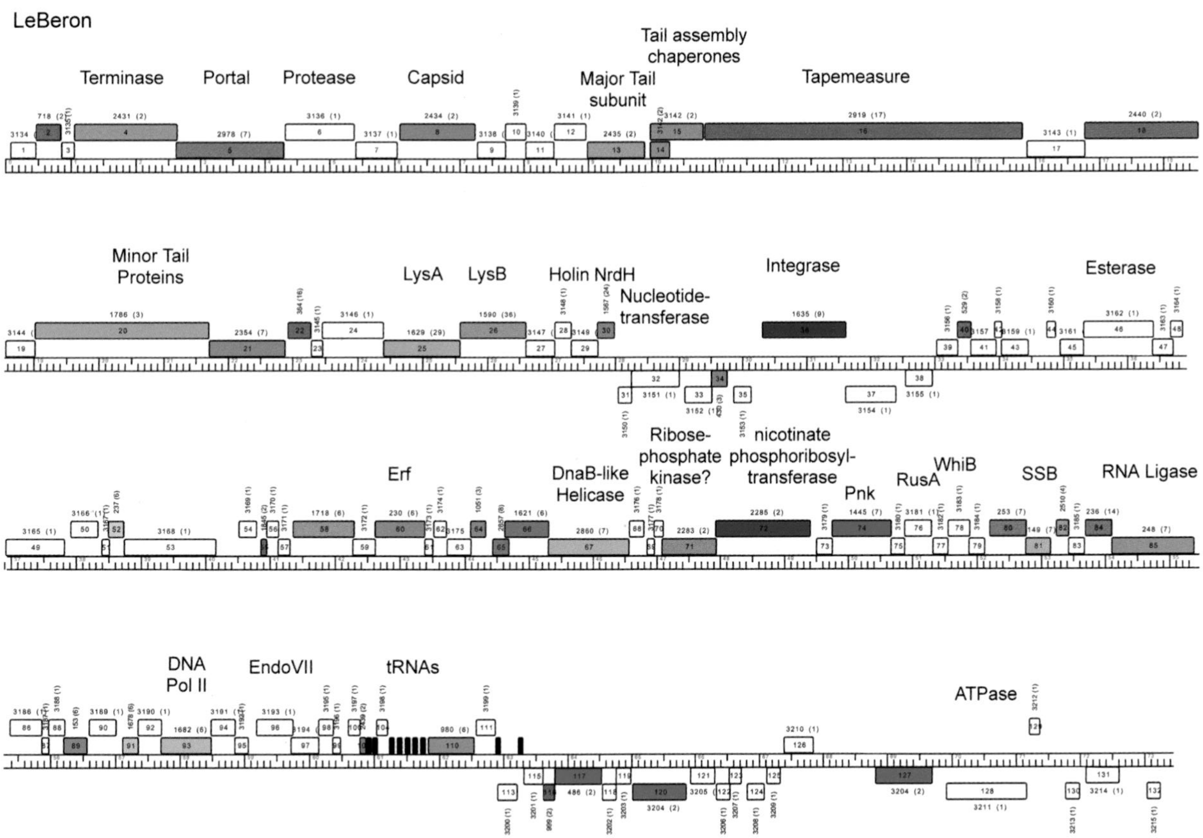

Figure 15, Graham F. Hatfull (See Page 233 of this Volume)

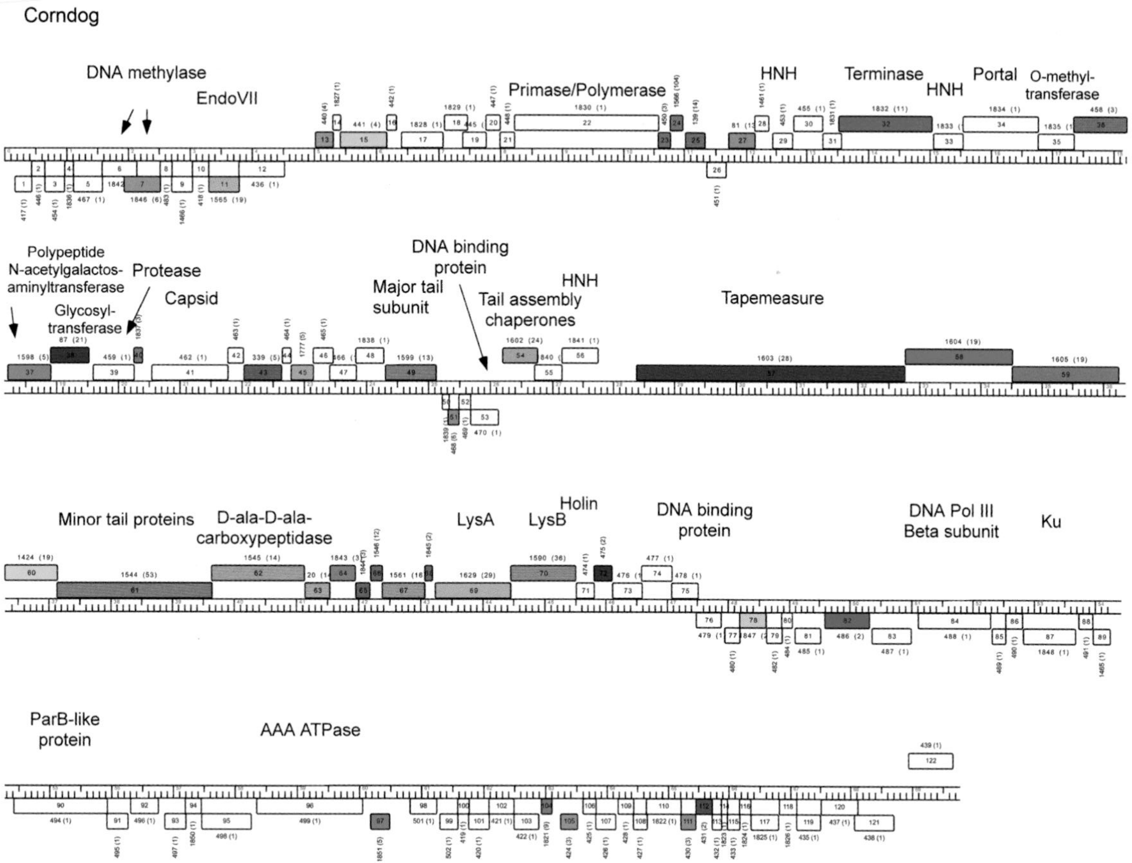

Figure 16, Graham F. Hatfull (See Page 235 of this Volume)

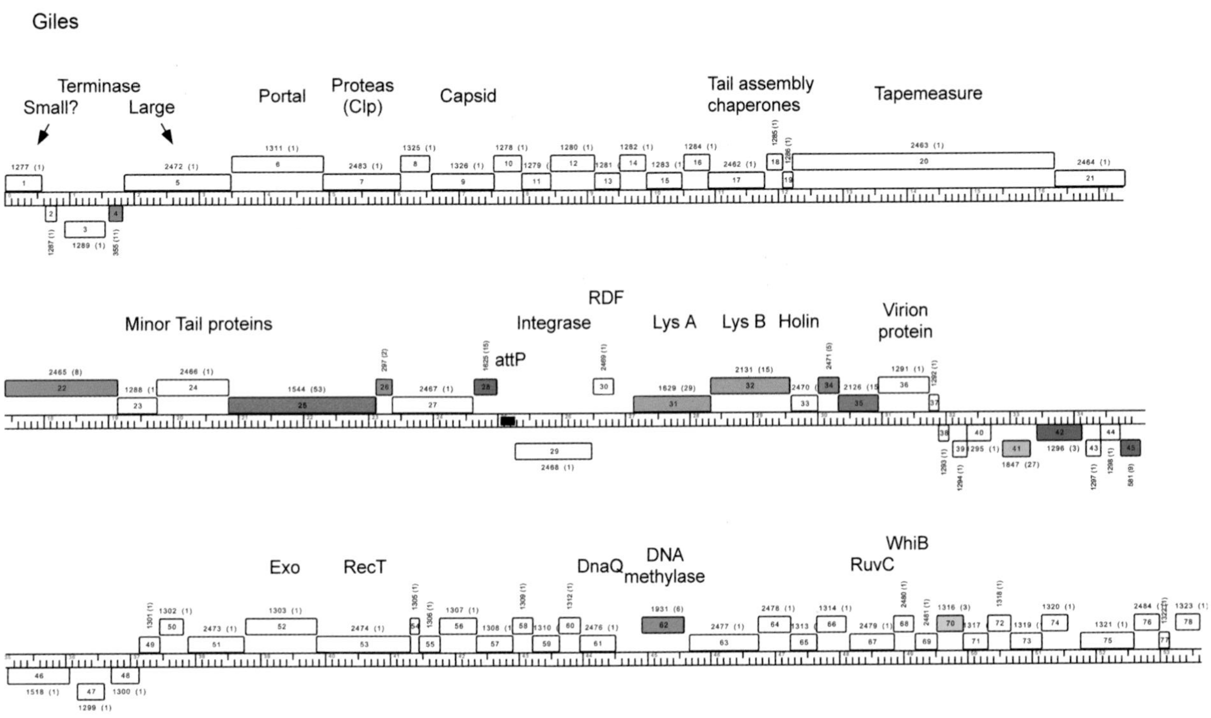

Figure 17, Graham F. Hatfull (See Page 238 of this Volume)

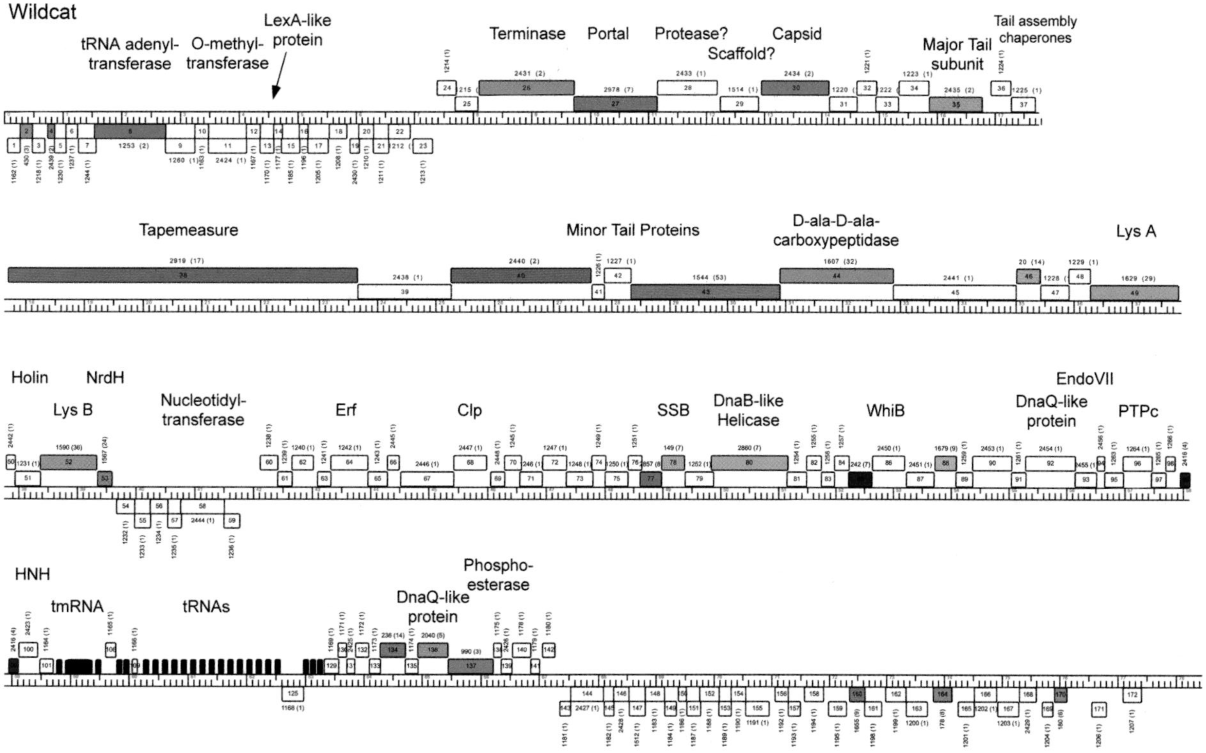

Figure 18, Graham F. Hatfull (See Page 240 of this Volume)

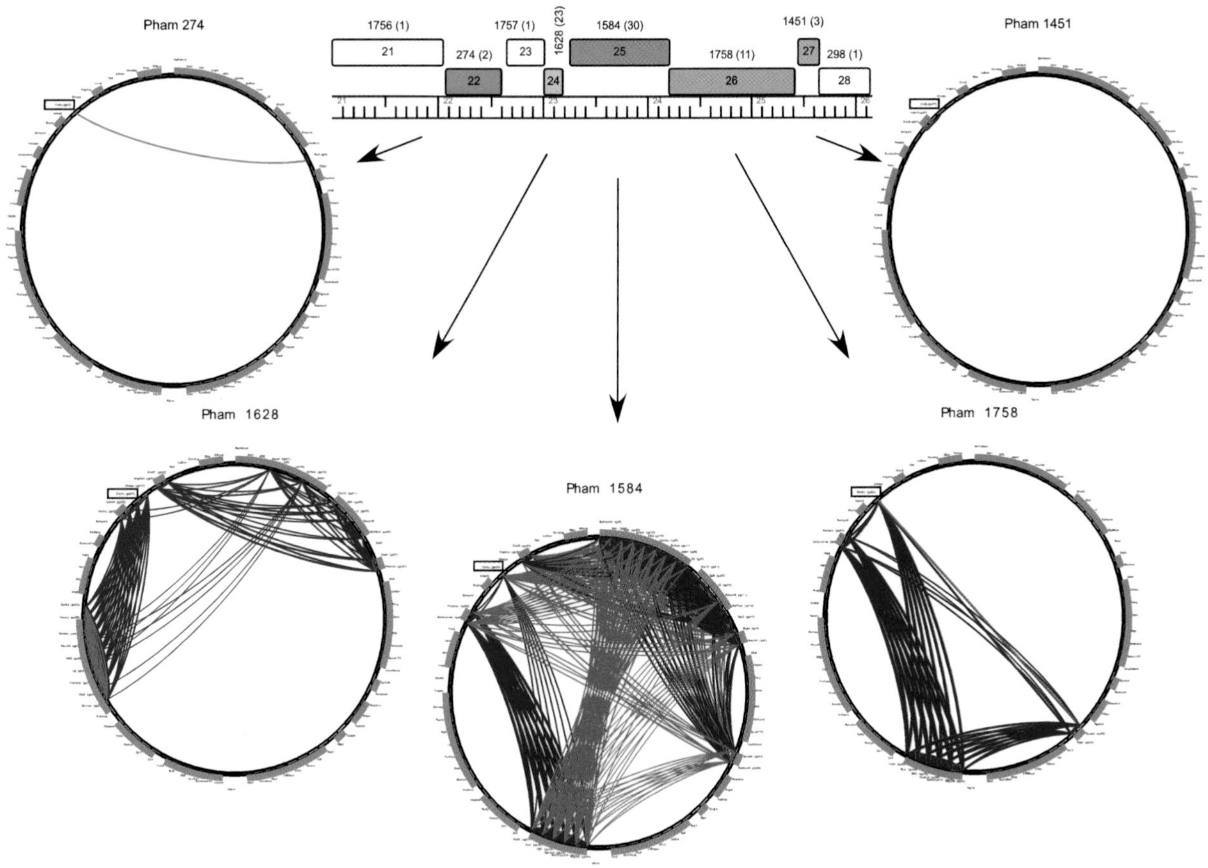

Figure 19, Graham F. Hatfull (See Page 244 of this Volume)

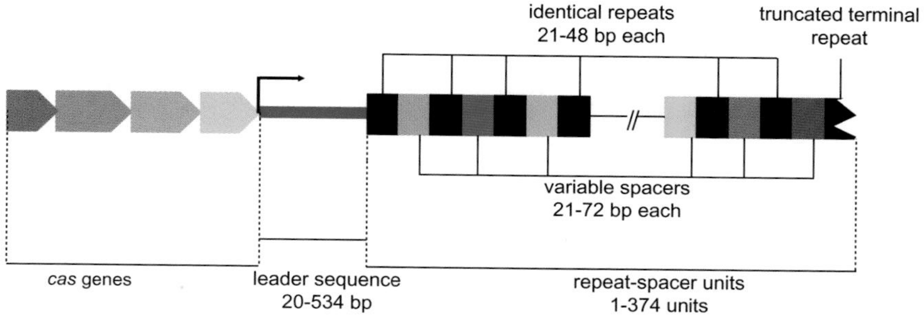

Figure 1, Agnieszka Szczepankowska (See Page 292 of this Volume)

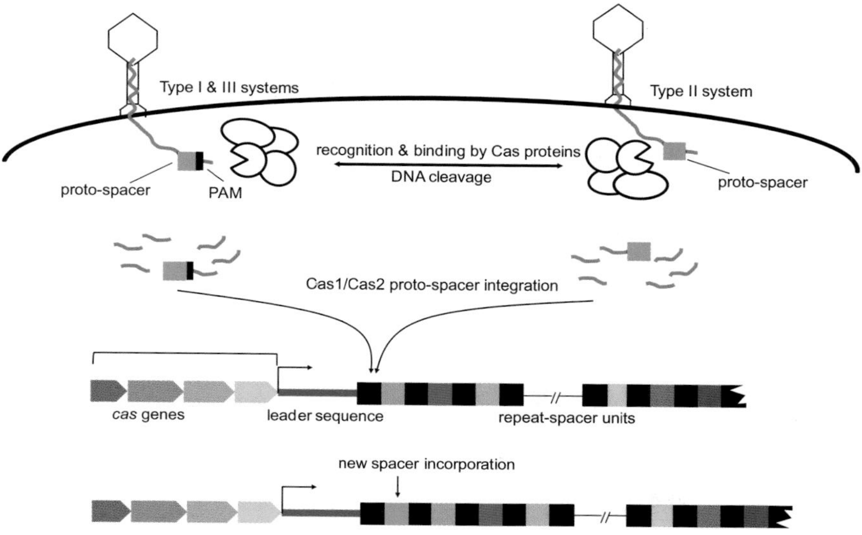

Figure 2, Agnieszka Szczepankowska (See Page 305 of this Volume)

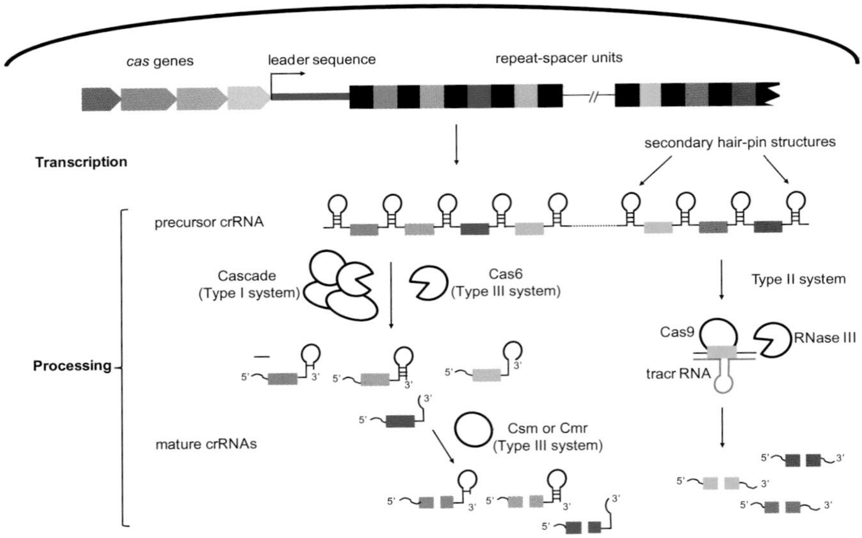

Figure 3, Agnieszka Szczepankowska (See Page 308 of this Volume)

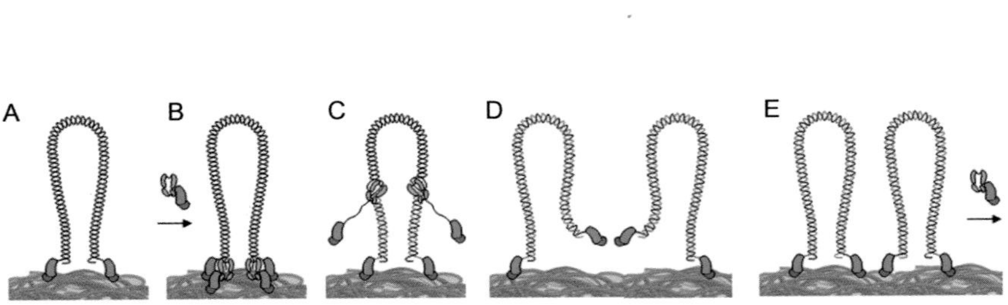

Figure 2, Daniel Muñoz-Espín *et al.* (See Page 357 of this Volume)